FUNGI

THE WORLD OF MUSHROOMS

Φ

FOREWORD

No organisms have captured the public imagination quite like fungi. Neither plant nor animal, fungi exist in their own kingdom. There are millions of species of fungi estimated to be in existence, and yet humanity has formally named only about 155,000 of them. Across the globe, fungi spread hidden as mycelium before bursting aboveground and into our visible world as mushrooms. Their fascinating forms, colors, and structures have inspired art, design, fashion, philosophy, science, and pop culture in every conceivable way—poetic, psychedelic, political, playful, imaginative, and informative. They enable trees to communicate, ecosystems to flourish, and fairy tales to inspire.

This book is arguably one of the most comprehensive tomes created that considers the wide gamut of fungi and mushrooms in visual culture. The works in *Fungi* were chosen by a panel of international experts, all with a special understanding of the mushroom in art, science, and history. The works included cover a breadth of techniques, from photography to textiles, botanical illustration to decorative art. These are the works that demonstrate fungi's indelible role in culture, across time and place.

What emerges is how our fascination with fungi has existed for millennia and conceptually shifted over time. Some of the oldest images in the book feature Mesoamerican stone sculptures, ancient Roman mosaics, medieval manuscripts, and Chinese and Japanese scroll paintings. There are even depictions of cave paintings from the Selva Pascuala in Spain, reflecting post-Paleolithic ritual use of psychoactive mushrooms. A seismic shift becomes apparent with the rise of botany and mycology as a specialism throughout the nineteenth and twentieth centuries, with deeply detailed scientific illustrations. Contemporary artists have explored the subject since the mid-twentieth century, with iconic names from John Cage to Yayoi Kusama, Andy Warhol to Martin Kippenberger, all approaching the topic. Fungi emerge in almost all types of media—painting, collage, drawing, sculpture, photography—not to mention decorative and design objects, including hats, brooches, dresses, lamps, buildings, shoes, teapots, ceramics, and more. Almost anything can be reimagined in fungal form.

What is so surprising is how endlessly fun and fresh mushrooms are to look at. Their delicate gills or spines and their endlessly varied cap shapes, even when simplified to an abstract outline, raise a smile or invite closer inspection. Fungi were often seen as emblems of decay until they emerged as one of the strongest tropes in children's book illustration. The influence of *Alice's Adventures in Wonderland*, Disney's *Fantasia*, and *The Smurfs* on our emotional relationship to fungi should not be underestimated.

Besides this rich visual history, what make mushrooms so intriguing are their possibilities for the future. Scientists are diving into the potentials mycelium and fungi provide for the development of cheaper, more natural building materials, textiles, and plastic alternatives. Other researchers are exploring the positive effects they can have on our minds and bodies. Fungi are natural bedfellows to innovation and technology.

But, perhaps, more profound are the ways in which fungi can inform our modes of thinking and relationship with the world around us. Working in symbiosis with nature, fungi may be the most inventive, and unexpected, exemplar of how humanity can reconsider how it exists within, and become better connected to, the natural world.

FRANCESCA GAVIN
Author, curator, and
Editor in Chief of *EPOCH Review*

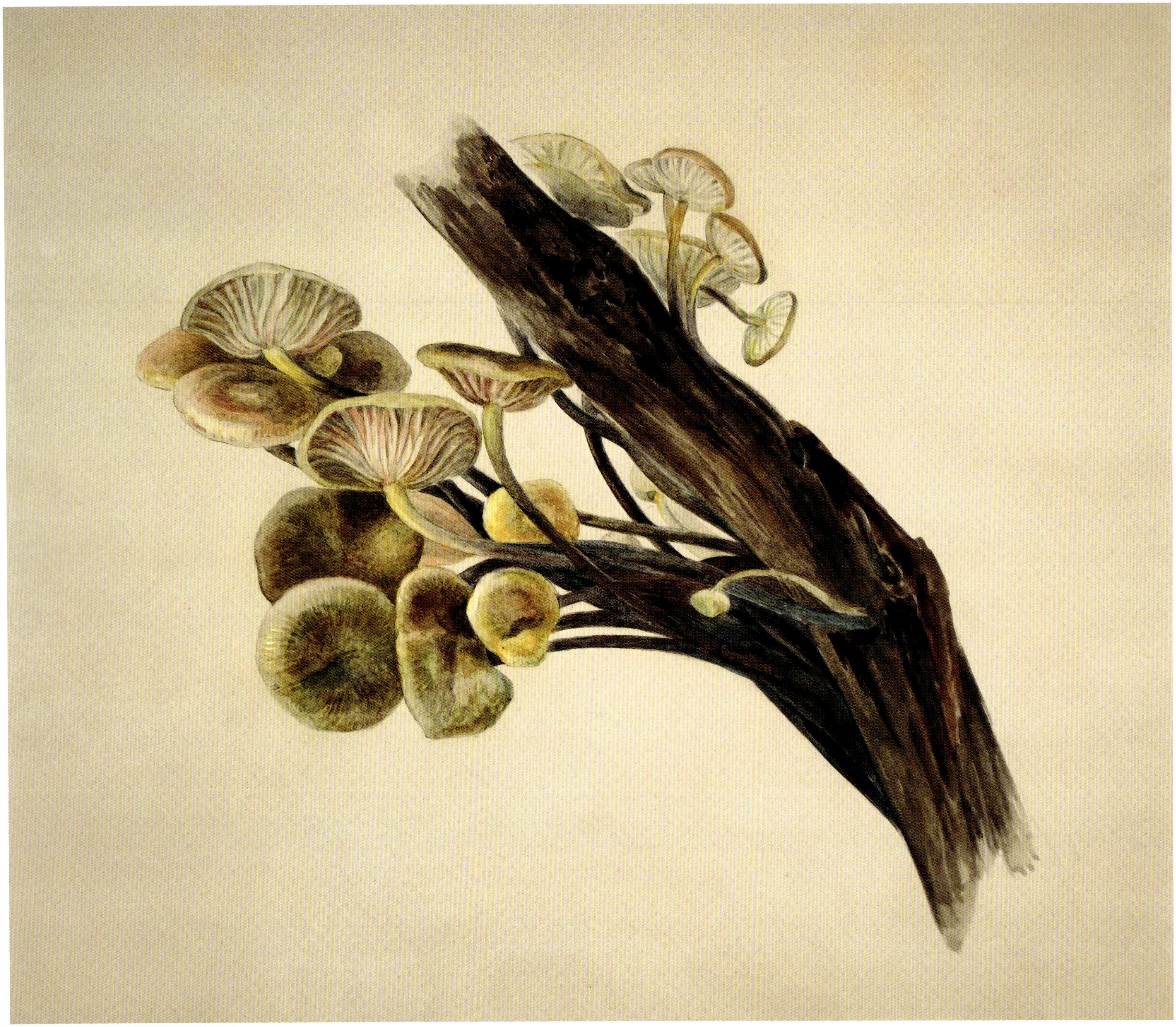

BEATRIX POTTER

Flammulina velutipes, 1892
Watercolor on paper, 8⅜ × 11 in. / 21.4 × 28 cm
Armitt Museum and Library, Ambleside, England

Tubular, yellow velvet shank mushrooms sprout from a torn branch in this scientific illustration by a young Beatrix Potter (1866–1943), best known as one of the most iconic children's authors and illustrators of the twentieth century. Despite Potter's literary prowess, mycology was her first love, and she studied lichens and fungi seriously from 1895, producing more than 350 watercolors and sketches, recording the subjects with impressive scientific accuracy. The self-taught mycologist even attempted to present a paper to the botanical Linnean Society of London in 1897, "On the Germination of the Spores of Agaricineae," accompanied by her astonishingly detailed illustrations. At the time, many academic and scientific institutions were exclusively open to men, and Potter's research was sadly rejected. Today, recognized for both their artistic nature and scientific importance, many of her detailed drawings are now held in London's Victoria and Albert Museum, as well as the Armitt Museum in Ambleside, England, near Potter's home. In 1901 Potter self-published her first children's book, *The Tale of Peter Rabbit*. The first edition sold out, and she was swiftly invited to join publisher Frederick Warne & Co., for which she wrote more than thirty books centered around animal characters such as Squirrel Nutkin and Jemima Puddle-Duck. The love of nature and artistic skill manifested in Potter's stories undoubtedly emerged from her many years of studying mushrooms in the decade prior to the start of her career as an author and illustrator.

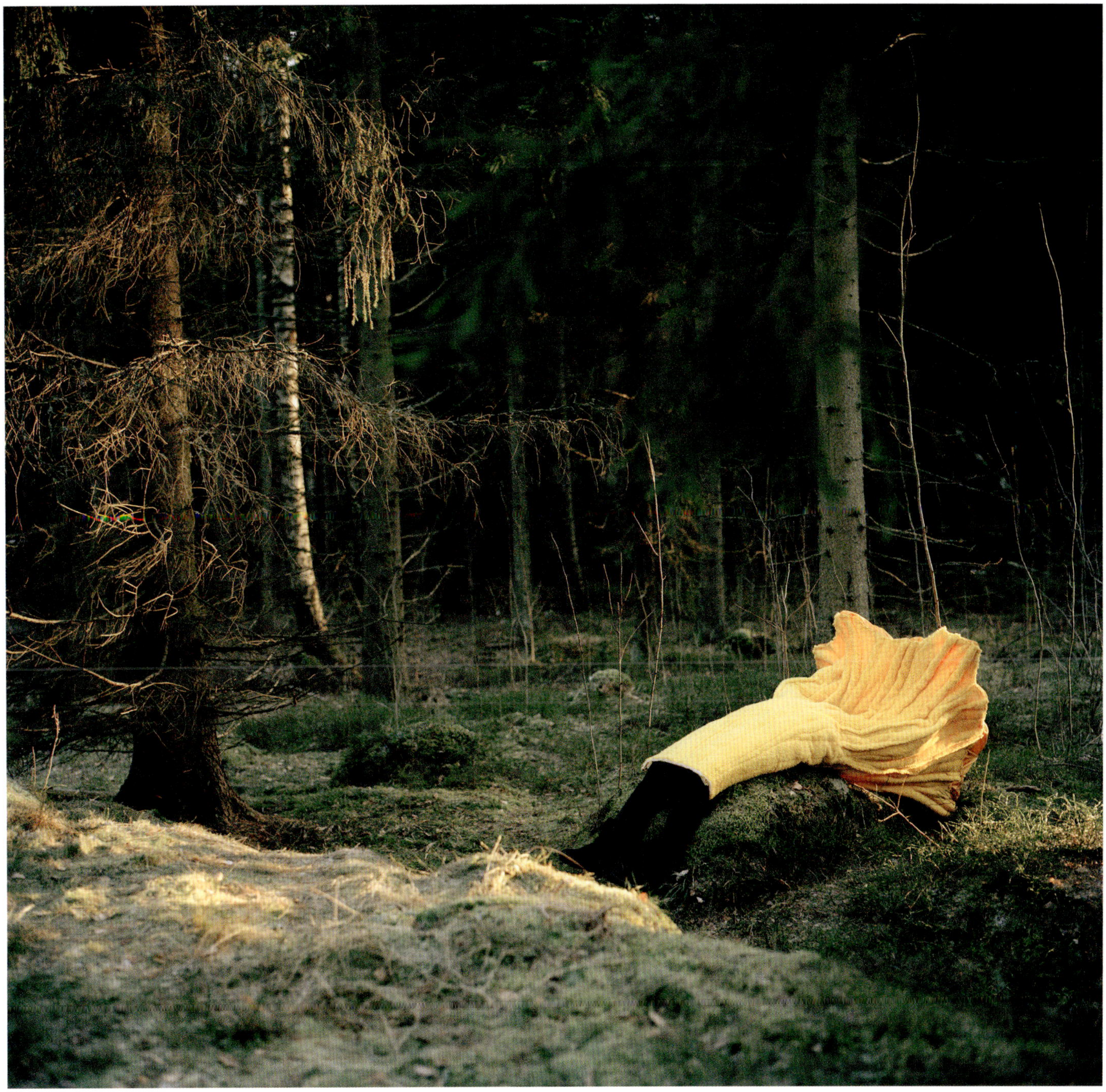

RIITTA IKONEN AND ANNI KOPONEN

Chanterelle, 2009
Photograph, 29½ × 29½ in. / 75 × 75 cm

Multidisciplinary Finnish artist Riitta Ikonen (b. 1981) is known for her performance-based art, collaborative projects, such as *Eyes as Big as Plates*, and her long-running and innovative *Mail Art* project, among many other works. Born in the "deep forests of eastern Finland," Ikonen was educated at the Royal College of Art in London, but she returns often to the woods of her homeland in her work, as seen here in *Chanterelle*. Working with photographer Anni Koponen, *Chanterelle* is part of a series featuring oversize pieces of wearable art Ikonen has constructed to reflect humans' interaction with nature. Here, a human chanterelle mushroom lies quietly among the fir trees, its bright yellow flesh ablaze against the muted green of the forest floor. In other tableaus, Ikonen costumes her models as reishi, stinkhorn, and lion's mane mushrooms, all photographed in their "natural" woodland habitats. The series was inspired by mycologist Paul Stamets's TED Talk "6 Ways Mushrooms Can Save the World," which discusses how fungi can clean up pollution, generate fuel, and fight viruses, among other virtues. Ikonen's exploration of humans' connection to their natural environment is more than just aesthetic; it's an urgent call to action in the face of drastic ecological change. Ikonen has received numerous awards for her work and has exhibited in solo and group shows around the world, including at the Victoria and Albert Museum in London, the Museum of Contemporary Art Kiasma in Helsinki, and SEIBU Shibuya in Tokyo. Showcasing the whimsical side of her passion for all things fungi, she is also a frequent favorite at the Telluride Mushroom Festival's costume parade (see p.98).

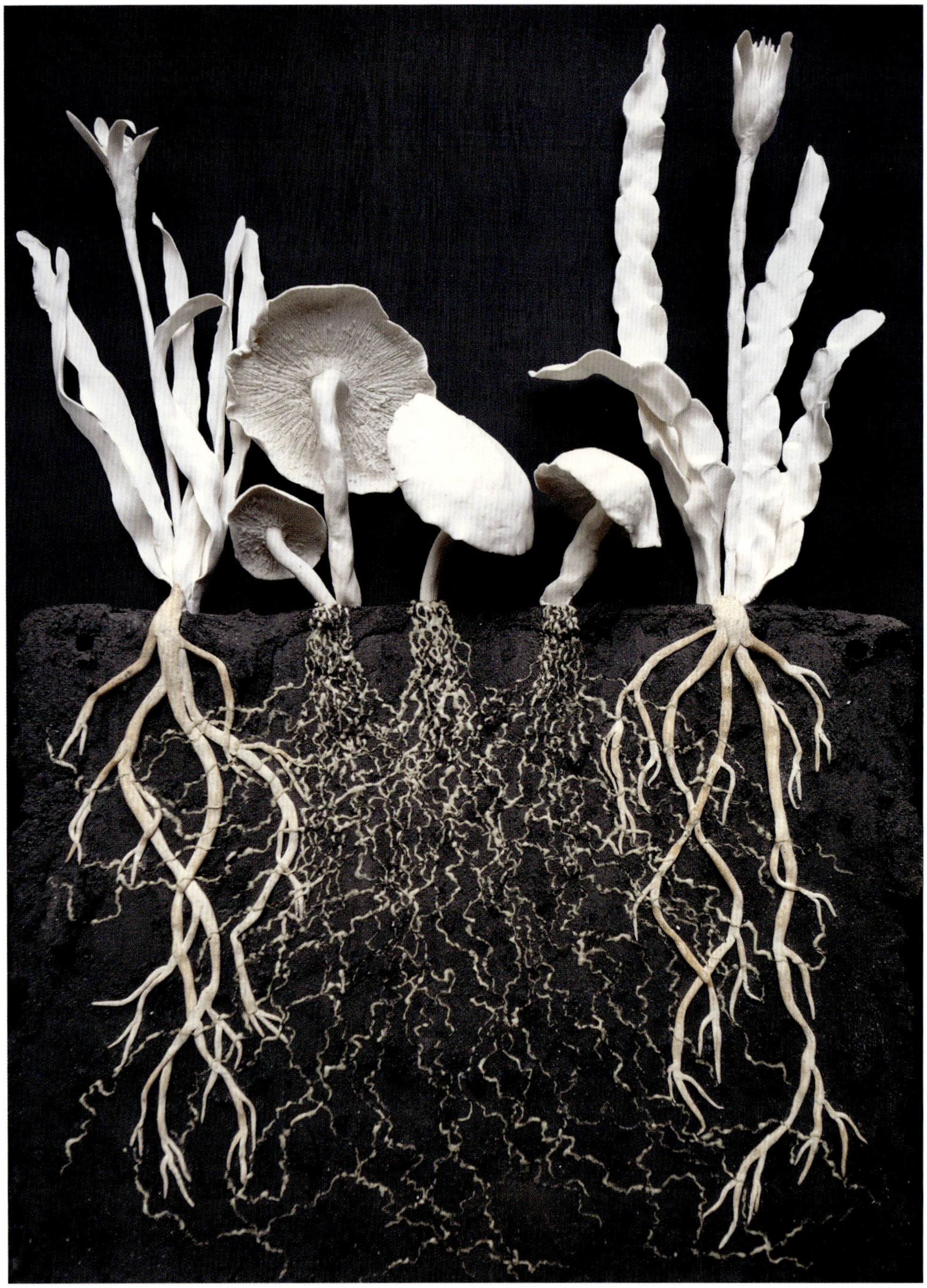

JO PEARL

Unearthed Mycelium, 2023
Ceramic, 23⅝ × 15¾ in. / 60 × 40 cm

British ceramicist Jo Pearl (b. 1968) traveled a circuitous route before becoming an artist. Educated at the London School of Economics, she worked in public affairs television and public relations before attending Central Saint Martins, where she earned a degree in ceramic design. Pearl—an activist artist who works in both sculpture and animation—maintains that the soil that composes her chosen medium, clay, is the foundation and the inspiration that undergirds so much of her work. In her series of sculptures *Unearthed* (2023), she plumbs the often beautiful and wildly diverse world that lives underneath our feet. From microscopic bacteria to insect life, Pearl celebrates the rich biodiversity of this mysterious realm, bringing underground life into sharp relief through her highly detailed and sometimes playful works.

Here, in stark white ceramic, *Unearthed Mycelium* depicts the familiar caps and stems of mushrooms growing between two flowers, a common enough sight in a garden or on the forest floor. But more intriguing is what lies beneath: a complex system of plant roots and mycelium that twine together to form a complicated, interconnected web, breathing life into the soil and every living thing in it. This work demonstrates both the symbiotic relationship found in this slice of topsoil and—with its resemblance to the structure of the human body's circulatory system—how humankind and the earth depend on a delicate balance of conditions in order for each to remain healthy.

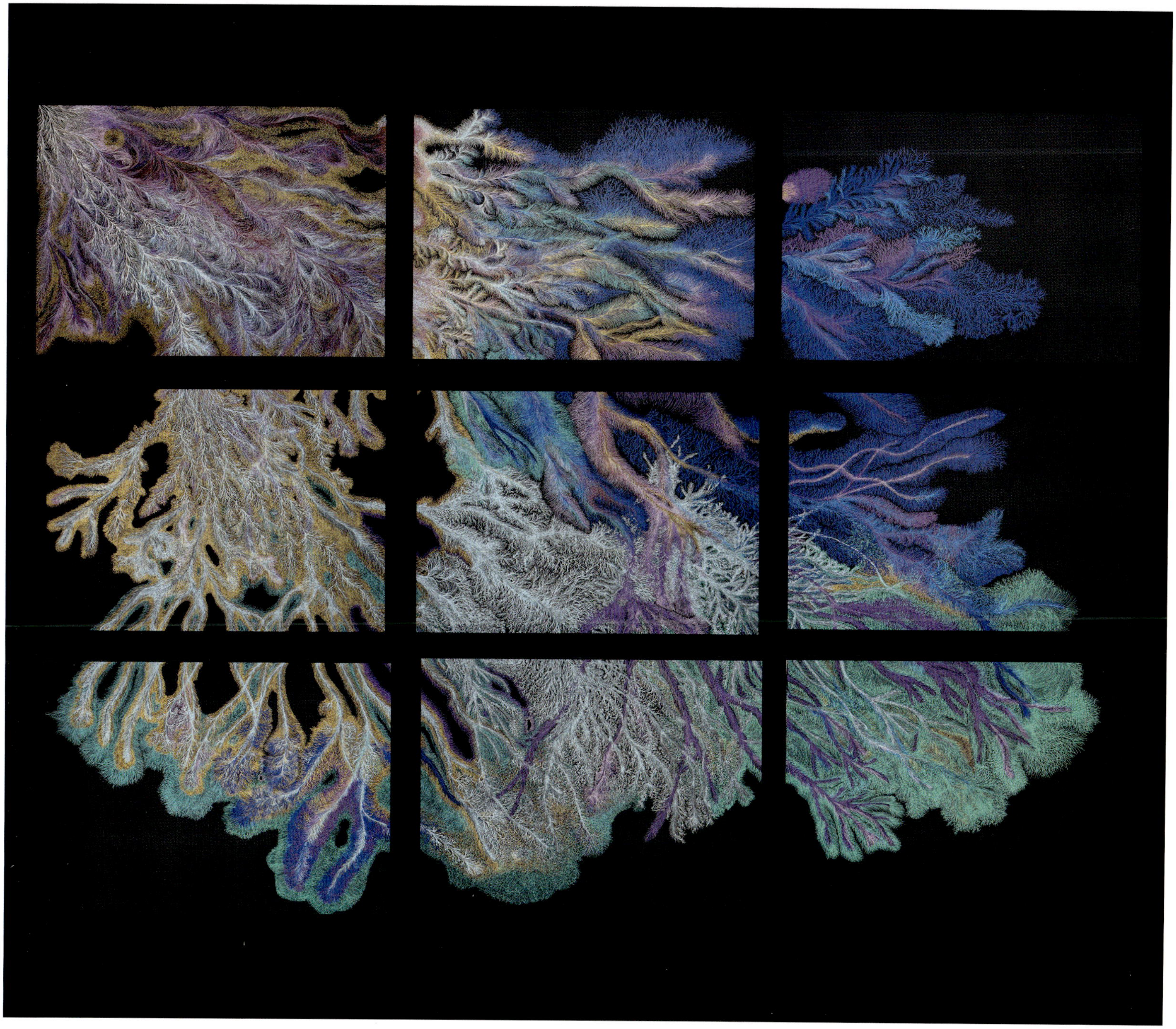

SARAH HEARN

Mycelium Network Polyptych, 2023
Gel pen on black paper,
overall 66 × 90 in. / 167.6 × 228.6 cm
Private collection

In this dazzling nine-panel drawing, Sarah Hearn (b. 1978) reveals the underground system of mycelium, the intricate network of fine, rootlike structures (hyphae) that forms the main body of a fungus. Recalling the vascular systems of plants and animals as well as those of the human body, the feathery, branching patterns are rendered with a vibrant array of blue, teal, purple, white, and earthy orange gel pen against a jet-black background—a vivid, larger-than-life reminder of the interconnectedness of Earth's living organisms. Subterranean mycelial networks are mostly microscopic, although those of some fungal species can grow to enormous sizes—the largest is thought to be that of an *Armillaria ostoyae* in the Malheur National Forest in Oregon, which covers an estimated 2,384 acres (965 hectares). Sometimes hyphae join with plant roots to form symbiotic mycorrhizal networks, taking sugars from plants in exchange for moisture and nutrients. A self-described citizen scientist, Hearn is driven by curiosity, research, and investigation with a view to encouraging a greater understanding of and respect for the natural world. Her drawings, photographs, and sculptures hover between empirical studies and the realm of science fiction, transporting viewers to fantastical worlds as she explores biological phenomena that cannot be perceived without specialized technology. While many of her works are made in collaboration with scientists and are grounded in the exploration of microbiology, she will often take artistic liberties. Here, she lets her imagination take hold to visualize an aspect of the mycological world that is usually hidden from view.

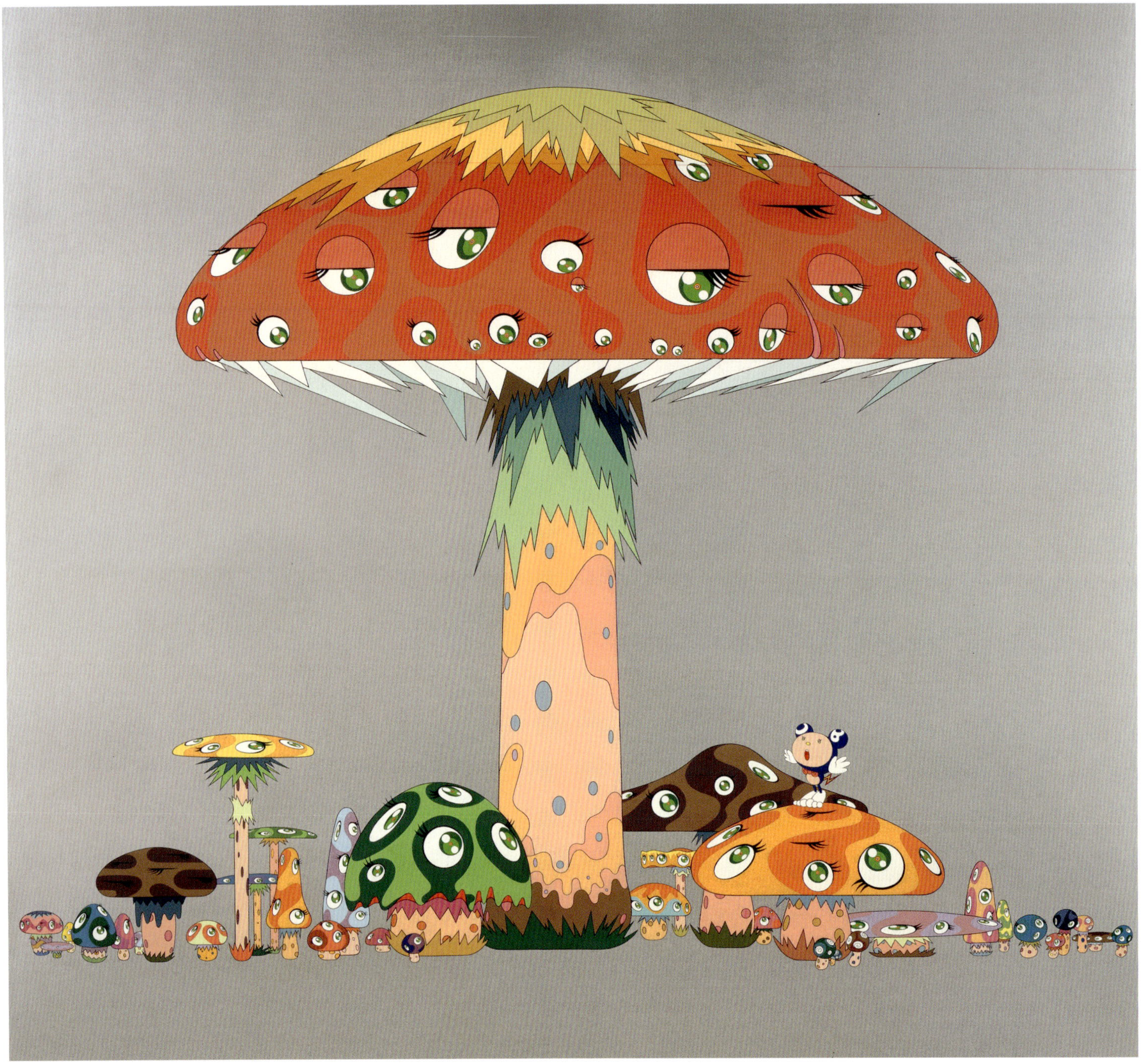

TAKASHI MURAKAMI

DOB in the Strange Forest, 1999
Acrylic on canvas mounted on board,
70⅞ × 70⅞ in. / 180 × 180 cm

In this colorful, cartoonlike print, a colossal eye-covered mushroom towers above a panoply of smaller yet similarly psychedelic fungi with bulbous eyes. Gazing anxiously at the imposing fruiting body is Mr. DOB, the iconic alter ego of Takashi Murakami (b. 1963), an ever-evolving character appearing in many of the Japanese artist's paintings, sculptures, and prints. Working in his trademark Superflat style, which blends traditional Japanese art with contemporary Japanese pop culture, Murakami is known for his anthropomorphic subjects, from smiling flowers and playful pandas to surreal mushrooms. Yet beneath his whimsical aesthetic lie darker currents. This scene is partly inspired by the fifth chapter of Lewis Carroll's *Alice's Adventures in Wonderland*, in which Alice meets a caterpillar on a mushroom who tells her that taking a bite from one side of the mushroom will increase her size, while the other side will cause her to shrink (see p.79). But the composition also speaks to Japan's collective trauma since the atomic bombings of Hiroshima and Nagasaki in 1945. Here, the giant fungus references the sobering sight of a billowing mushroom cloud that follows a nuclear explosion. Murakami has a very personal connection to the attacks: his mother's hometown of Kokura, Japan, was originally the primary target for the second atomic bomb to be dropped by US forces, but because the sky was too overcast on the day of the air strike, the city was spared and Nagasaki was bombed instead.

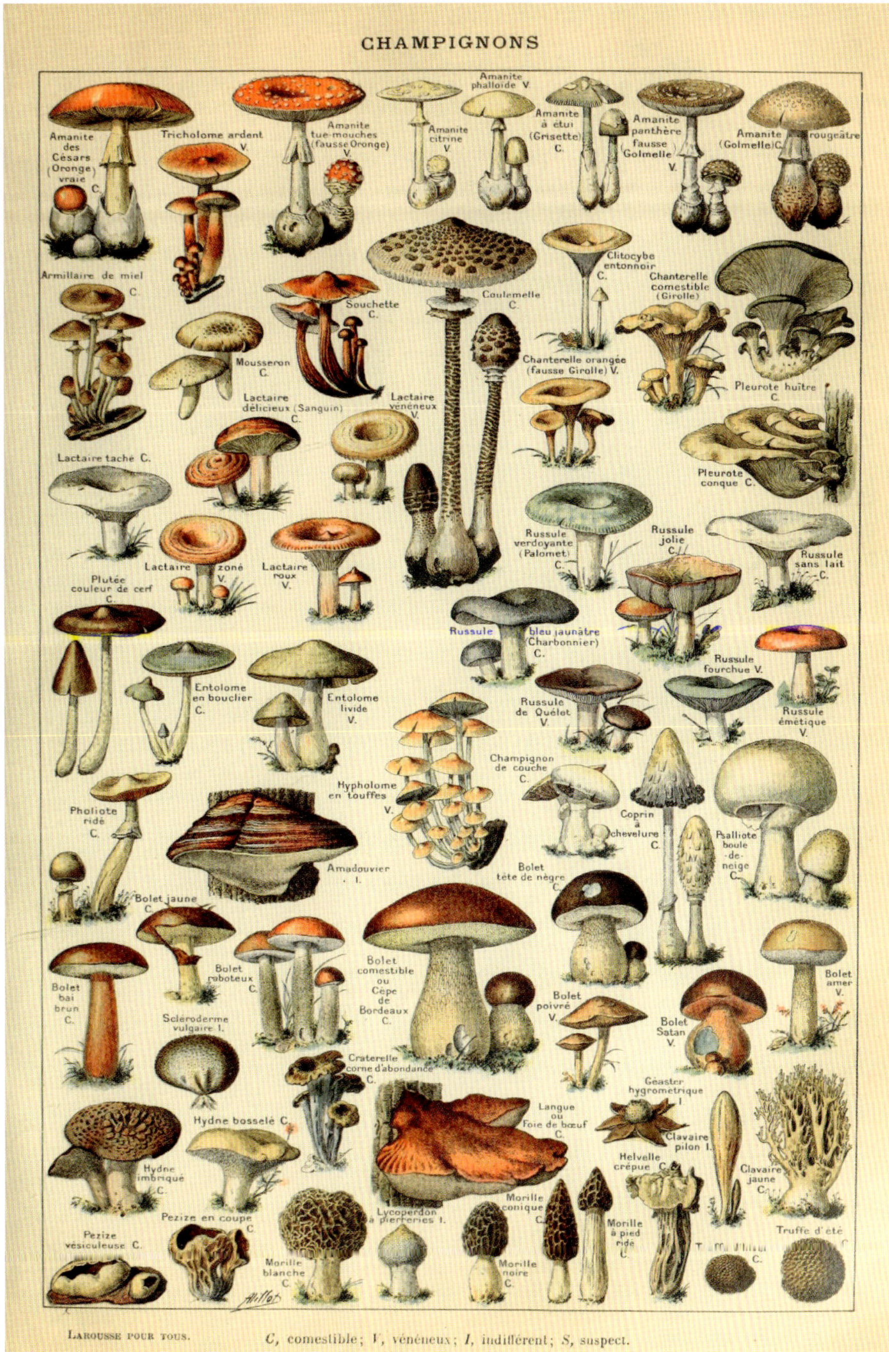

ADOLPHE MILLOT

Champignons, from *Le Petit Larousse Illustré*, c. 1900
Lithograph, approx. 7¾ × 5⅜ in. / 19.7 × 13.5 cm
Private collection

Forming a plate from the French encyclopedic dictionary *Le Petit Larousse Illustré*, more than sixty species of European fungi are juxtaposed on the page in a colorful array by French painter, lithographer, and entomologist Adolphe Millot (1857–1921). A senior illustrator at the Muséum national d'Histoire naturelle in Paris, Millot contributed a number of plates to the hugely popular encyclopedia, primarily focused on the natural world. Depicted in impressive detail, with all their key features shown, each species of fungus here is annotated with its French common name, and letters keyed at the foot of each plate indicate whether the mushroom in question is edible, poisonous, indifferent, or suspect. *Amanita* in the top row, which are mostly toxic, include the familiar vibrant fly agaric (*Amanita muscaria*), with its white-spotted red cap, alongside its relatives, notably the very poisonous death cap (*Amanita phalloides*). Standing tall at the upper center is the majestic parasol mushroom (*Macrolepiota procera*), with, to the right, the equally edible creamy yellow chanterelle (*Cantharellus cibarius*), with its characteristically undulating cap. Farther down, a field, or meadow, mushroom (*Agaricus campestris*) sits alongside another edible species, the shaggy ink cap (*Coprinus comatus*). Below these is a collection of boletes, with, center stage, the famous cep (*Boletus edulis*, also known as the penny bun or porcino), prized by many a gourmand. At the bottom, we find morels (genus *Morchella*) with their honeycomb, pitted caps, and, at right, truffles (genus *Tuber*), the highly aromatic subterranean species famously used to complement a variety of culinary dishes.

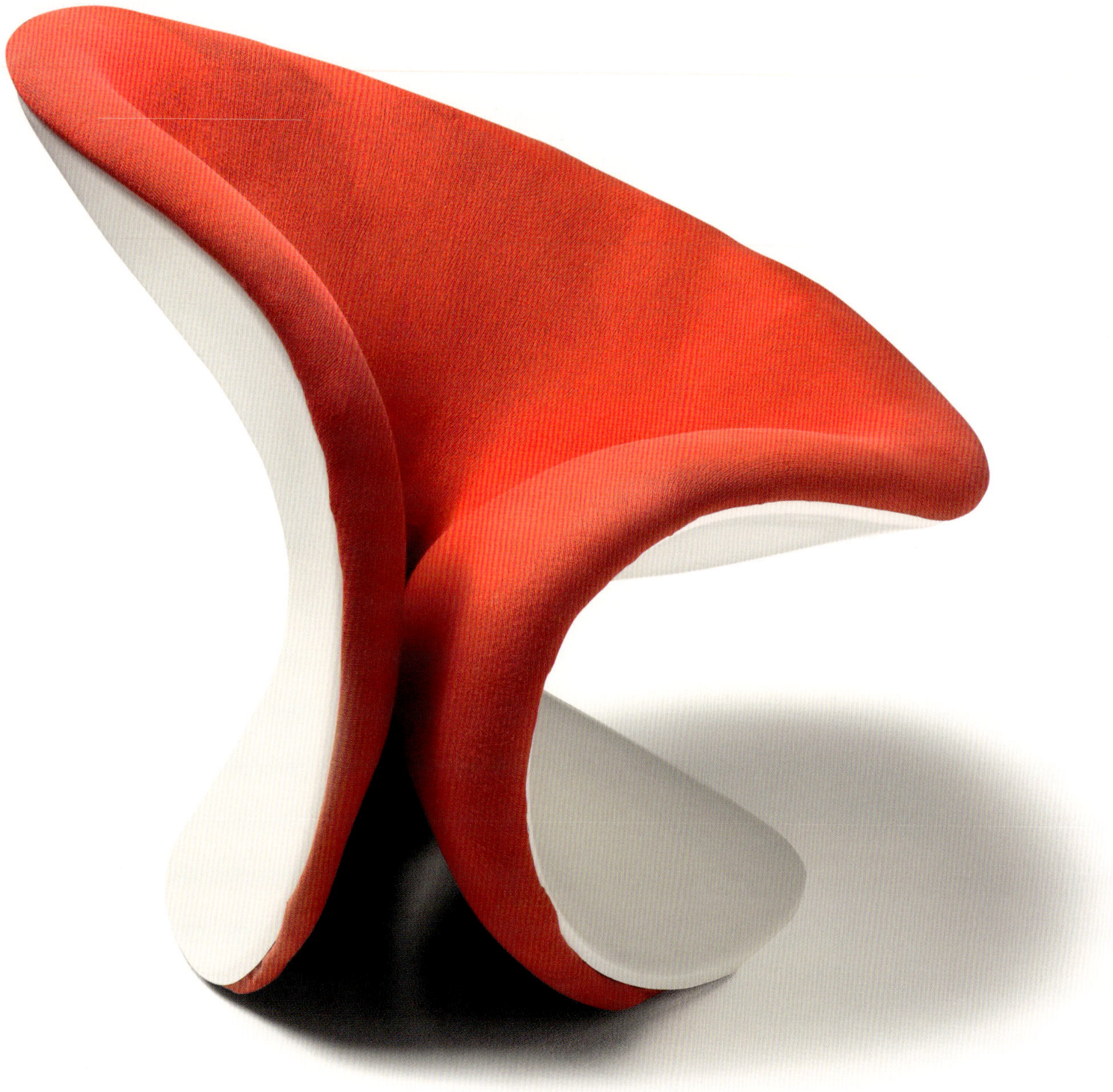

JEAN-PIERRE LAPORTE

Fauteuil Girolle (Chanterelle Armchair), 1969
Fiberglass, polyester, foam, and pink jersey fabric,
28¾ × 27½ × 23¾ in. / 73 × 70 × 60.5 cm
Musée National d'Art Moderne, Centre Pompidou, Paris

Born and raised in Bordeaux, French designer Jean-Pierre Laporte (b. 1944) was introduced to the process of furniture-making at an early age. His father worked for Thonet, one of Europe's most esteemed furniture manufacturers. It was there that, in 1961, Laporte was introduced to modernist designer Pierre Paulin, who would be instrumental in influencing Laporte's career aspirations. Laporte attended the École Boulle and then graduated from the École nationale supérieure des Arts décoratifs (National School of Decorative Arts) in 1965. Laporte began taking design commissions before establishing his own design studio in Paris in 1969. That same year he designed the *Fauteuil Girolle*, an armchair inspired by the shape of a chanterelle mushroom. Thonet produced the chair, which appeared at a time when industrial design was making drastic leaps forward. Modernism was a distinct rejection of the old ways, embracing simplicity in form, material, and concept. It was about minimalism instead of florid detail, and employing construction materials such as glass, steel, and plastics to evoke sleek sophistication. Much like Arne Jacobsen's *Egg Chair* and Eero Saarinen's *Tulip Table*, both introduced a decade before, Laporte's chair interprets its influence from the natural world through a distinctly modernist lens, reducing the form to its essence. It embraces the fluidity of the organic—no sharp angles, no straight lines—and employs inorganic materials that were then on the cutting edge of technology. Constructed from fiberglass with bright pink fabric, the pedestal unfurls into a graceful trumpet-like form, just like its mycological namesake.

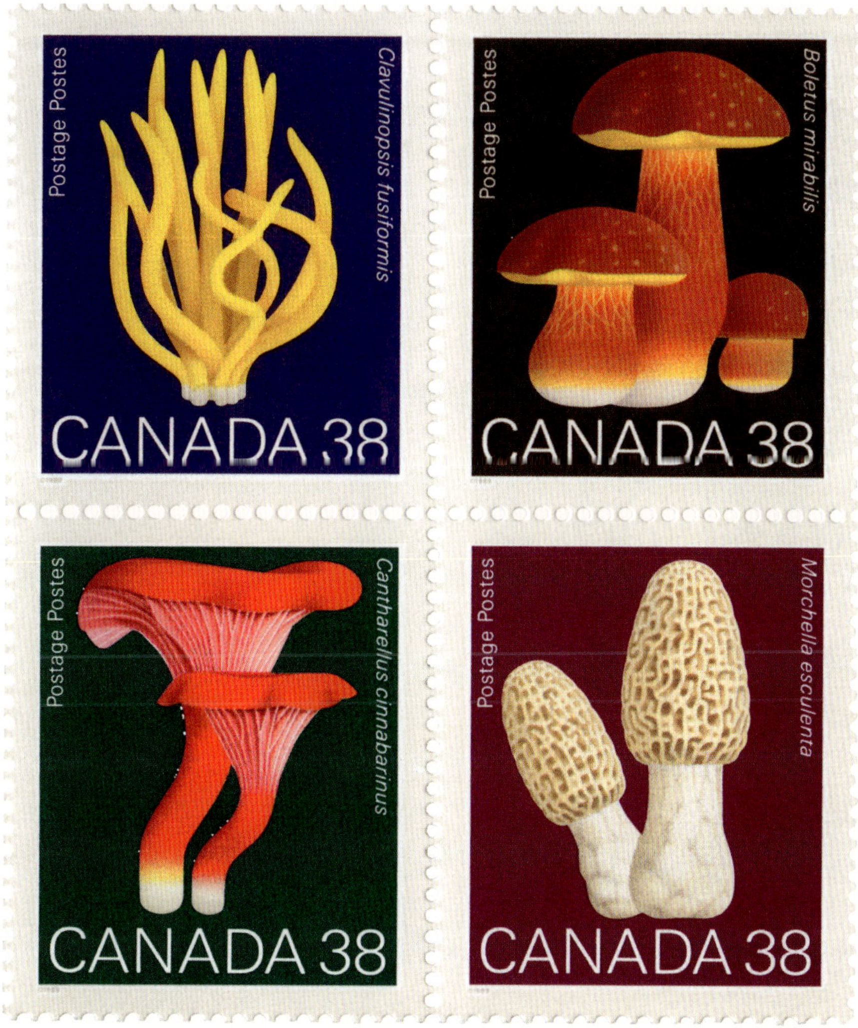

ERNST ROCH

Mushrooms of Canada, 1989
Stamps, block 2⅞ × 2⅜ in. / 7.2 × 6 cm
Canada Modern Archive

Set against monochrome backgrounds are four richly colored and stylized depictions of edible species of mushroom found across Canada: from the cylindrical yellow forms of golden spindles (*Clavulinopsis fusiformis*) and the velvety, reddish-brown caps of admirable bolete (*Aureoboletus mirabilis*, formerly *Boletus mirabilis*) to the distinctive flamingo pink funnels of cinnabar chanterelle (*Cantharellus cinnabarinus*) and the spongelike textures of the common morel (*Morchella esculenta*). This striking set of themed postal stamps was designed by Ernst Roch (1928–2003) and issued by Canada Post in 1989, coinciding with the conventions of both the Mycological Society of America and the American Institute of Biological Sciences taking place in Toronto that summer. Roch's gouache compositions combine artistic verve with scientific accuracy. The designs are representative of the Yugoslavian-born designer's clean and modern graphic aesthetic, which was rooted in his training at the Staatliche Schule für Angewandte Kunst (State School of Applied Arts) in Graz, Austria, which emphasized formal clarity and rational design. It was an approach that came to be known as the International Style of graphic design, of which Roch was a pioneer in Canada from 1953. These stamps were commissioned soon after he was awarded an honorary doctorate in fine arts from the Nova Scotia College of Art and Design in recognition of his considerable professional and artistic contributions to Canadian graphic design. In an echo of Roch's original project, Canada Post returned to the theme in 2025, issuing a new series of fungi-themed stamps illustrated by Emily S. Damstra.

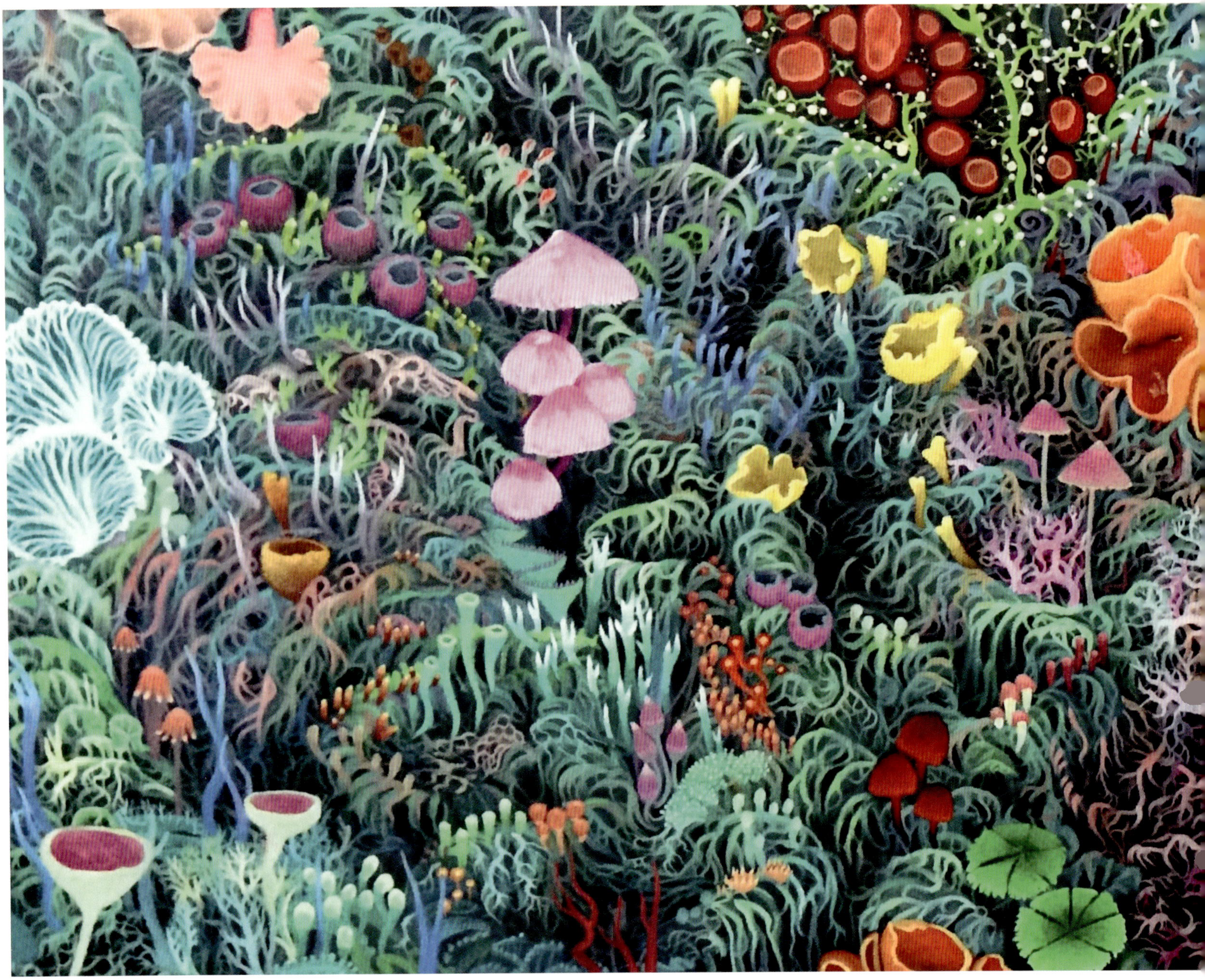

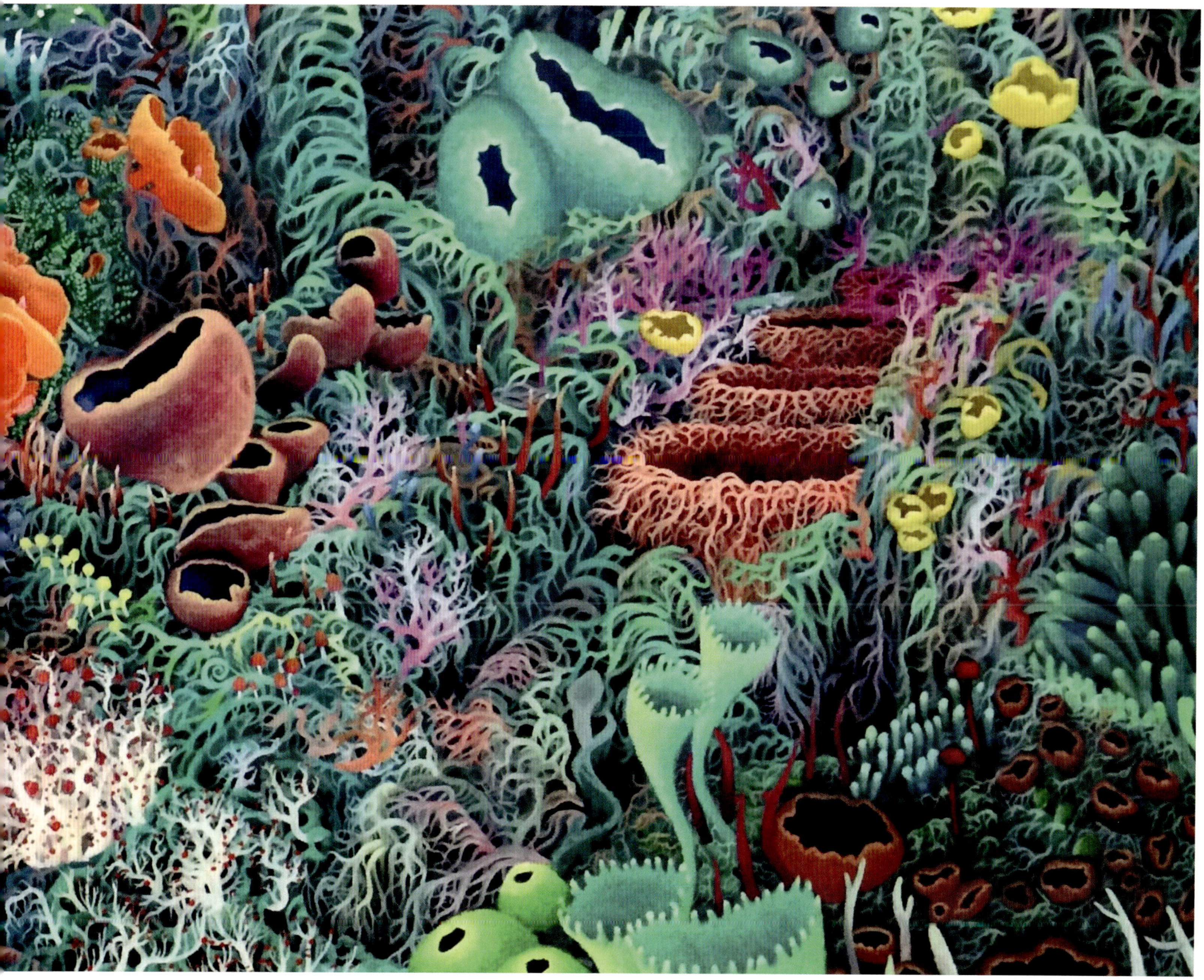

GREG ALLEN

Fruiting Bodies, 2017
Acrylic on canvas,
18 × 36 in. / 45.7 × 91.4 cm
Private collection

A dense assortment of brightly colored fungi of all shapes and sizes covers the ground alongside a multitude of lichen in this exuberant painting. The iridescent fungal landscape is one of a series of canvases, *Mycelium Matrix*, that launched the career of Canadian artist Greg Allen (b. 1982). Now living and working in Barú, Costa Rica—the rich colors of his work perhaps pay homage to the jungles of his adopted home—Allen's work has been compared to that of American artists Doze Green and Alex Grey (see p.80). All use psychedelic artwork to explore both the meaning of life and the role fungi plays in that exploration. The endless micro-universe of Allen's fungal world was partly inspired by the self-taught American mycologist Paul Stamets, who calls mycelium—the underground network of fungi—the "neurological network of nature." Allen wants his work to inspire further investigation into the fungi kingdom, and as with many fungi fans, he is awed by the longevity and importance of the mushroom in the evolutionary tree. Allen's canvas harks back to sixty-five million years ago and the second major extinction of life on Earth, when an asteroid strike created the Gulf of Mexico and wiped out more than 75 percent of all living species, including the dinosaurs. Fungi were notable survivors, thriving and proliferating on the newly dark planet. Allen celebrates this survival by depicting the intricate mycelium that remains as the "brains" of the planet.

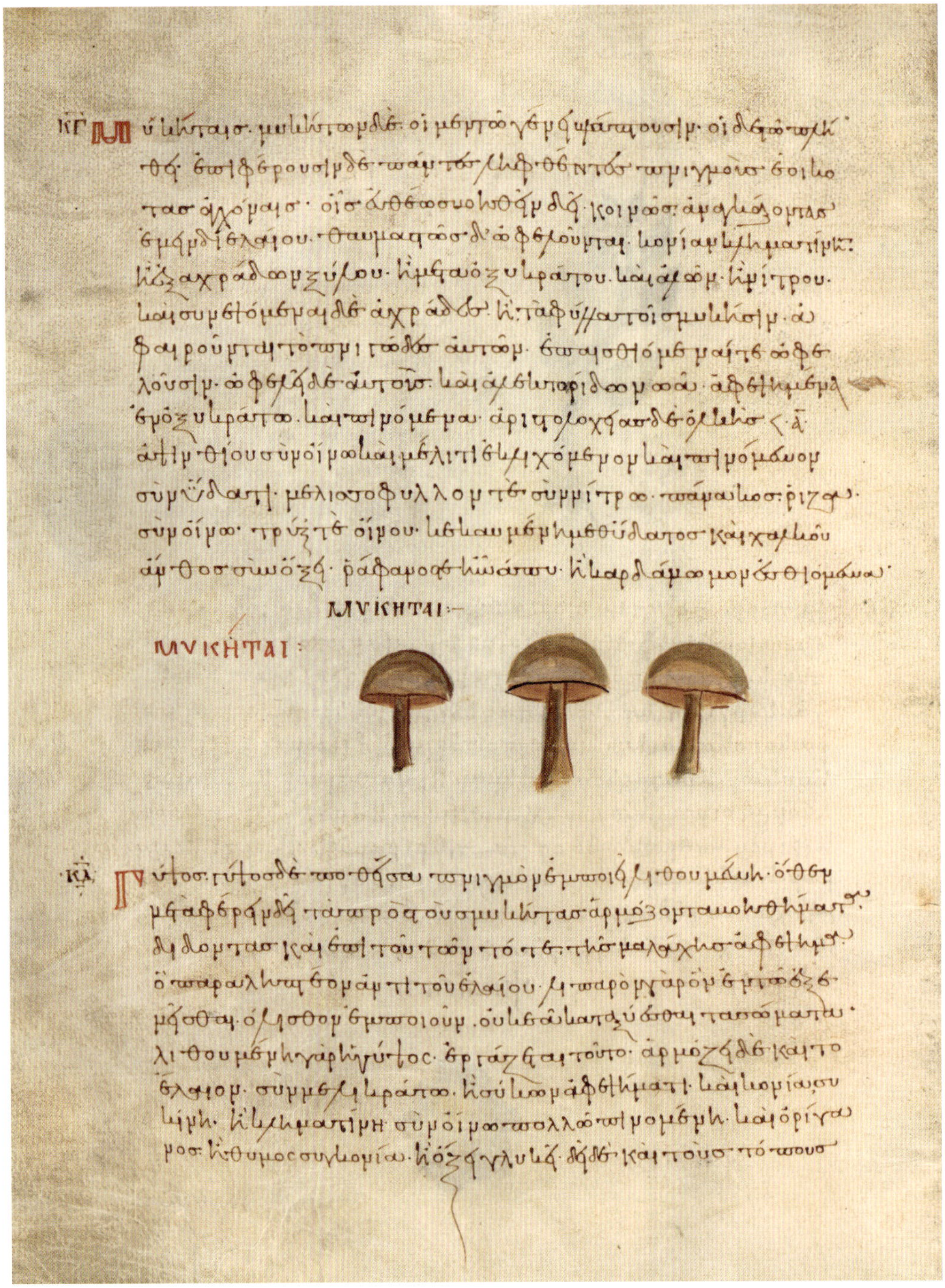

PEDANIUS DIOSCORIDES

Myketai (Fungi), from *De materia medica*, mid-10th century
Ink on vellum, 15½ × 11⅞ in. / 39.5 × 30 cm
Morgan Library & Museum, New York

A trio of mushrooms, standing upright and painted in isolation, comprises the sole illustration to appear on this page from the groundbreaking encyclopedia *De materia medica,* the work of the Greek physician, botanist, and pharmacologist Pedanius Dioscorides (c. 40–90 CE). Consisting of five volumes and written in Greek, *De materia medica* (its Latin title) covers herbal medicine and the different plants used in pharmacology, and it remained the go-to resource for more than 1,500 years after its publication. While the example here is simply labeled as *myketai* (Greek for *fungi*), Dioscorides wrote of mushrooms as being both poisonous and edible. He attributed the harmful varieties to their nefarious growing places, including rotten rags, and also cautioned that edible mushrooms spoil quickly after being picked and should be eaten in moderation. The inclusion of mushrooms indirectly paved the way for the study of mycology centuries later. Having studied medicine in Tarsus in present-day Turkey, Dioscorides references plants that were almost exclusively found in the Greek-speaking eastern Mediterranean. Written around 70 CE, his study became the foundational text of all modern pharmacopeias (official lists of medicinal drugs, their effects, and how to use them) and was translated into seven languages during the medieval period, including both Latin and Arabic, and was still in use in the nineteenth century. The illustration shown here is typical of the 769 that appear throughout the five volumes of this edition of Dioscorides's masterwork, which was written and illuminated in Constantinople (present-day Istanbul) during the middle of the tenth century.

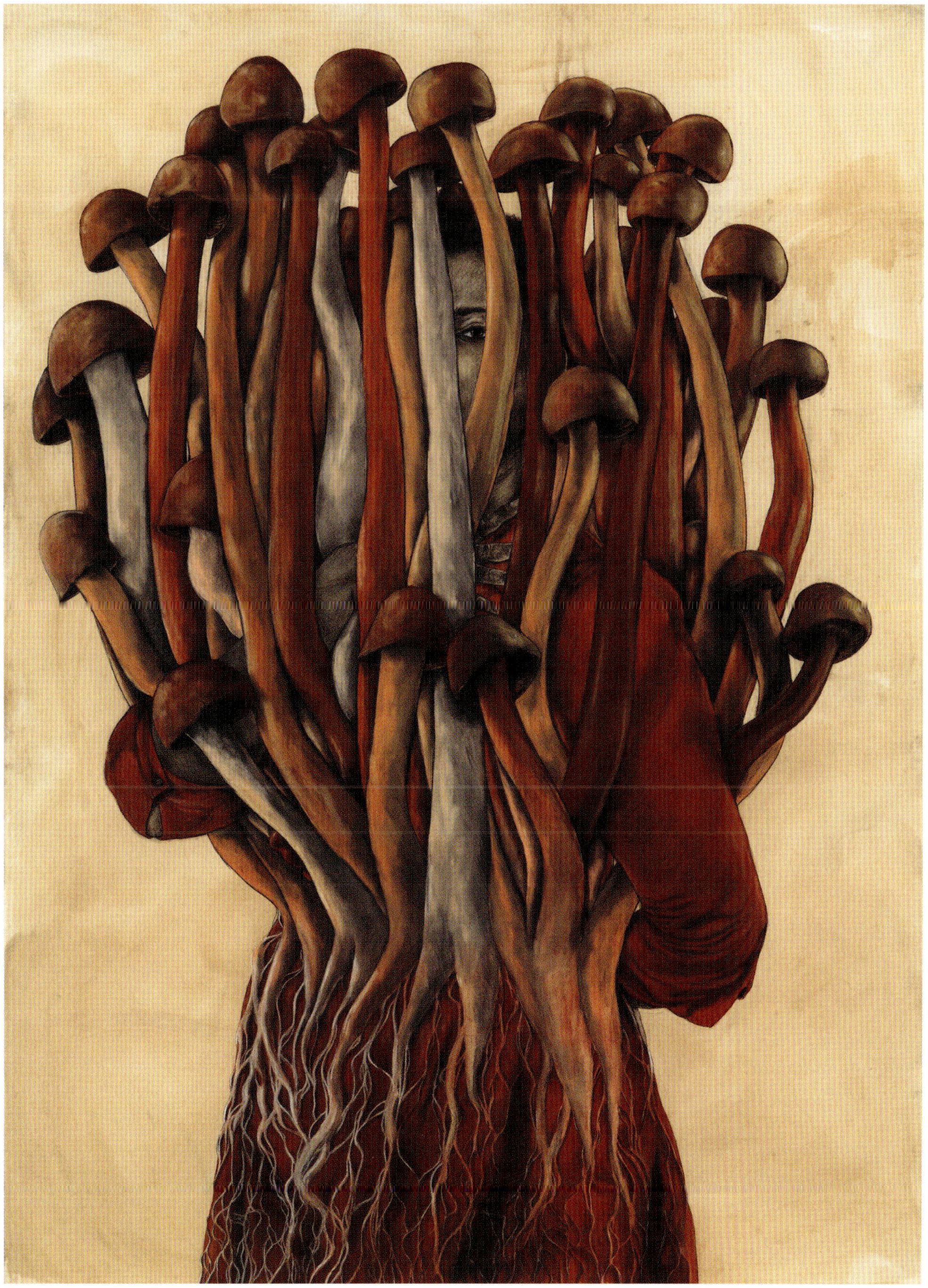

ROBERT PRUITT

Eve Hiding in the Garden of Eden, 2024
Charcoal, conté, pastel, and coffee wash on paper, 84 × 60 in. / 213.4 × 152.4 cm
Private collection

Eve, the first woman according to the Bible, is concealed amid a thicket of enormous black poplar mushrooms (*Cyclocybe aegerita*), one of her eyes staring at us from the oversize reddish-brown and cream stems that envelop her. At the bottom of the image, the treelike structure of the dense, branching mycelium appears to be growing on her clothing. This surreal drawing by American artist Robert Pruitt (b. 1975) reimagines the scene from the third chapter of Genesis, in which Adam and Eve attempt to hide from God after eating the forbidden fruit of the Tree of the Knowledge of Good and Evil. Eve is portrayed here as a Black woman, while the trees of the Garden have become enormous hostile mushrooms, and Adam is absent altogether. Uniting the physical and imagined realms, its masterful draftsmanship typifies Pruitt's monumental conté and charcoal drawings that explore African American perspectives on theism, transcendence, and mythology. Tapping into his experiences growing up in a Black community in Houston, his works weave together references from science fiction, religion, comic books, African culture, and art history. The subject of Adam and Eve hiding in the Garden of Eden is less common in art history than artwork portraying the couple's temptation and consequent expulsion from paradise, although Pruitt's approach subverts all of these representations. His fantastical and mystical interpretation challenges the canon of white, Western-dominated art history, alluding to the Black experience of alienation and marginalization in the United States.

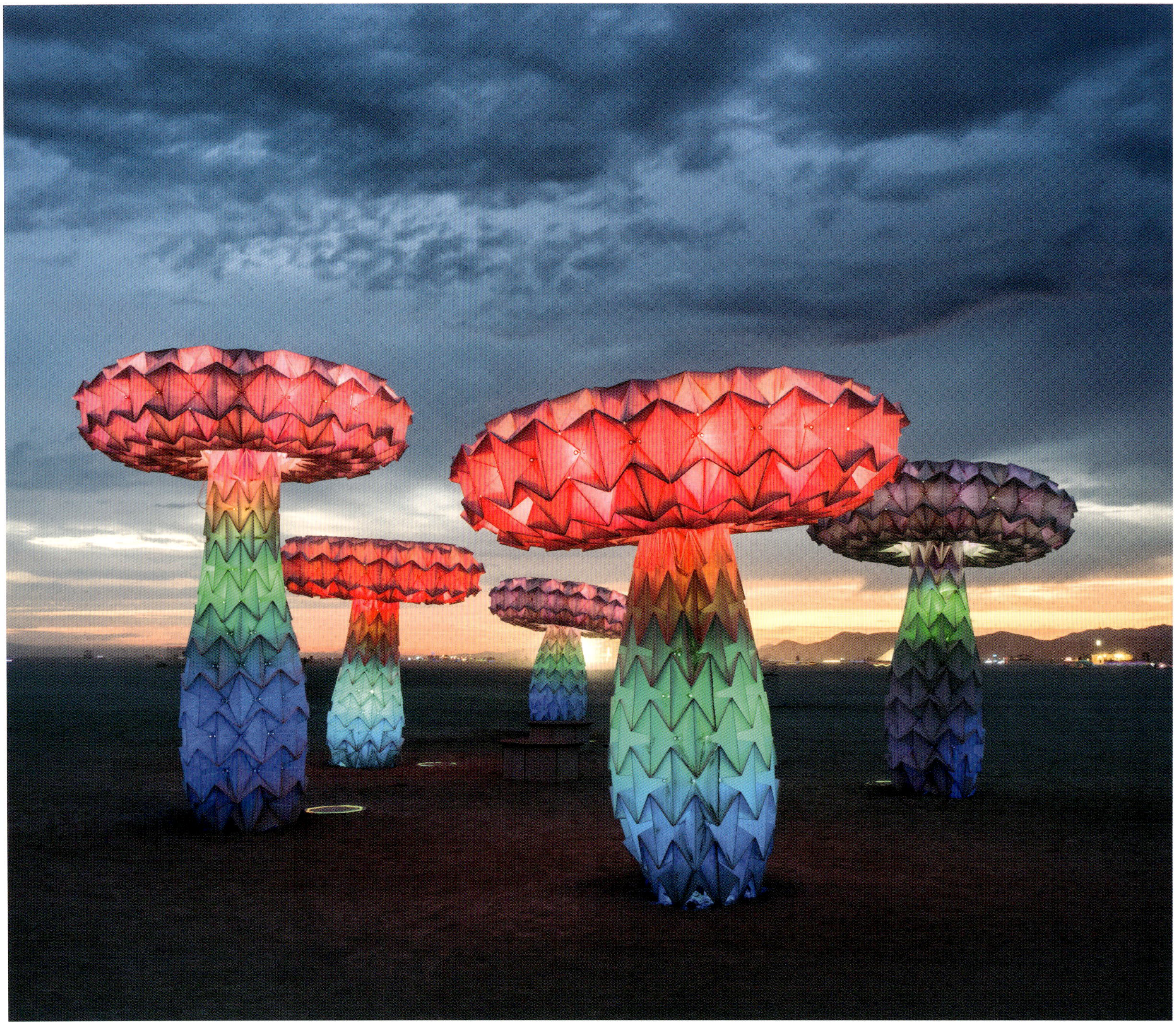

FOLDHAUS COLLECTIVE

Shrumen Lumen, 2016
Aluminum, steel, polypropylene, interactive electronics, 11–17 × 9–12 ft. / 3.3–5.3 × 2.8–3.7 m
Private collection, San Francisco; and Kiewit Luminarium, Omaha, Nebraska

A garden of giant mushrooms blinks kaleidoscopically against the night sky at Burning Man, but this is no hallucination. The mushrooms are the artistic brainchild of FoldHaus, a San Francisco Bay Area art collective dedicated to the intersection of art, engineering, and technology to create never-before-seen innovations in the public art space. FoldHaus premiered its first installation in 2010 and is currently led by designer Joerg Student and kinetic artist Jesse Silver. In *Shrumen Lumen*, the group constructed five giant fungi from eighteen sheets of corrugated plastic totaling more than 893 square feet (83 sq. m). Bent in origami-esque folds, the creases create an elastic form that expands and contracts—the caps of the mushrooms extend as much as 12 feet (3.7 m) in diameter. An interior motor powers an umbrella-like mechanism that creates undulating motion from within. Indeed, these shrooms are built for interaction, with pressure pads housed outside the forms to let onlookers activate the display. Also, inside each installation is a network of LED bulbs, allowing for a dizzying array of lights that blink rhythmically and hypnotically. The kinetic sculptures were funded in part by a Kickstarter campaign and installed for the 2016 Burning Man festival in Nevada's Black Rock Desert. It has since traveled around the world, from Dubai to New York City, among many other destinations, delighting viewers of all ages with its dazzling illumination, massive scale, and satisfying interactivity.

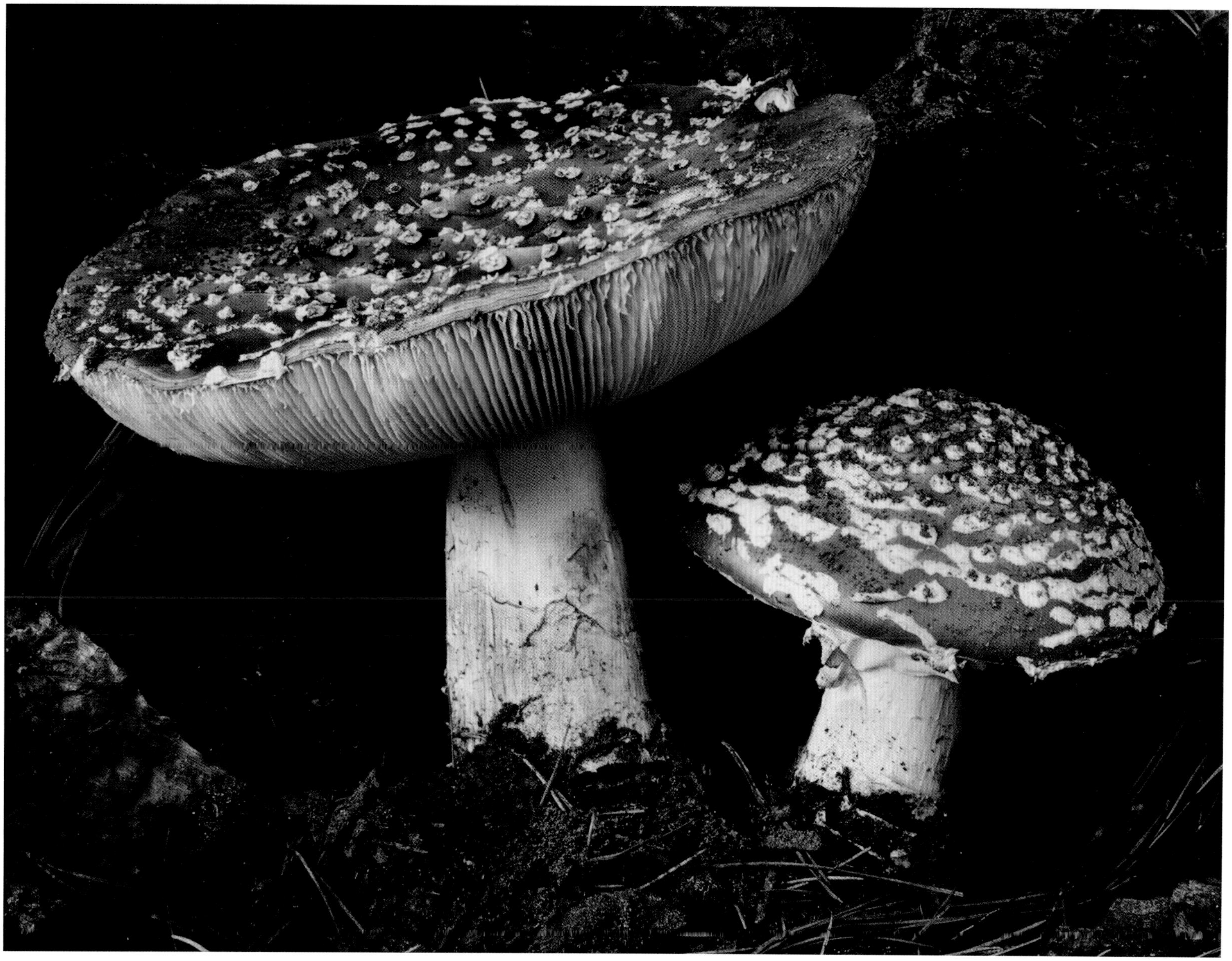

EDWARD WESTON

Toadstool, 1934
Gelatin silver print,
7½ × 9½ in. / 19 × 24.2 cm
Private collection

Edward Weston (1886–1958) was one of the most important American photographers of the twentieth century, particularly of the West. He was born in Highland Park, Illinois, but moved to California when he was only twenty-one. By that point he had already developed an acute interest in photography, after receiving his first camera for his sixteenth birthday and having one of his Chicago photographs featured in *Camera and Darkroom* magazine in 1906. But what he had not done yet was cultivate the detail-forward style that would come to define his work. From 1906 until 1922, Weston immersed himself in photography, apprenticing with a few photographers, shooting thousands of negatives, and eventually opening his own studio. While visiting his sister in Ohio in 1922, he took six photographs of the Armco steel mill that would forever change his style. These images eschewed the soft-focus approach he had been using in favor of a bolder, clean-edged look. He recognized this change in his own work and cultivated it. This photograph of two mushrooms, probably fly agaric (*Amanita muscaria*), perfectly typifies Weston's photographs from this period onward, from its dark background to its fine detail and hard edges. From the mid-1920s, Weston's career was marked by accolades, including many books and exhibitions, and in 1937 he became the first photographer to be awarded a Guggenheim Fellowship. Ravaged by Parkinson's disease, Weston took his last photographs in 1948, after which he spent the last decade of his life cataloging his images and overseeing the printing, publication, and exhibition of his work.

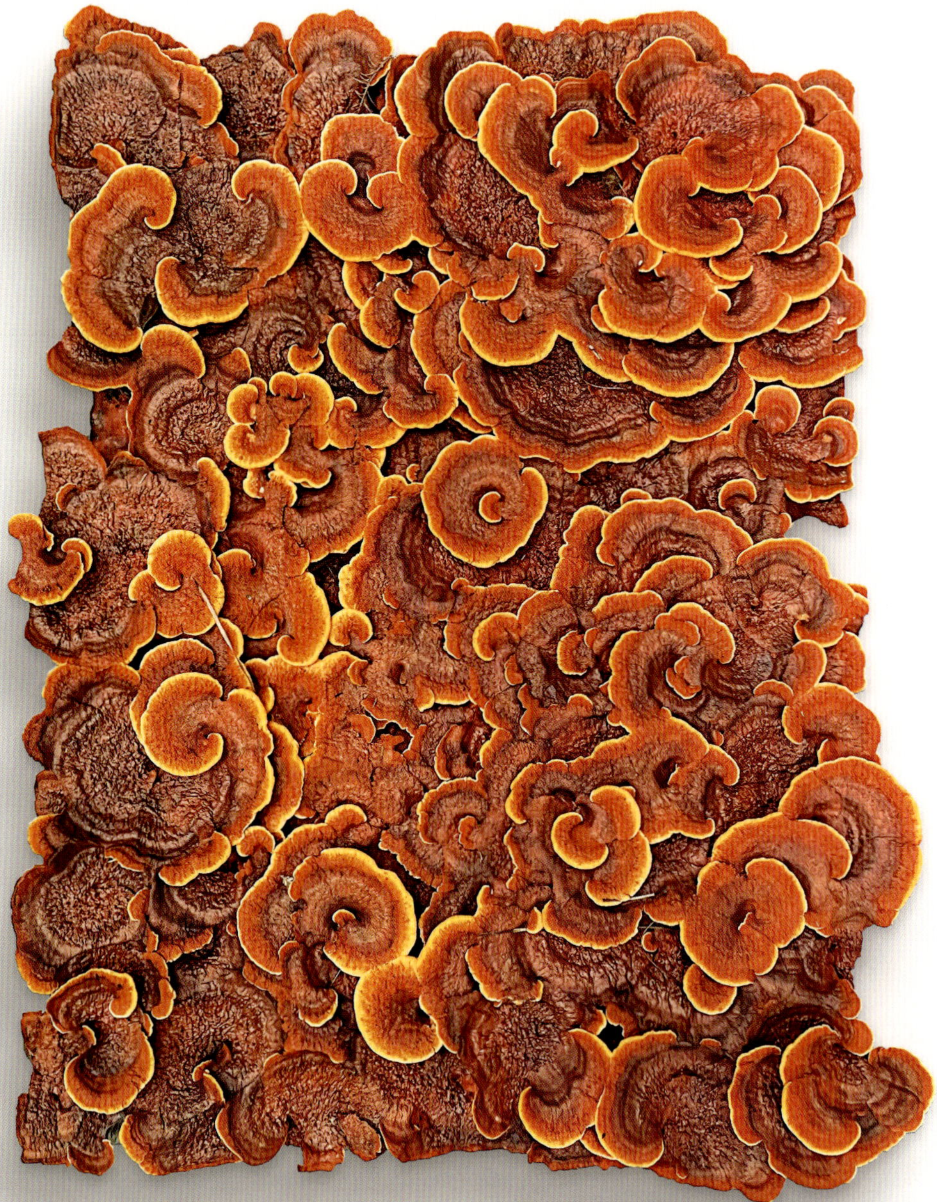

CHRIS RITSON

Mushroom Picture, 2022
Live fungus and sawdust,
20 × 14 in. / 50.8 × 35.6 cm

Fiery orange and yellow whorls of mushrooms comprise this living sculpture by American multidisciplinary artist Chris Ritson (b. 1985). Raised in O'ahu, Hawaii, and trained at the San Francisco Art Institute, Ritson works in installation as well as video, but it is his series of *Mushroom Picture* compositions that has garnered attention for its unique approach, combining contemporary art and mycology. Here, Ritson cultivated a thriving cluster of cinnabar bracket, or blood red bracket (*Trametes sanguinea*) in sawdust and promoted its growth until the mycelium spanned the whole canvas. The result is a spectacular display created by the red-orange caps, or conks, of the mushroom's fruiting bodies, giving the effect of molten lava in its luminosity. A species commonly found in the artist's home state, cinnabar bracket grows in tropical and subtropical regions on dead hardwood, such as oak and elm. The fungus is inedible but has myriad other uses: a pigment extracted from the caps is used in textile dyeing processes, and it has been applied widely in medicine for the treatment of everything from arthritis to toothaches, and continues to be researched. Ritson's artistic practice explores man's fraught relationship with nature, seeking to dampen humans' feeling of being at odds with the earth through his biogenerative works, which create instead of destroy, and are ultimately composted. He has also created natural works of art from algae and crystals, demonstrating that art can be more than just sustainable: it can sometimes create life itself.

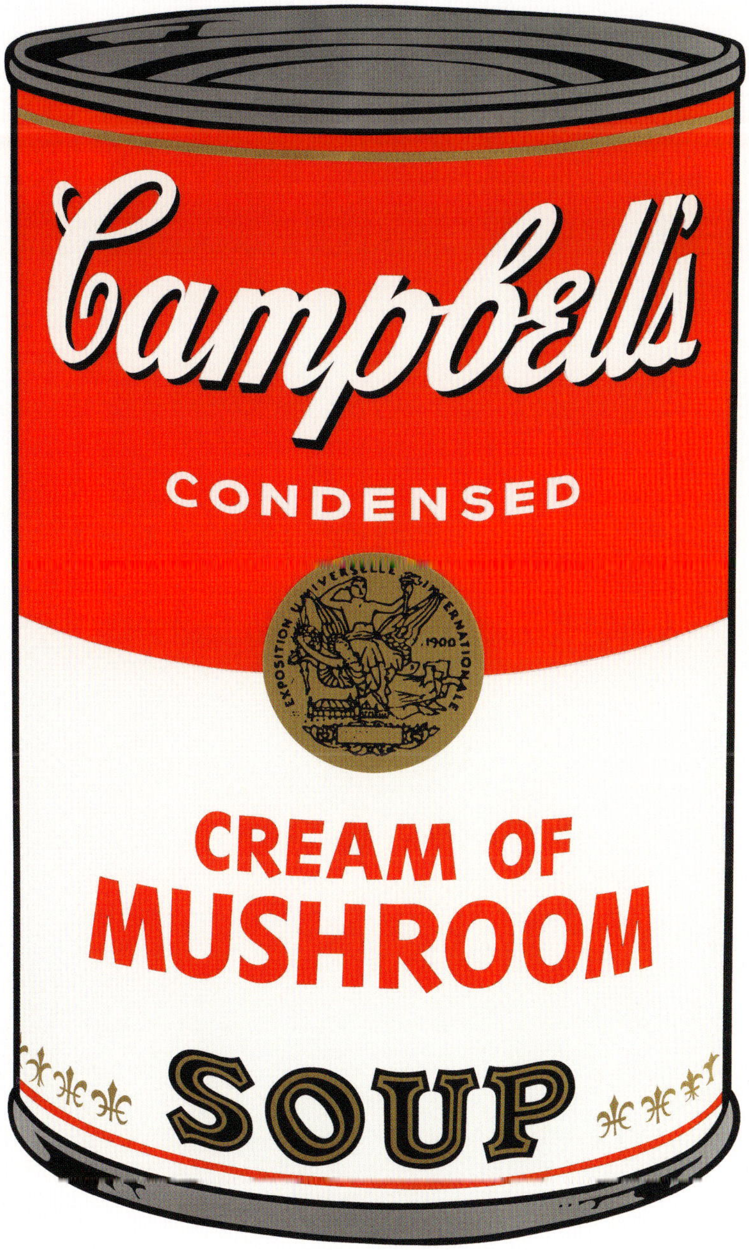

ANDY WARHOL

Cream of Mushroom, from the portfolio *Campbell's Soup I*, 1968
Screenprint, 35 × 23⅛ in. / 89.1 × 58.7 cm
Whitney Museum of American Art, New York

Transforming a mundane can of Campbell's Cream of Mushroom soup, the graphic red-and-white label is positioned boldly against a plain white background in this screenprint by Pop artist Andy Warhol (1928–1987). This elevation of the everyday to iconic status was one of Warhol's great contributions to twentieth-century art. Originally emerging from an illustration background, Warhol began to experiment with the idea of mass-produced, commercial products as an inspiration for both the process and subject of high art around 1960. His soup cans became one of the primary catalysts for the invention of Pop art. In 1962 Warhol created thirty-two silk-screen paintings of these cans and exhibited them in Los Angeles's Ferus Gallery. Originally, each 20-by-16-inch (50.8-by-40.6-cm) canvas was painted by hand, but over time Warhol reproduced them as screenprints. His process intentionally echoed the printing process used in advertising and signage. After World War II, the Campbell's brand had grown into one of the world's largest food companies, and its soup cans were everyday food found on most supermarket shelves. Warhol's popular images—which also included other soups, such as Tomato and Chicken Noodle—only made the brand more famous. Warhol's works were intentionally banal. He demonstrated how consumption and mass production were at the heart of the American dream, and how art could transform the experience of daily life. For many, looking at a can of Campbell's soup on a supermarket shelf would never be the same again.

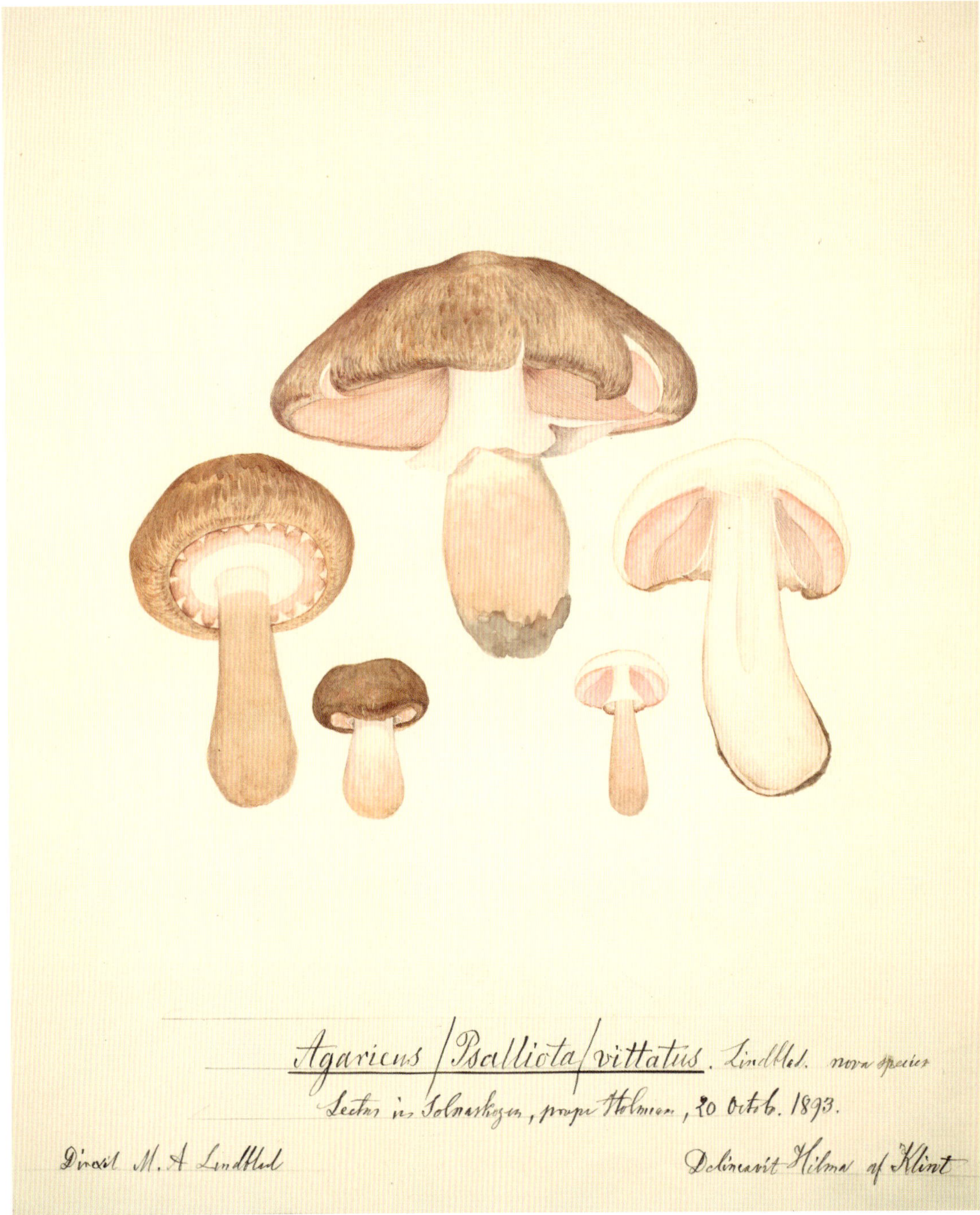

HILMA AF KLINT

Probably Agaricus bisporus (Cultivated mushroom), 1893
Watercolor, gouache, pencil, and ink on paper,
16⅜ × 12 in. / 41.6 × 30.3 cm
Naturhistoriska riksmuseet, Stockholm

Although enigmatic and visionary Swedish painter Hilma af Klint (1862–1944) was a working artist throughout her lifetime, it was only posthumously that she gained renown. Upon her death she left behind a massive trove of work—more than 1,300 paintings and 124 notebooks totaling about 26,000 pages—with the stipulation that none of it be shown for twenty years. A lifelong proponent of occultism, af Klint believed that she had the ability to communicate with beings on the spiritual plane, allowing her to create what are now credited to be among the first major abstract paintings in Western art, wildly original works that remained the focus of her practice for decades. Early in her career, however, af Klint worked in the precise, delicate art of scientific illustration. This botanical study of a humble *Agaricus bisporus*, the common cultivated mushroom, is one such work. Part of a series of seven mushroom compositions, each depicting a different species, this field study capturing the early brown and white tones of the mushroom was commissioned by mycologist M. A. Lindblad in 1893, but it was only first displayed publicly in the twenty-first century. After turning to abstract art in the late 1890s, af Klint would not pick botanical illustration back up until decades later. From 1919 to 1920, she produced nearly fifty botanical drawings, which she called *Nature Studies*, revealed to the public for the first time at New York's Museum of Modern Art as part of the groundbreaking 2025 exhibition *Hilma af Klint: What Stands Behind the Flowers*; this mushroom series was featured alongside them.

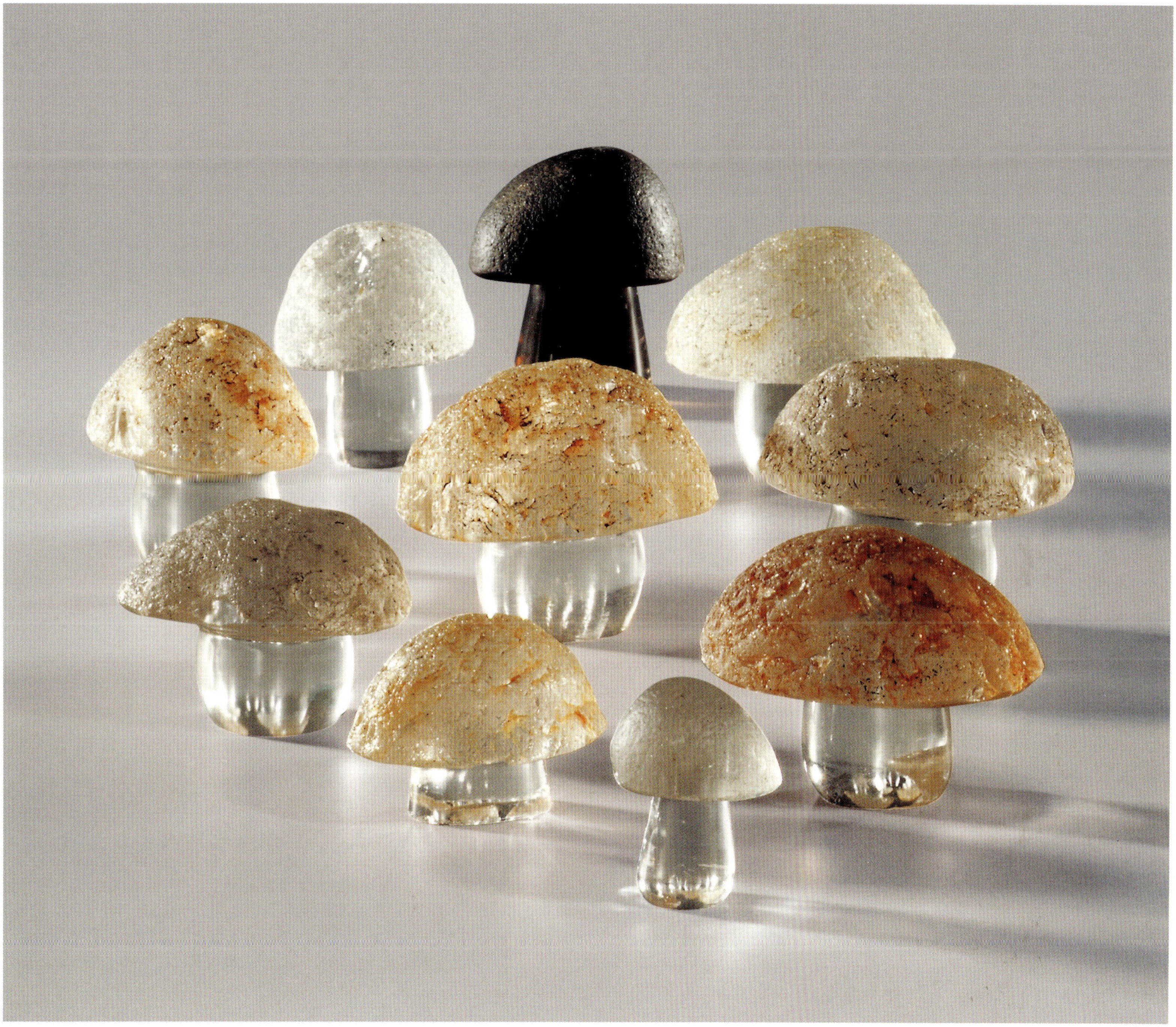

FABERGÉ

Seals carved in the shape of mushrooms, c. 1890
Rock crystal, dimensions variable
Private collection

Carved out of rock crystal, each of these ten mushrooms is unique. Thanks to its brittleness, the translucent quartz material from which they are carved is notoriously difficult to work with, giving these seals a naturalistic appearance, with indentations and spores on some of the caps, and variance in size, color, and shape. Designed by the House of Fabergé, the Russian jewelry firm helmed by Peter Carl Fabergé (1846–1920), these fungi might seem an unexpected creation from a house best known for its highly ornate enamel eggs created for the Russian imperial family. In fact, Fabergé worked with several different materials, including rock crystal from Russian mines, to craft intricate objects for Europe's leading families. These mushrooms are seals used to authenticate important documents by stamping a design in soft wax. Each bears a different cypher on its base: a crowned *A* and crowned *V*, as well as the interwoven initials *EA* and *AA*. They were made for Prince Valdemar of Denmark, third son of King Christian IX, who became a customer of Fabergé after his older sister, Princess Dagmar, married the future Czar Alexander III. It was as an Easter surprise for Dagmar—known in Russia as Maria Feodorovna—that Alexander first commissioned Fabergé to design an egg in 1885. The choice of mushrooms for Valdemar is apt: fungi have long been part of the culture in Denmark, where around nine hundred edible species grow today.

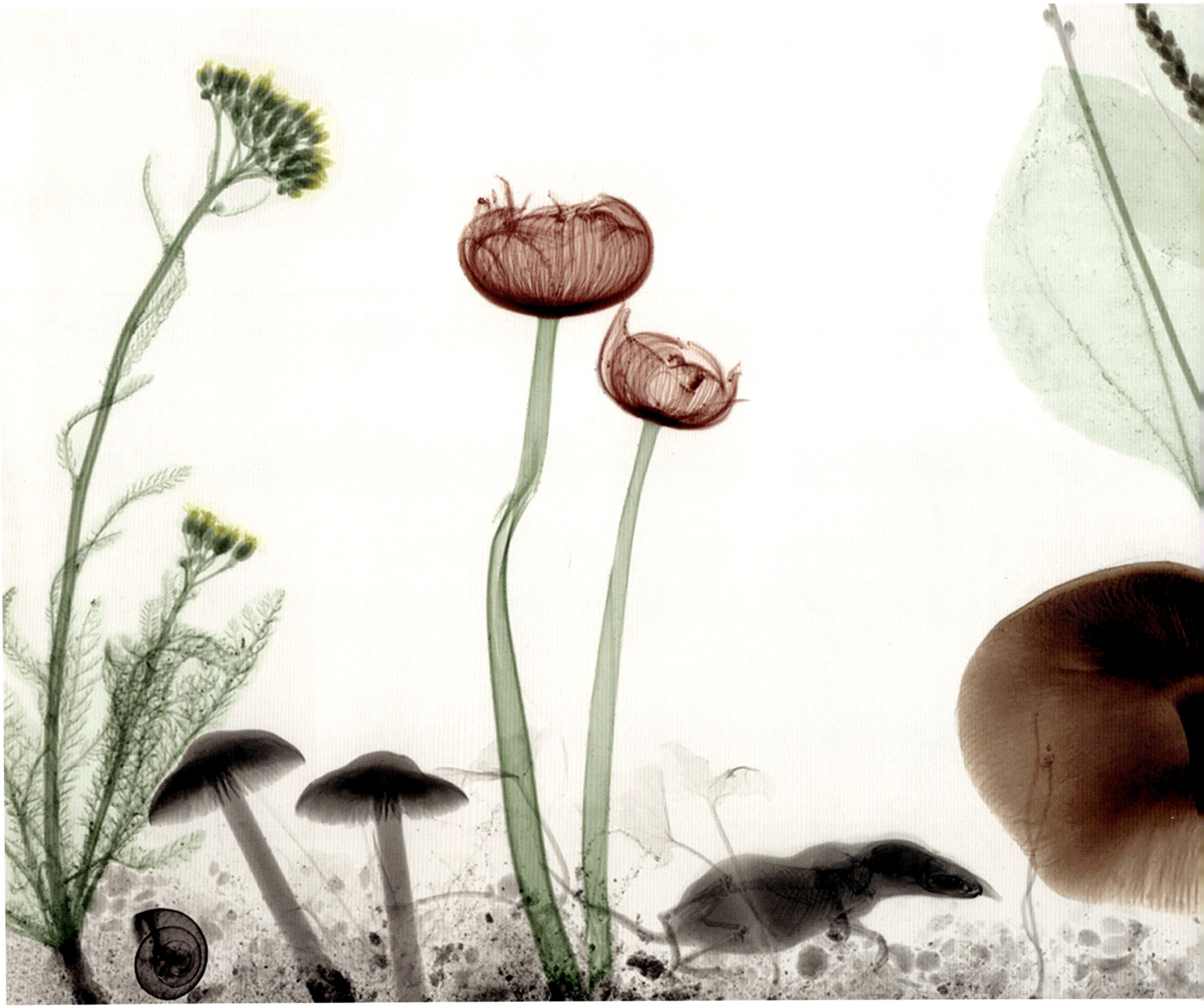

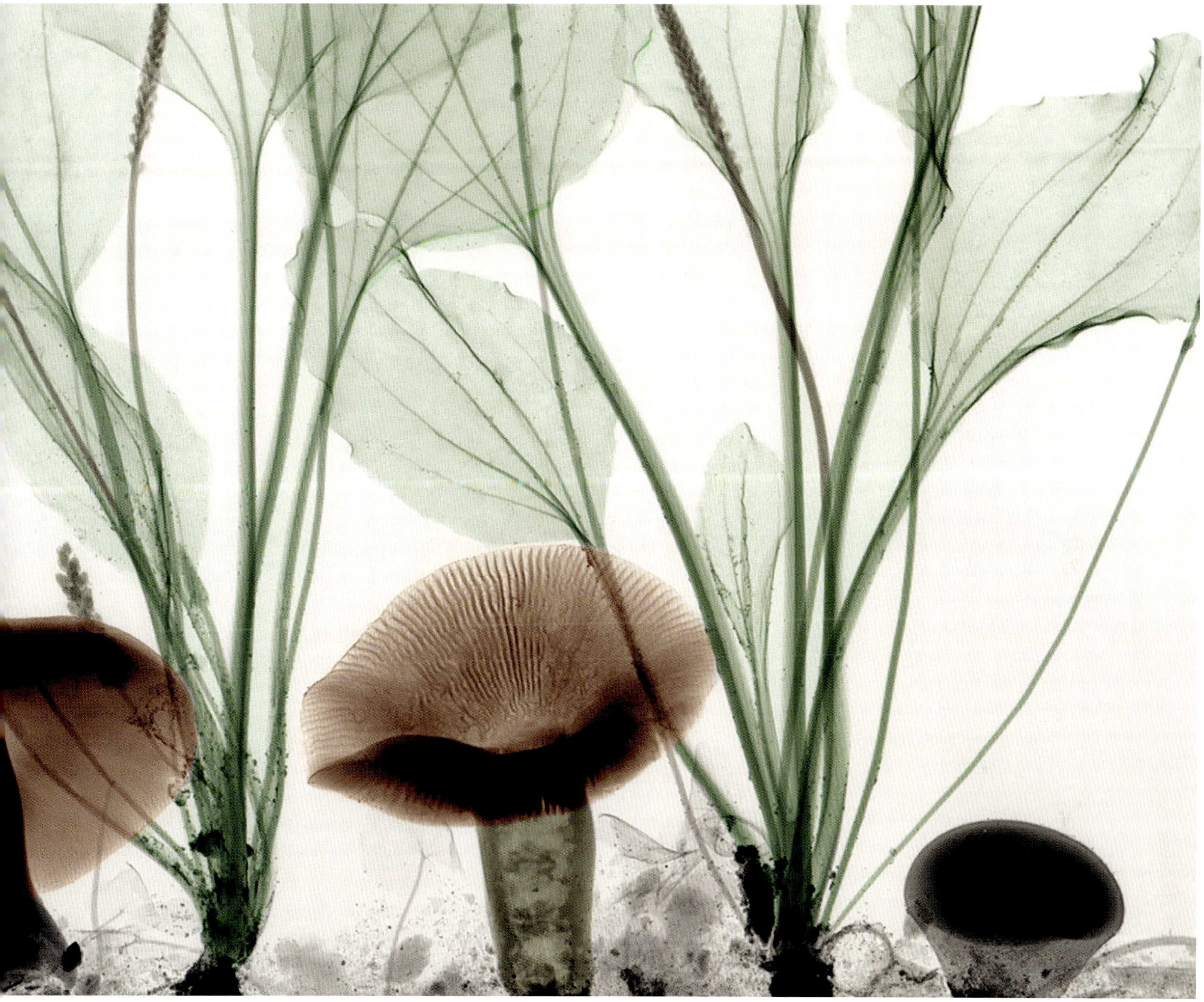

ARIE VAN 'T RIET

Shrew Among Mushrooms, 2014
Digitized colored X-ray, dimensions variable

Growing up in what is called the Green Heart of the Netherlands—a largely rural area in the western part of the country—Arie van 't Riet (b. 1947) was enamored by the many plants and animals around him. Wanting to know everything about this natural world, including the parts his naked eye couldn't perceive, he gravitated toward biology. This later led to a degree in radiation physics and a career as a medical physicist in hospital radiation oncology departments. Little did he know, this was all preparation for his second act as an artist. Science met art under Van 't Riet's gaze by chance, when he started to digitize and improve some flower X-rays he had shot as practice for his day job. Finding them visually compelling, he began to add more plant life and eventually animal specimens (all collected humanely) to mirror the coexistence of plants and animals in the natural world. He calls these narrative environments biological dioramas, or "bioramas." To make them, he slides an analog X-ray negative measuring 13 by 16 inches (33 by 41 cm) under each biorama and shoots the X-ray life-size. A biorama often needs to be adjusted and reshot, but Van 't Riet never uses any kind of stacking or digital layering in the creation of his art. Once he is happy with a biorama X-ray, he frequently adds color digitally to flowers and plants only, as seen in the mushrooms and greenery in this work. He never colors animals, as he feels it is impossible to do so in a way that looks natural.

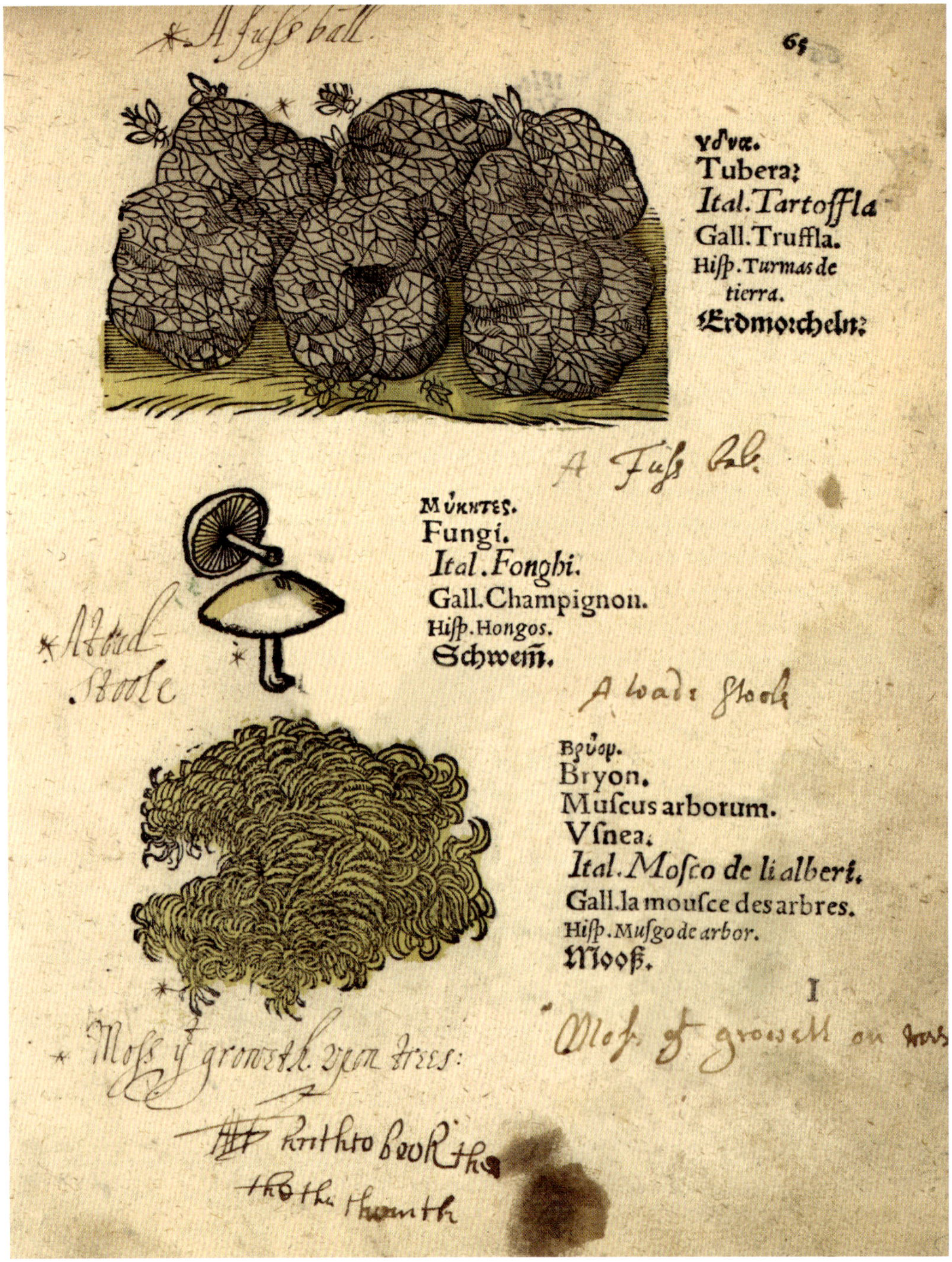

ADAM LONICER

Page from *Kreuterbuch (Herbal)*, late 16th century
Hand-colored woodblock engraving on vellum,
7⅝ × 6 in. / 19.5 × 15 cm
Private collection

This hand-colored woodcut is one of hundreds that appeared in the *Kreuterbuch*, or herbal, by botanist Adam Lonicer (1528–1586), which chronicled the growing and medicinal uses of plants and fungi, alongside the documentation of various animals and mythical creatures, as well as the creation of gardens. A man of many interests, Lonicer was a professor of mathematics at the Lutheran University of Marburg, Germany, and also studied medicine, later becoming the City Physician of Frankfurt. His true passion, however, was herbal medicine and botany, and he published several editions of his *Kreuterbuch* between 1557 and 1577. The publication became a bestseller, and editions were still being produced in Germany more than two hundred years later, making Lonicer a household name. (The honeysuckle genus *Lonicera* was also later named in his honor.) In fact, this page comes from an atlas edition of illustrations only, published in the late sixteenth century. The woodcut features three subjects—truffles, a generic fungus, and tree moss—each accompanied by captions in Greek, Latin, Italian, French, Spanish, and German. *Tuber melanosporum*, also known as the black or Périgord truffle, is the epitome of gourmet indulgence, and today is one of the most expensive culinary fungi in the world. Boars and the larvae of the truffle fly (*Suillia tuberiperda*), which are delightfully depicted in the illustration, eat the fruiting bodies, then distribute the truffle by excreting its indigestible spores. The interior of the truffle is black, with strong notes of chocolate and molasses, and the aroma is likened to that of a forest after a heavy rainfall.

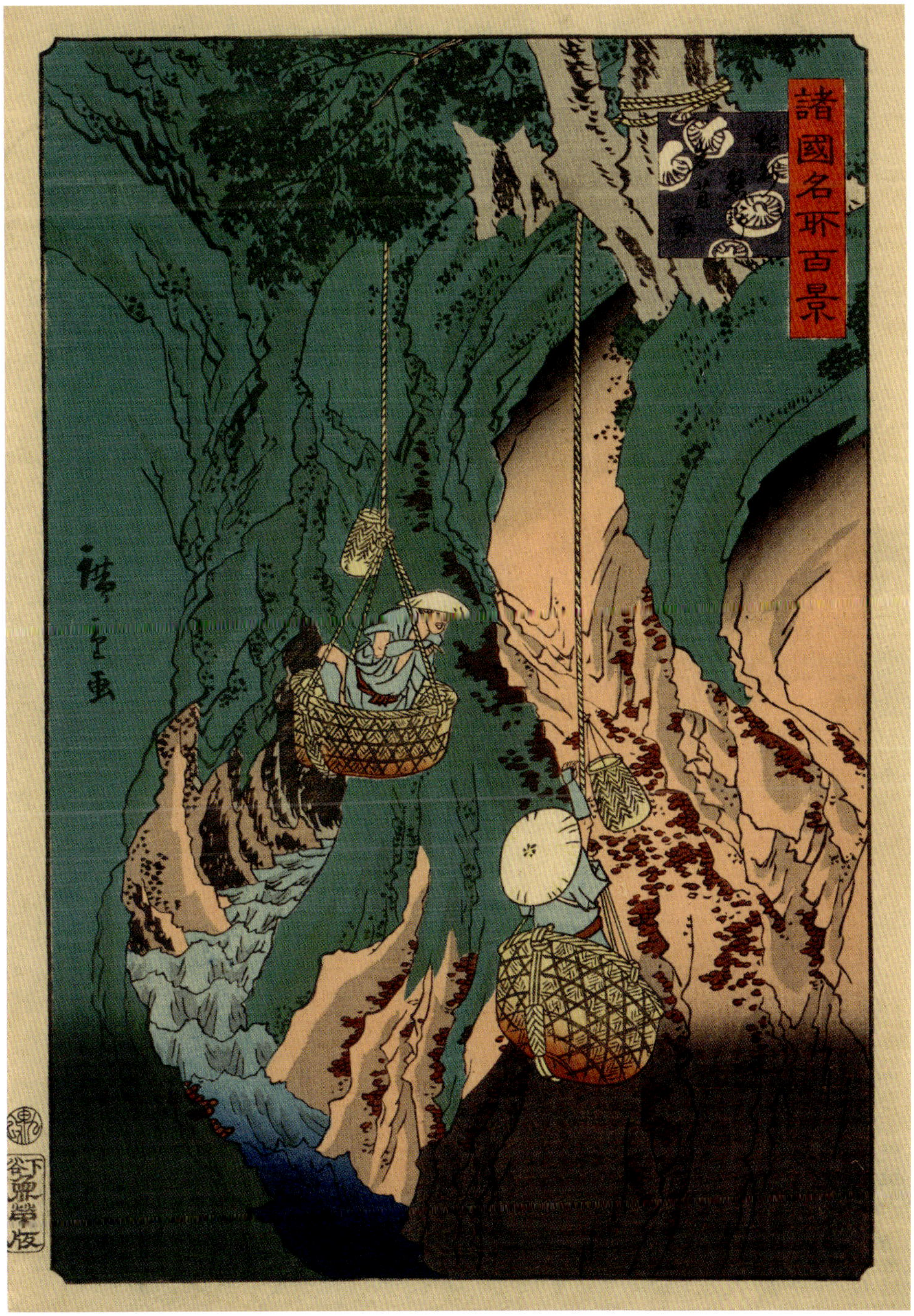

UTAGAWA HIROSHIGE II

Iwatake Mushroom Gathering at Kumano in Kishi Province, from the series *One Hundred Famous Views in the Various Provinces*, 1860
Woodblock print; ink and color on paper, 14⅛ × 9⅜ in. / 35.9 × 23.8 cm
Private collection

Two men in wicker baskets, dangling from ropes tied to a nearby tree, gather mushrooms off the side of a precipitous cliff, seemingly unphased by the sheer drop beneath them. This woodblock print by Japanese artist Utagawa Hiroshige II (1826–1869) belongs to a larger series, *One Hundred Famous Views in the Various Provinces*. Hiroshige II was the most accomplished of all the students of the Edo-era Japanese artist Utagawa Hiroshige, who is best known for his woodcut series *One Hundred Famous Views of Edo*, which shows different scenes of the imperial capital, Edo (now Tokyo). Hiroshige II took the name of his master and developed a style so similar that the two men's work is often confused. This print was made in 1860, a couple of years after the death of the original Hiroshige, when his student started to sign his prints as Utagawa Hiroshige II. Hiroshige II shows the two men picking *iwatake*—or rock tripe (*Umbilicaria esculenta*)—an edible lichen that grows on bare rock at high altitudes. Despite a bitter taste, the lichen is a delicacy in East Asian cuisines and can be eaten, usually in salads or deep-fried, once soaked and boiled in water several times. The scene gives an insight into nineteenth-century rural life in the Japanese mountains: both men seem at ease despite the danger, with such a plentiful supply of mushrooms being too enticing to ignore.

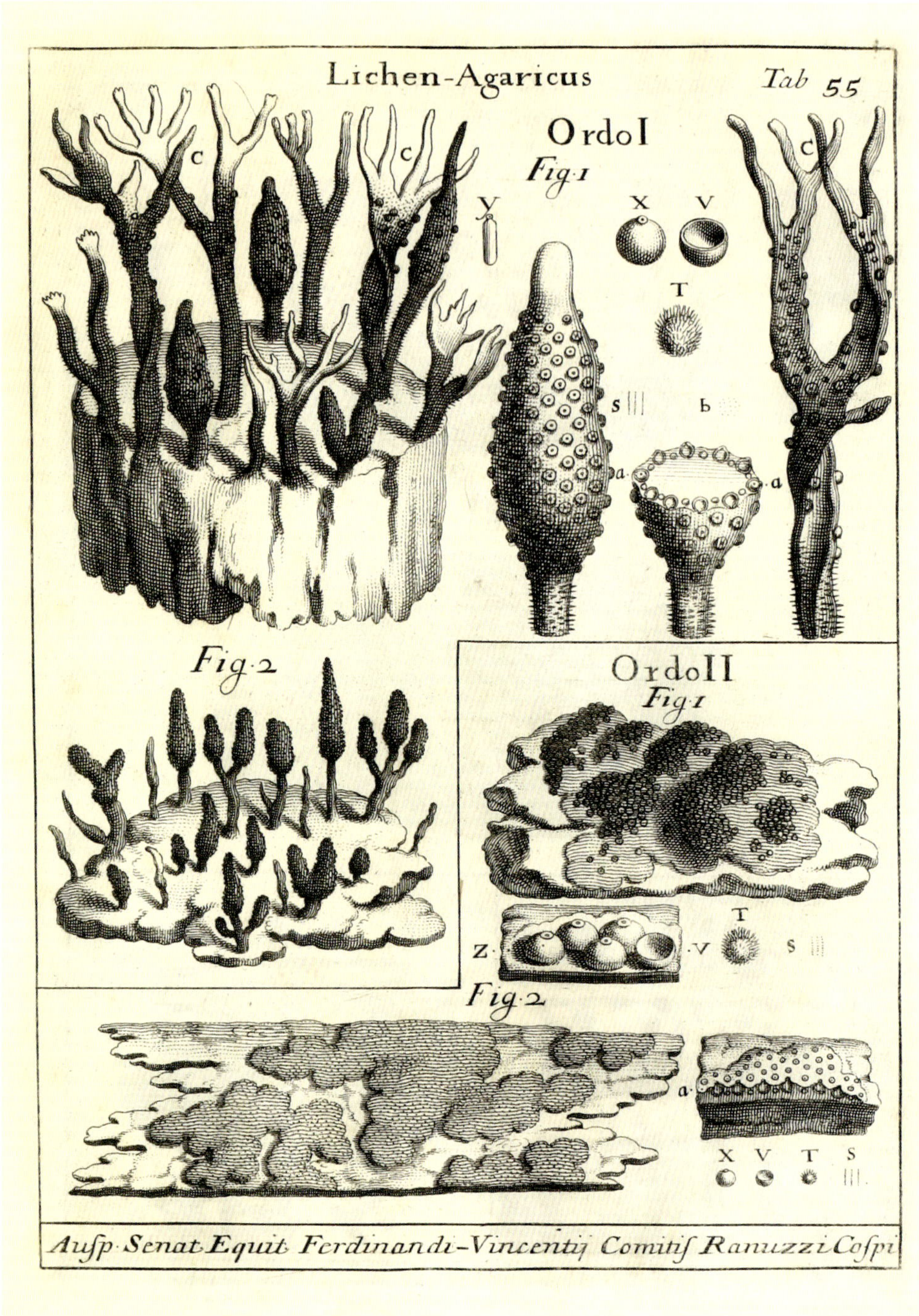

PIER ANTONIO MICHELI

Lichen Agaricus, from *Nova plantarum genera*, 1729
Engraving, H. 12¼ in. / 31 cm
University of Illinois Urbana-Champaign

Devoid of color but no less dramatic in their display, the unnatural fingers of lichen in this plate rise like zombies from the grave. Coral-like and crusted specimens grow upward from dense substrates or spread laterally in layered sheets, relying on finely engraved linework and crosshatching to communicate texture, density, and volume. Alongside the illustrations, magnified details emphasize tiny nodules and cups dotting the surfaces, and, unusual for botanical illustrations, shadows accentuate the forestlike growth. This is one of 108 plates from *Nova plantarum genera* by Pier Antonio Micheli (1679–1737), which revolutionized the study of fungus. The largely self-educated Italian was curator of the Florence botanical garden, founder of the Società Botanica Fiorentina, and botanist to the grand duke of Tuscany. As the first to carry out scientific studies on fungi, he is considered by many to be the founder of scientific mycology, and his experiments proved that fungi could be predictably grown from spores, flying in the face of popular theories that mushrooms spontaneously arose, for example, from moist, rotting matter or lightning strikes. Balancing order and wonder, this image is representative of Micheli's experimental work and supports his radical observations about fungal reproduction. These lichens, arranged in clearly delineated and categorized groupings, with labels and accompanying descriptions, are illustrated with care and precision, yet Micheli has also chosen to emphasize their otherworldly forms and textures. Here, Micheli positioned the fungi, which had previously been subjects of mere curiosity, as central actors in the unfolding story of science.

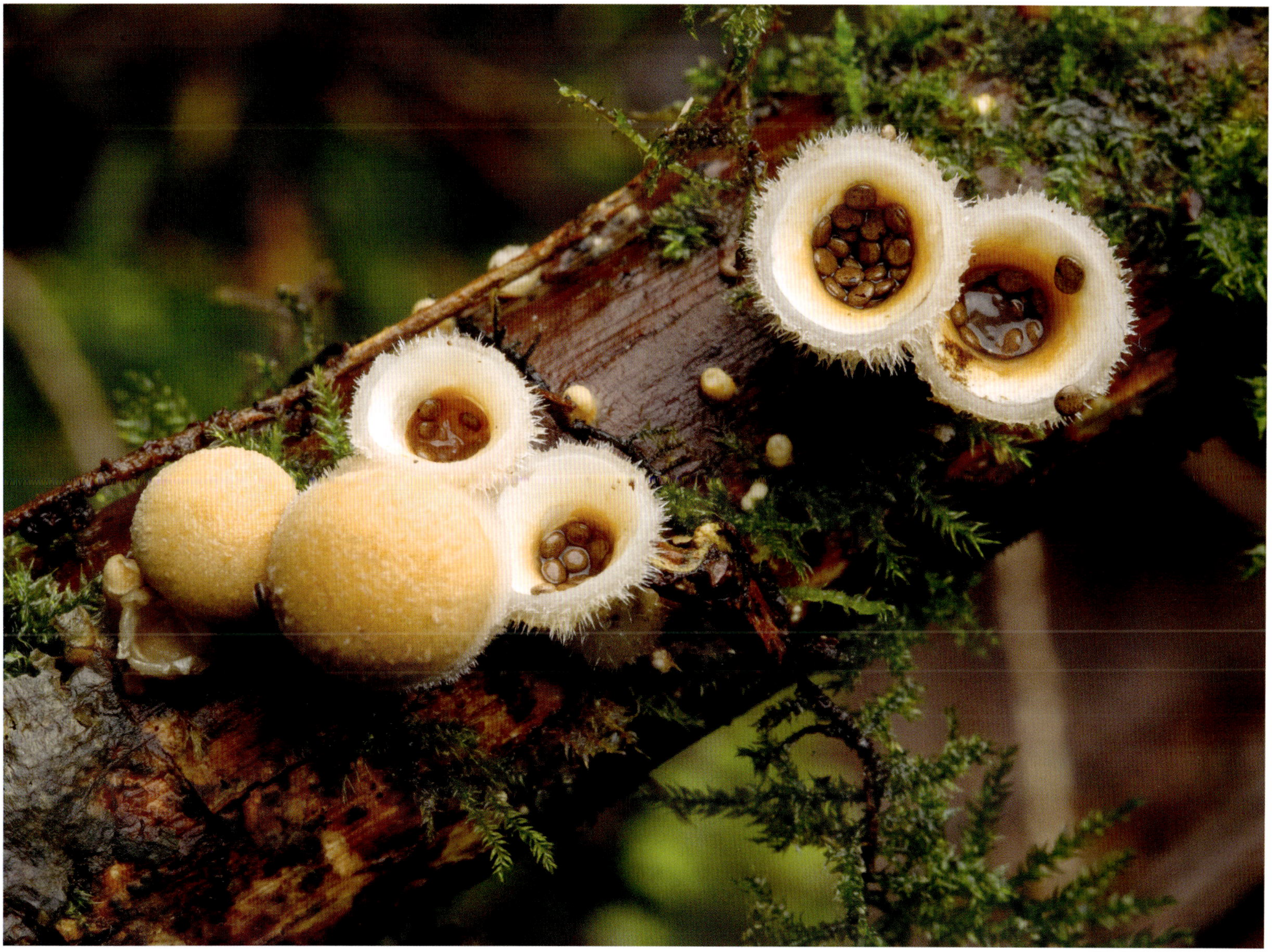

JAYMI HEIMBUCH

Bird's Nest Fungus (Nidulariaceae), Oregon, 2018
Digital photograph, dimensions variable

These tiny saucers might resemble a cozy bird's nest with a clutch of miniature eggs, but they are actually living organisms—and their form is entirely functional. Meet the bird's nest fungi, a group of around 150 to 200 species in the family *Nidulariaceae*. Each fruiting body grows in a concave shape commonly known as a splash cup, meant to house the "eggs," or peridioles, which each contain thousands of individual spores. The peridioles sit snugly inside until rain or another disturbance dislodges them. Incredibly, the force of impact can launch these spore-containing capsules more than 3 feet (91 cm). When each one lands, it releases a cord that, with any luck, tethers itself to a blade of grass or a twig using a super-sticky end called a hapteron. And there it dangles until it is dislodged, broken, or most ideally eaten and passed through an animal's excrement, further increasing the distance of the fungus's dispersal and chances for survival. There are five different genera of bird's nest fungi, but all are quite small, with the cups typically growing no wider than half an inch (15 mm). They are also gregarious—in the scientific sense—growing individually but adjacent to one another in groups. Bird's nest fungi feed off decomposing organic matter, such as fallen trees—as can be seen here in this photograph of a charming cluster captured by Jaymi Heimbuch in the woods of Oregon.

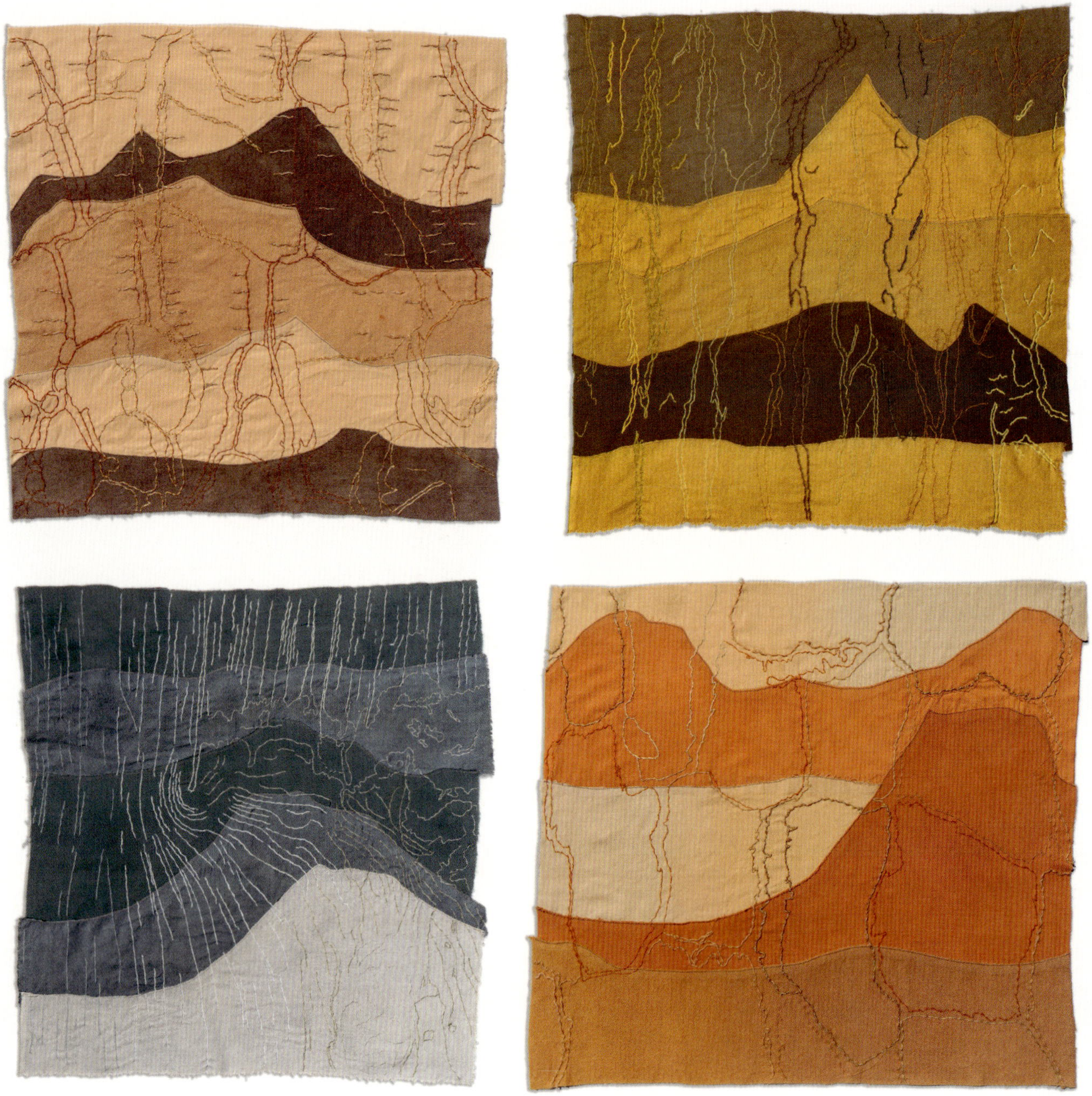

JULIE BEELER

Fungi Symbiosis, 2021
Mushroom-dyed wool and embroidery thread,
overall 49 × 49 in. / 124 × 124 cm
Private collection, Boston

A self-described alchemist, American textile artist, educator, and mycology enthusiast Julie Beeler (b. 1970) works exclusively with natural dyes, predominantly made from fungi, transforming fabrics into art. The textiles she employs are made from wool, which she says has been "cut, sewn, stitched, and constructed to reveal patterns, textures, and color that are a record of a place and time." The four different color palettes seen in *Fungi Symbiosis*, which recall the four seasons, have been extracted from different mushrooms foraged in her local forests at the foot of Mount Adams, in Washington State's Cascade Range. Overlaying her stitched, dyed fabric, Beeler adds a series of embroidered lines to represent mycelium networks—the vast underground of fine fungal threads that remain invisible. Her art starts with the fruiting body of a foraged mushroom, which she processes to produce pigment. Decades of observation and experimentation—she first foraged for mushrooms in forests in her native Oregon and around nearby Mount St. Helens as a child—have taught her that fungi change continually: the color a mushroom produces will vary with the season and the point at which it is picked in its lifespan. Beeler's ongoing experimentation with the kaleidoscope of color produced by different fungi species is recorded in her book, *The Mushroom Color Atlas* (2024), which reveals the chromatic world of fungi. Each of the 825 pigments in the book is identified with the fungus from which it is produced, including boletes, polypores, and tooth fungi, among many others.

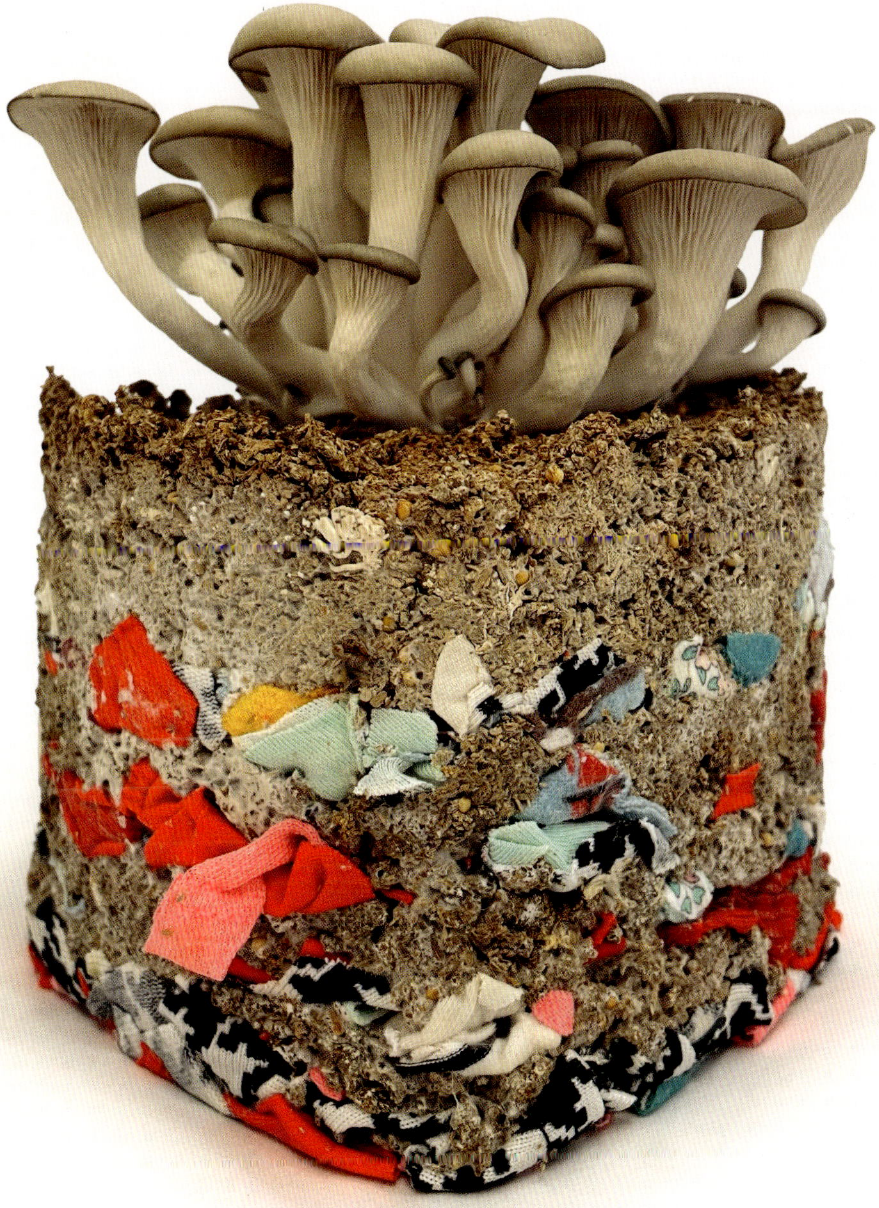

MINGA OPAZO

Re-Dress, 2021
Recycled clothing and oyster mushroom mycelium, 15 × 10 in. / 38.1 × 25.4 cm

For Chilean-born artist Minga Opazo (b. 1992), fungi represent an exciting solution to a modern problem: the permanent disposal of human-generated waste that lives in landfills around the world. Trained as a textile artist, Opazo, who now lives and works in Los Angeles, began investigating textile waste, which led to her desire to contribute to solving the problem as well as representing it in her work. Working alongside fungi researcher Danielle Stevenson—whose studies into mycoremediation (the degrading of contaminates in soil) has shown how soil can be cleaned by mushrooms—Opazo creates a series of sculptures that use mushrooms as waste management, including *Re-Dress*. Within the layers of shredded textiles and organic material is the mycelium of the oyster mushroom (*Pleurotus ostreatus*), known for its capability to digest any organic material, which over time ultimately breaks down the fabric into soil. At the top of the cube-shaped mound, the fruiting bodies of the mushrooms make visible their presence within the sculpture. Opazo's passion for fungal sculptures is fueled by the statistic that 85 percent of clothing ends up in a landfill. The Atacama Desert in her native Chile is just one location where discarded textiles from Europe and the United States form huge mountains of waste. Ongoing research shows that fungi are saprophytes, meaning they can break down plastics and other fossil fuel–based materials while decontaminating the soil and returning it to a clean state. Opazo's sculptures combine art and science in an arresting way, affording the viewer an opportunity to reconsider the hidden superpowers of the humble mushroom.

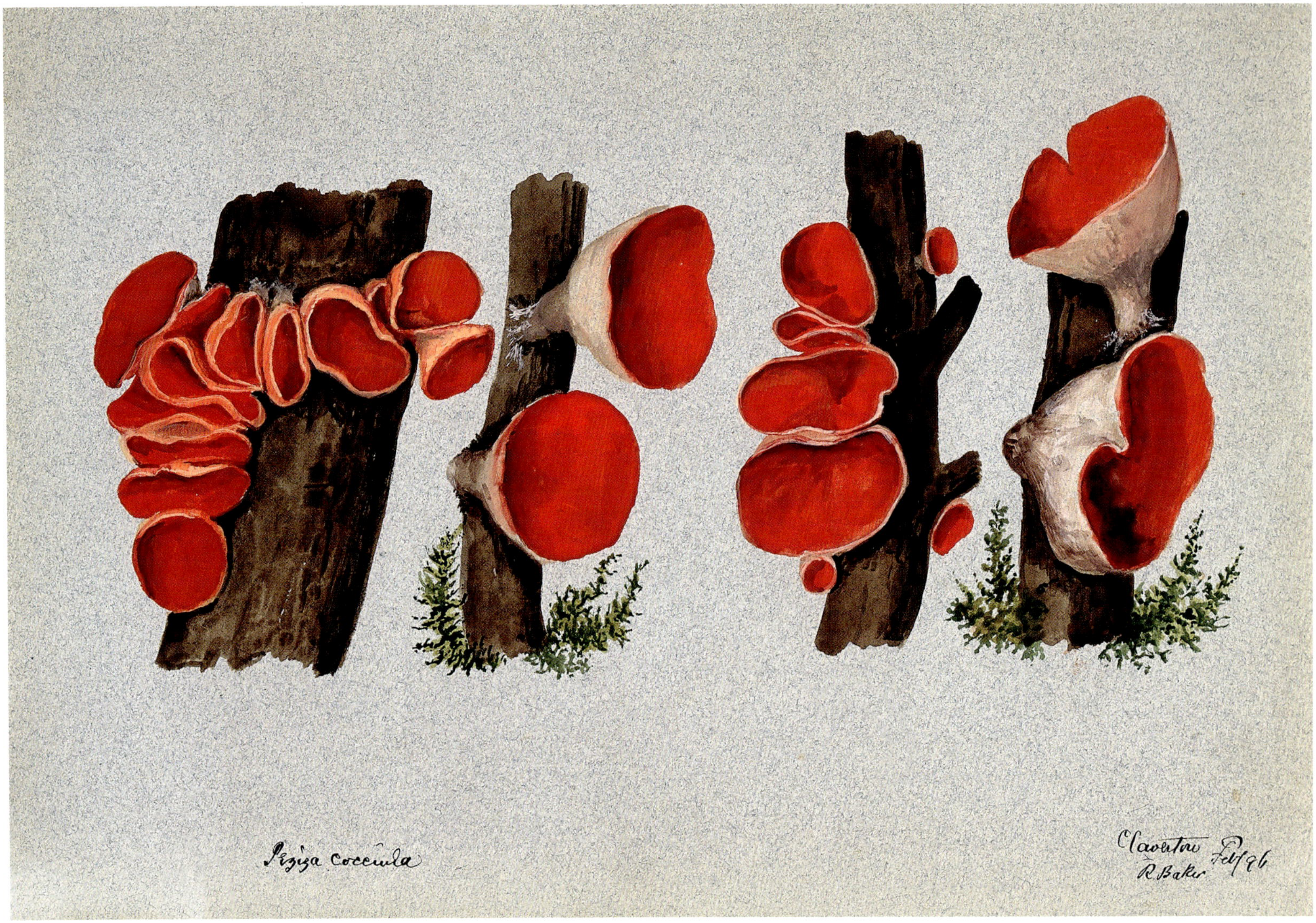

R. BAKER

Peziza coccinea, 1896
Watercolor on paper,
7½ × 10¾ in. / 19.2 × 27.2 cm
Wellcome Collection, London

Found across the world in shady, damp deciduous woodlands, the ruby elfcup (*Sarcoscypha coccinea*), which is often mistaken for the nearly identical scarlet elfcup (*S. austriaca*), is easy to spot with its large bright red cups and short stems. The fungus typically appears during late winter and early spring, growing on dead branches and twigs of elm, hazel, and willow trees and under fallen leaves. Edible mushrooms, they can grow up to about 2½ inches (6 cm) across, making them twice as wide as they are tall. As they mature, their caps grow to resemble deep cups—hence their name—but over time the cups become shallower, and their edges often split. When they release their spores, they make a puffing sound as the spores explosively disperse. This mushroom's distinctive form is captured here in a watercolor by a little-known British amateur artist identified only as R. Baker, who painted at the end of the nineteenth century. It is thought that this drawing was made in the west of England near the River Severn, possibly for an art class, which were popular in Victorian Britain. While nothing more is known about the artist, Baker's artistic ability is clear to see. Each of the four twigs is adorned with scarlet elfcups of varying sizes: those on the far right show just how big they can grow, while those on the far left are much smaller and clustered together.

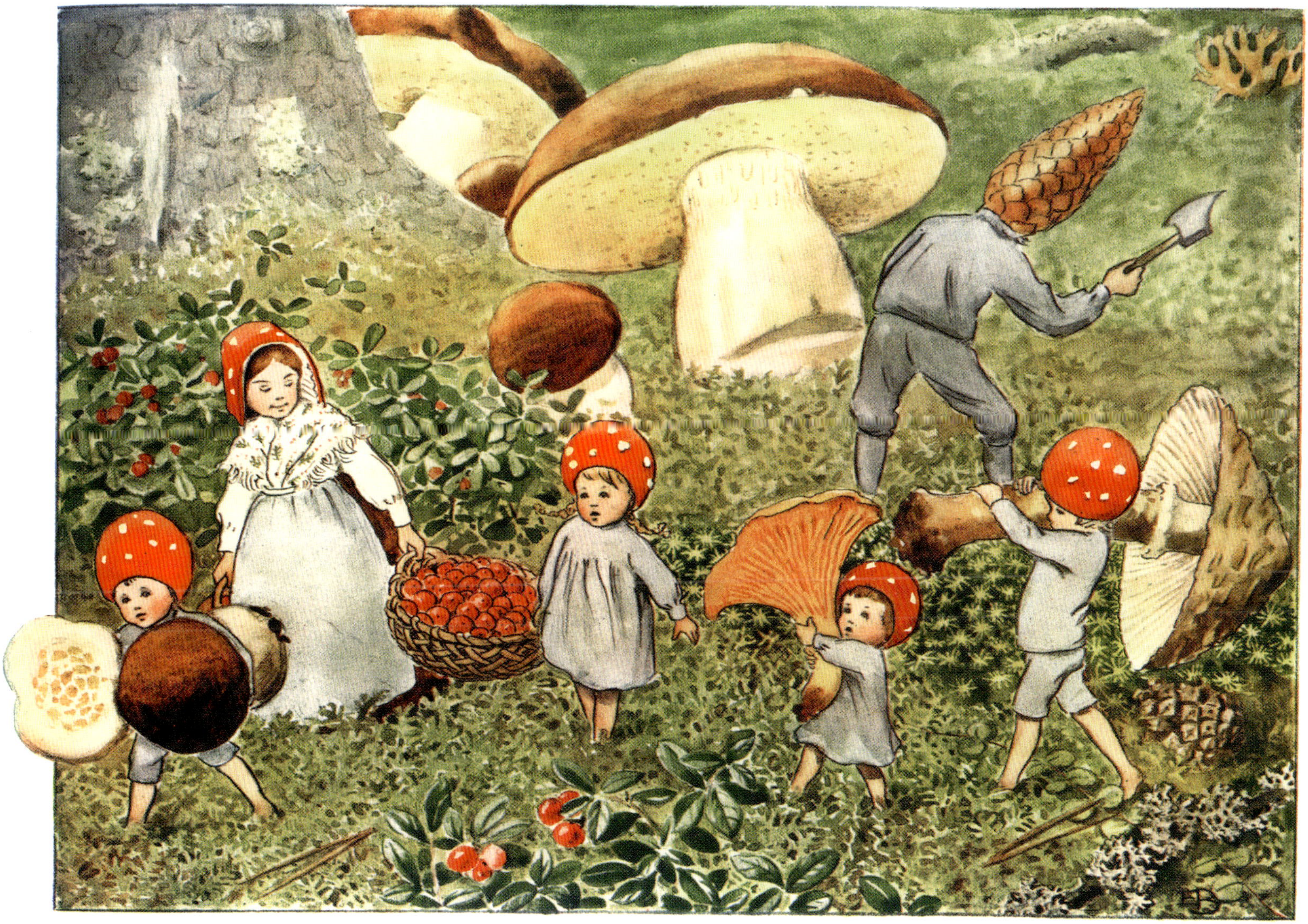

ELSA BESKOW

Page from *Children of the Forest*, 1910
Ink and watercolor on paper,
7½ × 10¾ in. / 19.3 × 27.2 cm

Swedish author Elsa Beskow (1874–1953) wrote and illustrated more than forty books in her lifetime, her images and stories celebrating the nature and freedom of childhood. This timeless scene of four elfin figures and their parents gathering huge edible mushrooms and fruit forms a page of Beskow's classic children's book *Children of the Forest*, published in 1910. A contemporary of Beatrix Potter (see p.6), Beskow likewise created a cunning blend of fairy-tale fantasy and naturalism: the father figure, wearing a pinecone headdress, is slashing at a porcino, also known as a cep or penny bun (genus *Boletus*); one girl carries a fluted yellow fungus, probably a chanterelle (*Cantharellus*), while the boy on the right shoulders a parasol mushroom (*Macrolepiota procera*). The children and their mother are crowned with large helmets of white-dotted, bright red fungi resembling fly agaric (*Amanita muscaria*). Other pages of the book show further woodland adventures in different seasons. The children live deep in the roots of an old pine tree, play with squirrels and frogs, and shelter under toadstools when it rains. Born Elsa Maartman in Stockholm, Beskow was partly brought up by her aunts, who no doubt inspired her other popular tale, *Aunt Green, Aunt Brown and Aunt Lavender* (1918). She studied art education at the Konstfack, part of Stockholm's technical school, married minister and social worker Natanael Beskow, and had six sons. She published her first book in 1897, and for more than a hundred years since, children have grown up with Beskow's books, bringing a small piece of the forest's magic into their homes.

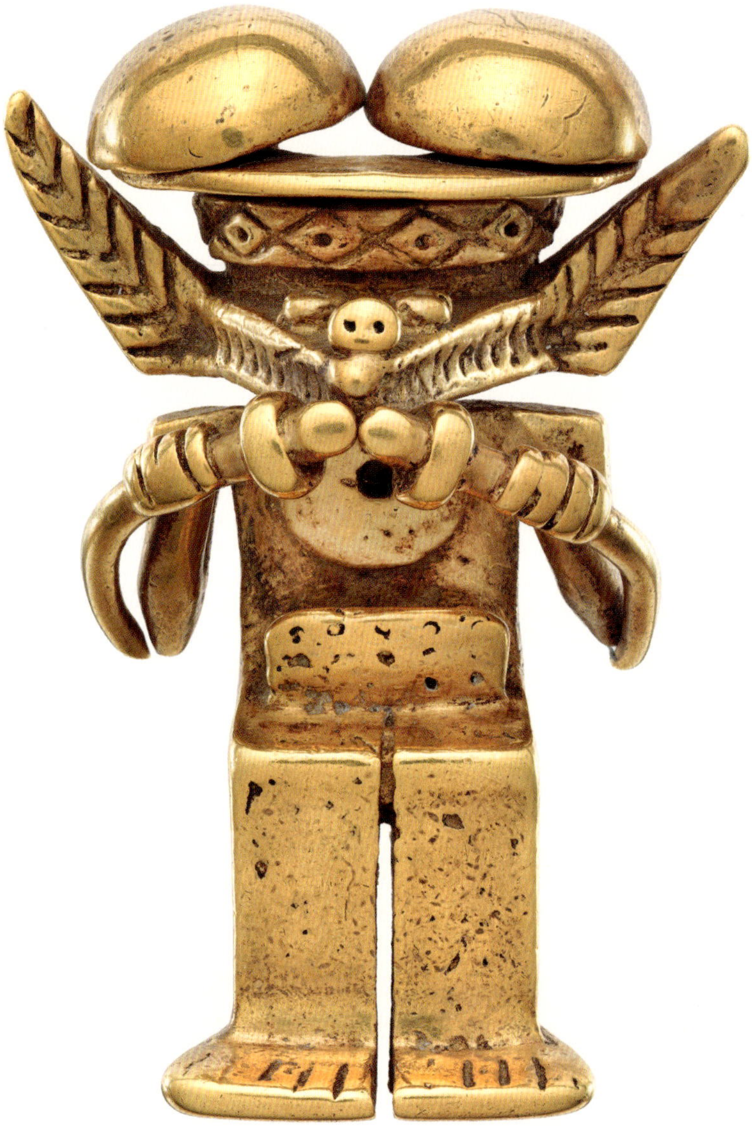

ANONYMOUS

Figure pendant, 1st–7th century
Gold, 2½ × 1¾ × ⅞ in. / 6.4 × 4.4 × 2.2 cm
Metropolitan Museum of Art, New York

Made from solid gold, this figure pendant is topped with two dome shapes that have, among other interpretations, been thought to represent mushrooms. While much about this enigmatic figure remains a mystery, the body is that of a human and its facial features are that of an animal, with a long, pointed nose and leaves or feathers near the ears. The flat cap is decorated with a lozenge pattern, and the two domes on top have been variously interpreted as mushrooms, gourds, or ceramic vessels. What is known is that the pendant was made in present-day Colombia sometime between the first and seventh centuries CE. Many similar pendants have been discovered in the Tolima and Calima regions, from the southwestern highlands to the Caribbean coast. The pendants were made using a lost-wax technique, whereby molten gold was poured into a mold built around a wax figure; the wax melted, drained away, and was replaced by the molten metal. Figures such as this one were made for more than a thousand years, but what they were meant to represent still remains unclear. Some scholars think they represent shamans in a state of spiritual transformation, with bent knees indicating the position the shaman took while meditating or communing with the gods. In that interpretation, the fungal forms are representative of psilocybin mushrooms, which shamans used to induce hallucinogenic trances to help with their divine communication.

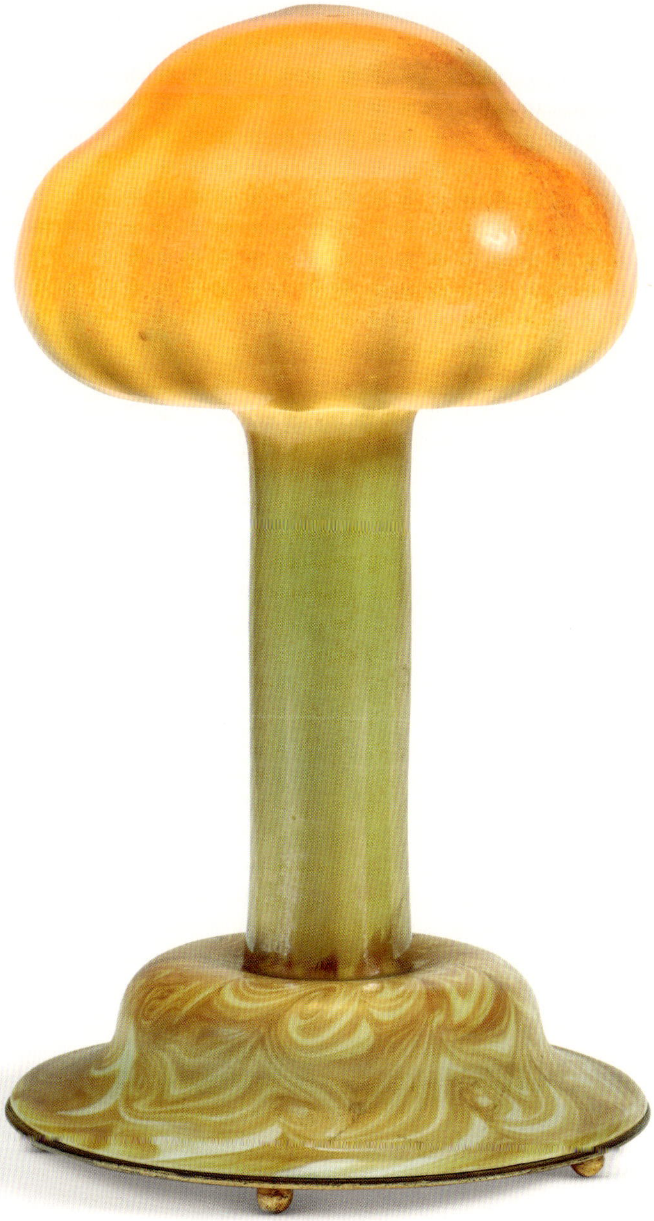

TIFFANY STUDIOS

Mushroom Lamp, c. 1902
Favrile glass and gilt bronze,
11¾ × 6½ in. / 29.9 × 16.5 cm
Private collection

Emitting a golden light from its bulbous cap, this attractive table lamp with its mushroom-shaped lampshade was made by the famed decorative arts house Tiffany Studios. Louis Comfort Tiffany (1848–1933), a key figure in the American Art Nouveau movement, founded the studio in Queens, New York, in 1902. While the small lamp may not be as instantly recognizable as the studio's iconic multicolored leaded-glass flower-and-butterfly lamps, it is made using Tiffany's signature favrile glass and patinated bronze. Favrile glass, which Tiffany developed and patented in 1894, is an iridescent glass in which the color is ingrained rather than being painted on, as was more typical of glass of the period. Tiffany coined the name *favrile* for his new glass because, he said, it sounded French and expensive, though it has been posited that the word is derived from the Old English *fabrile*, meaning "handmade." The glass would go on to become a key component of American Art Nouveau. While no artist has been attributed to the *Mushroom Lamp*, it is now known that many of Tiffany Studios' best-known designs were the work of Clara Driscoll (1861–1944), who worked at the company on and off from 1888 to 1909 and headed up the Women's Glass Cutting Department. The lamp was probably intended for a small side table and aimed at the midrange market, since its construction was not as labor-intensive as that of Tiffany's more famous leaded lamps.

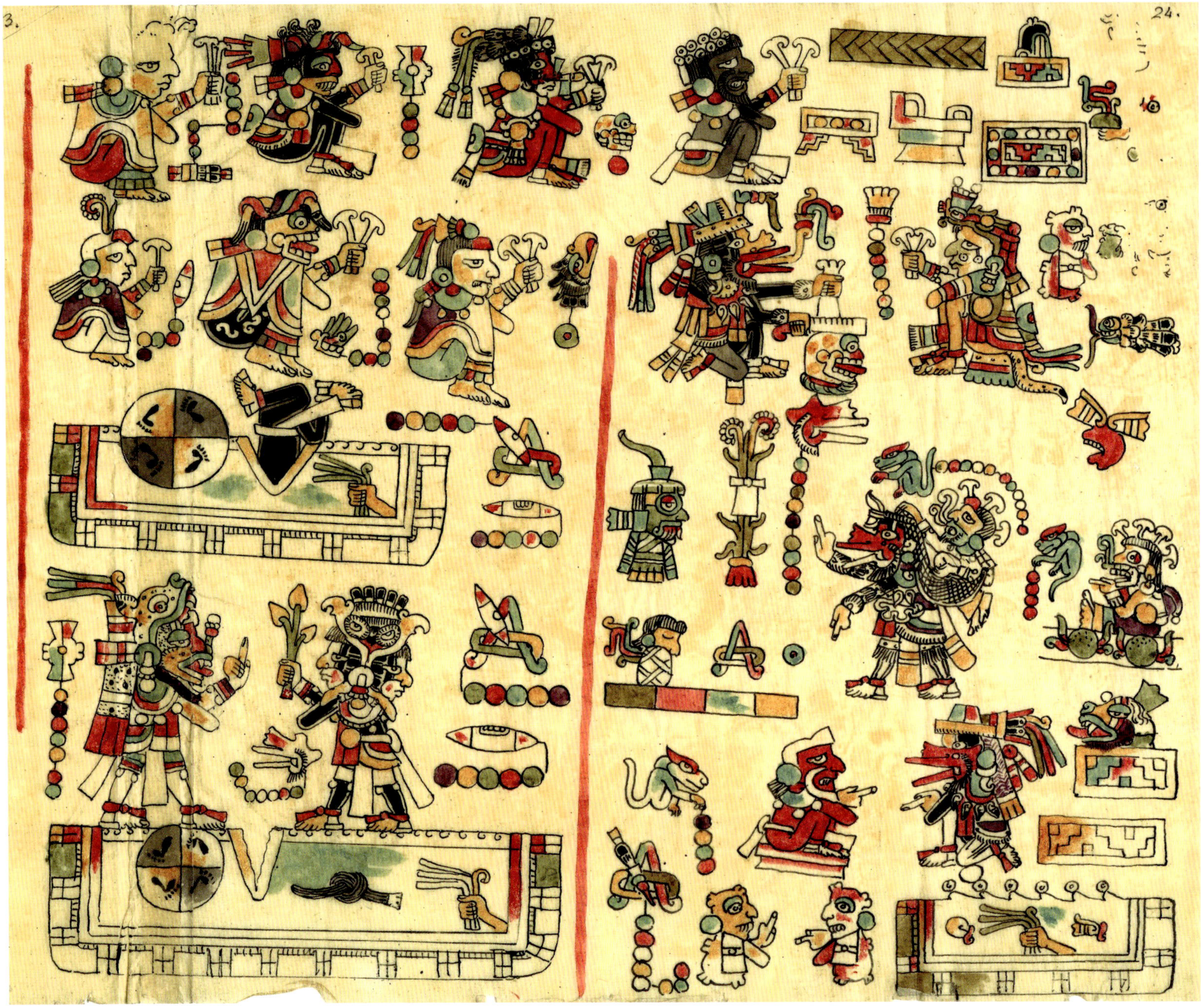

AGOSTINO AGLIO

Tracing of *Codex Vindobonensis Mexicanus I*, 1825–31
Ink and color on paper, H. 9 in. / 22.6 cm
British Museum, London

This nineteenth-century facsimile of a page from the *Codex Vindobonensis* is part of an account that describes the early history and origins of the ruling families of the pre-Hispanic Mixtec culture of what is now southern Mexico. Dating from about 1500, the *Codex* originally comprised fifty-two painted leaves describing life in the Mixtec kingdom, which occupied the present-day regions of Oaxaca, Guerrero, and Puebla. This page illustrates a mushroom ritual, which most likely took place when the calendar changed during a sacred period, and is intended to be read counterclockwise, starting from the bottom right corner. Mushrooms—particularly psilocybin "magic" mushrooms—played an important role in sacred Mixtec rituals, where their psychedelic qualities were thought to aid divination and communication with the gods. Mushrooms grew during the rainy season and were particularly associated, along with water and maize, with fertility. In this illustration, the lizard gods, Lady Eleven Lizard and Lady Four Lizard, are pictured at center right. Lady Eleven Lizard, with three mushrooms in her headdress, is carried by the God of the Wind, while Lady Four Lizard has four mushrooms in her headdress. Both represent the female spirit of sacred mushrooms. Following the page counterclockwise, they move to the left toward other deities, who hold two mushrooms in their hands to represent the duality of life. Written in the decades before the Spanish conquest of 1521, this document was later copied under the auspices of the conquerors, and the original is now housed in the Austrian National Library in Vienna.

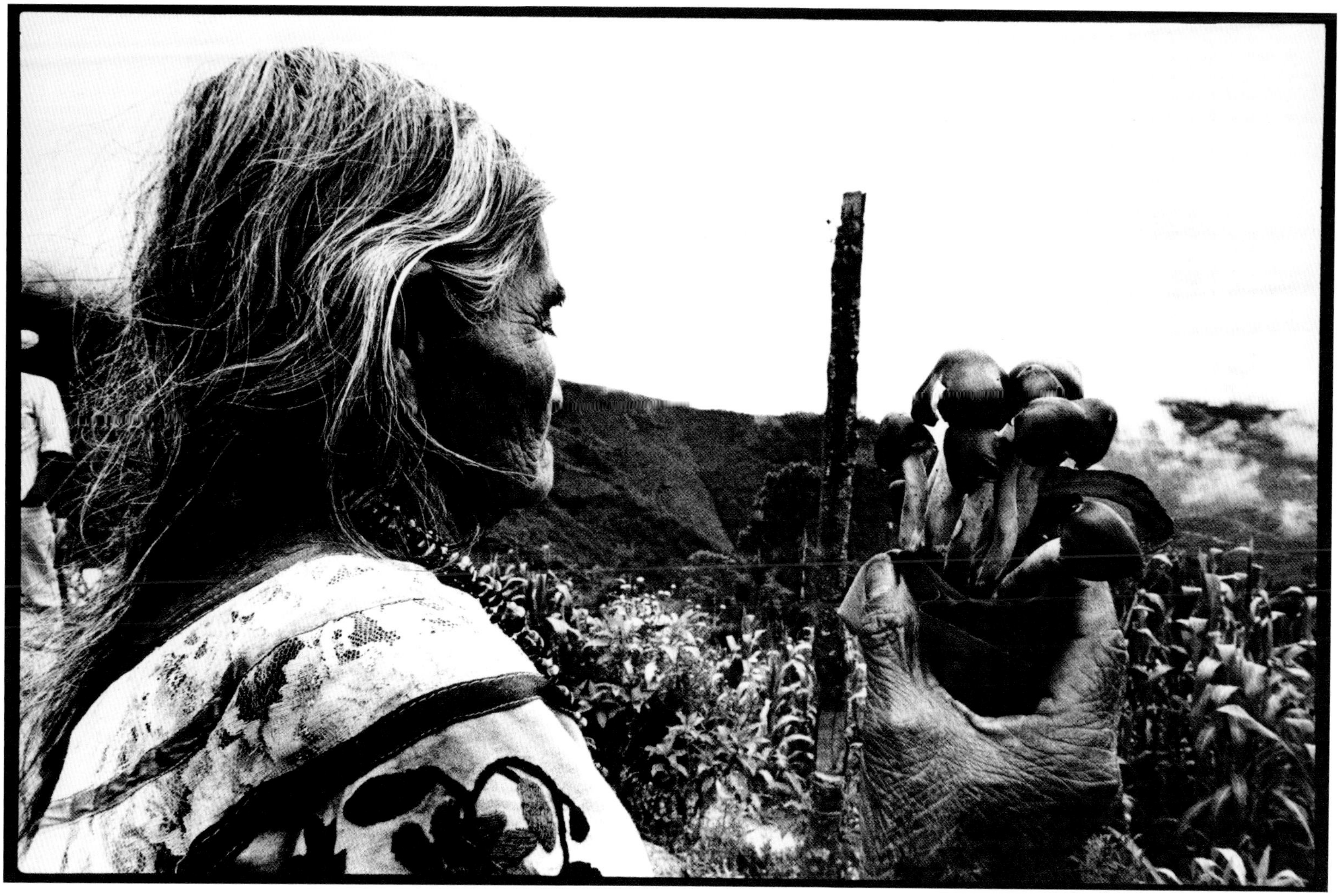

GUSMANO CESARETTI

Maria Sabina, Huautla de Jiménez, 1981
Photograph, dimensions variable

Mazatec *sabia* (wise woman) Maria Sabina stares off across the Huautla de Jiménez landscape, in southern Mexico, a cluster of psychedelic landslide mushrooms (*Psilocybe caerulescens*) brandished in her fist as one would wield a tool. And indeed, for her, these mushrooms were tools. Sabina used psilocybin in leading healing rituals called *veladas*, during which participants communed with the spiritual realm. After American ethnomycologist Robert Gordon Wasson deceived Sabina into allowing him to be the first Westerner to participate in a *velada*, and subsequently revealed her identity in a book, many hippies and other Westerners (including, it was rumored, several famous rock stars) began to seek out Sabina, asking for her psychedelic guidance. The Mazatec community resented the attention and blamed Sabina. She was ostracized, her house was burned to the ground, her son was murdered, and she was put in jail for a short time. She never recovered, and when she died years later, she was impoverished and malnourished. This photograph was taken by Gusmano Cesaretti (b. 1944) in 1982, only a few years before Sabina died. Cesaretti was born in Lucca, Italy, but has been a resident of Los Angeles since 1969. He has been integral to documenting East Los Angeles street culture since 1970, immersing himself in the Chicano movement. In 2014 he completed filming for a documentary about the South Central Los Angeles outlaw motorcycle club the Chosen Few, released in 2016 as *Take None Give None*. Cesaretti is still an active photographer and also lends his talents to Hollywood, helping to imbue films, especially those about Los Angeles, with an authentic look and feel.

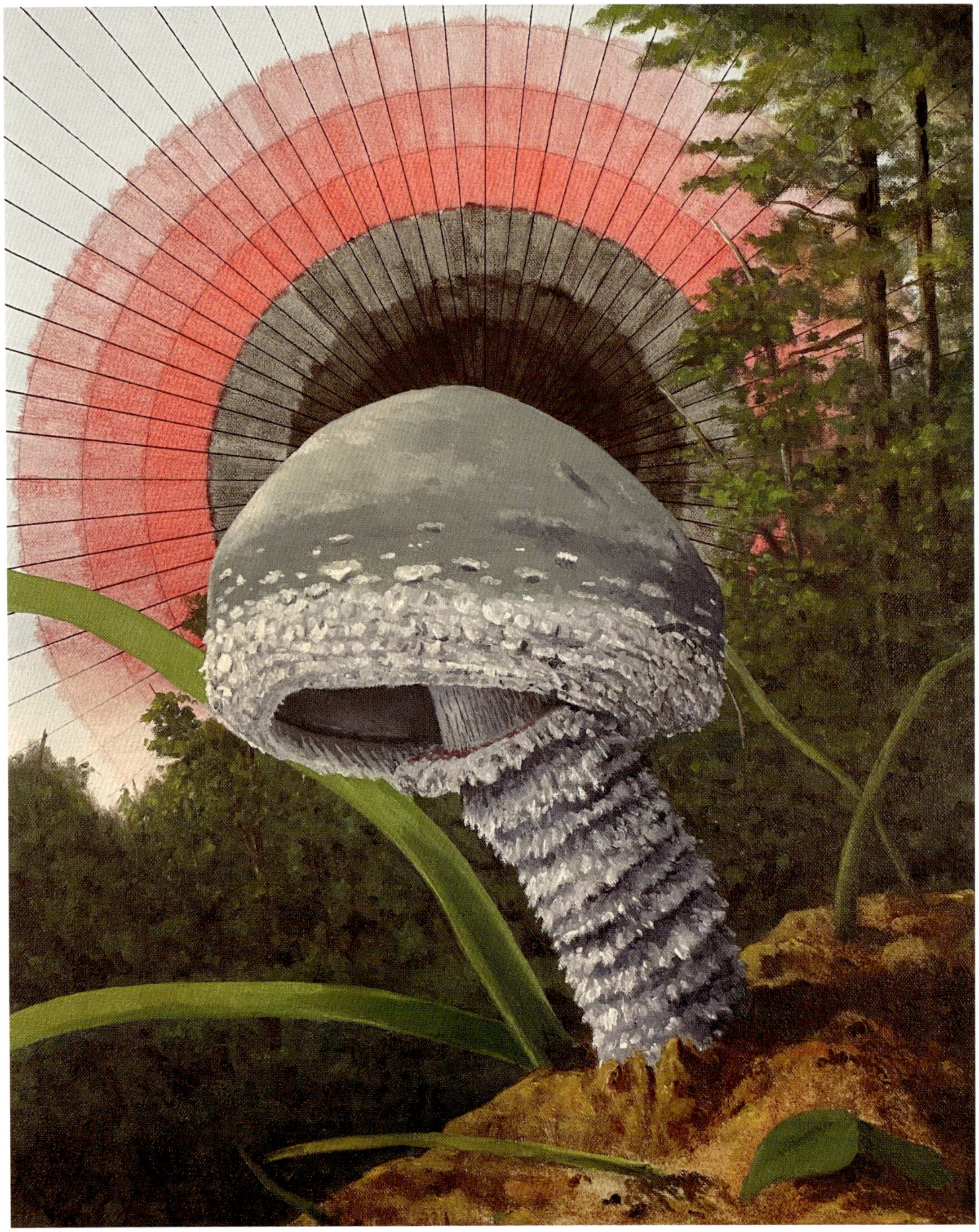

PEARL COWAN

Radiant Void, 2021
Oil and acrylic on canvas,
20 × 16 in. / 50.8 × 40.6 cm
Private collection

Framed by an abstract expanding pink aura, the ashy gray fruiting body of a mushroom grows out of tan soil in an idyllic, open forest landscape. This is the signature compositional approach to fungi in the work of contemporary American painter Pearl Cowan (b. 1982). Born in Texas and based in Massachusetts, Cowan graduated from the University of Texas at San Antonio in 2006 and has exhibited her intimate works on canvas and paper across the United States, as well as in *The Art of Mushrooms* at the Fundação de Serralves in Porto, Portugal. Her oil, gouache, and watercolor paintings depict detailed fungi in sublime landscapes surrounded by geometric forms. Cowan's *Radiant Void* series of works originated in 2019 as a response to her own loss of faith and renegotiation of ideas of the divine. The works mix both botanical illustration and experiments in color theory with diagrammatic forms that emerged at the start of the twentieth century around abstraction in the work of spiritual artists such as Emma Kunz and Hilma af Klint (see p.22). Cowan's take is more modern, however, almost veering toward photorealism. She has also used a similar approach for tropical flowers, again fusing botanical detail with shapes that imply spiritual enlightenment. The human figure is absent in these works. Instead, we are presented with something purely based in nature—the meeting place of flora, fauna, fungi, and the divine.

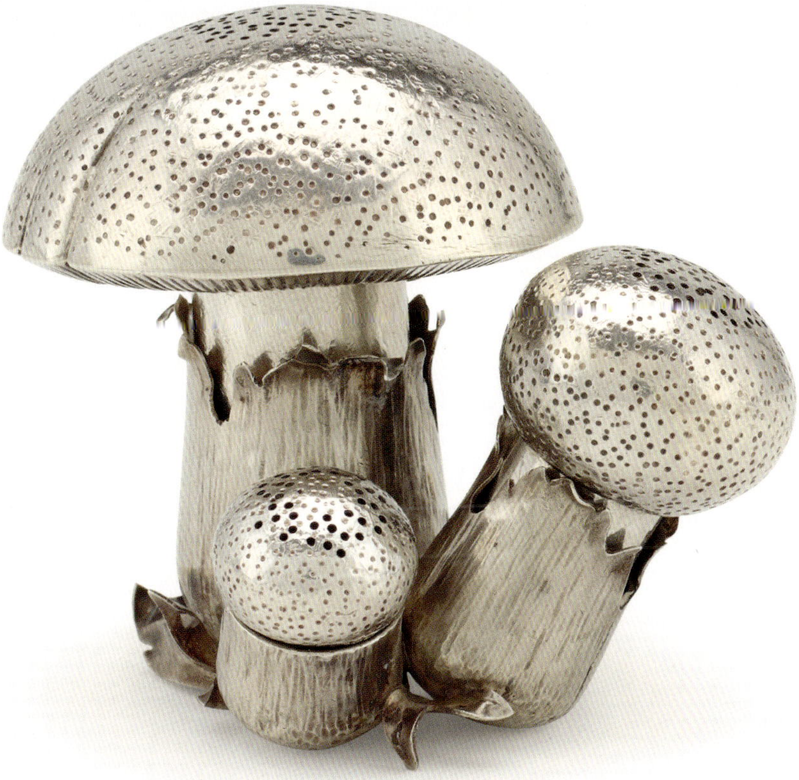

MARIO BUCCELLATI

Funghetti salt and pepper shaker set, 1965
.925 sterling silver, 3⅝ × 3⅛ × 3⅛ in. / 9.8 × 8 × 8 cm
Private collection

Upon founding his eponymous brand in Milan in 1919, Mario Buccellati (1891–1965) quickly established a reputation for creating intricately detailed, traditionally made jewelry and silverware. Trained in a Milanese workshop, Buccellati accumulated vast stores of knowledge on historical metalworking, championing methods originally popular in the Renaissance. He was drawn to how light interacted with engraved surfaces, developing techniques that created a textured effect and were the antithesis of the minimalist styles popular in Europe at the time. The jewelry and silver household objects he created went beyond ornamentation, striking an elegant balance between form and function. Fans of his work, including Pope Pius XII and the poet Gabriele D'Annunzio—who hailed Mario as the "Prince of Goldsmiths"—appreciated the rich textures and delicate finishing on the intricate accessories and ornaments. Buccellati was drawn to depicting nature, with flowers, insects, and fruit often decorating his designs. Mushrooms became popular in the late nineteenth and twentieth centuries, a reaction against the industrialized machine age, and their sculptural forms and whimsical silhouettes later became a popular part of Art Nouveau jewelry. Buccellati elevated the humble mushroom into silver salt and pepper shakers, finishing the mushroom caps with a lacelike pattern, softening the polished silver, making it warm and alive and imbuing the shakers with a precious character. Taken alone, each mushroom shaker serves a practical purpose; grouped together, they become a precious still life, underpinning Buccellati's feeling that dining should be a sensually rich experience.

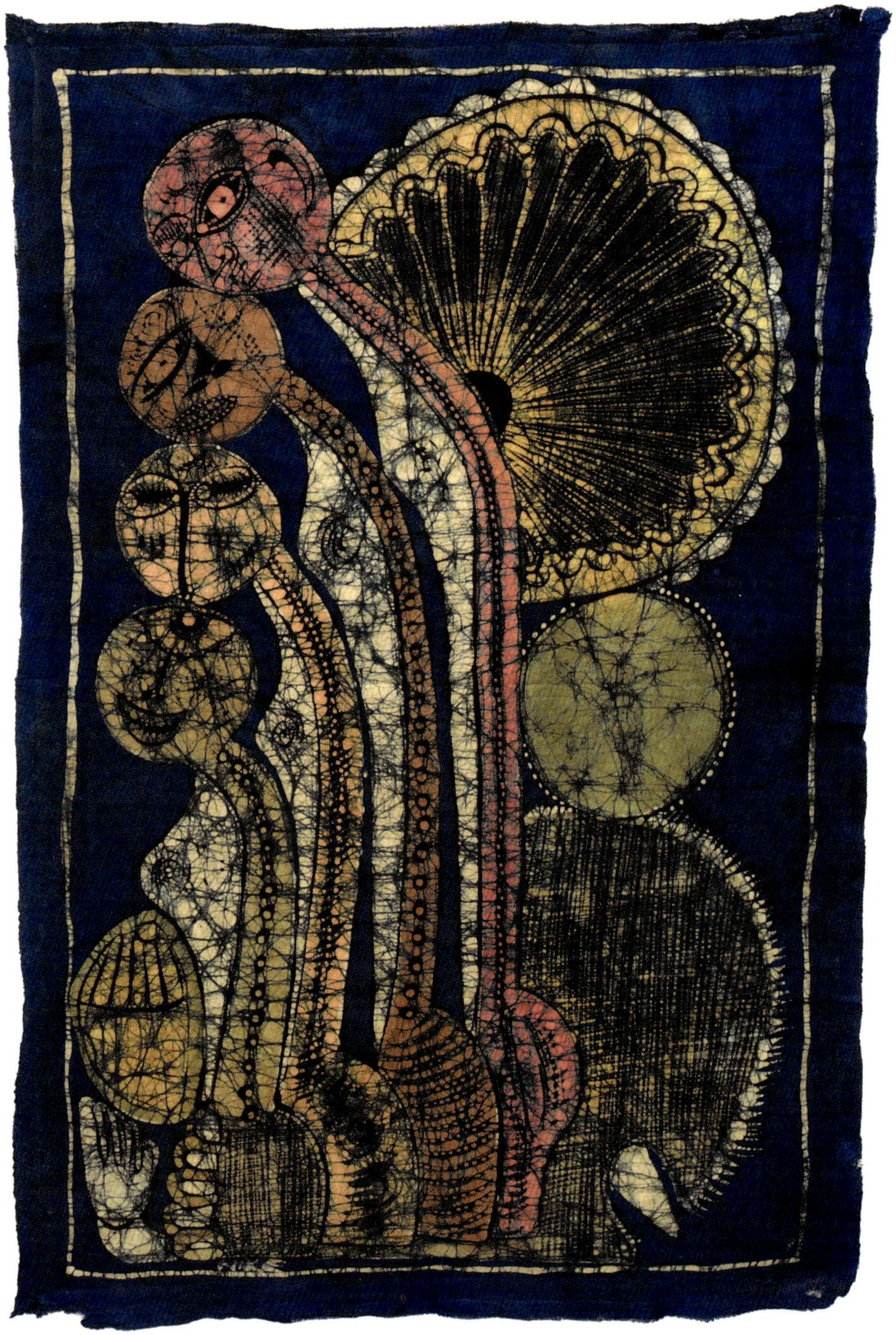

NIKE DAVIES-OKUNDAYE

The Mushroom of Life, 1977
Batik with natural dyes,
45 × 29 in. / 114.3 × 73.7 cm
Private collection

Celebrated Nigerian activist and textile artist Nike Davies-Okundaye (b. 1951) was a leading figure in the Osogbo artistic movement, which emerged after Nigeria's liberation from colonialist rule in 1960. Through public art projects and community outreach, the Osogbo School strove to promote and define the cultural identity of Nigerians after nearly a century of British occupation, combining the traditional methods of the Yoruba people with Western modernist aesthetics. Central to their practice was a symbolic depiction of Yoruba spirituality, folklore, and scenes of domestic life. Mushrooms feature prominently in Yoruba culture, where they are used to promote fertility, fight disease, ensure a good hunt, grant prosperity, and even inspire romance. Here, Davies-Okundaye reinterprets these Yoruba beliefs into the titular mushroom of life.

A prime example of traditional Nigerian dyed fabrics, Davies-Okundaye combines wax-resistant Batik dyeing techniques with the deep blue and bright patterns typical of *adire* cloth. Drawings and geometric motifs are laid upon the cloth using cassava paste, which resists the applied dyes, to create vivid tableaus like the one pictured here. A tireless artist and advocate for women's rights in her native country, Mama Nike, as she is affectionately known, has been a prolific textile artist and promoter of art for more than six decades. Her work is held in the permanent collections of New York's Museum of Modern Art; the Victoria and Albert Museum in London; the American Museum of Natural History in New York; and the Smithsonian National Museum of African Art in Washington, DC; among many other prestigious institutions around the world.

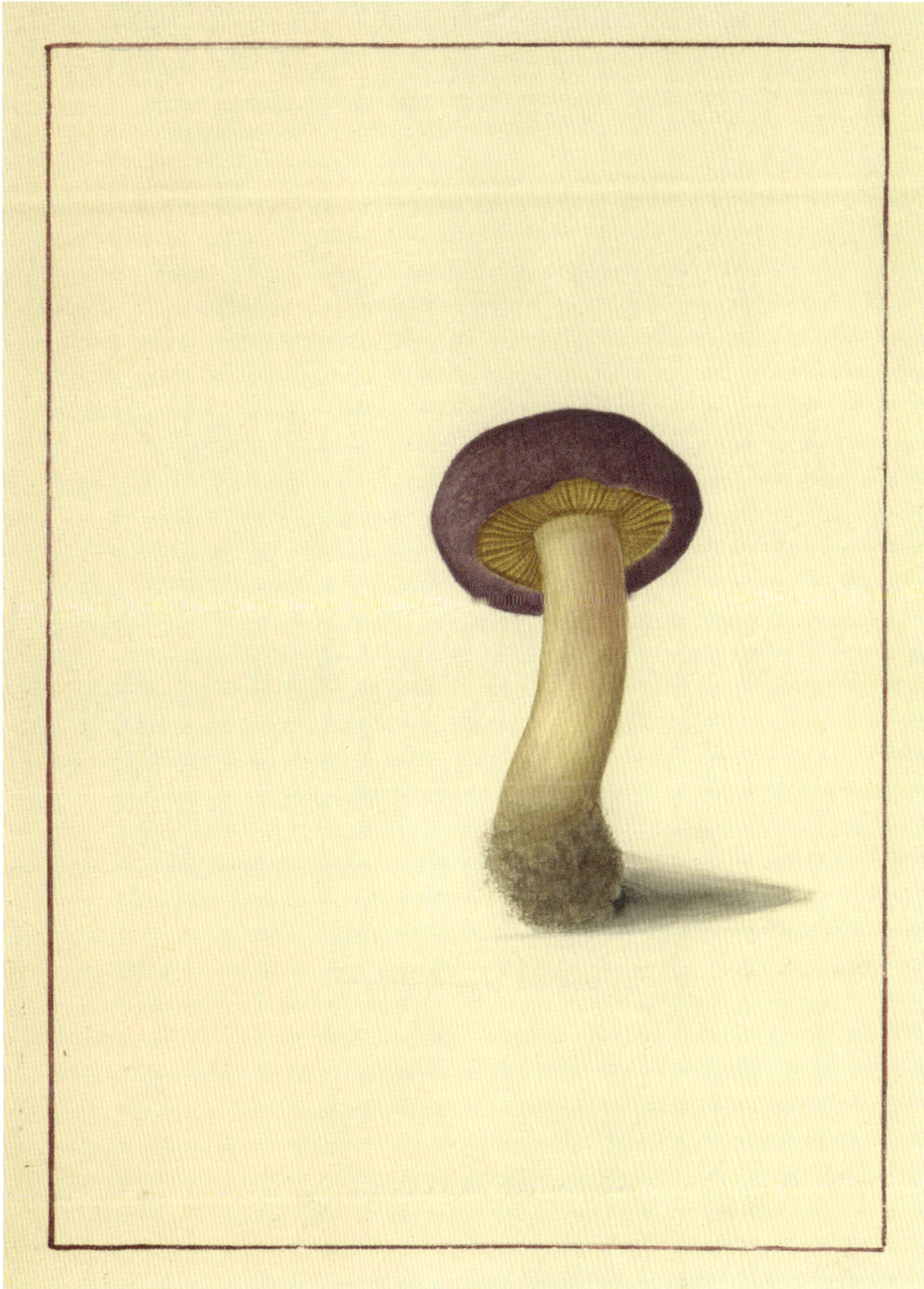

JAMES BOLTON

Agaricus serratus, from *Icons fungorum circa Halifax sponte Nascentium*, 1786
Watercolor drawing, H. 14⅛ in. / 36 cm
United States National Agricultural Library, Beltsville, Maryland

With traces of moss still clinging to the stem, a freshly picked purple-capped mushroom rises on the page, its delicate underside of gills rendered in meticulous detail by English naturalist James Bolton (1735–1799). The illustration is one of 242 plates that comprise *Icons fungorum circa Halifax sponte Nascentium* (*Illustrations of Fungi Growing Wild Around Halifax*), a collection of watercolors with accompanying descriptions by Bolton published in six volumes from 1784 to 1792. The son of a weaver, Bolton initially took up his father's trade but later became a self-taught art teacher. He developed a particular interest in fungi, collecting and illustrating wild specimens from around his home in the Halifax area of Yorkshire, England. This field research, along with his watercolors, developed into *Icons fungorum*, one of the first English-language works on fungi. Bolton identified this specimen as *Agaricus serratus*, or "serrate gill'd Agaric," but today it is known as plums and custard (*Tricholomopsis rutilans*), which is native to much of Europe. In his description, Bolton writes, "This beautiful and singular agaric I gathered under the beech trees in the park at Bramham . . . Aug 28, 1786. It differs from all the agarics that I have seen in that the gills are serrated; they are of a fine gold colour, and the pileus looks like cloth of a mixed purple colour. . . . This is the only specimen I ever met with." The mushroom is shown in an early stage, before the cap has expanded to its full width of up to 6 inches (15 cm).

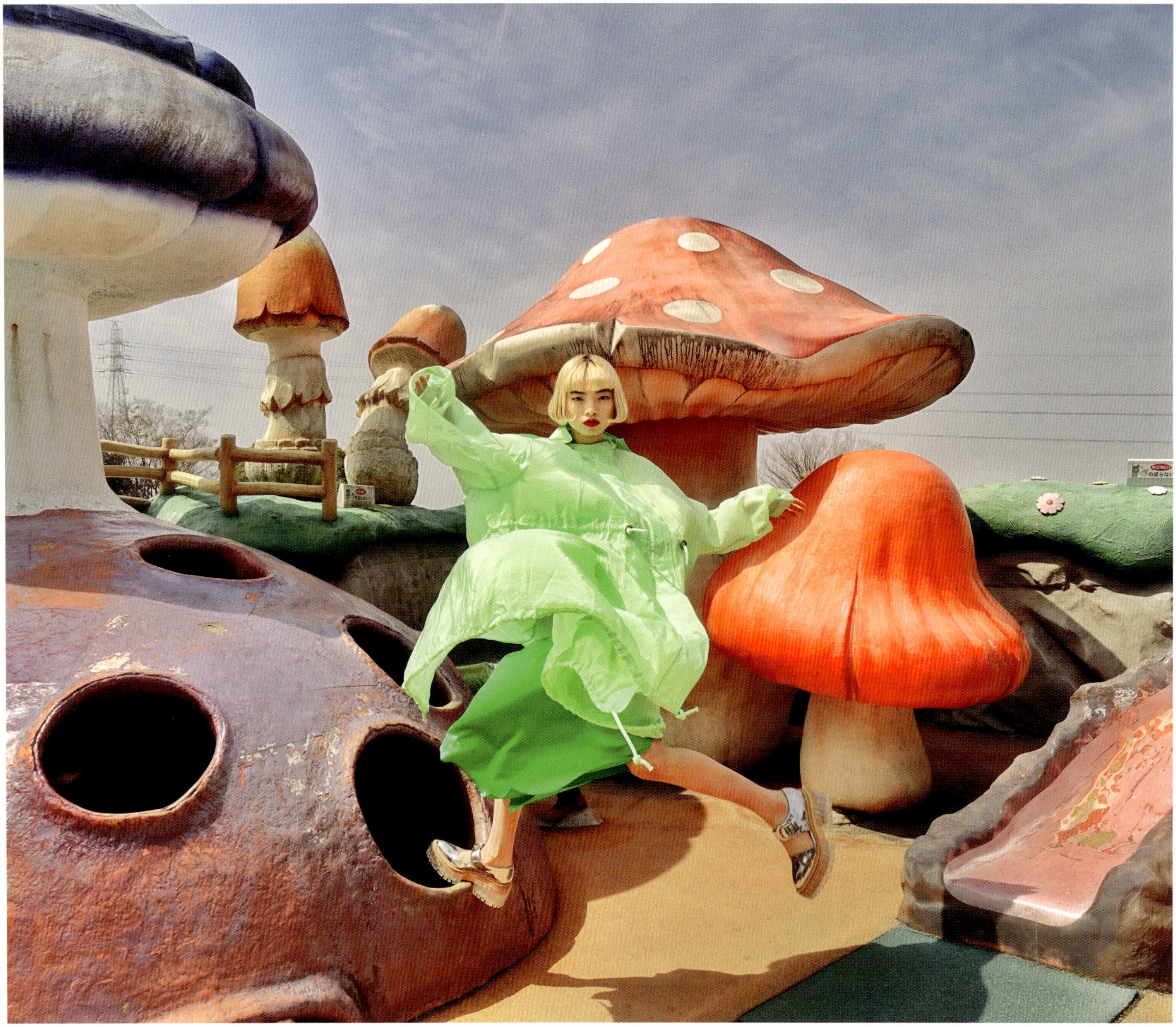

AOI YAMADA FOR STELLA McCARTNEY

Mushrooms Are the Future, Spring/Summer 2022 campaign
Photograph, dimensions variable

Japanese dancer Aoi Yamada (b. 2000) leaps across giant mushrooms in an outdoor park, her lime green, silk maxi dress and parachute jacket billowing outward. Backed by a soundtrack created while foraging mushrooms, Yamada recorded and edited the campaign video for British fashion designer Stella McCartney's 2022 Spring/Summer campaign, titled *Mushrooms Are the Future*. The collection brought serious attention to mushrooms, with textiles covered in hand-drawn prints, cutout shapes echoing mushroom gills, and colors reflecting fungi's earthy tones and phosphorescent neons. Alongside the campaign with Yamada, McCartney also released a psychedelic short film with Merlin Sheldrake (see p.223), and other fungi activations included a mushroom wallpaper called *Fungi Forest* with British heritage fabrics house Cole & Son and a catwalk show for the collection narrated by mycologist Paul Stamets. A year earlier, in 2021, Stella McCartney launched the Frayme Mylo handbag—the first luxury bag created from Mylo, a mycelium-based vegan leather. Mylo was created with long-term partner Bolt (formerly known as Bolt Threads), a lab researching biotech materials that are aligned with the environment. Part of Bolt's Mylo Consortium, McCartney was joined by Adidas, Kering, and Lululemon to invest in and explore new opportunities with the infinitely renewable material. Far ahead of the current trends for environmentally conscious materials, McCartney has been committed to not using leather, fur, feathers, or animal skin in her clothes and accessories since 2001.

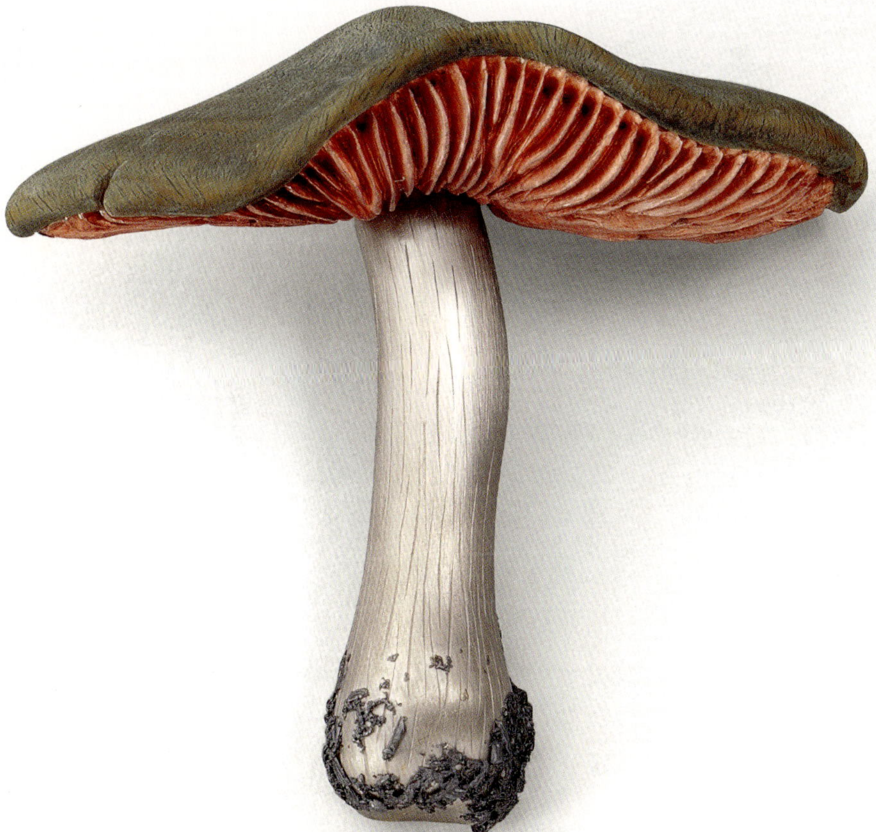

HEMMERLE

Mushroom brooch, 2004
Pock wood, copper, and white gold,
2⅞ × 3 in. / 7.2 × 7.5 cm
Private collection

Founded in 1893 in Munich, Hemmerle is a fourth-generation family-run jeweler known for creating unique, one-of-a-kind pieces that often juxtapose high-end jewels with unexpected, often "lower," metals and other materials. Hemmerle began as a goldsmith house specializing in providing medals of honor and gem-set adornments for the Bavarian court. While the house was continuously sought after for their impeccable technique and workmanship, it was in 1995 that their signature style emerged. Working on a client commission for a woman who "detested flashy gems" and wore only nineteenth-century Berlin cast-iron jewelry, Hemmerle crafted a piece that set a stunning diamond in a ring made of textured iron. With such a spectacular contrast, it was the beginning of the jeweler's design-driven, architectural, modernist creations that highlight daring and unexpected combinations of colors and textures, and high and low materials. One of Hemmerle's many high jewelry collections inspired by the natural world, *Mushrooms*, launched in 2004, embodies the house's unique approach. Comprising numerous mushroom brooches, the collection does not feature specific species, but instead uses a self-described magical realism to simultaneously capture and render abstract the fungal forms. The piece shown here—composed of pock wood (pockmarked lignum vitae wood), copper, and white gold—resembles many different species of mushrooms while also redefining the form. It represents one of only about two hundred pieces that Hemmerle produces annually, each made by a single skilled craftsman who often devotes hundreds of hours to creating just one piece.

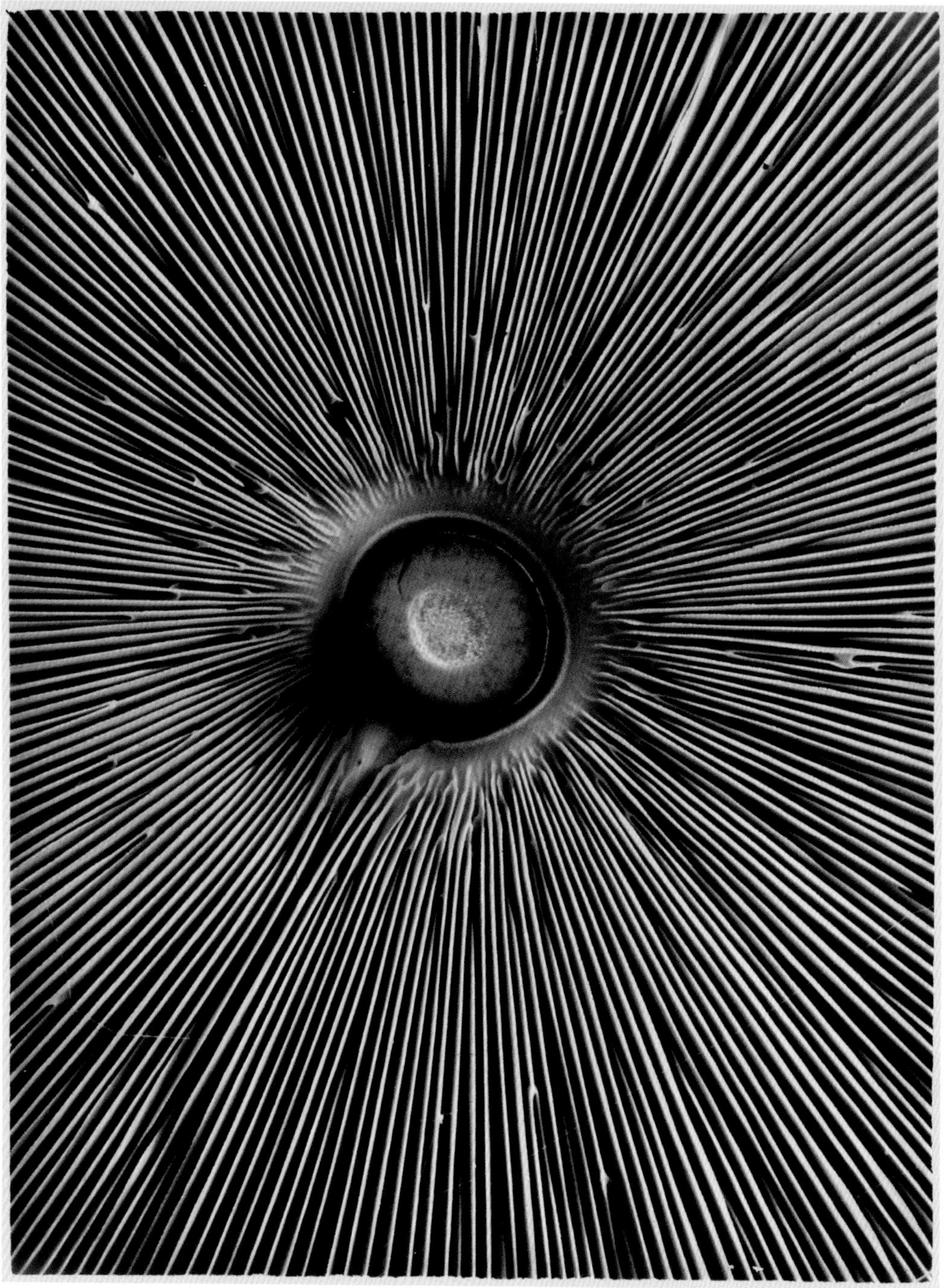

ALBERT RENGER-PATZSCH

Parasol Mushroom, c. 1930
Gelatin silver print, 9⅛ × 6⅝ in. / 23.1 × 16.7 cm
Metropolitan Museum of Art, New York

The delicate gills of a parasol mushroom (*Macrolepiota procera*) are the subject of this striking photograph, which captures their radial pattern in spectacular close-up detail. With its pin-sharp focus, close cropping, and use of high-contrast black and white, the image exemplifies the "pure photography" practiced by Albert Renger-Patzsch (1897–1966). Rejecting Pictorialism, which sought to imitate painting, and dismissing the avant-garde experiments of artists who used darkroom techniques to startling effect, the German photographer took an objective approach, emphasizing a clear and realistic depiction of his subjects without manipulation or embellishment. Instead of focusing on the distinctive parasol-shaped cap of this mushroom, he was drawn to the intricate form of its underside. The gills, also known as lamellae, are the parts of the mushroom that produce and release its spores. Millions of spores are typically produced and dropped from them, which are then carried away by wind currents that spread them far and wide. The edible parasol mushroom is a fairly common species, found solitary or in groups in woodlands and pastures throughout the northern hemisphere. Renger-Patzsch was acclaimed for his carefully composed images of plants, animals, landscapes, and industrial subjects, many of which he photographed in close-up to reveal the often overlooked details of the natural and man-made worlds. Driven by his interest in scientific inquiry, he believed that by embracing the camera's ability to faithfully reproduce an image of reality, he could unveil the true, inherent beauty of his subjects.

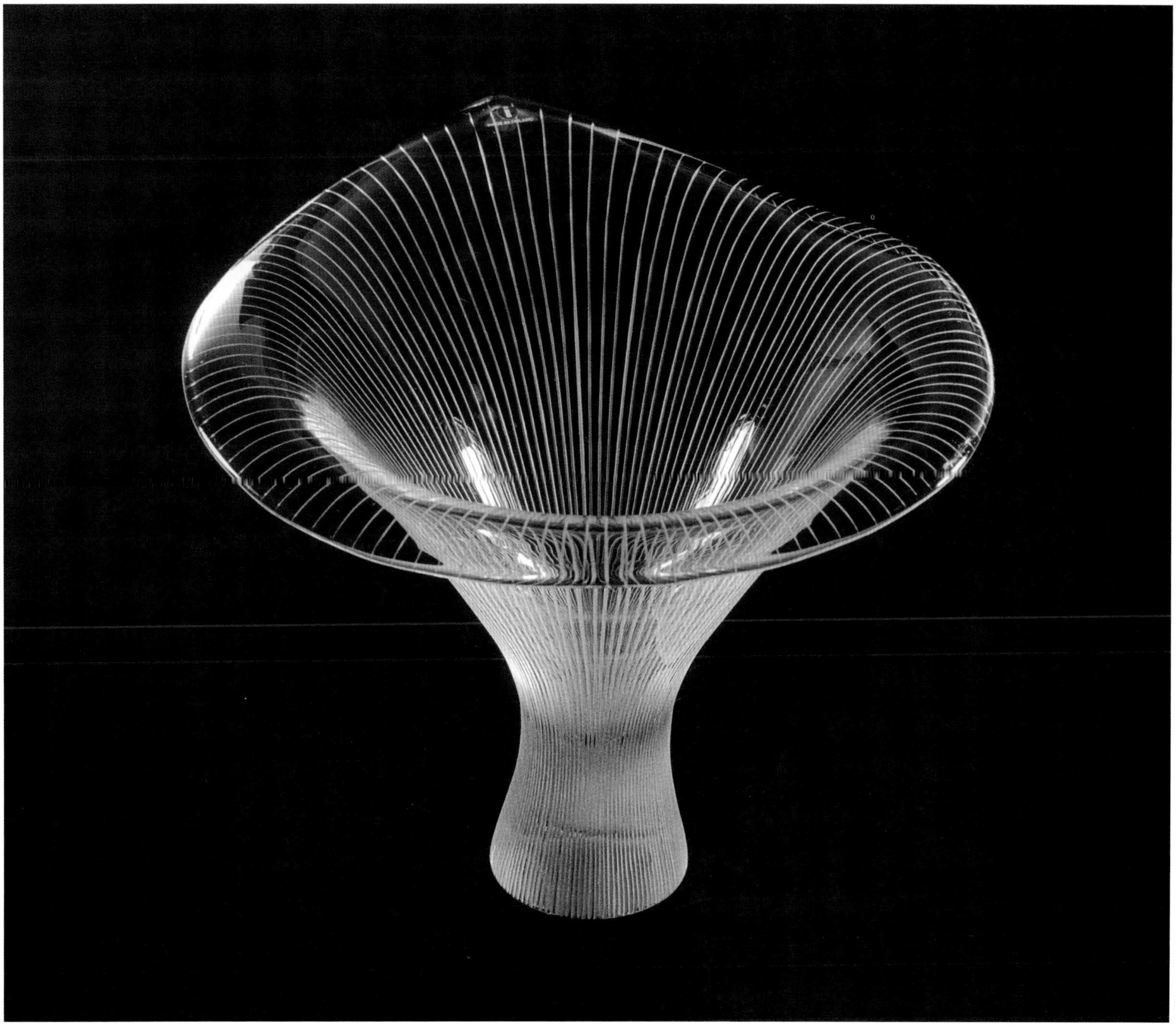

TAPIO WIRKKALA

Kantarelli vase, designed 1946
Mold-blown and wheel-engraved glass,
$7\frac{5}{8} \times 8\frac{1}{2}$ in. / 19.5 × 21.4 cm

Referencing its namesake, the simple, elegant form of the *Kantarelli* vase is inspired by the distinctive funnel shape of the chanterelle mushroom, the culinary delicacy that grows in the forests of Europe, North America, and parts of Asia. The piece is made of clear mold-blown glass, and artist Tapio Wirkkala (1915–1985) incised the outer surface of the vase with thin vertical lines to resemble the mushroom's characteristic gills. Born in Finland to a family of artists, Wirkkala was a prolific designer and sculptor who worked with metal, wood, ceramic, and glass. His designs were often inspired by the natural forms he encountered in the Finnish landscape, such as leaves and pebbles, and by natural processes such as melting ice. Describing his design philosophy, Wirkkala once said, "Take a piece of ice in your hand and find our culture in it; find it in a stone smoothed by the rapids, or in sand that waves have shaped." He conceived the *Kantarelli* vase in 1946 for Iittala—initially a glassware company and now a leading Finnish design brand specializing in tableware and cookware—and it quickly became one of their most popular products. Smaller, slightly modified versions were subsequently introduced, and the design remained in production until 1960. The vase won the Grand Prix at the Milan Triennale (an event showcasing architecture and industrial design) in 1951 and has long been considered an icon of Scandinavian design.

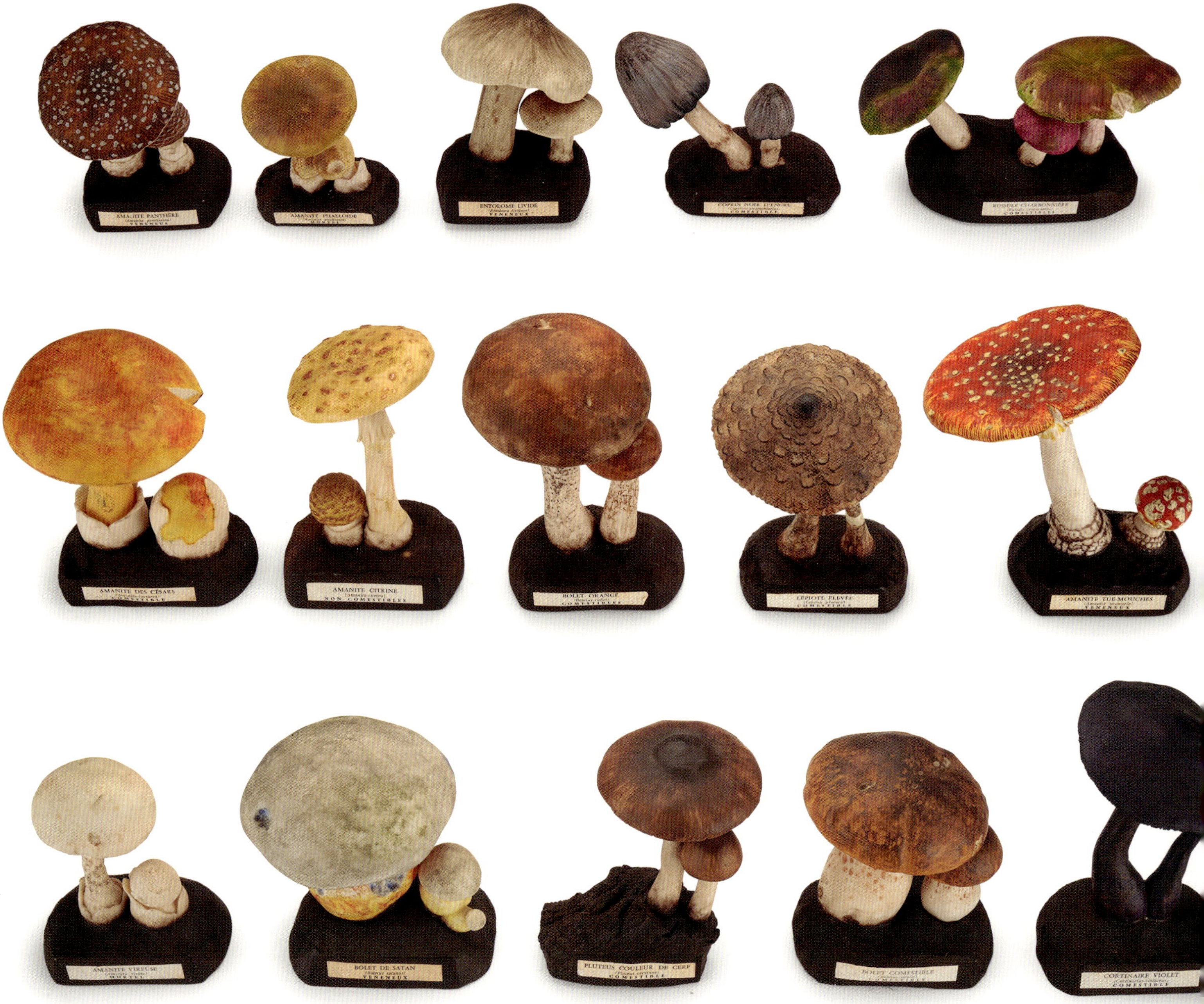
AMANITE PANTHÈRE
VÉNÉNEUX
AMANITE PHALLOIDE
MORTEL
ENTOLOME LIVIDE
VÉNÉNEUX
COPRIN NOIR D'ENCRE
COMESTIBLE
RUSSULE CHARBONNIÈRE
COMESTIBLES
AMANITE DES CÉSARS
COMESTIBLE
AMANITE CITRINE
(Amanita citrina)
NON COMESTIBLES
BOLET ORANGÉ
COMESTIBLES
LÉPIOTE ÉLEVÉE
COMESTIBLE
AMANITE TUE-MOUCHES
VÉNÉNEUX
AMANITE VIREUSE
MORTEL
BOLET DE SATAN
VÉNÉNEUX
PLUTEUS COULEUR DE CERF
COMESTIBLE
BOLET COMESTIBLE
CORTINAIRE VIOLET
COMESTIBLE

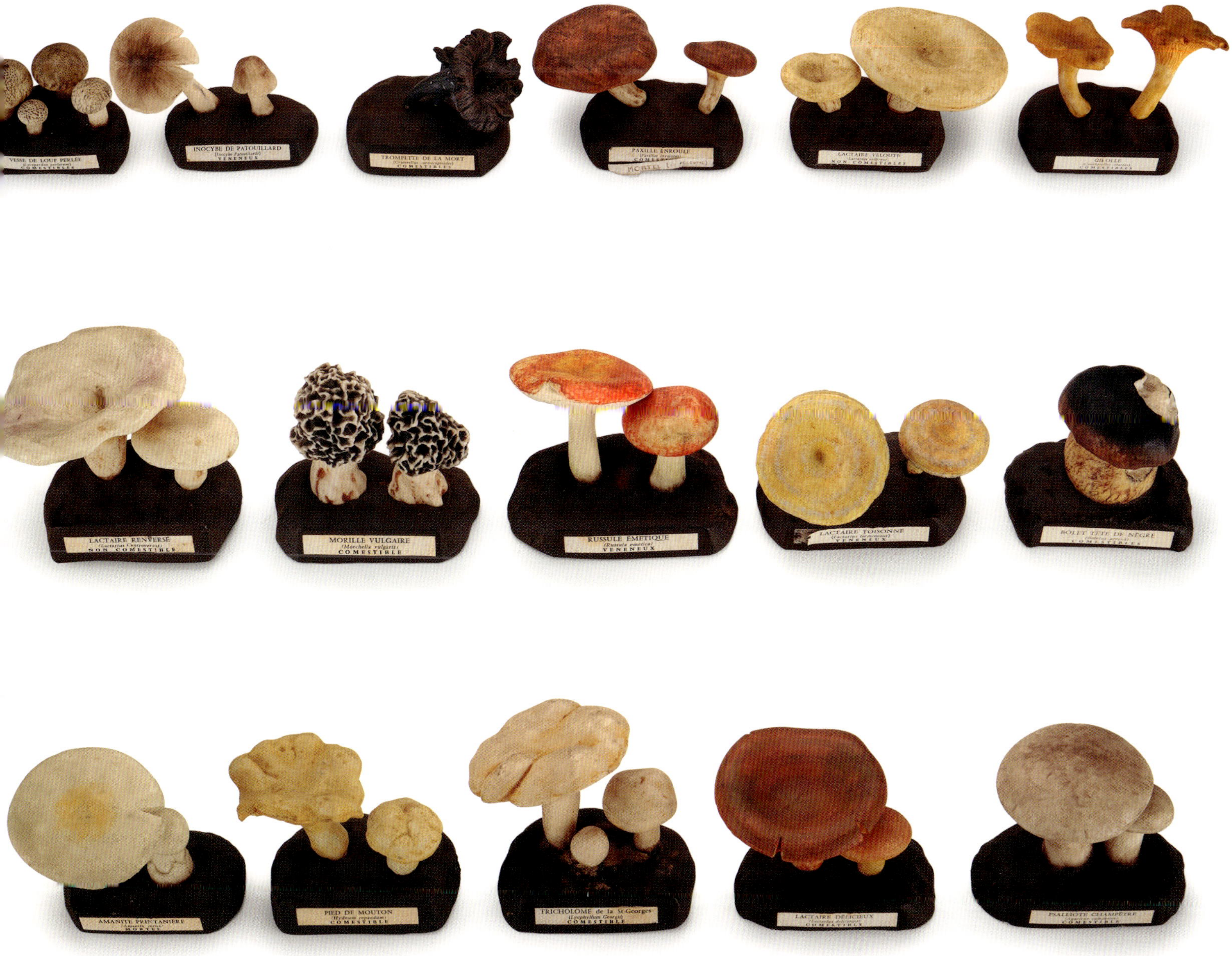

ANONYMOUS

Didactic rubber mushrooms, c. 1930
Rubber with paper labels, dimensions variable
Private collection

When discussing educational models, it is often in relation to anatomical models, including everything from the Visible Man, available at most toy stores, to more specialized models seen only in medical offices or hospitals, from a human eyeball the size of a bowling ball to a variety of movable joints. But there are many other types of educational models, all of which give access to lifelike examples in a controlled environment, allowing for the careful study of minute details, movements, or processes that would otherwise be difficult or impossible to apprehend. This collection of thirty-one rubber mushrooms cast from live specimens—each with its scientific name and whether it is edible, poisonous, or fatal printed on a small paper label—was made as a teaching tool. Whether this set belonged to an educator, an amateur forager, or a seasoned mycologist is unknown, but it served the important function of providing mushroom education without its owner having to locate numerous live specimens or handle a variety of poisonous, even fatal, mushrooms. Sets such as this one, which originates from the South of France from around 1930, appear to be most common in Europe. German company Somso—an elite name in the field of educational models—also made a similar set around the same time. These mushrooms—and educational models as a whole—are now highly collectible, not only as teaching tools but as curiosities and objets d'art.

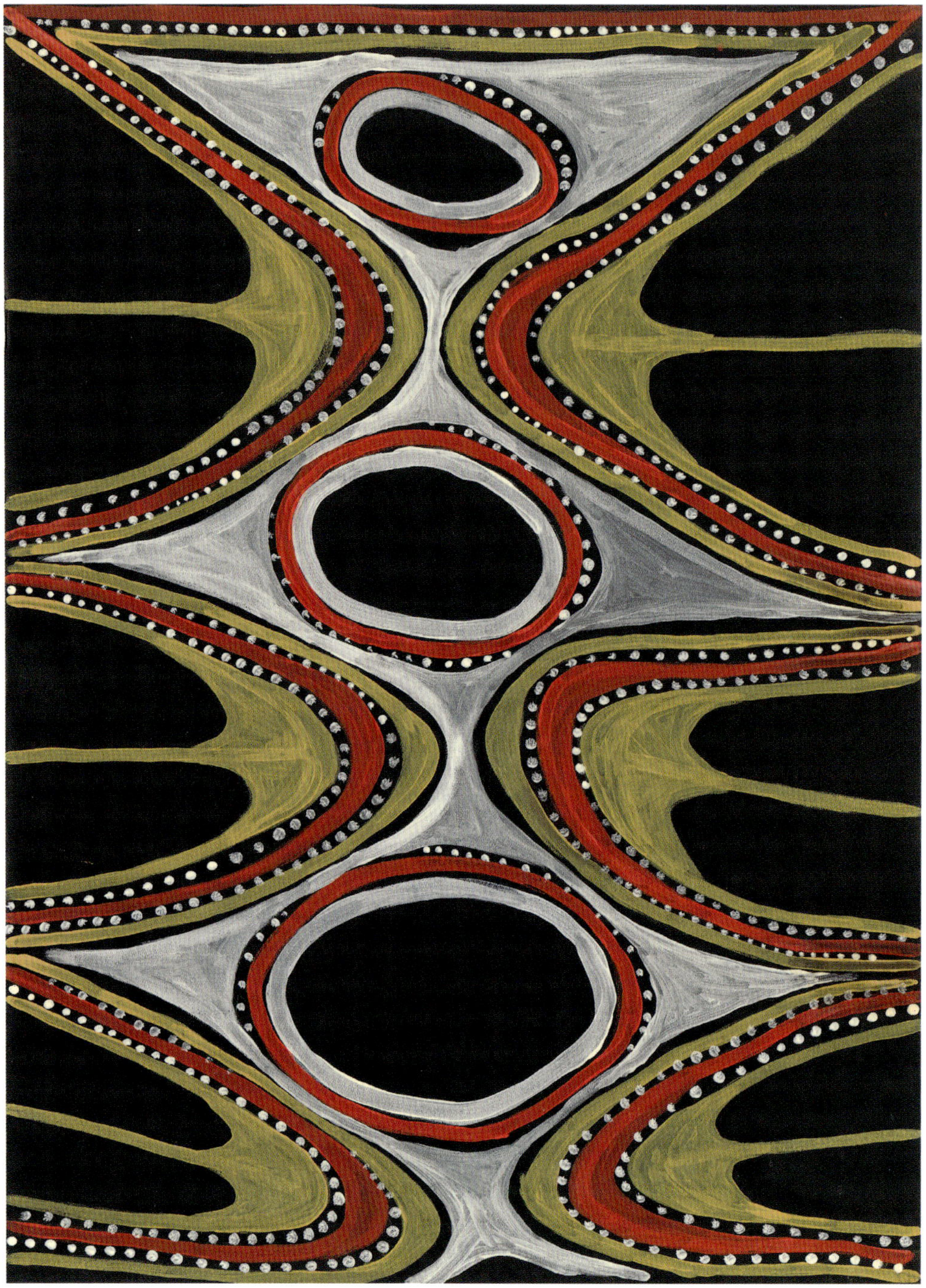

ABUNIGI

Hapakwinya, Mushroom, 1973
Gouache on black paper, 39¾ × 28⅜ in. / 101 × 72 cm
National Gallery of Victoria, Melbourne, Australia

Against a black background appears an abstracted series of rounded forms outlined with yellow and red lines, interspersed with lines of white dots. These shapes represent mushrooms and the mushroom spirit in a work by New Guinean artist Abunigi (b. c. 1935), painted with gouache on black paper. A member of the Kwoma people, Abunigi created this work in the early period of his artistic output, while his later paintings eschew paper for tree bark. The Kwoma live in northeast New Guinea in the Pacific Ocean archipelago of Melanesia, more specifically in the hills north of the Sepik River. Abunigi's work typically represents the spiritual and natural worlds of his people, focusing on subjects such as butterflies, forest spirits, centipedes, and, as here, mushrooms. The palette of red, white, black, and yellow—derived from natural pigments found in plants and minerals—reflects the ceremonial status of the artwork and points to the significance of mushrooms in Kwoma culture. As well as culinary mushrooms, Papua New Guinea is home to psychoactive fungi, such as boletes, which were initially blamed for an outbreak of "mushroom madness" among communities living in the highlands during the 1930s. In fact, the malady was found to be the result of tobacco overconsumption. Abunigi's use of abstract and geometric designs to portray different subjects is typical of the art of the vast island, although the precise meaning of specific shapes varies from clan to clan and from artist to artist.

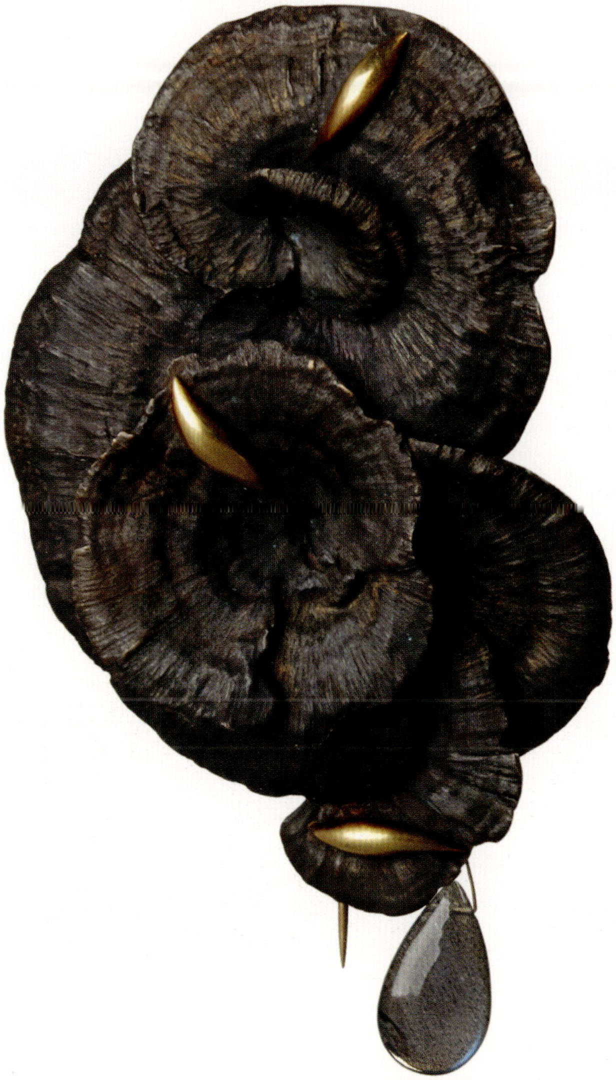

GABRIELLA KISS

Mushroom Brooch, 2018
Bronze, 18-karat gold, and labradorite,
3 × 2 in. / 7.6 × 5 cm

A cluster of bronze mushrooms adorned with three tiny gold slugs and a labradorite stone drop forms this nature-inspired brooch by Canadian-born American jewelry designer Gabriella Kiss (b. 1959). Kiss has built her career on fashioning beautiful pieces that represent underappreciated parts of nature, from insects and snakes to branches and seedpods, antlers, and, as here, fungi. Working with precious metals, such as gold and silver, to which she sometimes adds precious and semiprecious gemstones, she pays homage to often-overlooked flora and fauna. Her studio in New York's Hudson Valley, and the surrounding landscape, provide continual inspiration for Kiss that finds its way into pieces from rings and brooches to earrings and necklaces. Kiss studied sculpture at the Pratt Institute in Brooklyn, New York, and frequently uses the lost-wax casting technique she learned there to make her jewelry. She first carves a model of her subject in wax, which is then enclosed in a clay mold. When molten metal is poured into a hole at the top of the mold, the wax melts and runs out through another hole at the bottom of the mold, while the metal fills the sculpted shape and hardens, after which the mold is broken off. The method allows Kiss to replicate the tiniest textural details, giving the brooch here a sense of profound verisimilitude, as if the mushrooms are freshly picked, slugs and morning dew included.

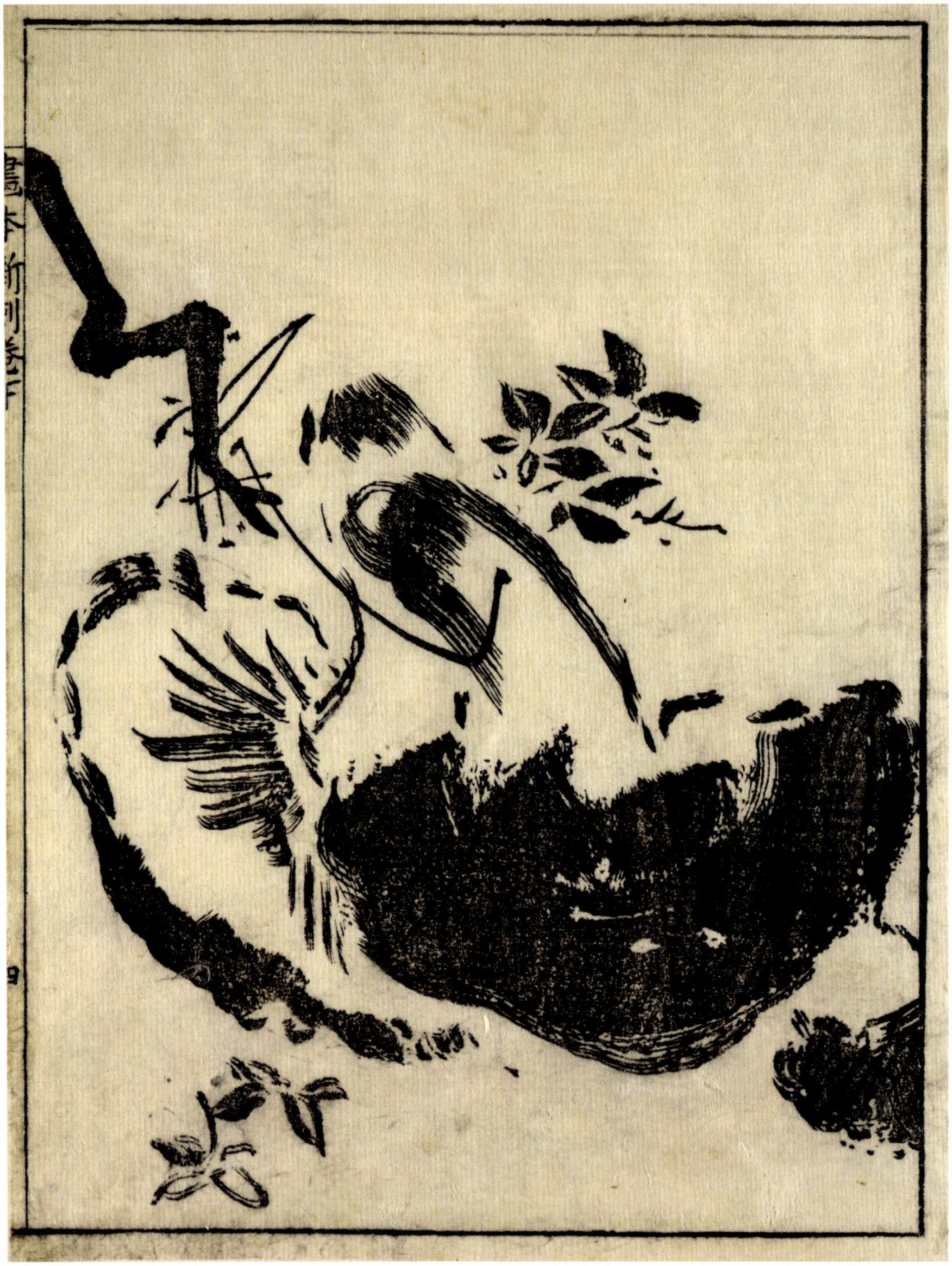

YAMAMOTO SHINKO

Mushrooms, c. 1701–30
Woodblock print, 8¼ × 6⅜ in. / 21 × 16 cm
Miriam and Ira D. Wallach Division of Art, Prints, and Photographs, New York Public Library

This striking monochrome print of two upturned shiitake mushrooms shows the jagged-edged dark cap of one, while the gills on the underside of the second are suggested with just a few deft lines. Small, scattered leaves set the scene, while a zigzag black line suggests the outline of a mountain. Startling in its apparent modernity, this is, in fact, a woodblock print made sometime in the early decades of the eighteenth century during Japan's Edo period (1603–1868). It is the work of woodcut artist Yamamoto Shinko (1650–1730), about whom little is known beyond the existence of a few surviving prints. Woodblock printing of ukiyo-e, or "pictures of the floating world"—the daily life and landscapes of contemporary Japan—was popular during the Edo period because the mass-produced prints were more affordable for ordinary people. By the 1760s, color prints abounded, including by artists such as Katsushika Hokusai, best known for *The Great Wave off Kanagawa* (c. 1830–32), and Utagawa Hiroshige, renowned for his horizontal landscapes. Yamamoto, however, belonged to an earlier generation who worked in black and white. At the time, advances in cultivation were making shiitake mushrooms more available—they had previously been found only in the wild—allowing them to become a staple for everyday Japanese families rather than a delicacy of the elite. Valued for their nutty taste, the winter shiitake (*donko*), with their thick caps—as seen here—could be dried to preserve them and were particularly sought after.

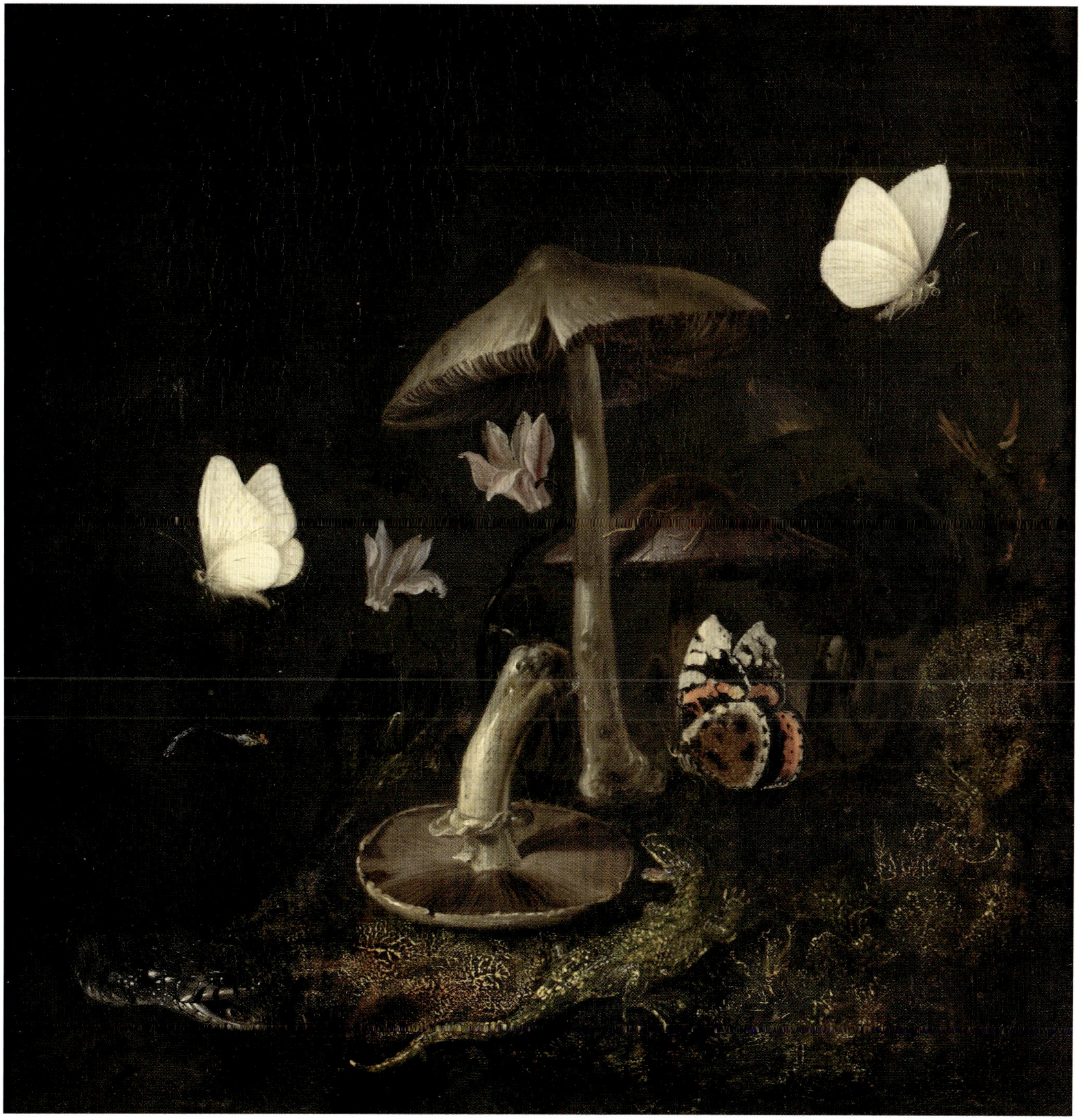

OTTO MARSEUS VAN SCHRIECK

Untitled, 17th century
Oil on canvas,
13¾ × 13⅛ in. / 34.8 × 33.3 cm
Private collection

Central to this somber painting is a cluster of mushrooms, one standing tall above another that lies inverted on the forest floor, clearly showing the brown gills under the pale cap and the membranous ring on the stem, while a couple more mushrooms recede into the gloom of the background. Painted by Dutch Golden Age artist Otto Marseus van Schrieck (c. 1613–1678), this work is a prime example of the *sottobosco*, or forest-floor still life, genre, of which he is widely regarded as the inventor. Van Schrieck's nickname was the Snuffler due to his habit of rummaging about in the garden searching for details of wildlife. According to his wife, he also kept lizards and snakes to study more closely. Indeed, reptiles and also insects appear in many of his works, usually alongside plants and fungi, as seen here. The brightest objects in this otherwise dark painting are two flowers on twisting black stalks, probably cyclamen, and a trio of butterflies composed of two whites and a red admiral. On the left floats a dragonfly with a blue body and red eyes. In the foreground, a green lizard rests on the mossy forest floor, its mouth open, probably alarmed by the snake slithering toward it from the left. While the fungi are the dominant subjects—probably in the genus *Agaricus*, which contains several edible species—van Schrieck has brought together other creatures from the forest ecosystem, albeit in a combination unlikely to occur naturally.

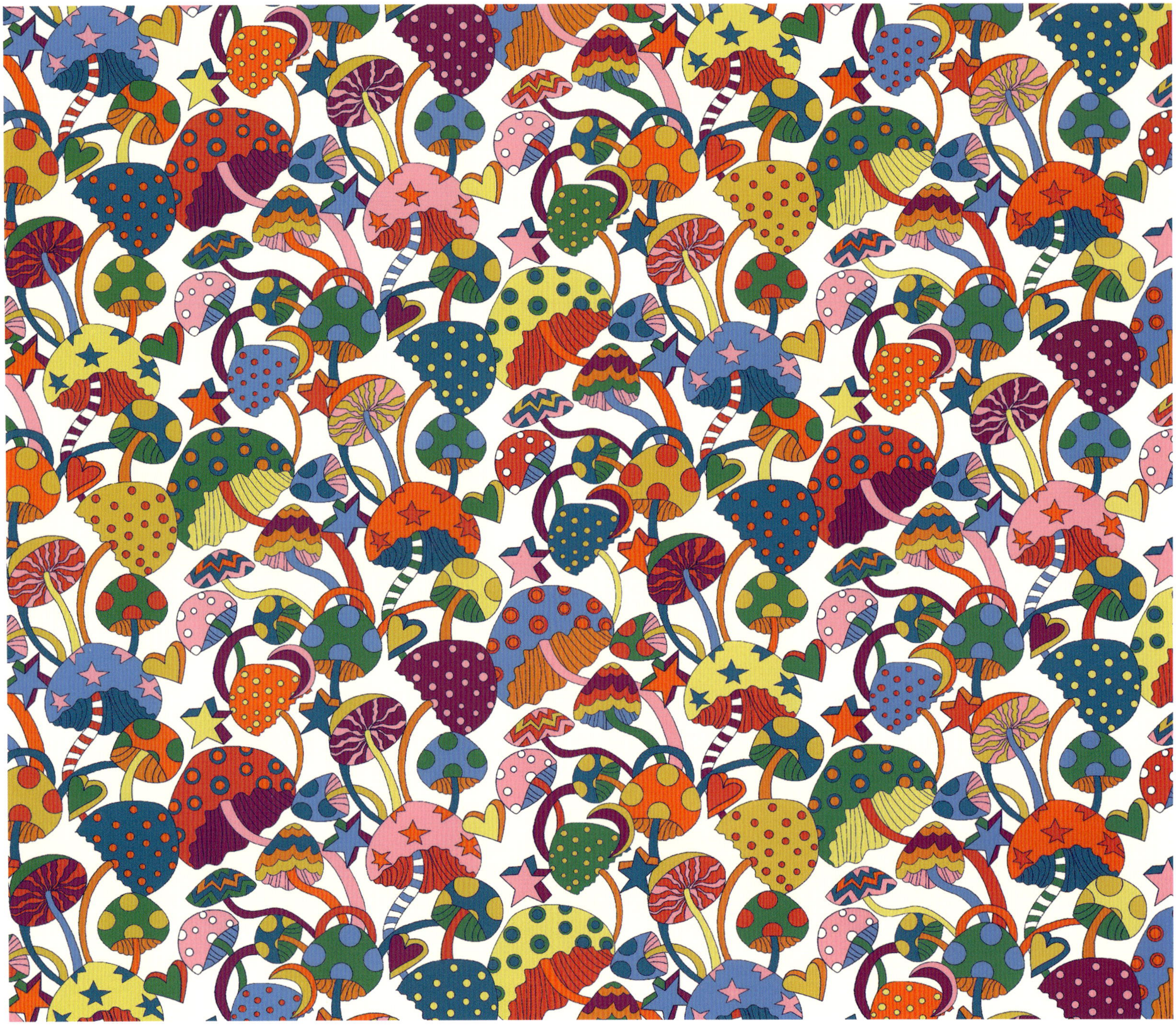

LIBERTY FABRIC

Magic, 2020
Tana lawn cotton,
dimensions variable

The venerable Liberty London has been an arbiter of refined taste and luxury since its inception in 1875. Founded by importer Arthur Lasenby Liberty, the store offered Londoners access to fine fabrics and objets d'art, particularly from the Far East. It quickly became *the* place to shop in London among its fashionable set (Oscar Wilde was an early patron and proponent). Since 1924, it has resided in West End on Great Marlborough Street in a location as iconic as the brand itself. The two-story, black-and-white Tudor Revival–style building was constructed from the wood of two decommissioned Royal Navy ships. Celebrating craftsmanship and forward-thinking design, Liberty soon became synonymous with the Art Nouveau movement and its celebration of aesthetics. In fact, the store's branding incorporates R. Beauclair's *très* Art Nouveau *Ianthe* print, which the store first made available as a wallpaper design in 1902. For more than 150 years, Liberty has offered a wide array of luxury goods. But it is perhaps best known for its fabrics, with more than fifty thousand patterns in its archives. Liberty's prints teem with vibrant colors and patterns that seem to jump off the bolt—from its earliest Art Nouveau motifs to the ditsy florals of the 1930s to the geometric patterns of the 1960s and 1970s and beyond. The psychedelic *Magic* mushroom pattern seen here is one such example: a dazzlingly colorful and densely populated tableau buzzing with vitality and possessed with the qualities that have made Liberty the choice of Londoners for generations.

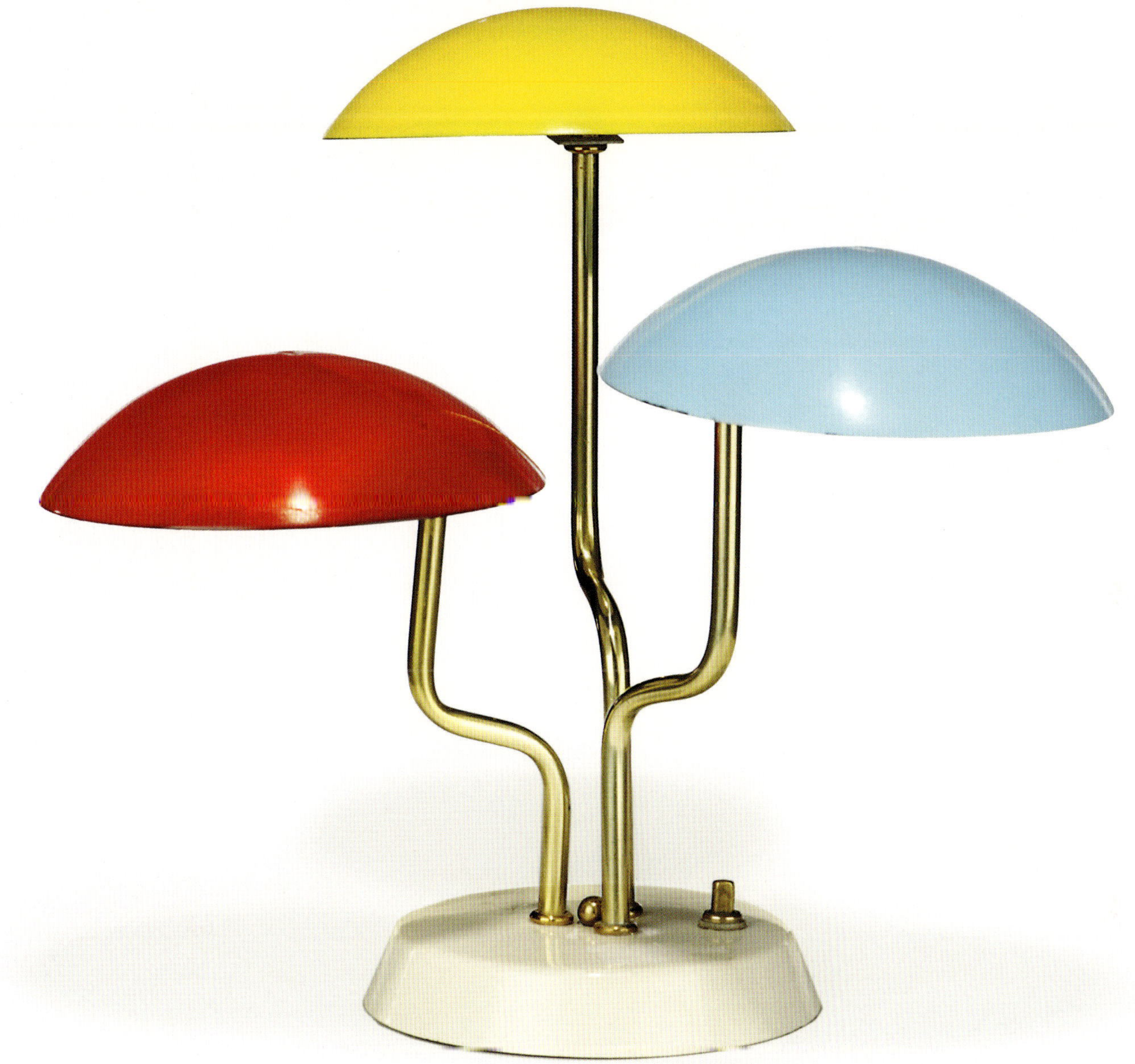

GINO SARFATTI

Table lamp, designed 1951
Enameled steel and brass,
H. 11¼ in. / 28.5 cm
Private collection

Born in Venice, Italy, to a Jewish father and a Catholic mother, as a teenager Gino Sarfatti (1912–1984) had an intense interest in studying the classics. However, by his late teens he had become more science-minded and pursued aeronautical engineering at the University of Genoa. By the end of university, however, his yen for aeronautics had waned. In 1935 he moved with his family to Milan, where he briefly sold art glass before finding his passion: the application of what he had learned studying aeronautical engineering to the field of lighting design. In 1936 he founded the lighting company Lumen along with partners Aldo Valcarenghi and Dino Mondolfi. For unknown reasons, in 1938 Sarfatti parted ways with Lumen, and in early 1939 he cofounded lighting company Arteluce along with his wife, Jolanda Marazza, and architects Maurizio Bolchini and Franco Buzzi. Several of the lamps Sarfatti designed for Arteluce borrowed the form of the mushroom—some standing alone, others in groups—including the one pictured here. In the 1940s, Sarfatti fled Italy for Switzerland to avoid persecution for his Jewish heritage, returning to the helm of Arteluce at the conclusion of World War II. In 1954 Sarfatti's Model 559 lamp won the inaugural Compasso d'Oro prize for industrial design excellence. After numerous other awards and accolades, the designer and his partners sold Arteluce in 1973. At the time of Sarfatti's death, more than seven hundred lighting designs were attributed to him.

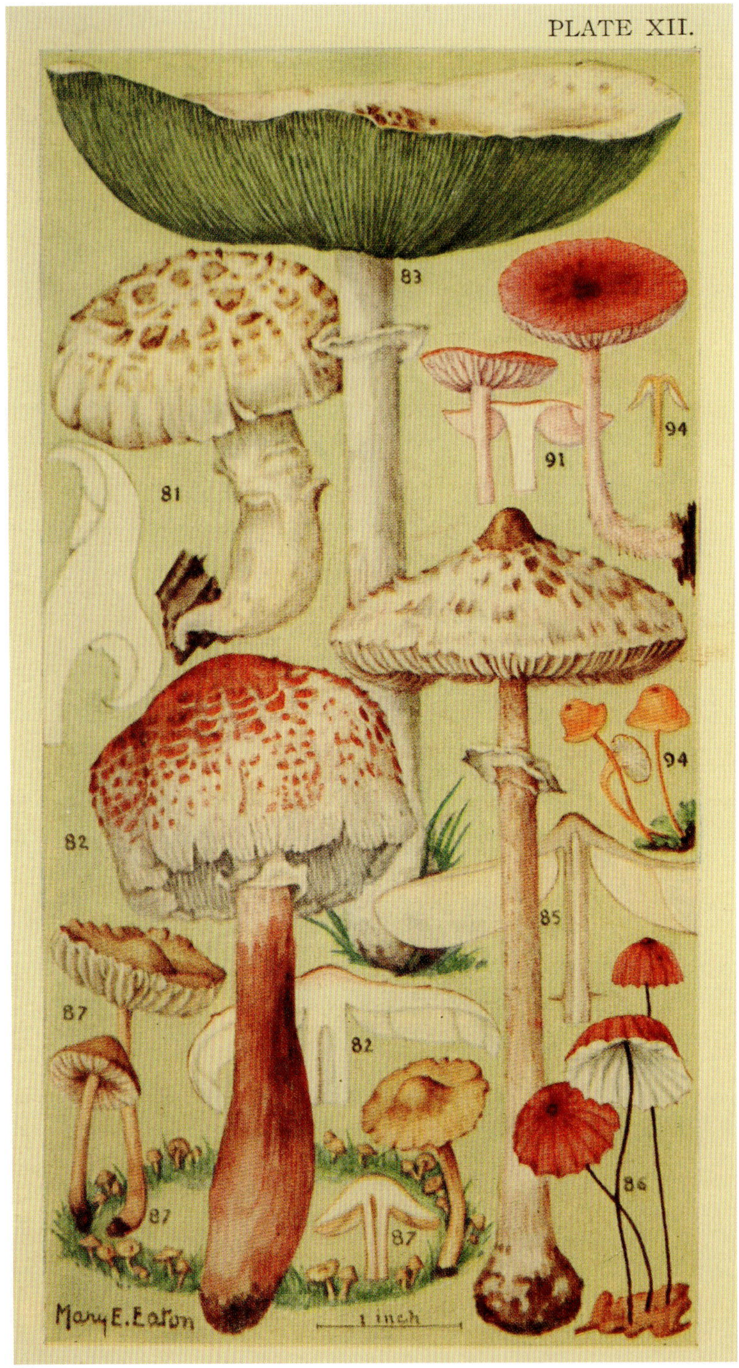

MARY E. EATON

Plate XII, from *Field Book of Common Gilled Mushrooms* by William Sturgis Thomas, 1928
Chromolithograph, H. approx. 7 in. / 18 cm
University of Connecticut Libraries, Storrs

William Sturgis Thomas's *Field Book of Common Gilled Mushrooms: With a Key to Their Identification and Directions for Cooking for Those That Are Edible* (1928) is a soup-to-nuts introduction to American fungi, chock-full of tips and tricks for the amateur forager. The bulk of Thomas's text, covering hundreds of species, focuses on each mushroom's defining features, cataloging its physical attributes, manner of growth, odor, spore characteristics, place and season of growth, and taste. The volume is particularly enhanced by sixteen stunning color plates painted by botanical illustrator Mary Emily Eaton (1873–1961). The vibrantly colored grouping here includes species from several families—including the central green-spored parasol (*Chlorophyllum molybdites*), with its upturned green gills, and the fairy ring mushroom (*Marasmius oreades*) at bottom right in its varying stages of growth—all painstakingly rendered in striking detail and at accurate scale. Born in England, Eaton was formally trained as an artist at the Taunton School of Art and South Kensington's Royal College of Art. At the age of thirty-six, she visited her siblings in Jamaica, where her passion for nature illustration was sparked; she started by painting the island's butterflies and moths before turning an eye to flora. Two years later, she landed a position in the United States as an artist for the New York Botanical Garden, where she remained for more than two decades. Best known for her sumptuous watercolor compositions of flowers and cacti, Eaton's work is held by the National Geographic Society, Smithsonian Institution, and British Museum, among other prestigious entities.

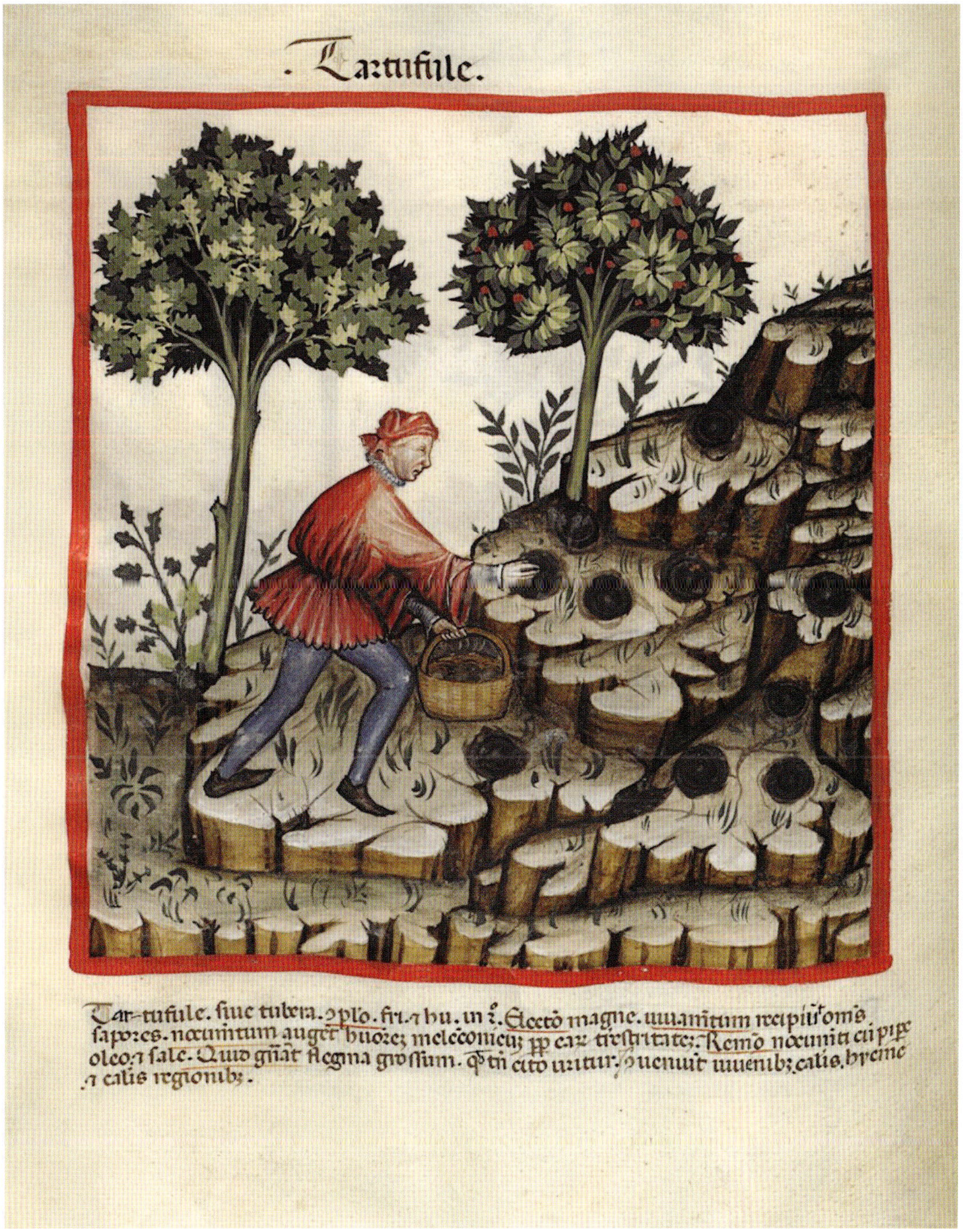

IBN BUṬLĀN

Tacuinum sanitatis, c. 1380–99
Ink on parchment, 13⅛ × 9 in. / 33.5 × 23 cm
Austrian National Library, Vienna

Basket in hand, a man forages for truffles on a hillside in this illustration from a late fourteenth-century Latin edition of *Tacuinum sanitatis,* a medieval guide to healthy living intended for the cultured nobility. Lavishly illustrated, this expensive handbook was based on an eleventh-century Arabic manuscript, *Taqwim al-Sihha bi al-Ashab al-Sitta* (*Rectifying Health by Six Causes*). Written by physician and philosopher Ibn Buṭlān (d. c. 1066), the manuscript went beyond the popular herbalist guides of the day to divide the components of a healthy life into six "causes," or parts, which it addressed in turn. The first part was concerned with nutrition. Ibn Buṭlān advised eating and drinking in moderation, and this illustration—one of many that show vegetables, fruits, and culinary and medicinal herbs being gathered—portrays a bucolic scene as the man collects truffles, which he considered necessary for a good life. The guide explains that most cap-and-stem mushrooms are deleterious to one's health but lists the truffle, which resembles a small black potato, as an exception. Although the guide includes the truffle as a vegetable, it is, in fact, the fruit of a subterranean fungus of the genus *Tuber* that is commonly found close to the roots of trees, such as the oak. While the *Tacuinum sanitatis* does not claim the truffle to be particularly nutritious, the fungus was a highly prized culinary treasure in the homes of the nobility, known for its distinct flavor and its rarity.

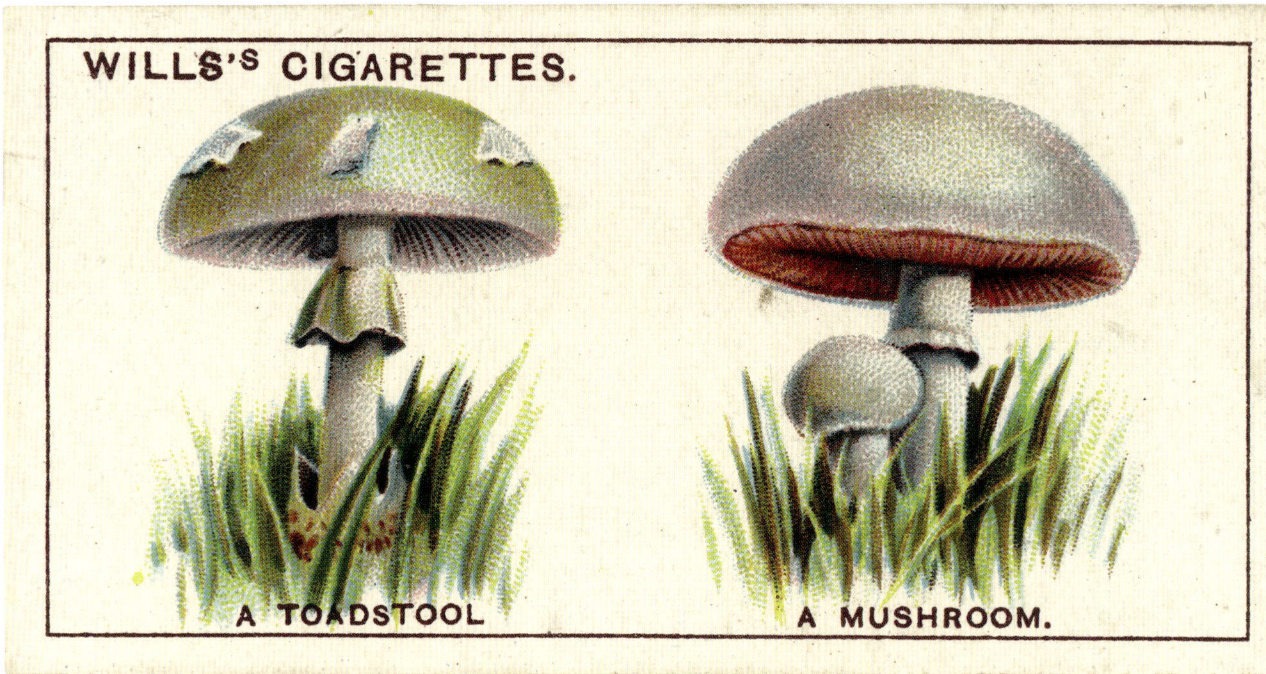

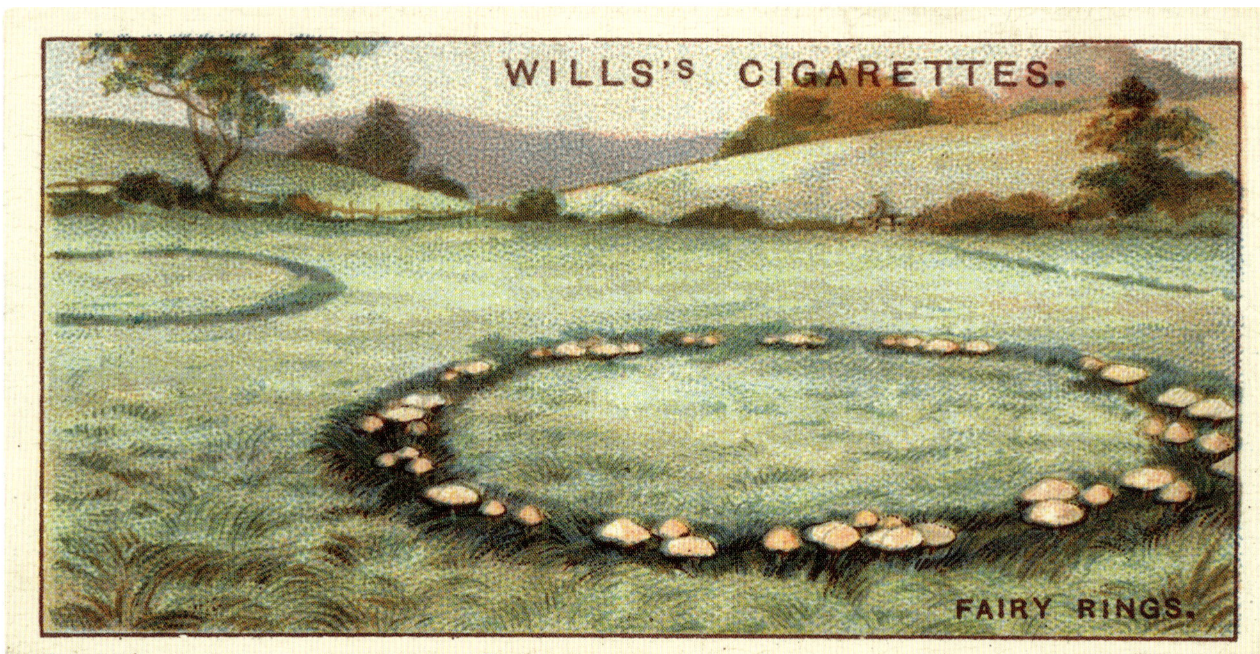

W.D. & H.O. WILLS

Wills cigarette cards, c. 1920–40
Chromolithograph, each approx.
1⅜ × 2⅝ in. / 3.5 × 6.7 cm
Private collection

British tobacco company W.D. & H.O. Wills was founded in Bristol, England, in 1786, although it did not take its name until 1830. It was one of the thirteen founding entities of the Imperial Tobacco Company, and the first British tobacco company to mass-produce cigarettes. The Wills company had a reputation for taking care of its workers, pioneering free medical care and paid holidays for employees, among other benefits. Starting in 1887, Wills became one of the first UK tobacco companies to include advertising cards (known as cigarette cards) with their packs of smokes. Later, in 1895, they began producing themed sets—*Ships and Sailors*, *Aviation*, *British Butterflies*, *Garden Flowers*, and *Air Raid Precautions*, as well as sport-centric sets—encouraging consumers to collect all the cards in a given set. In the 1920s and 1930s, they produced a popular set titled *Do You Know?*, which offered smokers interesting facts and corresponding illustrations about subjects such as nature, history, and the arts. The two cards pictured here are part of that series, the top one titled *Do You Know a Toadstool from a Mushroom?* and the bottom one titled *Do You Know the Cause of Fairy Rings?* While there is not, in fact, any scientific difference between a toadstool and a mushroom, at the time, edible varieties were colloquially referred to as "mushrooms" and poisonous ones as "toadstools." A fairy ring, on the other hand, is a naturally occurring circular ring of mushrooms caused by an underground mycelial network expanding outward from a central point.

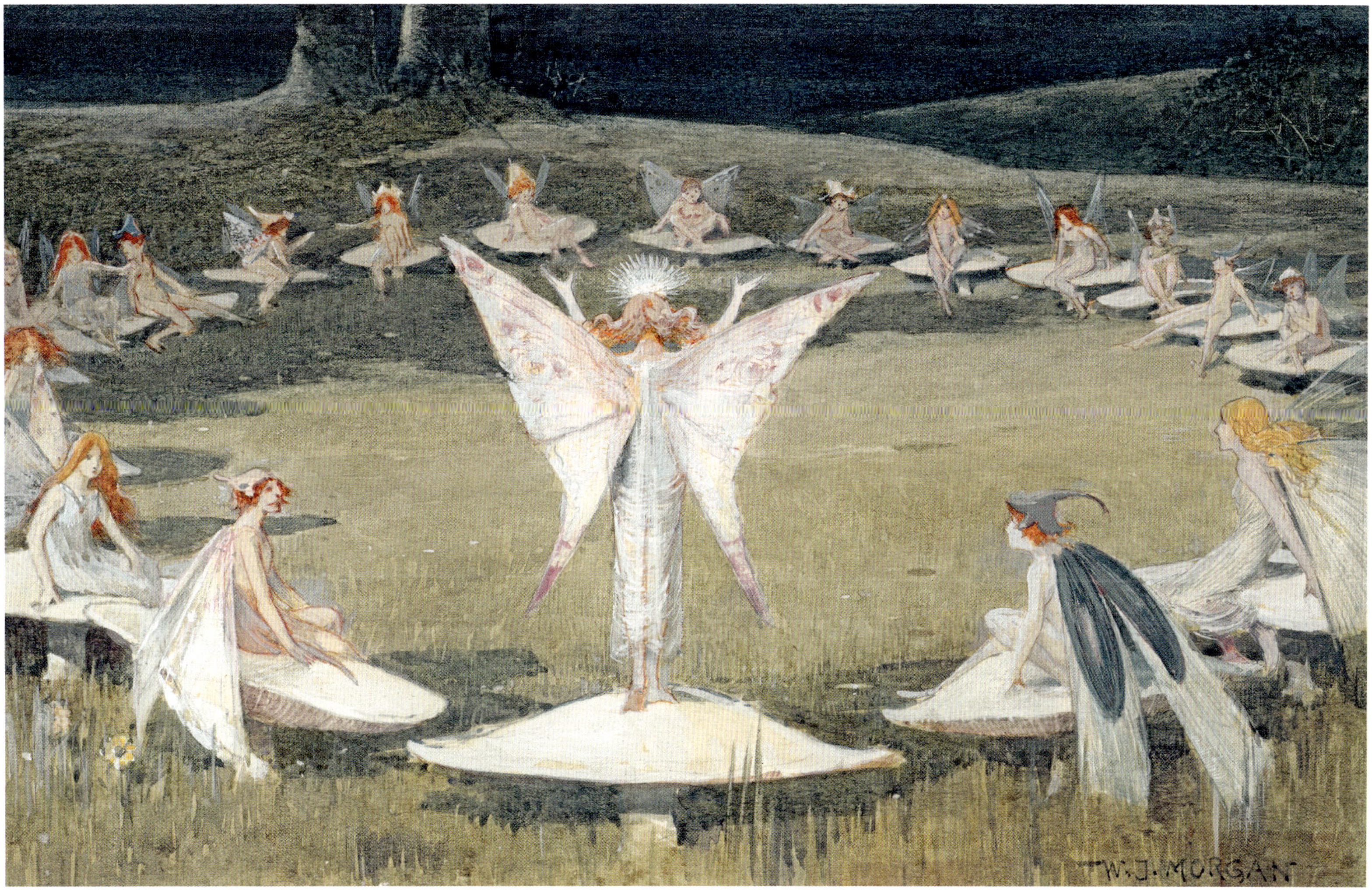

WALTER JENKS MORGAN

The Fairy Ring, c. 1870–80
Watercolor with white heightening on paper,
6⅛ × 9 in. / 15.5 × 23 cm
Private collection

In the latter half of the nineteenth century, all things supernatural—including mesmerism, occultism, illusionism, séances, and much more—were in vogue in England. This enthusiasm coincided with a rise in children's literature, which featured fairies and sprites in abundance, to create a veritable mania for magical creatures, with fairies foremost among them. The Victorian fairy paintings of artists such as John Anster Fitzgerald, Richard Dadd, and others, captured the zeitgeist of the era, often taking inspiration from Shakespeare's *A Midsummer Night's Dream* to create a distinctively British Victorian art form. Fairy paintings were a reaction to the rapid (and rapacious) effects of industrial and technological advances, as well as a rejection of the realism in the nascent art form of photography. Warm, whimsical, and enchanting, fairies were the embodiment of hope, celebration, and even hedonism. *The Fairy Ring* by Walter Jenks Morgan (1847–1924) is a beautiful example of the form, the lithe bodies of the sprites and verdant pastoral settings both unmistakably Art Nouveau. The scene here suggests a confabulation, the fairies gathered in a circle, each seated on his or her own mushroom. Painted in watercolor with white heightening, the central figure stands in the foreground with her back to the viewer. Yet her glittering crown and her stance convey the potency of her stature; she is unmistakably their leader. An accomplished lithographer, painter, and illustrator, and member of the Royal Society of British Artists, Morgan worked extensively for magazines, including *The Graphic* and *Illustrated London News*.

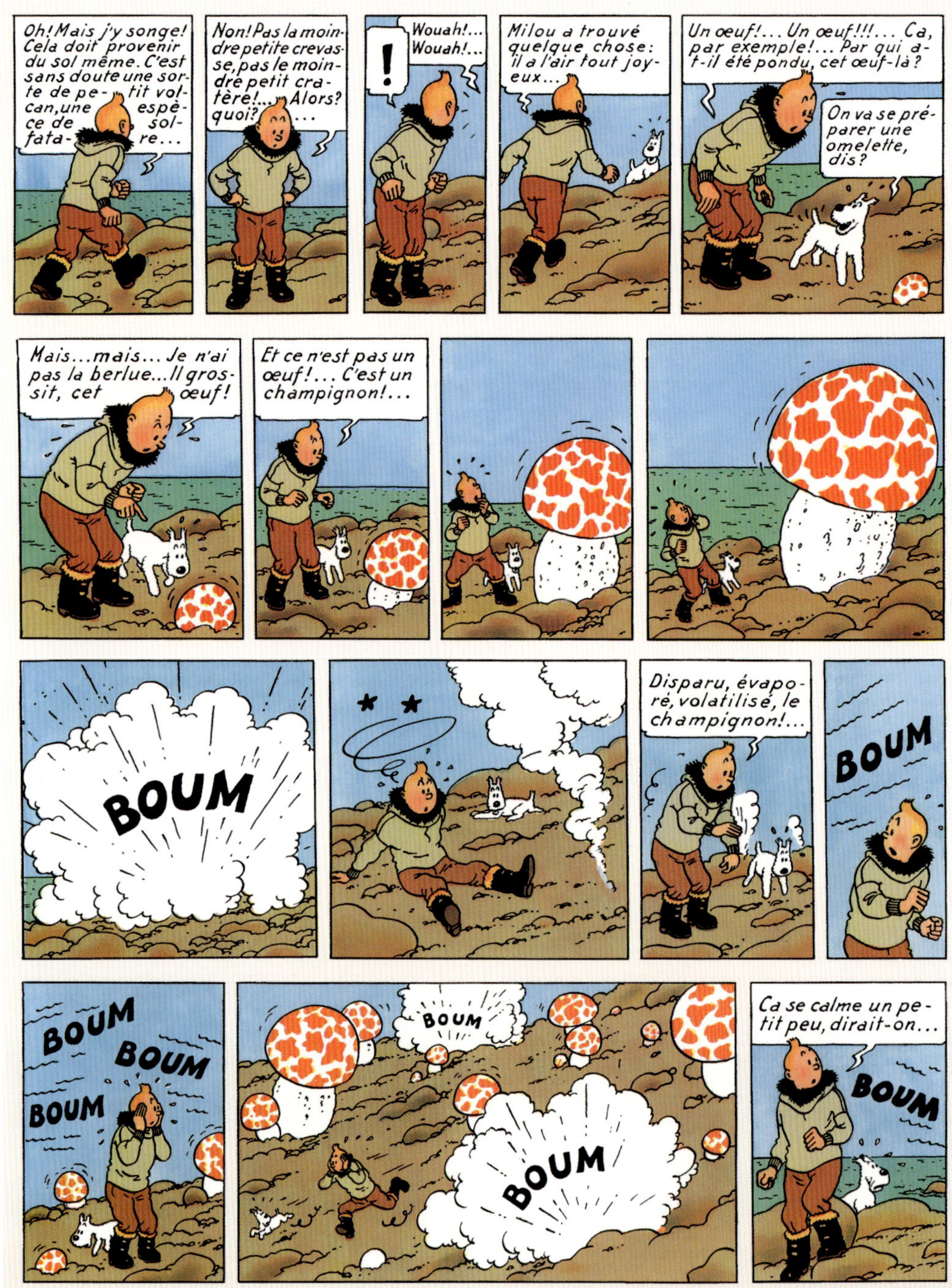

HERGÉ (GEORGES PROSPER REMI)

Les Aventures de Tintin: L'Étoile Mystérieuse (*The Adventures of Tintin: The Shooting Star*), 1942
Printed comic book, 11¾ × 8¼ in. / 29.7 × 21 cm

For generations of children who grew up reading comic-strip books about the adventures of the enterprising young boy reporter Tintin and his sidekick, Snowy, this page will be instantly recognizable. Taken from the 1942 book *L'Étoile Mystérieuse* (*The Shooting Star*), the comic strip shows giant red-and-white mushrooms growing and exploding out of the ground, the result—so the story tells—of an unknown substance found in a meteorite, or shooting star, that has crashed into the Arctic Ocean. Created by the Belgian comic-strip artist Georges Prosper Remi (1907–1983), better known by his pen name of Hergé, the stories about Tintin were originally published in the newspapers *Le Vingtième Siècle* and later *Le Soir* before being turned into books. This story was the first of Tintin's adventures that appeared in color. Each one of the twenty-four stories (translated into more than 110 languages) stands alone: Tintin and Snowy, aided by the whiskey-loving Captain Haddock, overcome villains, solve mysteries, and travel the world to places including the Arctic Ocean (as here), Tibet, and the Congo. Hergé was known for his meticulous research; here, the mysterious substance in the meteor (named phoslite after Professor Phostle in the book) was said to prefigure the mushroom cloud caused when the atomic bomb was dropped on Hiroshima on August 6, 1945. The enduring popularity of Tintin lies not just in Hergé's unique drawings but also in his storytelling: derring-do mixed with messages of morality about right and wrong.

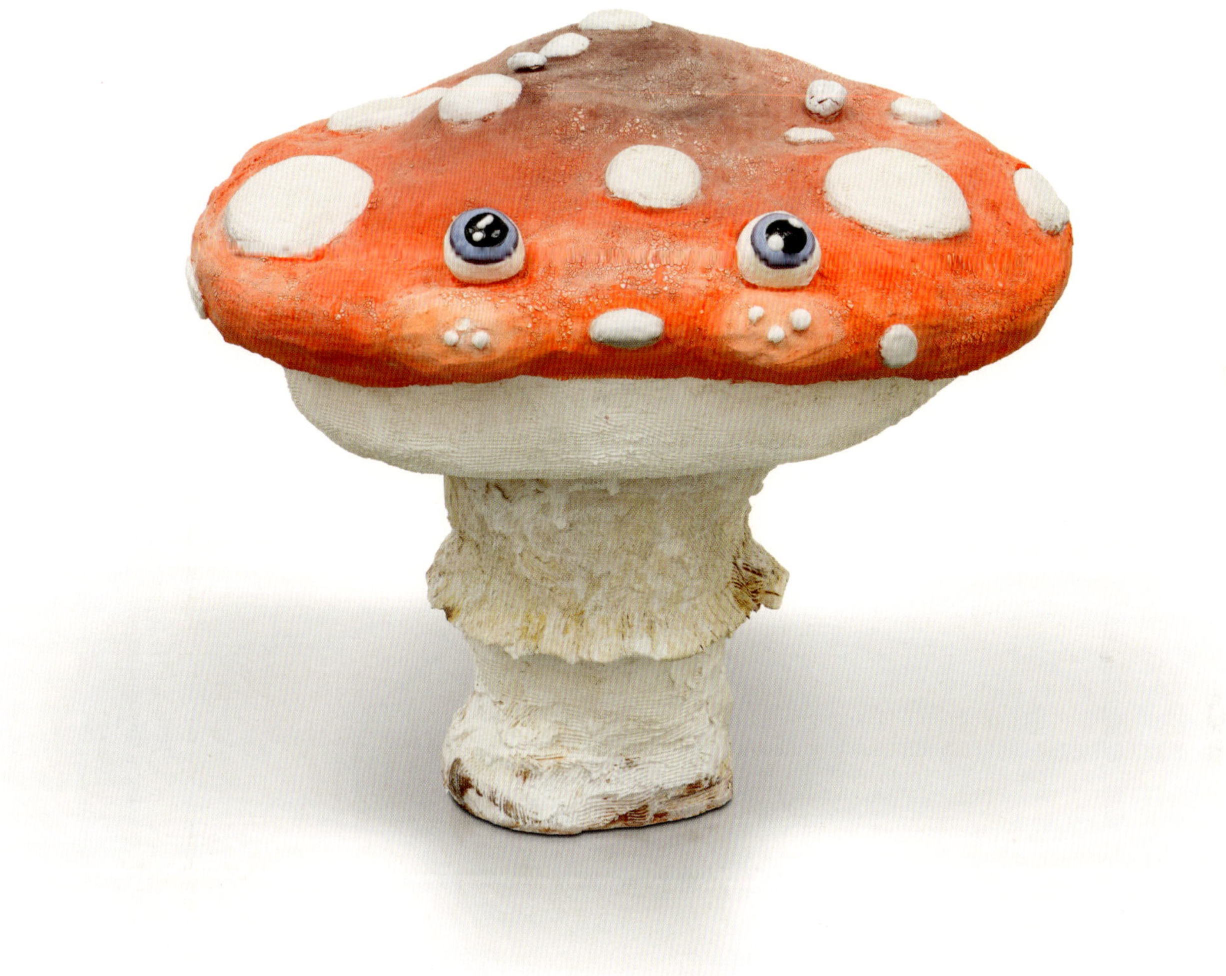

ALAKE SHILLING

I'm Just a Fun Guy, 2023
Glazed fired clay,
20½ × 19 × 17½ in. / 52.1 × 48.3 × 44.5 cm
Private collection

Recalling the toadstools of fairy tales and folk stories, with its bright red cap littered with white speckles, this oversize ceramic mushroom whimsically peers back at the viewer from two eyes perched atop its cap. The sculpture, cleverly titled to be a play on the word *fungi*, is by Black artist Alake Shilling (b. 1993). The Los Angeles–based artist is, by her own admission, fascinated by bugs and their associated flora and fungi. The toadstool reminds Shilling of her childhood and is part of a wider ceramic ecosystem that includes other mushrooms as well as ladybugs, butterflies, and worms. Her celebration of nature's tiny, and often unsung, heroes is inspired by her West African ancestry, particularly the traditional totems of Yoruba culture. These totems often reflect a personification of nature and play an integral part in how the Yoruba depict the importance of nature in the everyday lives of the people. Probably the most easily recognizable of all mushrooms, the toadstool after which Shilling has modeled her sculpture—known as the fly agaric or fly amanita (*Amanita muscaria*)—is both inedible and poisonous; its vibrant red umbrella-shape cap belies both those facts. As the mushroom grows, the white speckles on the cap appear to shrink, but this is just an illusion. Instead, the larger red surface only makes them appear smaller. When completely mature, they can reach 3 to 8 inches (8 to 20 cm) wide and are commonly found in northern Europe and Asia.

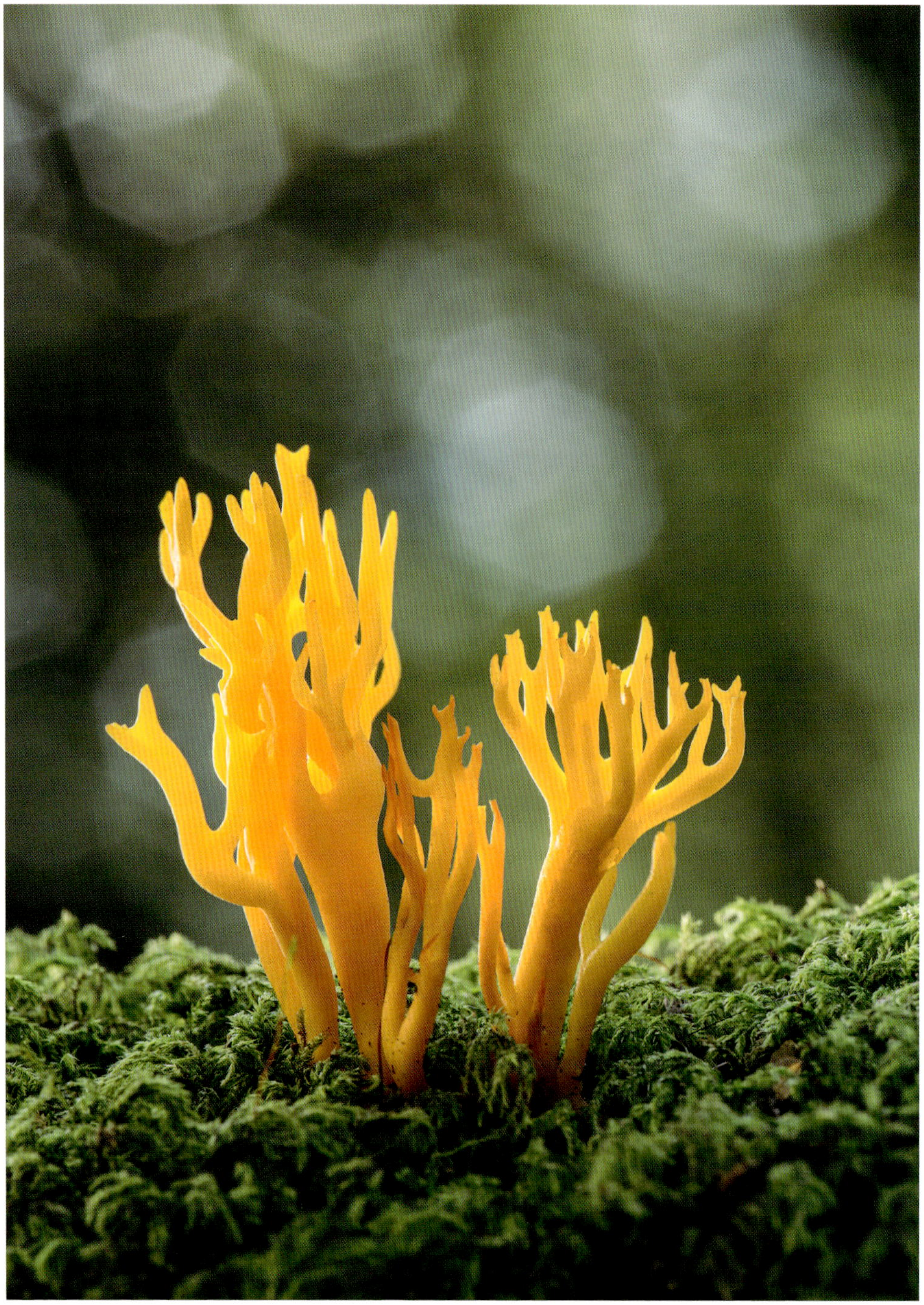

GUY EDWARDES

Yellow Staghorn Fungus (Calocera viscosa), New Forest National Park, Hampshire, England, UK, 2016
Photograph, dimensions variable

In this detailed macrophotograph, the bright, shining, slightly glutinous branches of a yellow staghorn fungus (*Calocera viscosa*) reach upward in pleasing contrast with the soft carpet of green mosses on the forest floor. The yellow staghorn is widespread in Europe, growing mainly on the decaying wood of coniferous trees. It is also found in parts of North America, where it is known as the jelly antler fungus. The fruiting bodies are a bright golden yellow, resembling some types of branching coral, or miniature antlers. This image was captured by landscape and nature photographer Guy Edwardes (b. 1974) in the New Forest National Park, in Hampshire, England. A protected area of ancient woodland and heath, it is a site well-known for its rich and diverse array of fungi, many of which grow on the rotting timber of fallen trees and branches. The New Forest is a local happy hunting ground for Edwardes, who is based in nearby Dorset. When composing his shots, he pays as much attention to the background as to the subject itself, taking care not to include elements that might distract. Here, this has been achieved by creating an unfocused backdrop (an effect known as bokeh), which highlights superbly the main brightly lit subject. Some of Edwardes's favorite subjects are fungi, and his portraits reveal their beauty in extreme close-up, achieved by combining technical expertise with great patience.

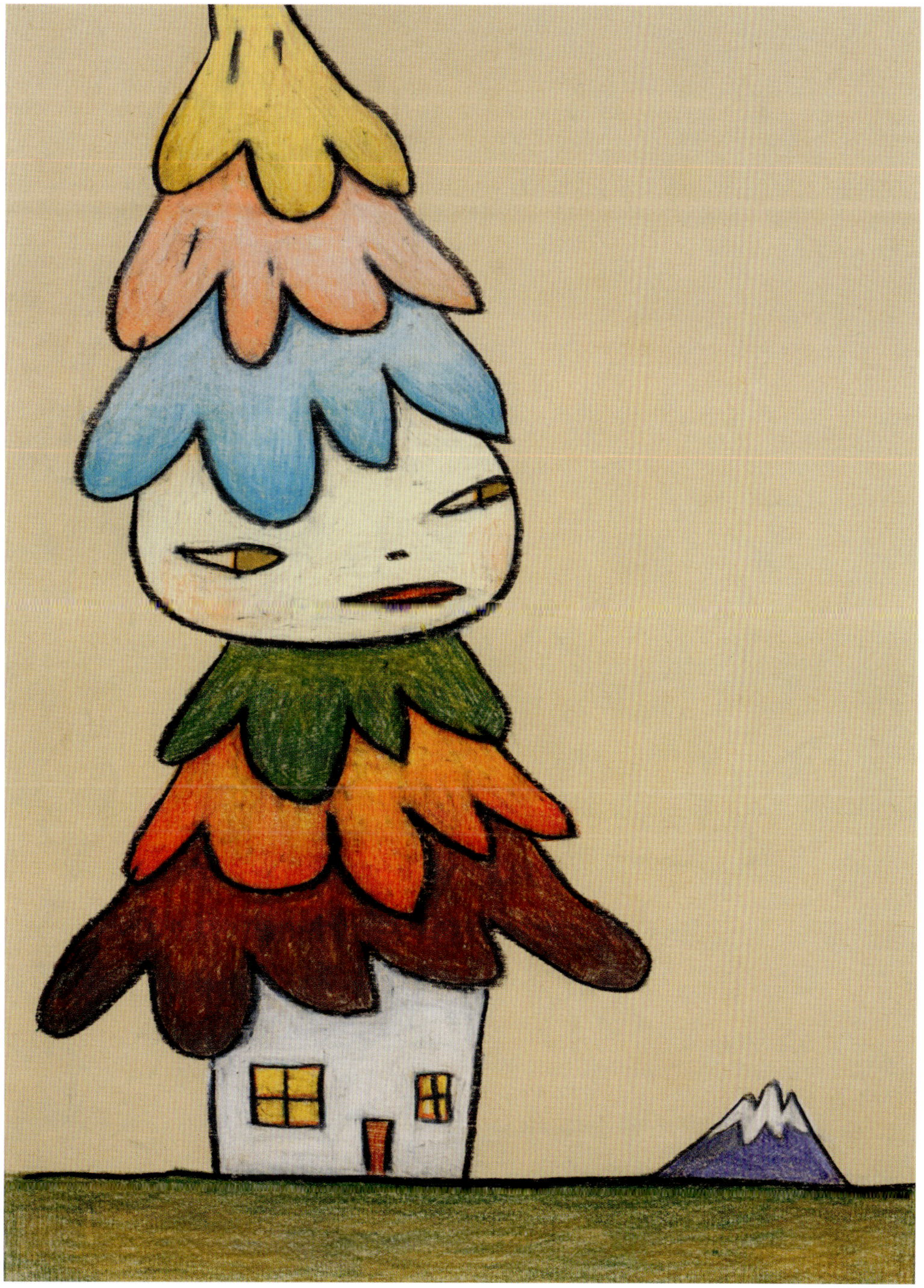

YOSHITOMO NARA

Mushroom House, 2011
Colored pencil on paper,
14⅜ × 10¼ in. / 36.5 × 26 cm
Private collection

Japanese contemporary artist Yoshitomo Nara (b. 1959) grew up in rural northern Japan, a region known for its volcanic terrain, evergreen forests, and snowbound winters. His artistic sensibilities were formed in this distinctive place, and his childhood memories remain at the core of his unique and varied art practice. Nara is renowned for his intriguing paintings of wide-eyed, large-headed children, who possess an air of innocence but with an edge of defiance and sometimes menace. However, drawing is central to his artistic production. Drawing allows for rapid execution, enabling Nara to capture his creative impulses spontaneously. *Mushroom House* exemplifies the bold outlines of his assured linear style and his use of vivid color carefully rendered in gradients. The drawing is reminiscent of a child's vision of a house—a square with a door and windows on either side—but a simple triangular roof has metamorphosed into fantastical, bulbous layers that mushroom vertically, reaching beyond the edge of the paper. Their shape brings to mind fir trees weighed down by heavy snowfall, the "snow monsters" experienced by Nara as a child, and a gigantic round head, set within the branches, recalls the artist's interest in animism—the idea that spirits inhabit the natural world—a belief rooted in Japanese religion and folklore. The head casts a sideways glance toward a distant, snow-capped mountain that recalls Mount Iwaki, a constant presence overlooking Hirosaki, Nara's hometown.

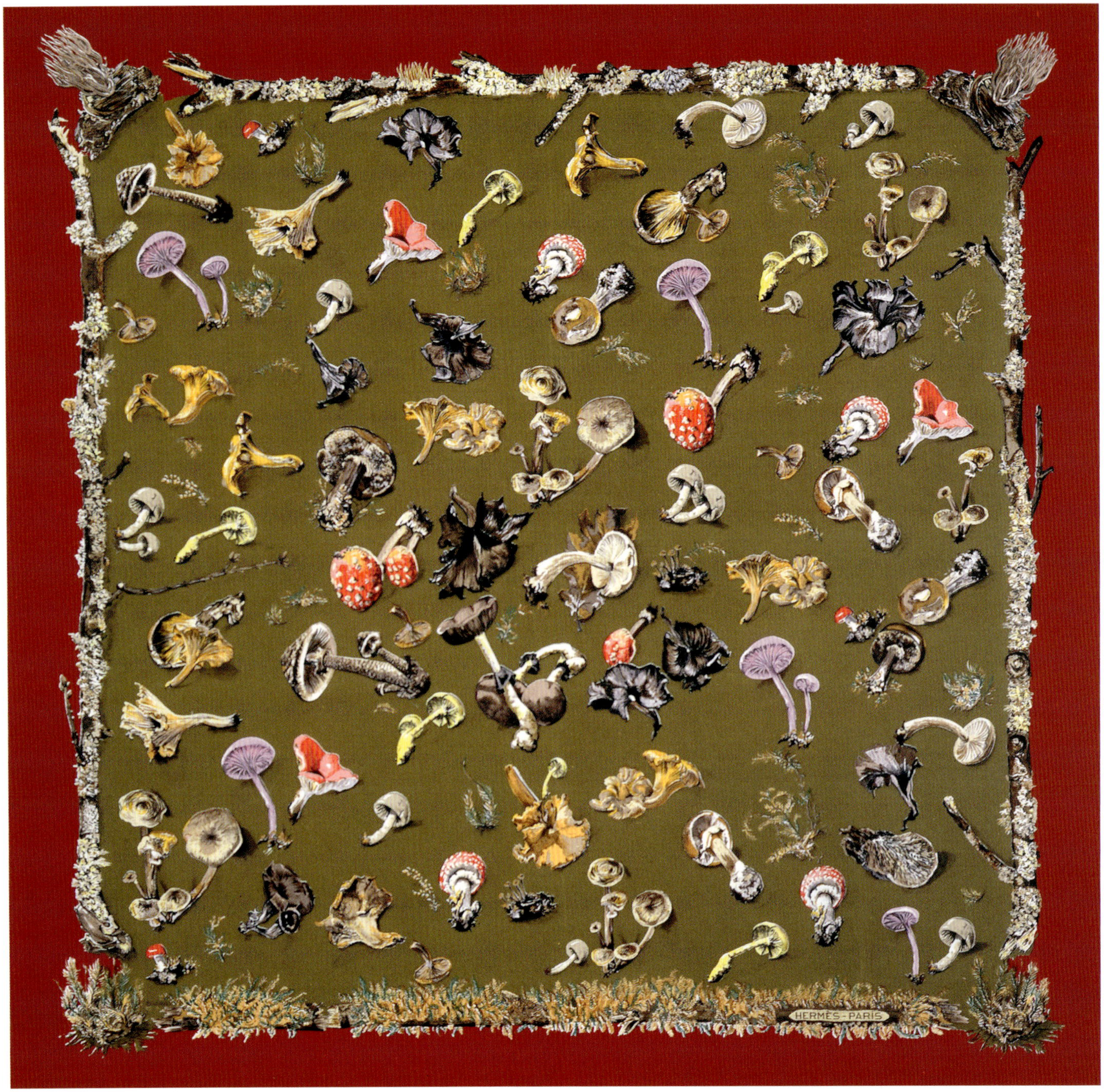

ANNE GAVARNI D'AUTRAY AND FRANÇOISE DE LA PERRIÈRE FOR HERMÈS

Champignons, 1960
Silk, 34⅞ × 34⅞ in. / 88.5 × 88.5 cm

Artfully arranged on a dark green background, an explosion of mushrooms decorates this silk scarf, from red-capped, white-speckled toadstools (*Amanita muscaria*) to violet webcaps (*Cortinarius violaceus*) and yellow flowerpot parasols (*Leucocoprinus birnbaumii*). Some of the mushrooms are still attached to their organic growing medium, evoking a sense of autumnal change, while the burgundy border edged with lichen, bracken, and berries further captures the season. The illustration is the work of the French designers Anne Gavarni d'Autray (1922–2014) and Françoise de la Perrière (active 1950s–mid-1990s) for the French luxury house Hermès. First issued in 1960, *Champignons* belongs to one of the more than two thousand *carré* ("square") designs that have been created by more than 150 contemporary artists for Hermès since 1937. The remit is often to showcase nature's beauty, from shells, flowers, and, as here, fungi to imagery from the equestrian world, horses being a key theme of the Hermès heritage. Accuracy of detail is paramount: the artists—a number having a background in botanical art—capture the unique subtleties of each individual species. During a long collaboration with Hermès, de la Perrière designed more than thirty scarves, and Gavarni, who began exhibiting her artwork in Paris in the 1950s under the pseudonym Mélusine, is best known for her vegetable, mushroom, and dewdrop designs. Made in limited numbers, the scarves—particularly ones from earlier decades—have become iconic collector's items.

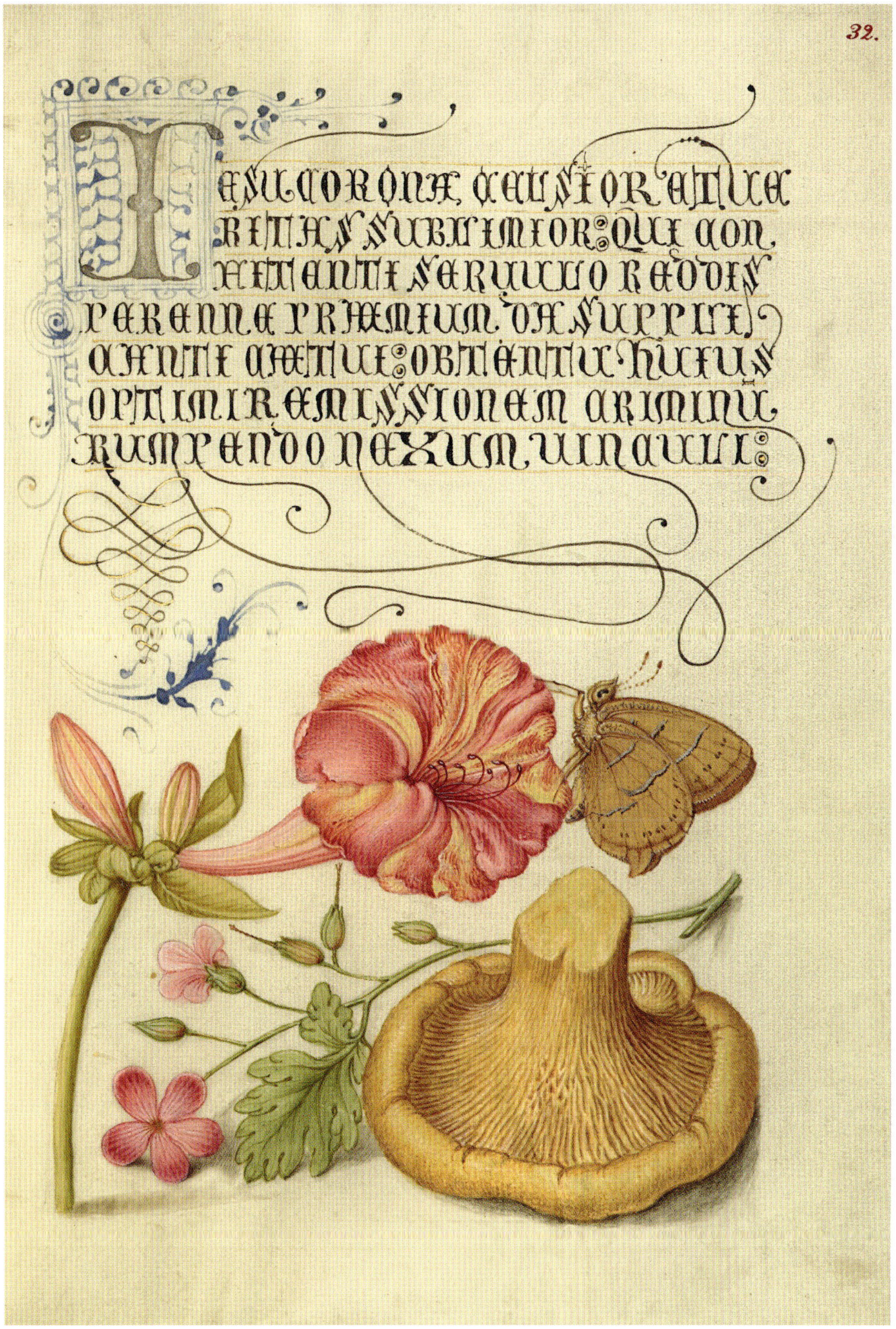

JORIS HOEFNAGEL

Four-o'Clock, Brown Hairstreak, Herb Robert, and Chanterelle, from *Mira calligraphiae monumenta* (*Model Book of Calligraphy*), 1561–62
Watercolor, gold and silver paint, and ink,
$6\frac{1}{2} \times 4\frac{7}{8}$ in. / 16.6×12.4 cm
Getty Museum, Los Angeles

Depicted in vibrant watercolor, an overturned chanterelle mushroom (*Cantharellus cibarius*) rests alongside a sprig of herb Robert (*Geranium robertianum*), while a brown hairstreak butterfly (*Thecla betulae*) perches upon a four o'clock flower (*Mirabilis jalapa*). Taken from a much larger work, *Mira calligraphiae monumenta* (*Model Book of Calligraphy*), this illuminated page is the work of self-taught Flemish artist Joris Hoefnagel (1542–1600). The curled cap and delicate gills of the golden yellow mushroom are rendered in realistic detail, and Hoefnagel included subtle shadows to add further verisimilitude to the image. However, the illustration was painted some thirty years after the text, which was written by a man Hoefnagel never even met. Between 1561 and 1562, Georg Bocskay, court secretary to Holy Roman Emperor Ferdinand I, wrote the book's text to demonstrate his calligraphy skills. Thirty years later, during the early 1590s, Ferdinand's grandson, Emperor Rudolph II, commissioned Hoefnagel to illuminate Bocskay's tome. Hoefnagel, who was one of the last manuscript illuminators, added illustrations of flowers, fruit, insects, and, as here, fungi. A true Renaissance man, Hoefnagel—as with many of his contemporary Netherlandish artists—had studied the natural world. His studies resulted in *The Four Elements*, a collection of three hundred of his exquisite watercolor drawings of various flora and fauna. Hoefnagel's illustrations were very influential on the practice of floral still-life painting that emerged in the Netherlands toward the end of the sixteenth century, with his botanically accurate illustrations becoming the benchmark for subsequent artists.

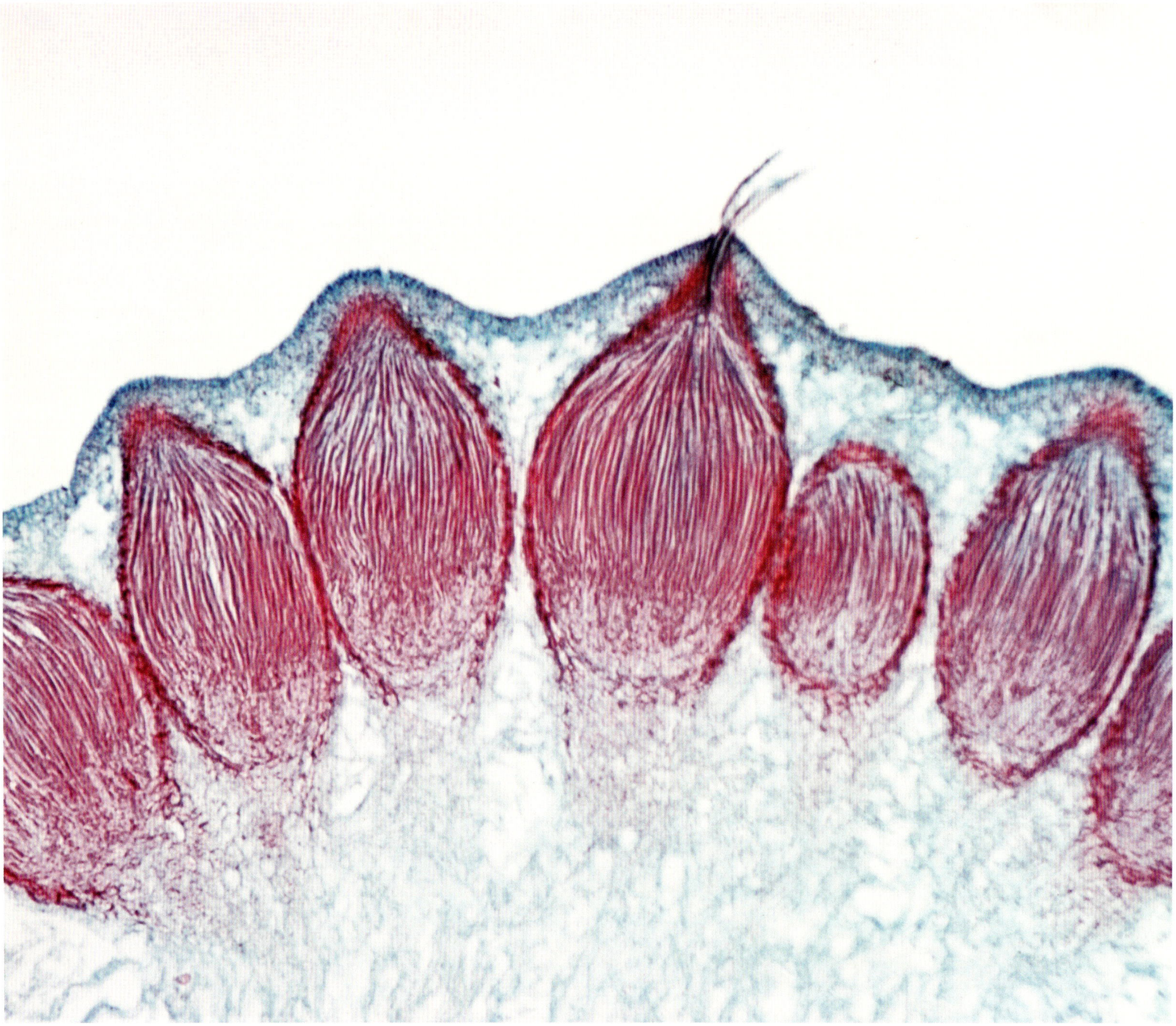

BIOPHOTO ASSOCIATES

Ergot (Claviceps purpurea)
Light micrograph, dimensions variable

Light micrography, or photomicrography, is essentially the process of taking a photograph through a microscope, allowing one to closely examine that which is not visible to the naked eye. The photomicrograph shown here depicts *Claviceps purpurea,* a parasitic ergot fungus that affects grasses and grains, especially rye. When infected crops are ingested, the fungus causes ergotism in humans. This condition is marked not only by crippling physiological manifestations, such as convulsions, gangrene, loss of limbs, and potentially death, but also psychological symptoms, including hallucinations and erratic, sometimes violent, behavior. In the Middle Ages, poisoning due to ingesting ergot-infected grain was common, and the resulting condition became known as Saint Anthony's fire, named for the monks of the Order of Saint Anthony, who worked to treat the affliction and ease the suffering of those in its grip. Numerous events of mass hallucinations and hysteria throughout history previously attributed to the occult—including the behaviors ostensibly exhibited by the women persecuted in the Salem witch trials—are now thought to have been manifestations of ergotism. The chemical lysergic acid can be derived from the alkaloid ergotamine, found in ergot fungus. In 1938 Swiss chemist Albert Hofmann used this lysergic acid to first synthesize the hallucinogenic drug lysergic acid diethylamide (also known as LSD or acid). LSD has shown incalculable influence within the arts, especially on visual art and music, and is enjoying renewed therapeutic interest within the field of psychiatry after a long period of dormancy in that arena.

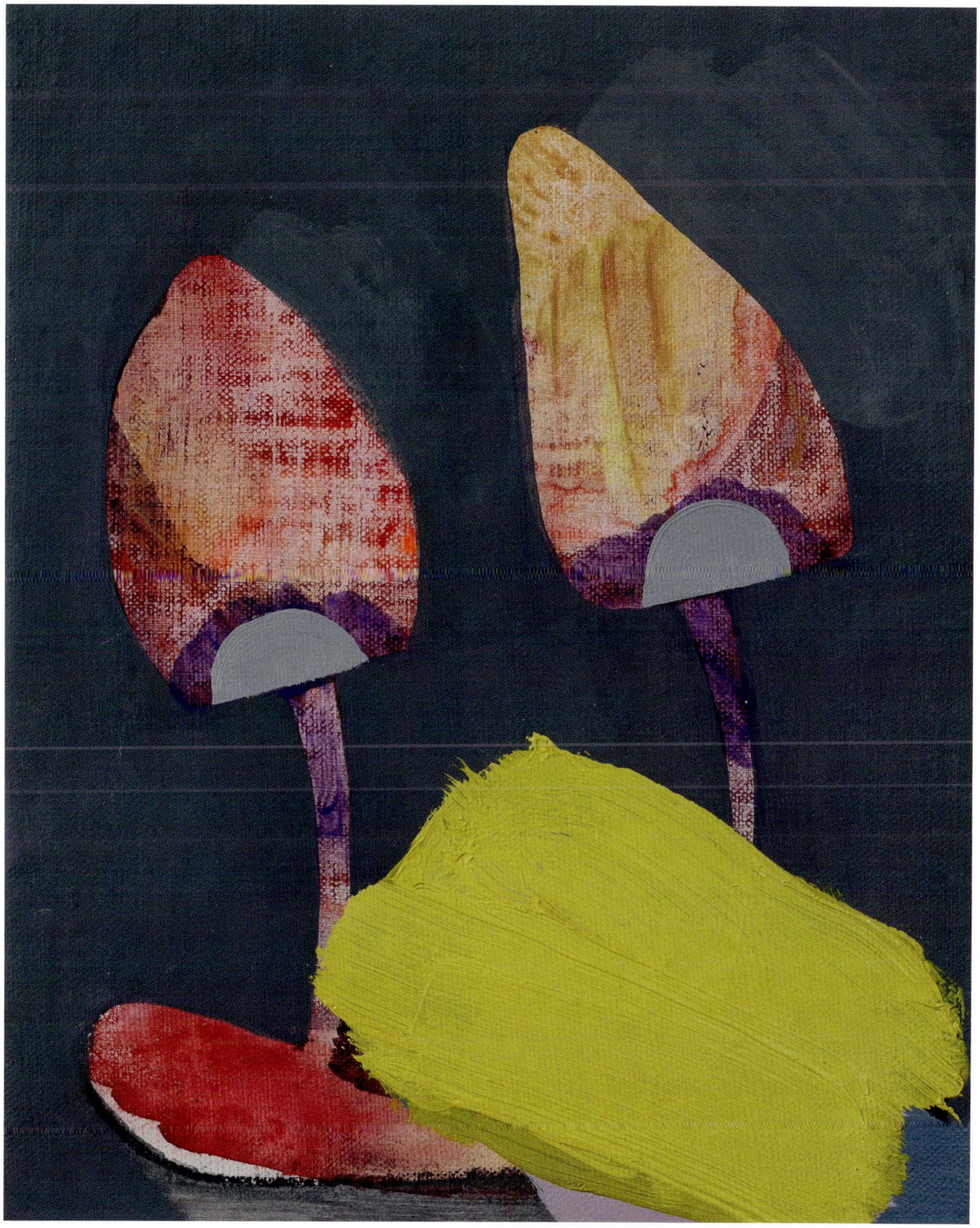

JOHN MILLEI

María Sabina #2, 2017
Oil on linen, 10 × 8 in. / 25.4 × 20.3 cm
Private collection, Los Angeles

The Abstract Expressionist paintings of Los Angeles–based painter John Millei (b. 1958) are best recognized by their pared-down forms and signature palette of secondary and tertiary colors. Millei works primarily in oil and flashe, piling great swaths of paint across his large-scale canvases. The muscularity of his brushstrokes in *María Sabina #2* underscores the power of the mushrooms, whose variegation and texture enliven the otherwise banal gray background. The titular subject, María Sabina, was a highly respected twentieth-century Mexican poet and spiritual leader (see p.37). Millei's composition is likely a reference to a mushroom *velada*—a nocturnal ritual led by a traditional *curanderos*, or healers (both physical and spiritual), such as Sabina, where participants eat psilocybe mushrooms, chanting and praying through the night. Through the ingestion of the fungi, practitioners are able to effectively seek divine guidance from both saints and ancestors in how to heal the ill and remedy other problems. The Mazatec people of the mountainous Sierra Mazateca in Oaxaca, Mexico, have been practicing *veladas* for centuries. But it was in 1955 that American businessman R. Gordon Wasson traveled to Oaxaca and deceived Sabina in order to participate in a ceremony. He absconded back to the United States with the sacred psilocybe spores, documented his experience in *Life* magazine in 1957, and exposed the traditional ritual (and later Sabina's identity) to the rest of the Western world—effectively opening the door to the wave of recreational psychedelics that would pervade the American counterculture in the following decades.

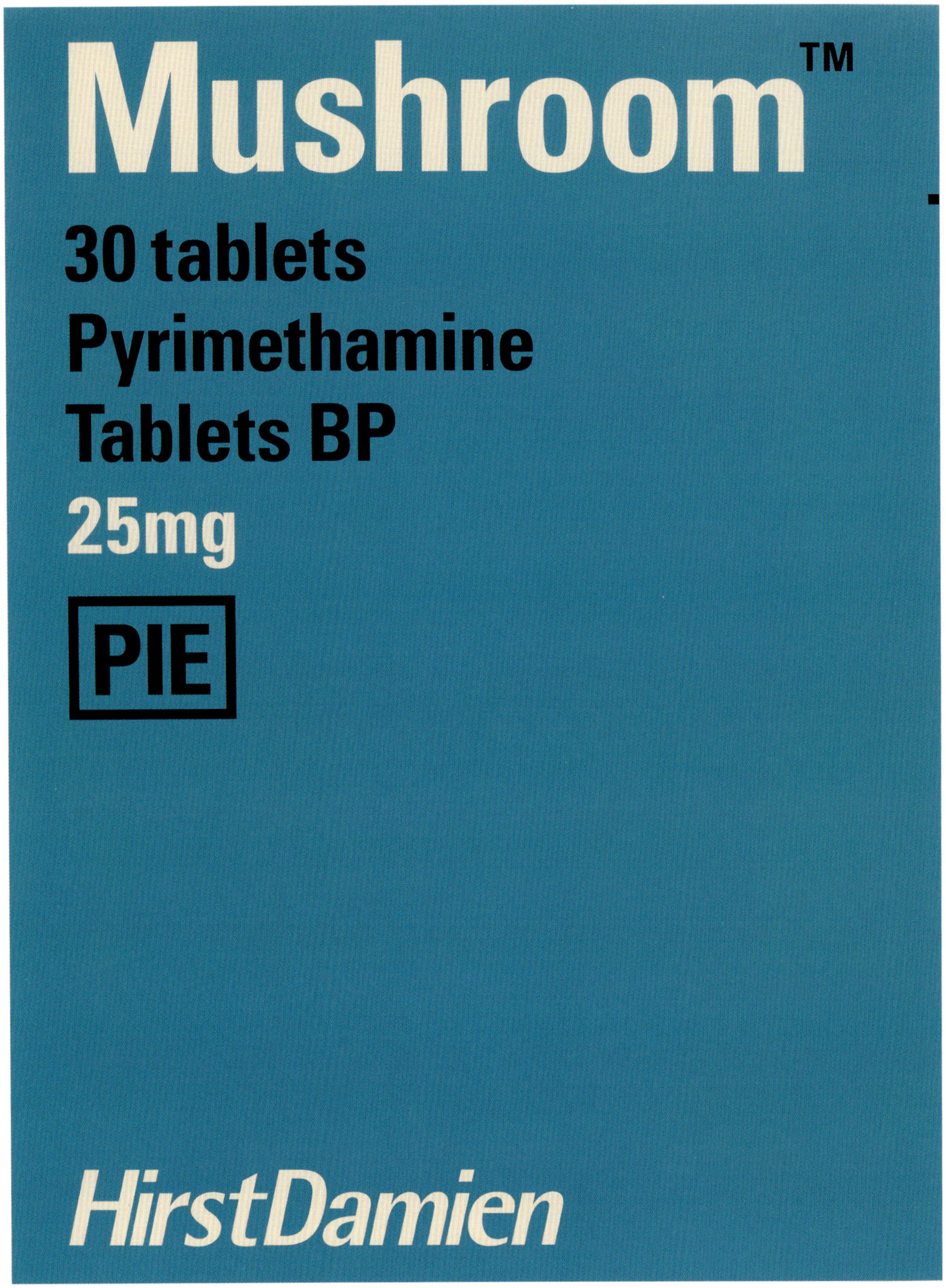

DAMIEN HIRST

Mushroom, from *The Last Supper* series, 1999
Screenprint, approx. 60¼ × 40 in. / 153 × 101.5 cm
Private collection

British artist Damien Hirst (b. 1965) rose to prominence in the 1990s, mastering a conceptual, often controversial style that is characterized by works that rethink the parameters of traditional art. Hirst explores themes of death, science, and religion through the lens of unconventional materials, using objects such as medical cabinets, formaldehyde, and pills to take a sideways look at the contemporary art scene. Hirst riffs off this tension between playful medium and serious content in the 1999 work *The Last Supper*, a series of thirteen screenprints. Its title, referencing Leonardo da Vinci's painting of the same name—where Jesus and his disciples share a meal before his crucifixion—strikes a subversive note by viewing familiar items, such as food and pharmaceutical drugs, through a religious lens. Each work, created in bold colors, with simple typography and a clean design, could be an advertisement or food packaging that would not look out of place on a grocery-store shelf. In the three prints *Dumpling*, *Sandwich*, and *Mushroom*, Hirst imbues the mundane items we consume every day with an unsettling context, twisting their comforting and nourishing properties into a look at the medicalized way through which we achieve comfort today. By including *Mushroom* in the series, Hirst continues this juxtaposition by referring to the natural healing character of mushrooms while also nodding to their occasionally poisonous and hallucinogenic properties. Mushrooms, like medicine, tread the line between harmful and healing—but which one it is becomes a question Hirst leaves unanswered.

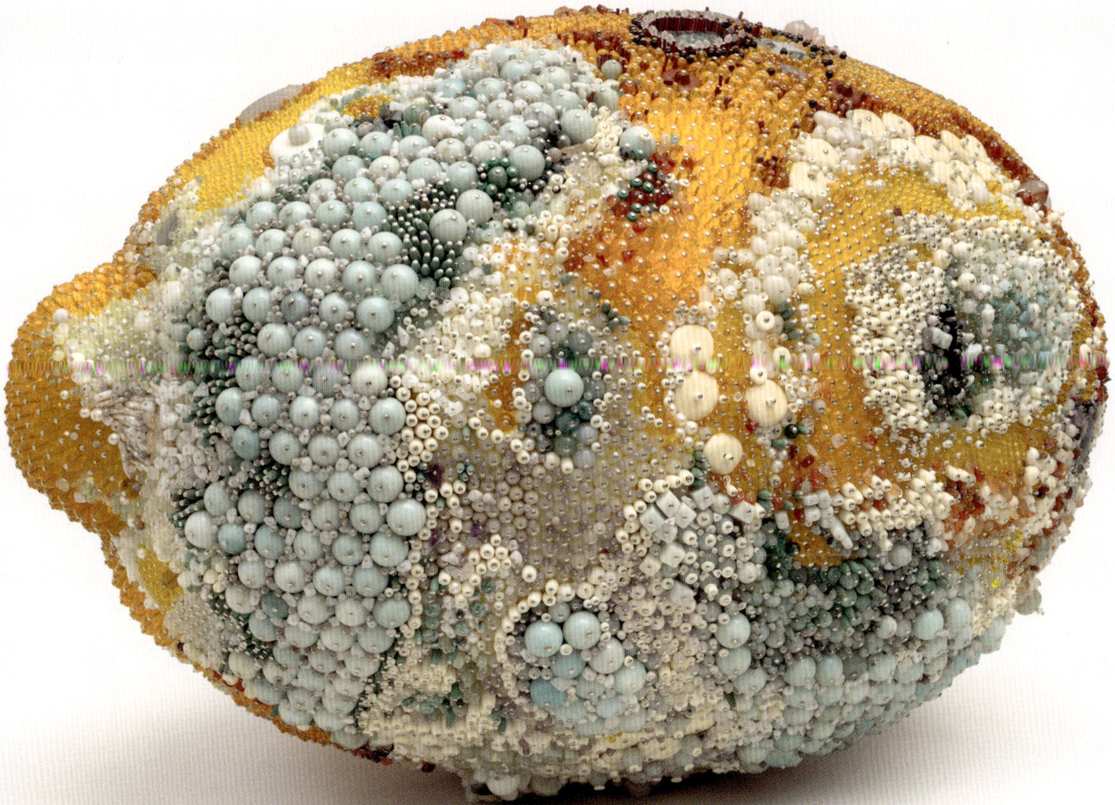

KATHLEEN RYAN

Bad Lemon (Creep), 2019
Amazonite, aventurine, black onyx, Italian onyx, turquoise, labradorite, carnelian, ocean jasper, sesame jasper, serpentine, fluorite, Ching Hai jade, snow quartz, magnesite, agate, brecciated jasper, rhodonite, rhodochrosite, red agate, garnet, tree agate, rose quartz, amethyst, lilac stone, limestone, marble, mother-of-pearl, bone, freshwater pearl, glass, and steel pins on coated polystyrene,
20 × 28½ × 20 in. / 50.8 × 72.4 × 50.8 cm
Hiscox Collection

While many fungi are cultivated for consumption, it is never a welcome moment when one grabs a piece of fresh fruit to eat, only to discover a fuzzy, powdery coating of mold on its skin. California-born artist Kathleen Ryan (b. 1984) plays with this revulsion response in her intricate sculptures of rotten fruit. On closer inspection of her artwork, the blooms of white, gray, and green "fungus" are revealed as meticulously placed, sparkling crystals and semiprecious gemstones, which sit alongside a uniform surface of tiny glass beads. In using the more valuable materials for the most repellent part, the artist inverts typical notions of value. Ryan also seeks to draw attention to the delicate beauty and buzzing energy of the fungus, explaining that the "fruit is dying, but mold is thriving." Such fungal invasions begin when airborne spores land on fruit skin and, feeding off sugars and moisture, grow as multicellular filaments called hyphae. Often spreading from a patch where the peel has been damaged, enabling the pathogen to access the nutritious flesh more easily, the interior is penetrated with fungal threads that take hold and thrive, sometimes rendering the fruit toxic to humans. Due to the high levels of acidity in their flesh and antifungal essential oils in their rinds, lemons are resistant to many common fungi that spoil other fruit. However, some, such as the bluish-green–and–white *Penicillium digitatum* and *Penicillium italicum*, have evolved to tolerate acidity, and therefore lemons are prone to some fungal growth, just as Ryan's sculpture depicts.

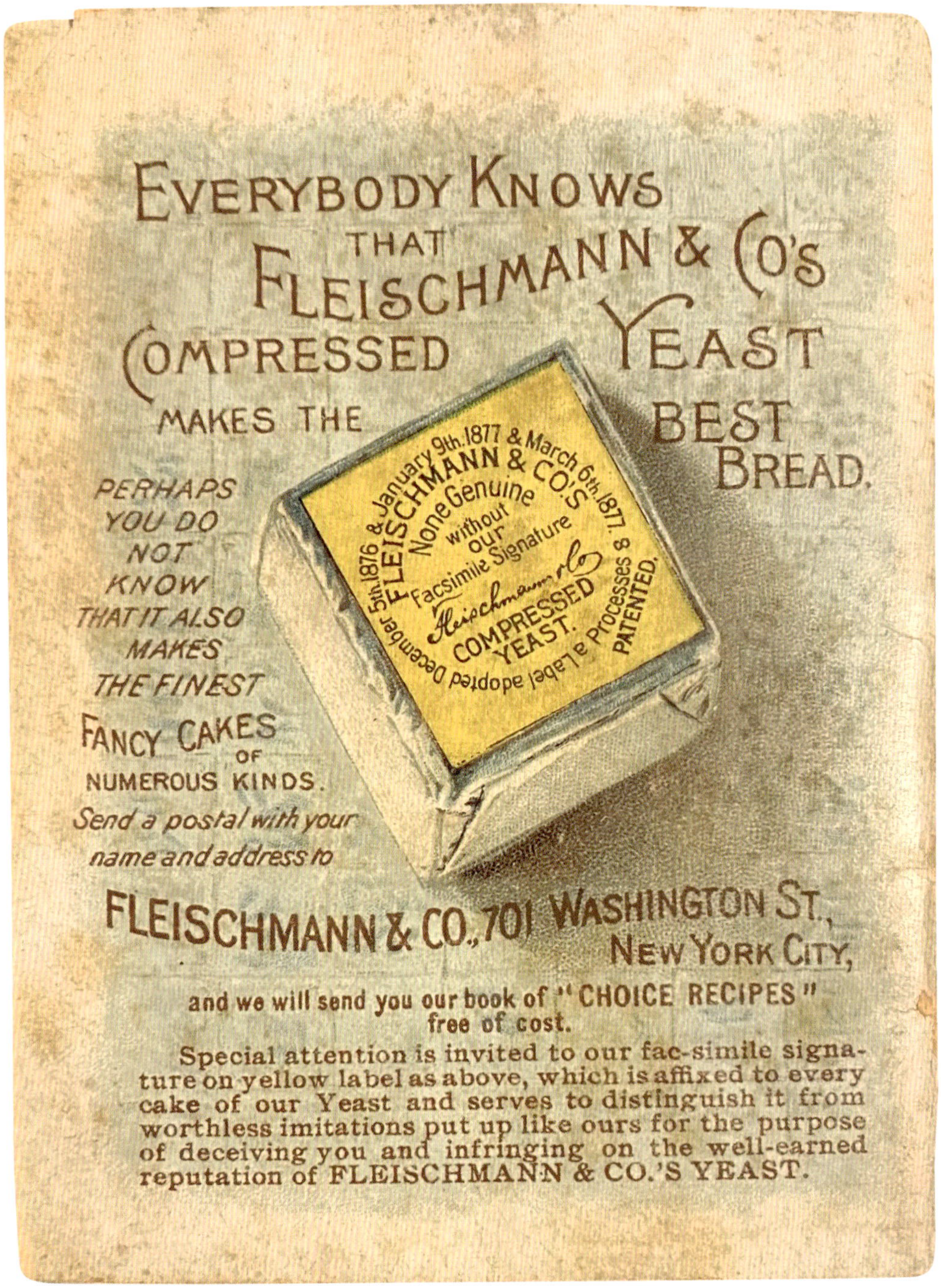

FLEISCHMANN & CO.

Advertising booklet, 1900
Chromolithograph,
4½ × 3¼ in. / 11.4 × 8.3 cm
Private collection

With its distinctive yellow label and printed warning about imitations, Fleischmann & Co.'s compressed yeast revolutionized American baking in the late nineteenth century. The company was founded by brothers Charles and Maximilian Fleischmann, who emigrated from Austria to Cincinnati, Ohio, in the 1860s. Disappointed by the quality of local baked goods, they sought to create better-rising bread as they had known in Europe. Partnering with American businessman and distiller James Gaff, the Fleischmanns built America's first yeast plant on the north bank of the Ohio River, where they produced and patented a compressed yeast cake with excellent leavening power. As America's first commercially produced yeast, it was demonstrated at Philadelphia's Centennial Exposition in 1876 with a range of freshly baked Viennese breads and pastries. To the delight of home bakers, each foil-wrapped block delivered consistent quality and made great-tasting bread and cakes. By the time this printed advertisement appeared in 1900, the brand was a household name and the Fleischmanns had made their fortune. Found naturally in soil, on plants, and even on human skin, yeast is a single-cell, living fungus crucial for fermentation. Baker's yeast is a strain known as *Saccharomyces cerevisiae*, which releases carbon dioxide that causes dough to expand and rise. While the original cubes of Fleischmann's yeast had a limited shelf life without refrigeration, this was overcome in the 1940s when new manufacturing methods allowed the development of dried yeast that could be stored for longer periods.

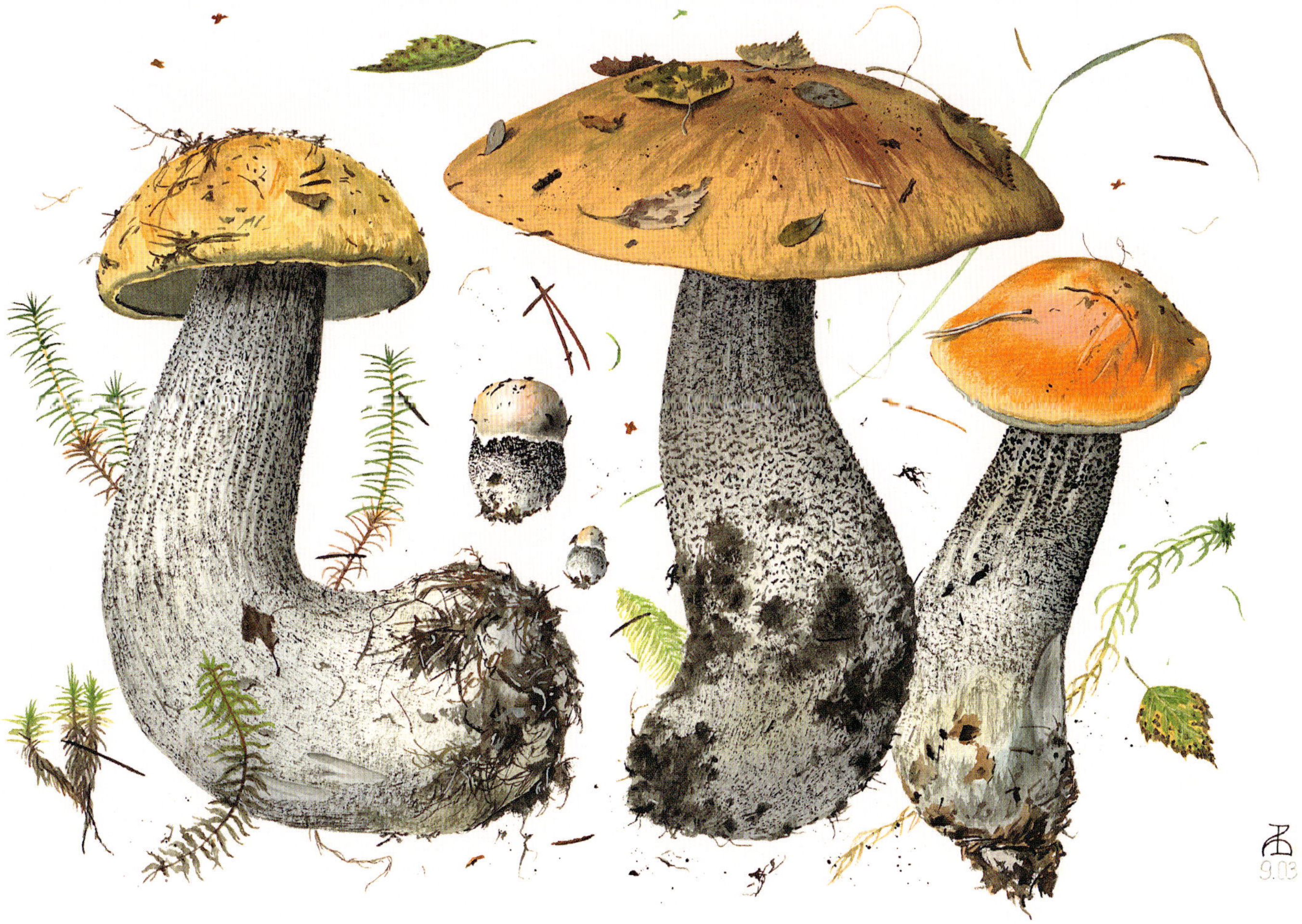

ALEXANDER VIAZMENSKY

Leccinum versipelle, 2003
Watercolor, 12 × 16 in. / 30.5 × 40.5 cm
Private collection

In this remarkably realistic painting by Alexander Viazmensky (b. 1946), three mature fruiting bodies of a mushroom are depicted in fine detail. Drawn from freshly gathered specimens, the remnants of mosses, grass, and fallen birch leaves still cling to the mushrooms' glutinous caps and stalks, a nod toward their woodland habitat. The species depicted is the orange birch bolete (*Leccinum versipelle*), a relative of the famous cep (*Boletus edulis*) that is also edible when cooked. Often harvested for food in Finland, the Baltic States, and Russia, they are, however, mildly toxic and must be adequately prepared by frying or boiling before consumption. Orange birch boletes are typically found under birch trees on sandy soils in forests or heathland. The small blackish scales along the stalk, so faithfully depicted here, are known as scabers and are a distinctive feature of this species. Viazmensky is one of the finest nature artists working today. Born in Saint Petersburg in 1946, he graduated from the Leningrad Academy of Art and since 1976 has worked as an independent artist. His paintings have been exhibited around the world, and he is a member of the American Society of Botanical Artists. Fungi are some of his favorite subjects, and his portraits of mushrooms are unsurpassed. His paintings of wild fungi display acute, close observation combined with an extraordinary artistic ability. Viazmensky lives simply, close to the forest, and often spends hours foraging for fungi, for both his kitchen and his studio.

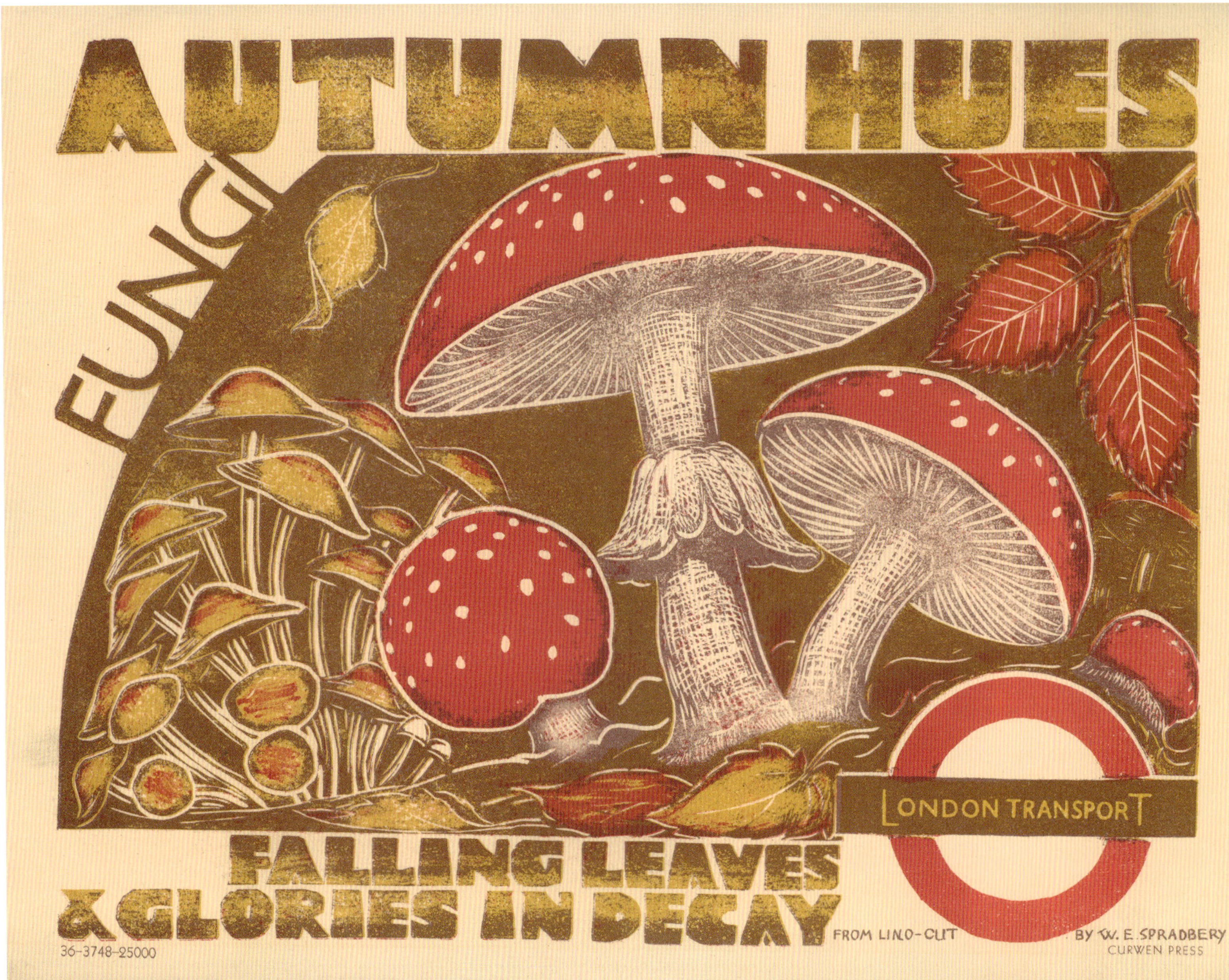

WALTER E. SPRADBERY

Autumn Hues, 1936
Color lithograph, 9⅞ × 12⅜ in. / 25 × 31.4 cm
London Transport Museum

Printed in warm shades of yellow, russet, brown, and crimson, this poster by Walter Spradbery (1889–1969) for London Transport was designed to encourage Londoners to take a bus or train to the countryside to enjoy the changing colors of autumn foliage and to explore the seasonal abundance of fungi flourishing among the fallen leaves. The central motif is a trio of fly agaric mushrooms (*Amanita muscaria*), a species commonly found in woods and forests. With their red caps spotted with white, these mushrooms are familiar in popular culture, from children's picture books and fairy tales to garden ornaments and Disney's 1940 film *Fantasia* (see p.213), but their visual appeal belies their true nature. They are poisonous, although not usually fatal; their common name derives from their one-time use as an insecticide. Seen to the left is a group of sulphur tuft mushrooms (*Hypholoma fasciculare*), another common woodland species (also toxic). This poster, which was paired with a companion, *Summer Garlands*, was one of many designed by Spradbery for London Transport in the early decades of the twentieth century. At the time he produced this design, Spradbery was living in Buckhurst Hill in Essex, at the edge of Epping Forest, and he would probably have seen these fungi on his local walks. The soft, uneven printing, which helps to conjure an autumnal mood of fading color and decay, was characteristic of linocut printing, the medium in which Spradbery originated the image, as indicated by the caption at the lower right.

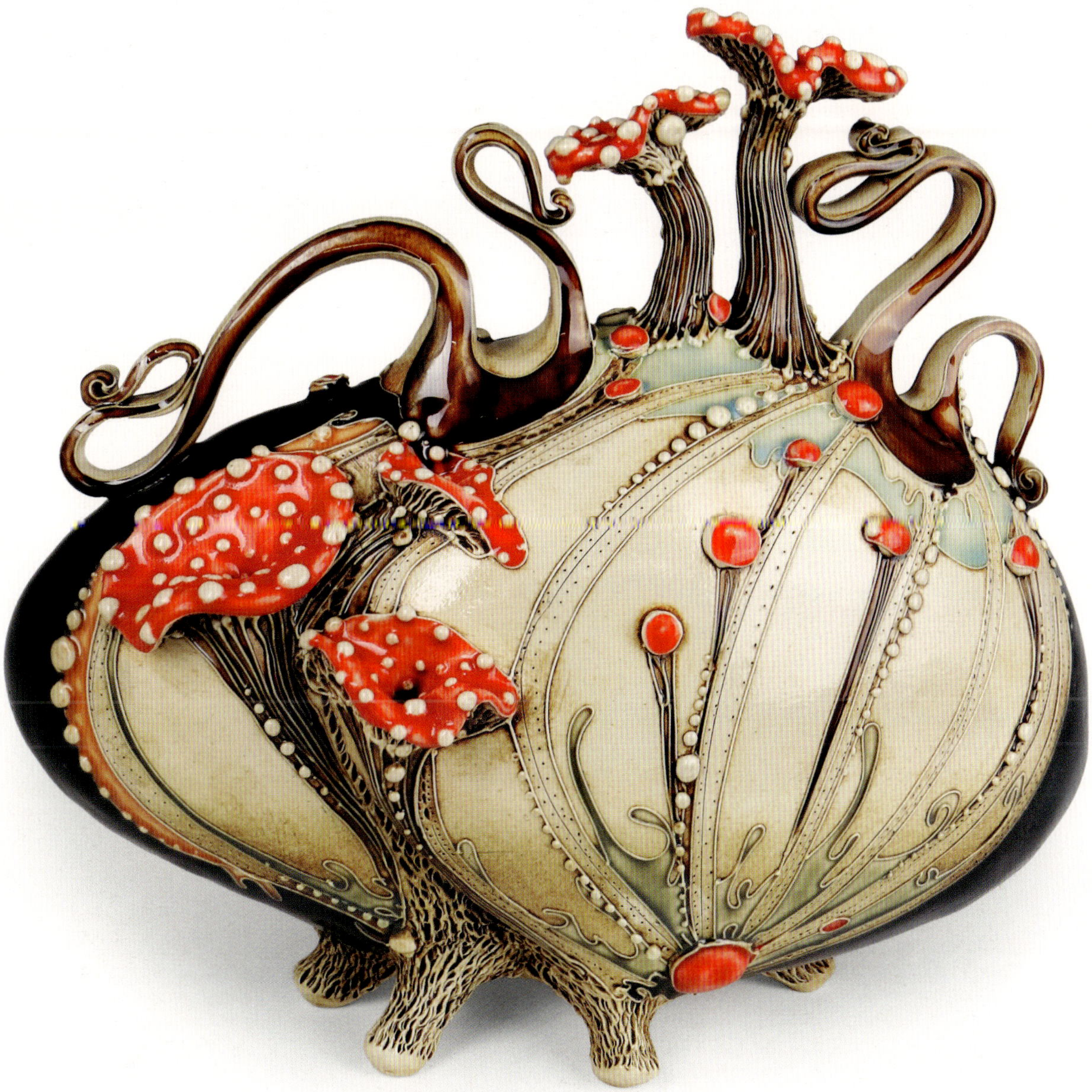

CAROL LONG

Red Mushroom Bottle, 2022
White stoneware fired with cone 5 oxidation,
10 × 11 × 5 in. / 25.4 × 27.9 × 12.7 cm
Private collection

Raised on a farm in Kansas, ceramic artist Carol Long (b. 1965) grew up exploring in nature, marking the beginnings of her lifelong fascination with, and love for, all manner of plant and animal life. Her ceramics practice began in earnest in high school and deepened in college—she jokes that she spent so much time in the ceramics lab it's a wonder she passed any of her other classes. With more than a nod to Art Nouveau, Long makes colorful, elaborately formed (and intentionally misformed) pieces that are richly decorated with flora and fauna, including butterflies, birds, flowers, and fungi. Her sculptures, tiles, vases, bottles, boxes, and other functional and strictly decorative pieces often feature exaggerated proportions and surrealistic patterns that evoke an idea of growth and motion. Much of her work also showcases, among its copious surface detail and ornamentation, her masterful use of slip trailing—the use of white lines to separate different colored glazes, an effect similar to stained glass and cloisonné, an enameling technique in which thin strips of wire separate colors for decorating metalware. Long describes her artistic process as an intuitive one; she rarely has an end point in mind when she starts a piece, but instead lets it develop as she works. This bottle, the shape of which resembles a stomach or other organ, seems to pulse and undulate, with mushroom caps, stems, and lamellae sprouting on its surface, from its feet to its crown.

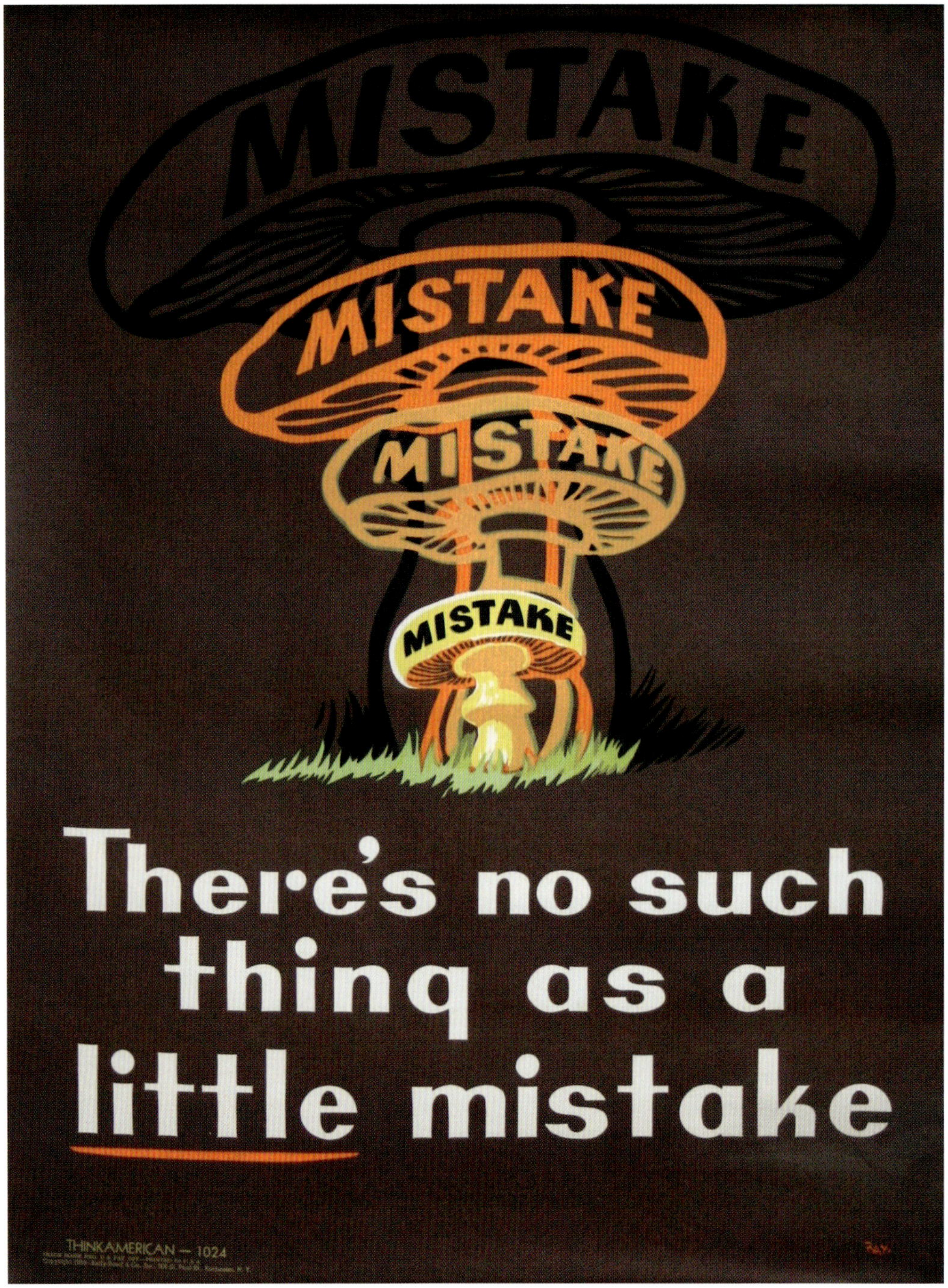

CHESTER RAYMOND MILLER

There's No Such Thing as a Little Mistake, 1959
Printed poster, 27 × 20 in. / 68.6 × 50.8 cm
Private collection

In the late 1930s, as the Axis powers began to advance across Europe and the possibility of world war threatened to reach American shores, a group of American businessmen in Rochester, New York, founded the Think American Institute. Created to promote American values, individualism, and industry at a time of increasing global instability, they sought to combat anti-American and anti-industrialist propaganda. The institute was spearheaded by William G. Bromley, president of design firm Kelly-Read & Co. Beginning in 1939 and continuing until 1961, the group published more than one thousand posters through its Think American series, which were distributed to businesses, schools, and financial institutions across the country. The vast majority of them were designed by Kelly-Read art director Chester Raymond Miller (1898–1968), such as in this stunningly rendered example from 1959 cautioning about the toll of nuclear weapons. The unmistakable form of a mushroom cloud is cleverly incorporated into the typography of its graphic message, set against a stark background that suggests the weapon's potential for annihilation. The ravages of nuclear bombing would certainly have been familiar to the masses, having seen the images of destruction in Japan just more than a decade earlier. A rebuke of Communist ideology and nationalized industry, this poster is a stark reminder of the ever-present specter of nuclear war during the Cold War era.

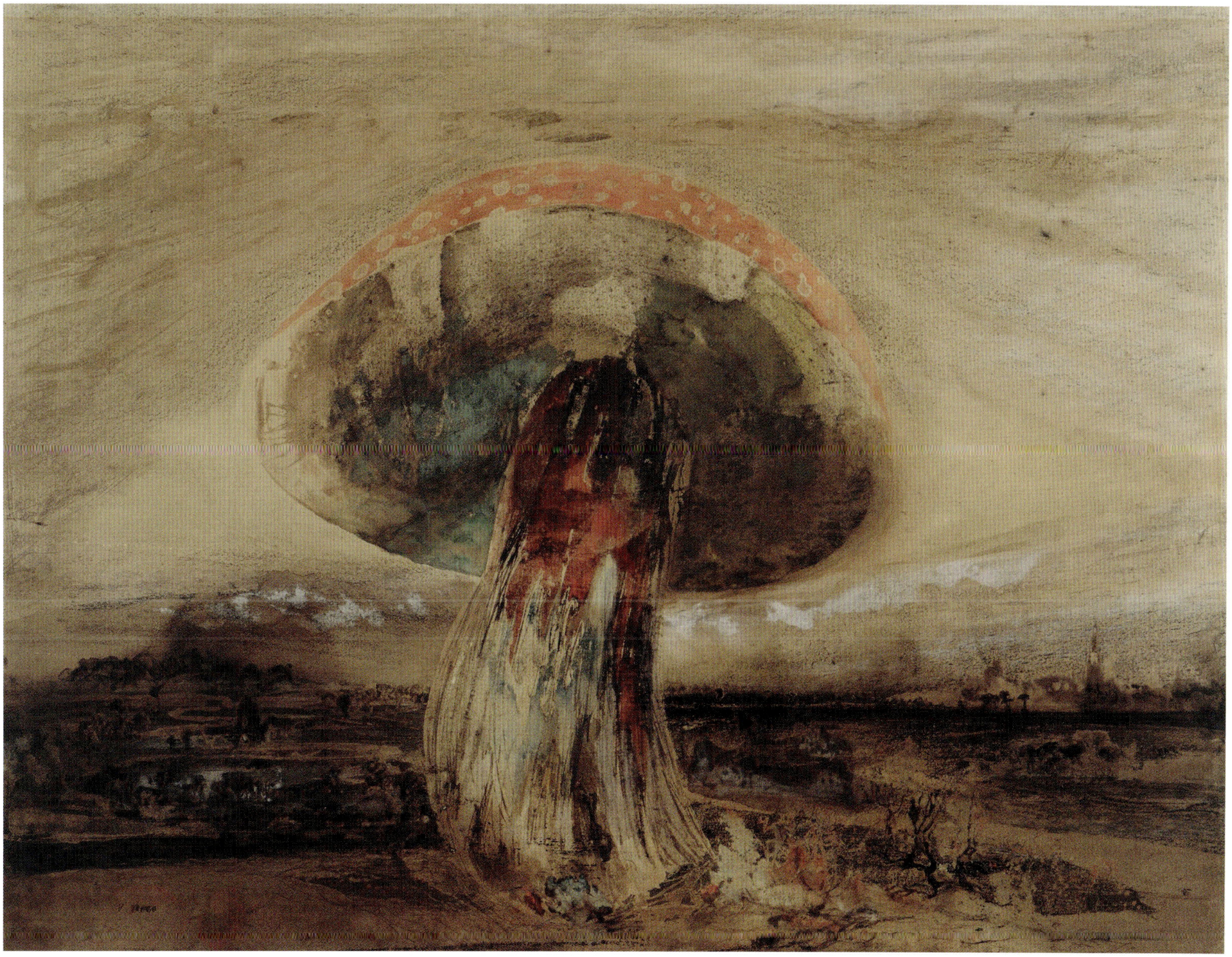

VICTOR HUGO

Champignon, 1850
Pen, pencil, wash, charcoal, ink, and gouache on paper, 18⅝ × 24 in. / 47.4 × 60.8 cm
Hauteville House, Guernsey, Channel Islands

Victor Hugo (1802–1885) was one of the most consequential figures in nineteenth-century France, not just as a writer but as a public figure. The author of enduring literary classics, including *The Hunchback of Notre-Dame* (1831) and the revolutionary masterpiece *Les Misérables* (1862), he was also an outspoken politician and critic. Yet despite his high profile, little was known of his proclivity for art, much less his artwork, which remained largely unseen until after his death. He created most of his works of art from 1851 to 1870 while in exile from his home country for his political views. In the more than three thousand works on paper that survive, Hugo's unconstrained imagination runs wild. He was highly experimental, using inkblots, lace and leaf stencils, and unconventional materials, such as coffee grounds and soot, among myriad other mediums. Rendered using a wide array of techniques and media, the central mushroom in *Champignon* dominates the frame, a sinister presence overtaking the desolate landscape. Look closer, where the stem meets the cap, and an anguished human face emerges, further underscoring the foreboding mood. Hugo's unrestrained work was as provocative as his writing. And to some, it presaged an artistic sea change: Pablo Picasso and André Breton, both ardent admirers, praised him as a forerunner of Surrealism.

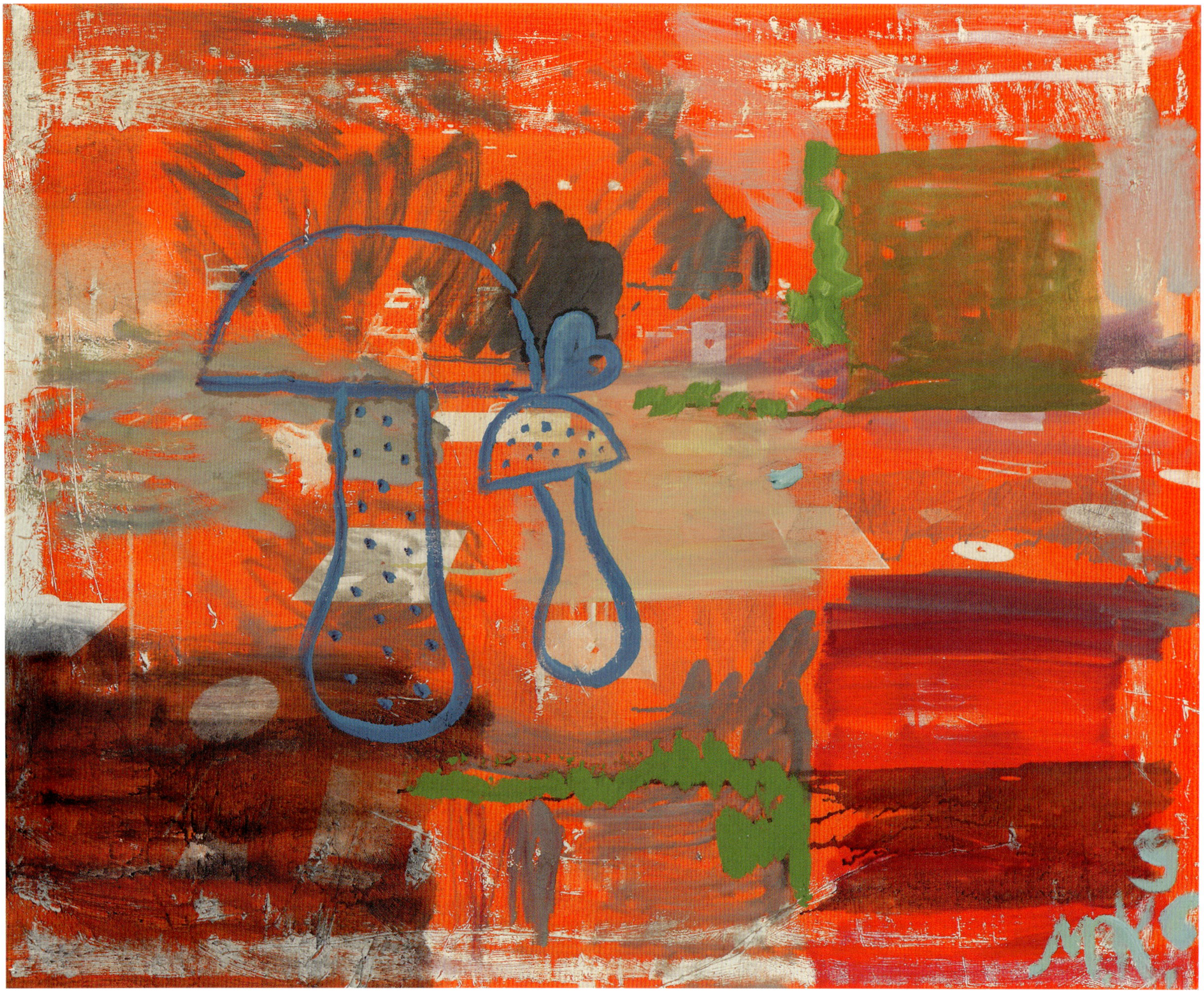

MARTIN KIPPENBERGER

Untitled, 1996
Mixed media on canvas,
29½ × 35⅜ in. / 75 × 90 cm

Rendered in bright, electric blue paint, the simplified outlines of two cartoonlike mushrooms float against a vivid orange ground layered with white, green, and red patches. Vigorous gray scribbles and loose brushwork add to the dynamism of this painting, which combines Abstract Expressionist and Pop art stylings typical of the eclectic approach favored by Martin Kippenberger (1953–1997). The larger of the two fungi sports a speckled stem, while the smaller has a spotted cap and is topped with a blue heart—a symbol perhaps signifying a love of mushrooms or maybe an ironic gesture to the contrary by the German artist, who was known for his subversive and playful sense of humor. A tireless self-promoter and one of the most influential artists of his generation, Kippenberger consistently challenged art-world norms by refusing to limit himself to a single medium, theme, or style, which resulted in a prolific and varied oeuvre encompassing sculpture, painting, works on paper, photography, installations, prints, and ephemera. From the mid-1970s until his untimely death at the age of forty-four, he created enigmatic works that were simultaneously absurd and profound, comical and bleak, often taking inspiration from popular culture, art history, architecture, music, politics, history, and his own biography. His imitation of earlier modern art styles in this composition reflects his belief that, as the twentieth century drew to a close, it had become virtually impossible to produce artworks—especially paintings—that were either original or authentic.

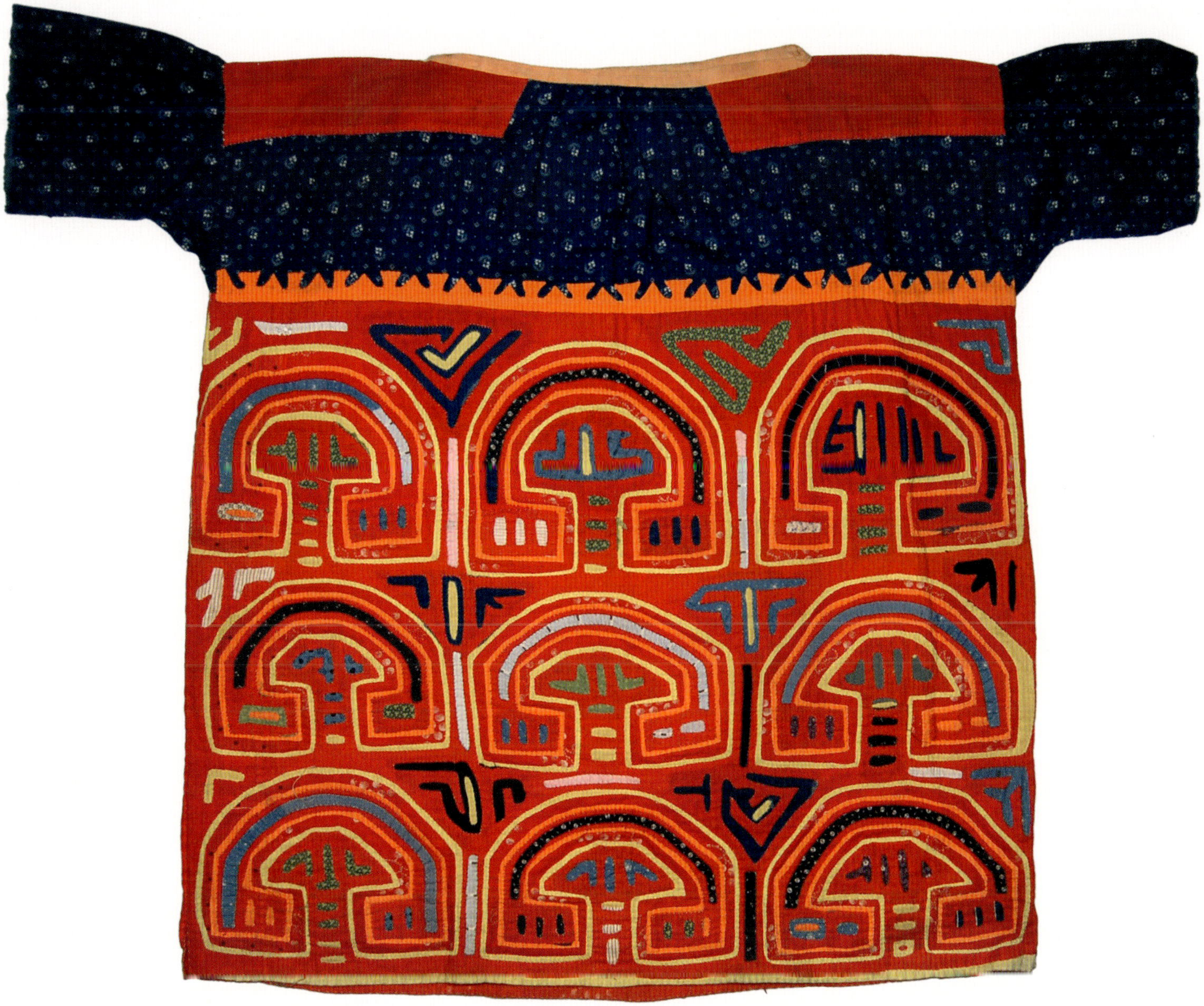

ANONYMOUS

Mola (blouse), c. 1900–24
Printed cotton with appliqué,
30⅜ × 34½ in. / 77 × 87.5 cm
British Museum, London

For the women who belong to the Cuna culture living in the San Blas Archipelago on the Caribbean coast of present-day Panama, this *mola*, or blouse, represents more than an item of clothing. The word *mola* comes from a Cuna word that can mean "cloth," "bird plumage," or "to cover," all of which are appropriate for a brightly decorated blouse. Made of cotton, the *mola* consists of two decorated panels that serve as the front and back. It can take as long as six months to complete each blouse, all of which are made using a technique called reverse appliqué, in which layers of fabric are stacked up, cut, and then sewn to reveal different designs. Typically, a *mola* is decorated with designs taken from nature, such as butterflies, birds, or stylized animals. Here, the back and front are covered in what appear to be mushrooms, the shapes of which are outlined in narrow, hand-stitched strips of colored cotton. It could be that they represent psychoactive mushrooms, such as *Psilocybe yungensis*, which are valued for their hallucinogenic properties. While mushrooms can grow in the tropical humidity of the islands, growing on decaying timber and organic matter, there is little recorded information of mycology playing a significant role in Cuna culture. However, the *mola* occupies an important function in Cuna life; Cuna women started to make the blouses in the late nineteenth century, drawing on fabric instead of the centuries-old tradition of decorating human skin.

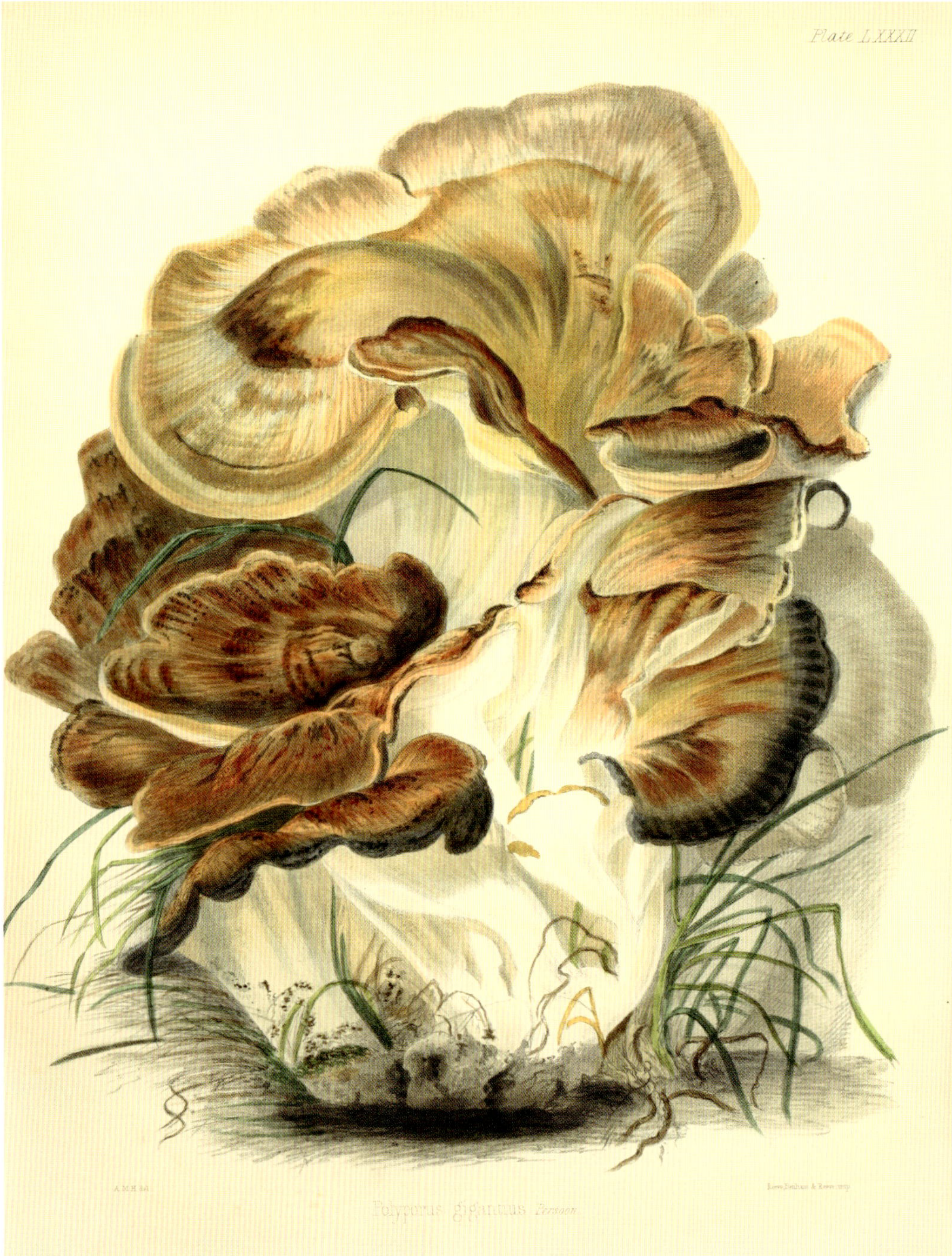

ANNA MARIA HUSSEY

Polyporus giganteus, from *Illustrations of British Mycology*, 1847
Hand-colored lithograph, H. 12⅝ in. / 32 cm
LuEsther T. Mertz Library, New York Botanical Garden

Anna Maria Hussey, née Reed (1805–1853), grew up at a time when women were discouraged from involvement in science, but this did not deter her from producing an authoritative and detailed account of local fungi, illustrated beautifully by herself and her sister Frances Reed. *Illustrations of British Mycology* was published in two volumes between 1847 and 1853, containing 140 colored plates with descriptive text and field notes, including this one of *Polyporus giganteus*. The giant polypore, now named *Meripilus giganteus*, is a common fungus with fan-like fruiting bodies, usually found growing at the base of broad-leaved trees such as beech. In her notes to accompany the plate, Hussey states that the fungus "was first observed at Hayes in 1847, recurring in 1848 on the stump of an acacia, buried in the turf. . . . The grass beneath the vigorous plant from which our portrait was taken, was loaded with the pale ochraceous spores, so copiously scattered that the green colour was entirely lost beneath them." Hussey had begun working as a botanical illustrator in the 1840s and corresponded with experts, including the mycologist Rev. Miles Joseph Berkeley, with whom she was able to discuss aspects of fungal classification and ecology. Her husband was rector of Hayes in Kent, England, not far from Down House, the home of Charles Darwin, with whom she shared a passion for natural history. The resulting book is a true fusion of art and science, and Hussey is now widely regarded as Britain's first woman mycologist.

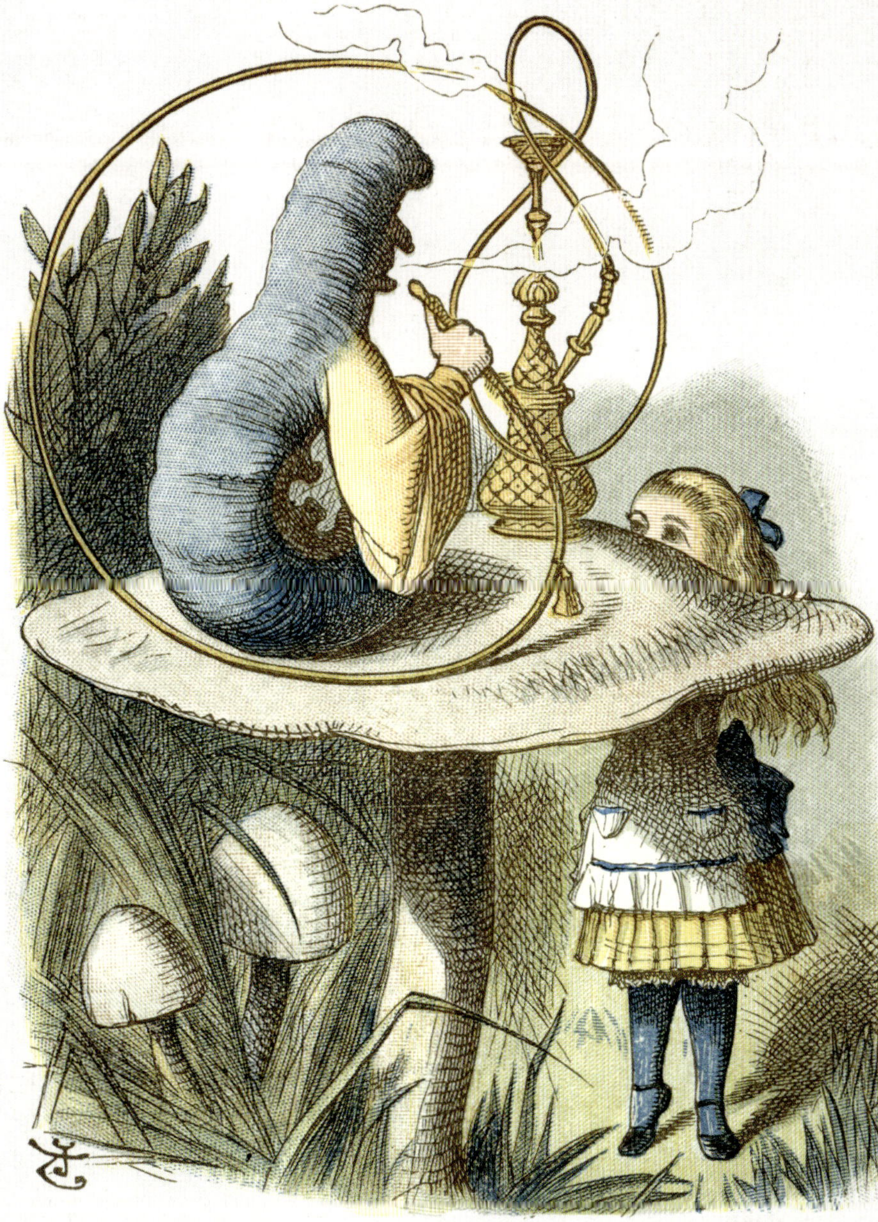

JOHN TENNIEL

Alice Meets the Blue Caterpillar, from
The Nursery "Alice" by Lewis Carroll, 1890
Color photoengraving, 10 x 7¾ in. / 25.5 × 19.5 cm
British Library, London

Lost in Wonderland and shrunk to a minuscule size, Alice stares up with astonishment at the blue Caterpillar, an anthropomorphic insect smoking a hookah and lounging on a giant pale mushroom. Alice begs the Caterpillar to help return her to her normal height, but the Caterpillar, unperturbed, does not immediately answer, yawning and puffing smoke instead. Finally, he hops off his fungus seat, lazily inviting Alice to eat one side of the mushroom to grow taller or the other to get even smaller. Through trial and error, she eventually regains her size and continues her journey out of Wonderland. This illustration is by British cartoonist John Tenniel (1820–1914) and was included in Lewis Carroll's children's book *The Nursery "Alice"* (1890), an abridged version of his 1865 classic *Alice's Adventures in Wonderland*, with all the original artwork now in color. Tenniel's interpretation of Alice and the characters in Wonderland have become indelible, especially the smoking Caterpillar and his mushroom. His fungi is unnamed, but its dull color and gilled underside echo the appearance of a humble meadow, or field, mushroom (*Agaricus campestris*), while the red-and-white-topped fly agaric mushroom (*Amanita muscaria*) had also become a whimsical good-luck symbol in Victorian England, so perhaps that was another influence. Whatever its fungal inspiration, this image was later adopted by the counterculture movement in the 1960s as a metaphor for psychedelic experimentation, widely influential in shifting the cultural perception of mushrooms away from decay and toward hallucinatory empowerment and childlike wonder.

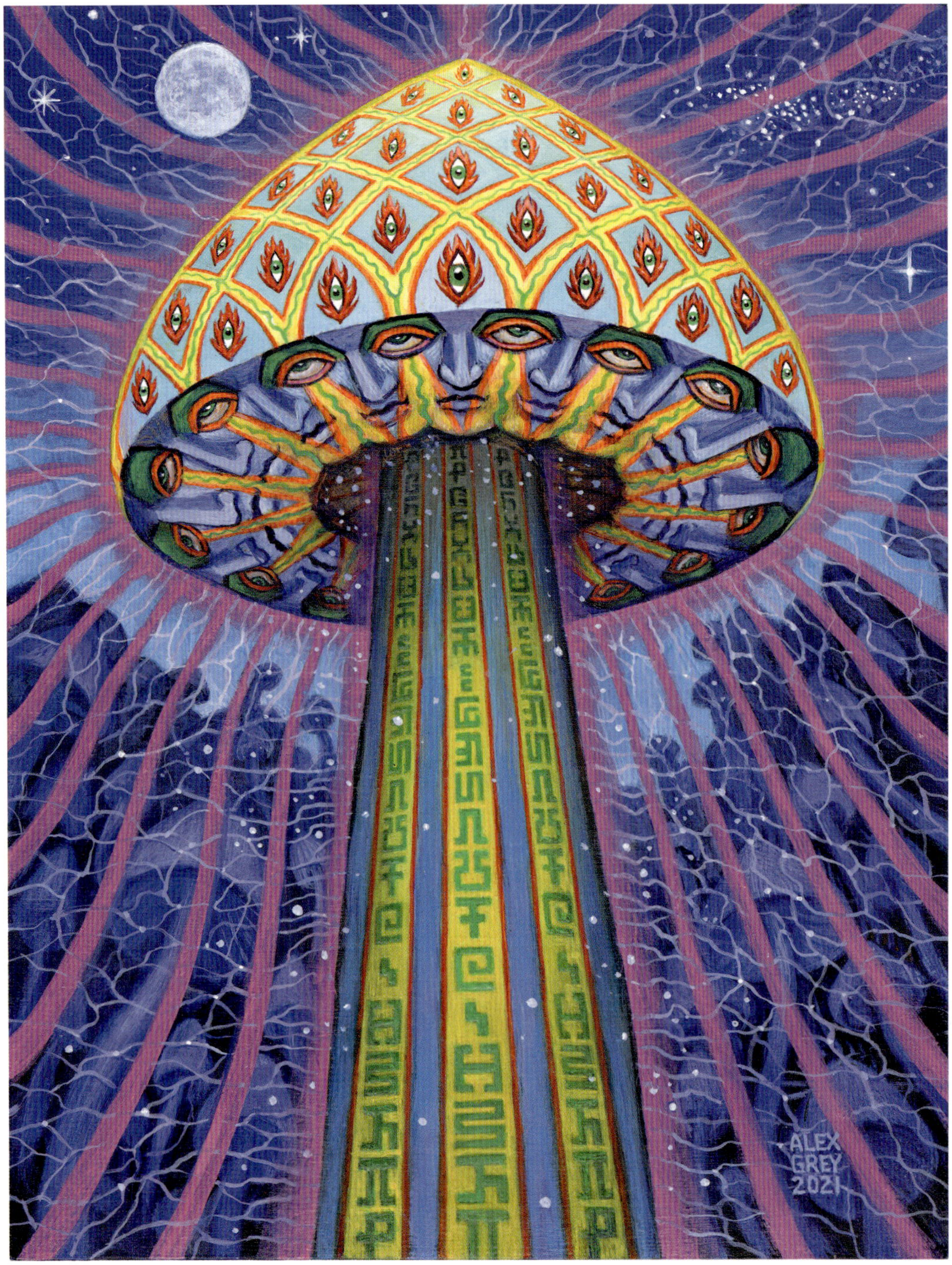

ALEX GREY

Blue Moon Mushroom, 2021
Acrylic on wood, 12 × 9 in. / 30.5 × 22.9 cm
Private collection

A giant mushroom looms large in the night sky beneath a small blue moon, its cap decorated with repeating patterns of all-seeing eyes sitting atop a stem covered in coded writing. This painting is the work of American visual artist, teacher, and author Alex Grey (b. 1953), and the secret symbols were invented by his wife and creative partner, Allyson. Grey is best known for his spiritual and psychedelic artwork, particularly his *Sacred Mirrors* series of twenty-one paintings and his artwork for album covers for rock bands, such as Nirvana, Beastie Boys, and TOOL. The inspiration for this work came, Grey said, during a blue moon in August, when "a colorful mushroom-being appeared in [his] imagination." A practicing Vajrayana Buddhist who believes it is possible to reach enlightenment in one's lifetime, Grey has further stoked his imagination through his experiences with psychedelic drugs, such as LSD and psilocybin mushrooms. For this image, Grey imagined the mushroom cap as a temple dome and the mushroom spores as stars. While much of his artwork features the human form—he spent five years working in the anatomy department at the Harvard Medical School—mushrooms are a recurring theme in his work because he considers them sacred and credits them for his prodigious output. In addition to his artistic practice, he and Allyson created the Chapel of Sacred Mirrors in Wappinger Falls, New York, as a retreat and center for the psychedelic community.

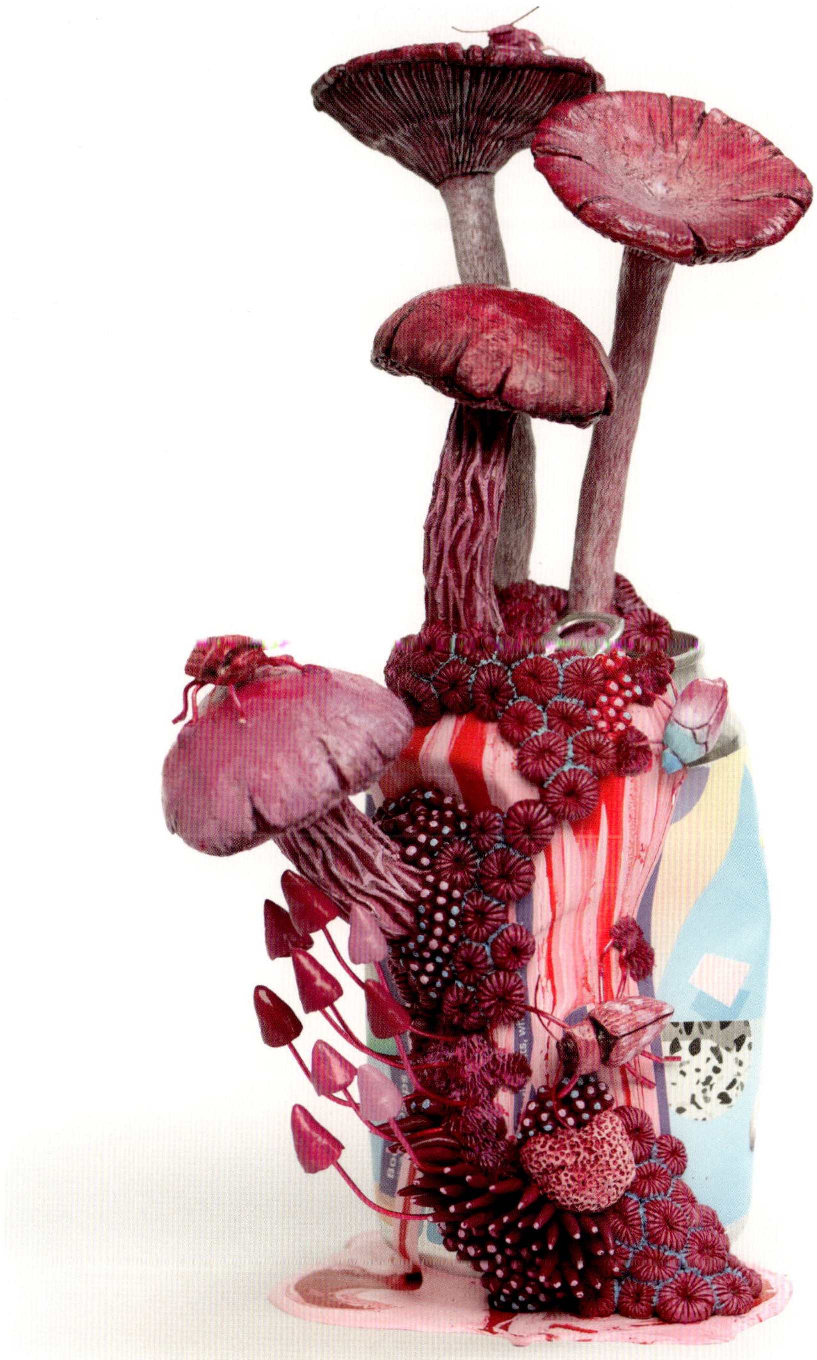

STÉPHANIE KILGAST

La Vie en Rose, 2019
Epoxy clay and acrylics on tin can,
9½ × 4¾ × 4 in. / 24 × 12 × 10 cm
Private collection

Emerging from an empty drink can complete with decoration and dents, a variety of mushrooms large and small are all painted in shades of bright pink. Made from epoxy clay and acrylic paint, this is "cheerful post-apocalyptic art," in the words of its creator, German-born, France-based mixed-media artist Stéphanie Kilgast (b. 1985). Kilgast describes her work as "an ode to life," which is inspired by the bond between humankind and nature. She sees it as a reflection on nature's ability to rebound, no matter the circumstances. *La Vie en Rose* belongs to a long-running series of sculptures, *Discarded Objects*, which Kilgast began in 2017 with the intention of creating colorful organic objects out of discarded manufactured pieces to show what happens when nature is allowed to flourish. Featuring a diverse range of animals, birds, and insects, Kilgast's sculptures are often colonized by mushrooms. They can be seen growing out of bottles, milk cartons, food tins, and—as here—drink cans in a variety of shapes, sizes, and rich colors that border on unnatural: vibrant blues, oranges, purples, greens, and pinks dominate. Upcycling and transforming trash into cheerful objects that remind the viewer of the wonders of nature—and, more specifically, the fungal world—lies at the heart of Kilgast's philosophy. She wants her art to remind people every day of the imbalance and waste that currently lie at the heart of humanity's dialogue with nature and to encourage a reexamination of that fundamental relationship.

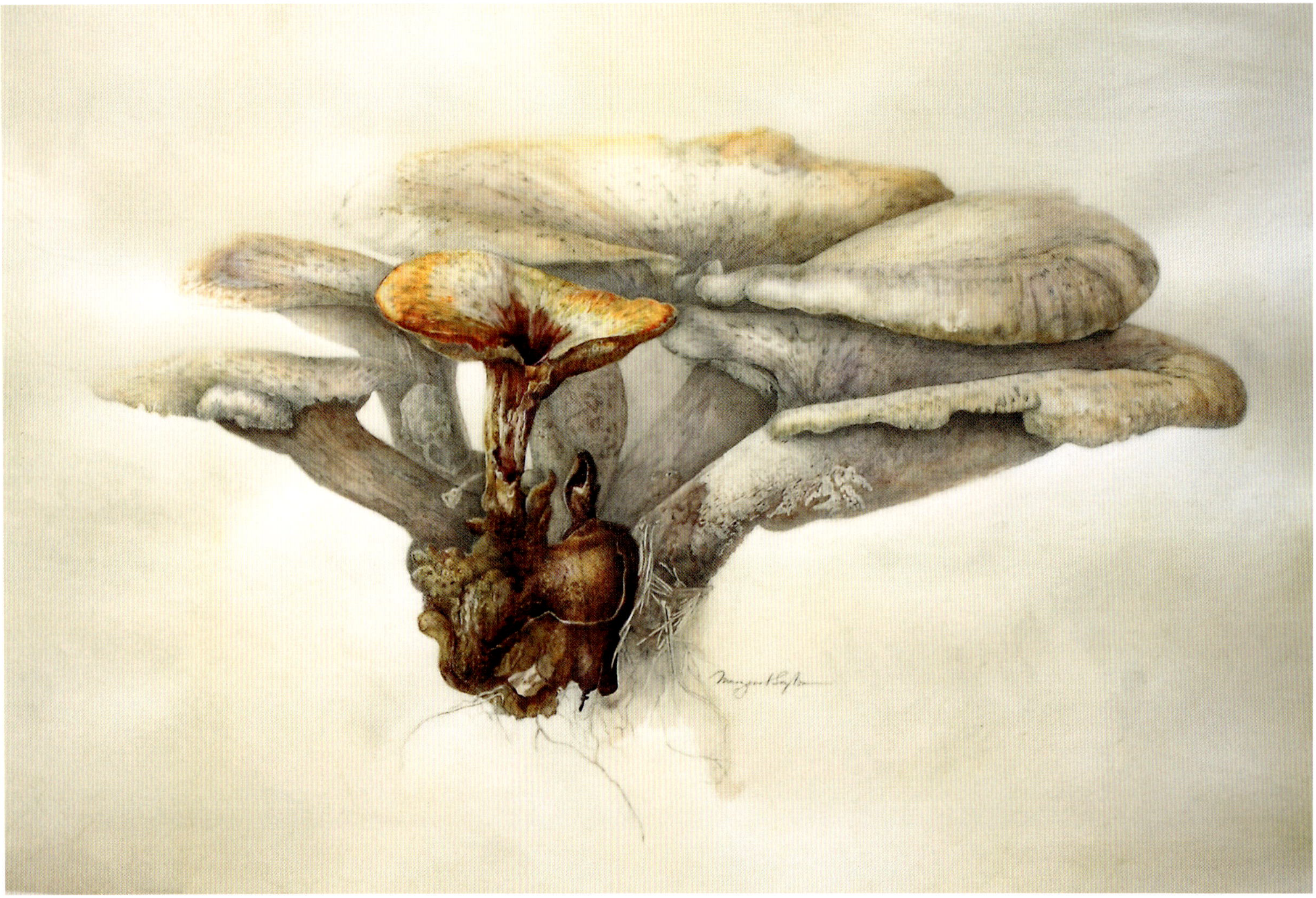

MARGARET SAYLOR

Dryad's Saddle, Polyporus squamosus, 2013
Graphite and watercolor on calfskin vellum,
12 × 18 in. / 30.5 × 45.7 cm
Private collection

Executed with finesse and precision, this beautifully observed mycological illustration depicts the large and attractive fungus known as dryad's saddle (*Polyporus squamosus*). Combining graphite and washes of watercolor on calfskin vellum, the highly detailed and colorful foreground captures the texture of the fungus, while the middle ground and background incorporate the grainy composition of the substrate to form an atmospheric fade. Margaret Saylor (b. 1966) is a noted botanical illustrator specializing in depicting fungi, wild mushrooms, and woodland plants. Her process is complex and labor-intensive, building up highly detailed images with many layers as she seeks to accurately render the shapes, forms, structures, and colors of her subjects. Although she regularly forages for specimens in the woods of her native Berks County, Pennsylvania, and throughout the US East Coast, she had never seen dryad's saddle before, finding it growing close to the bottom of a tree after a friend alerted her to its presence. Seeking a less obvious viewpoint, she decided to paint the fungus from behind, revealing a perspective hidden as it grew. Appearing in tiered, overlapping brackets on broadleaf trees, such as sycamore, willow, and poplar, the mushroom is typically devoured by insects as it ages. Its name refers to its shape while evoking the dryads of Greek mythology: tree nymphs or nature spirits that were said to use the fungus for a saddle as they rode through the forest.

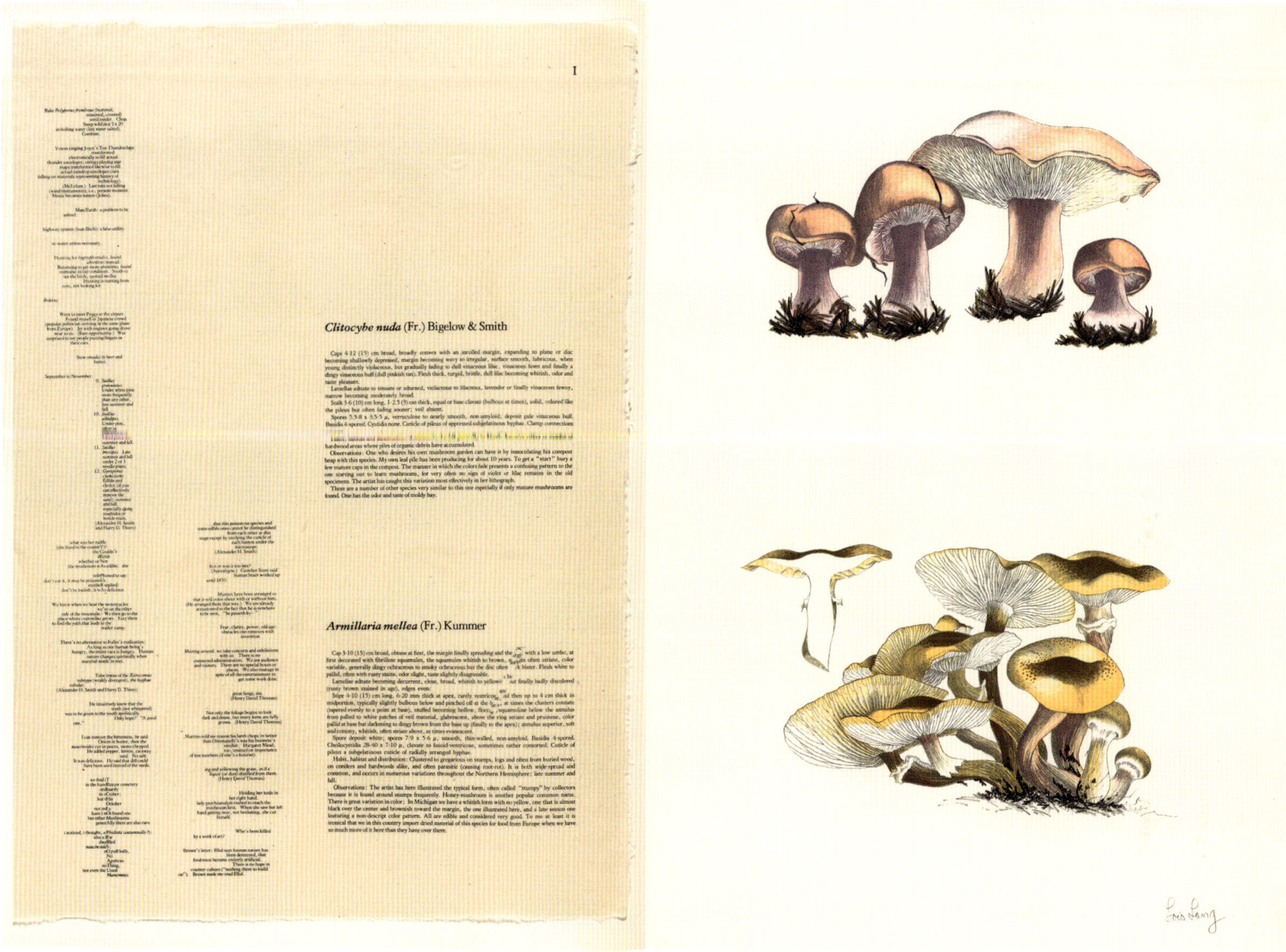

LOIS LONG AND JOHN CAGE

Mushroom Book, 1972
Portfolio of twenty lithographs on Arches paper, overall 22 × 29⅞ in. / 56 × 76 cm
National Gallery of Art, Washington, DC

A delicate lithograph of blewits (*Clitocybe nuda*) and honey fungus (*Armillaria mellea*) sits alongside paragraphs of scientific description and a column of free-form poetic notes. This is one of twenty spreads from *Mushroom Book*, a publication created as a collaboration between composer, artist, and poet John Cage (1912–1992) and artist Lois Long (1918–2005) in 1972. Printed at Hollander Workshop in New York State—operated by master lithographer Irwin Hollander—in an edition of seventy-five, the rare, oversize portfolio depicts a total of fifteen species of mushrooms. Long's illustrations show each specimen life-size, and the lithographs are interspersed with translucent sheets of tracing paper printed with scrawled notes and observations. Some pages include botanical descriptions of the depicted fungi by American mycologist Alexander H. Smith. Both Cage and Long were founding members of the New York Mycological Society. An amateur mycologist, Long's interest in mushrooms largely began when she moved to Upstate New York in the 1950s, where she was introduced to Cage. Cage's fascination with mycology, however, emerged in the Great Depression in the 1930s, originally when seeking sources of food. His love of foraging developed over the coming decades to reflect a strong interest in Zen Buddhism. Cage supported himself not only by his creative practice but also by foraging rare fungi for restaurants and is famous in Italy for his appearance on a television quiz show as a mushroom expert in 1959. Mushrooms also featured in his compositional work, such as his 1983 spoken word performance called *Mushrooms et Variationes*.

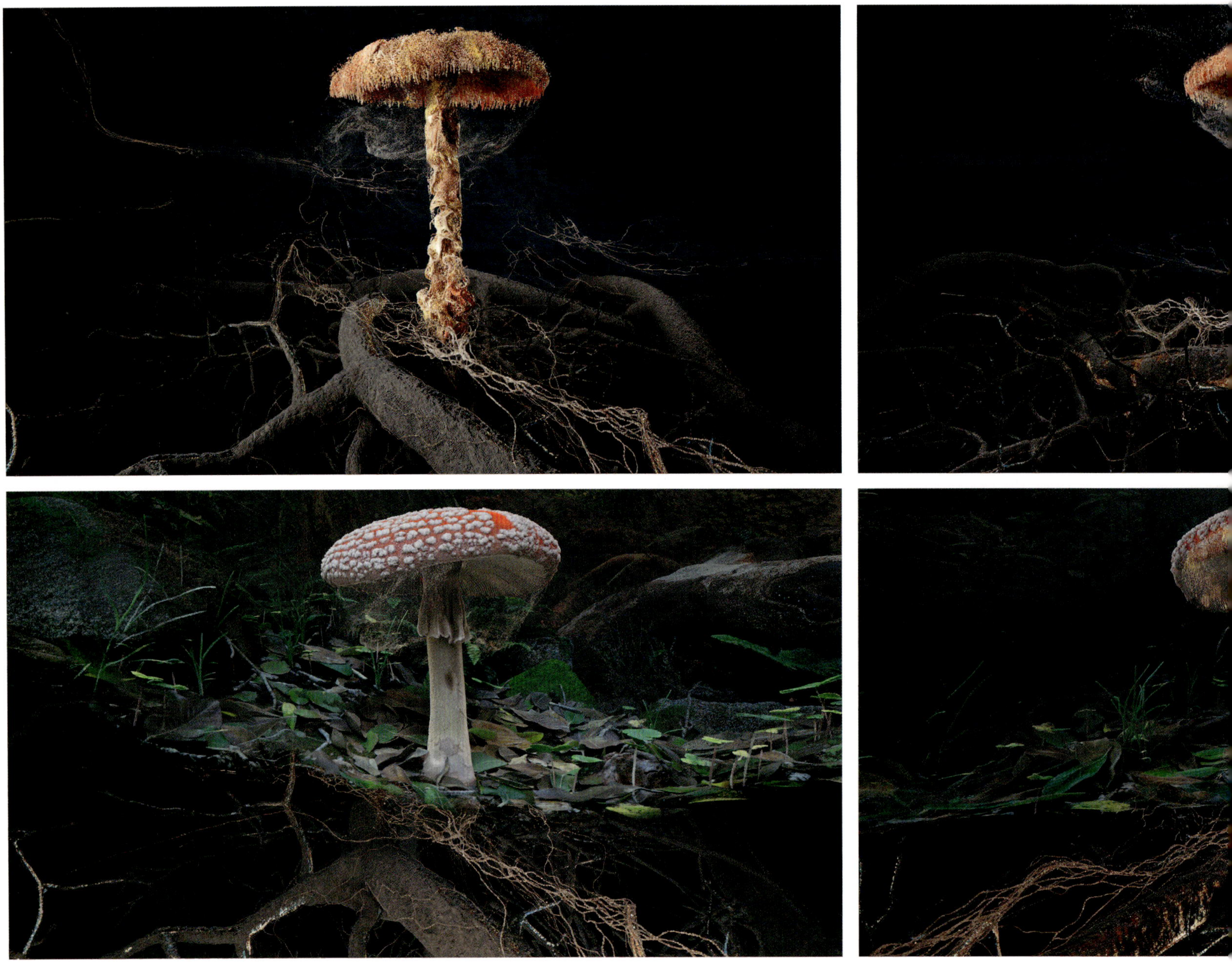

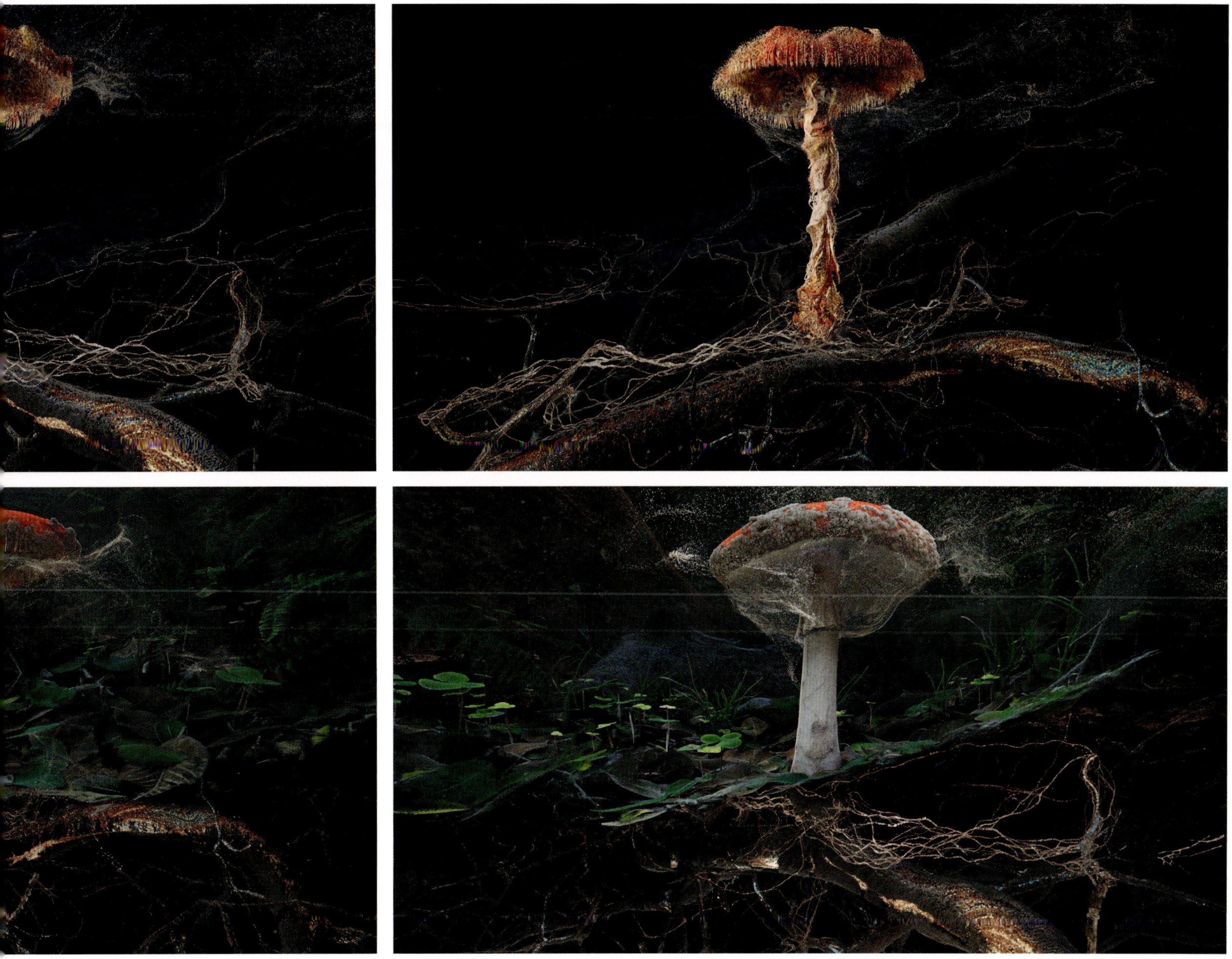

MARSHMALLOW LASER FEAST

Poetics of Soil: Fly Agaric I, 2025
Single channel digital video, color, and sound installation, dimensions variable
Commissioned by Somerset House, London

In a series of six stills from the audiovisual installation *Poetics of Soil: Fly Agaric I*, a mushroom releases a cloud of minute spores, while beneath, an underground complex world of intertwined roots and fungal mycelia emerges, pulsating with life as nutrients are transported. Immediately recognizable as the titular fly agaric (*Amanita muscaria*), with its white-spotted red cap, in a woodland setting, the form is reinterpreted within a digital dreamscape. Created by experiential artist collective Marshmallow Laser Feast (MLF), the immersive installation, first presented in the exhibition *SOIL: The World at Our Feet* at Somerset House in London, allowed viewers to appreciate the interconnected complexity of fungal systems. Founded in 2011, MLF is led by artists and directors Robin McNicholas, Ersin Han Ersin, and Barnaby Steel, and has since grown to collaborate with specialists from a range of disciplines to produce multisensory installations that fuse together information and imagination. Seeking to shine a light on the hidden kingdoms of life underground, the swirling artwork is narrated by mycologist and author Merlin Sheldrake (see p.223), and explains how the world of fungi extends beneath the soil into a living network of mycelial threads, including intimate associations with the roots of trees in mutually beneficial mycorrhizae, or symbiotic relationships between fungi and plant roots. Fungi are a major part of nature's system of nutrient recycling, breaking down dead material and enriching the soil to give life and energy to forest plants and animals. *Fly Agaric I* seeks to create a window into this unseen world, strengthening our connection with these ancient, elemental systems.

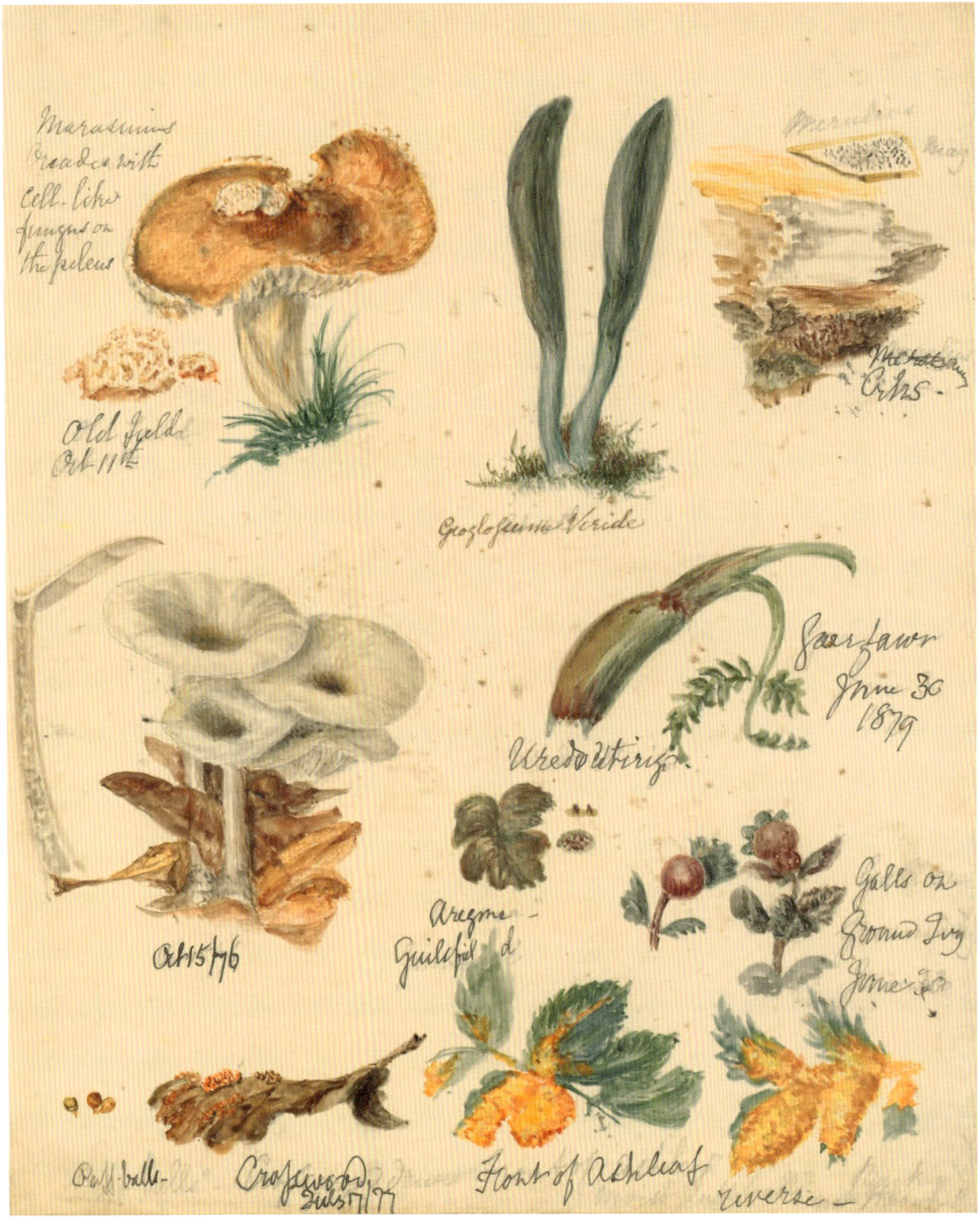

M. F. LEWIS

Page from *Fungi Collected in Shropshire and Other Neighbourhoods*, vol. 1, 1860–1902
Watercolor, ink, and pencil on paper, H. 12½ in. / 32 cm
Cornell University Library, Ithaca, New York

These exquisite drawings and watercolors belong to a much larger work: three volumes of more than one hundred pages of botanically accurate illustrations of different fungi species. They are the creations of amateur nineteenth-century English mycologist and accomplished artist M. F. Lewis. In Victorian Britain, women were largely excluded from scientific endeavors, and so Ms. Lewis recognized the value of using her initials to conceal her gender. Forced to teach herself about the "lower taxa" of fungi, lichens, algae, and mosses, Lewis belongs to a tradition of self-taught female mycologists that includes Beatrix Potter and Mary Elizabeth Banning (see pp.6, 168). Lewis initially illustrated mushrooms in her native Shropshire, an English county on the Welsh border, but she soon expanded her area of study to the neighboring Welsh hills and the English lowlands, drawing hundreds of mushrooms between 1860 and 1902. Most of the sheets include multiple specimens, which were drawn in pencil and then inked over and colored using watercolors. Each illustration was accompanied by the specimen's scientific name and the date and place it was found. Lewis herself largely remains a mystery, although her artistic skill and scientific knowledge are evident on every page. The three volumes of botanical illustrations—each with a hand-painted title page, two of which are signed "M. F. Lewis"—are not only a testament to her dedication to the discipline of mycology but also a lasting legacy.

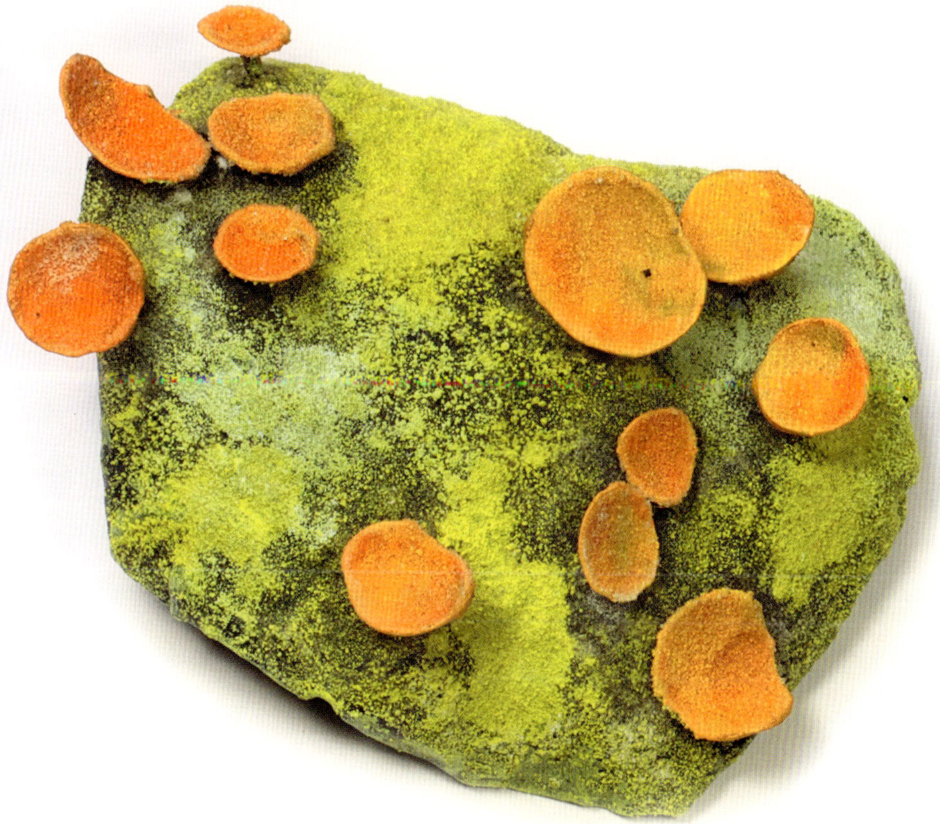

MYUNG URSO

Flora Brooch, 2016
Sterling silver, sand, marble dust, and pigment,
4 × 3 × 1½ in. / 10.2 × 7.6 × 3.8 cm
Private collection

Yellow mushroom caps emerge from a raised, unevenly shaped base speckled with lichen and moss to create this vibrant nature-inspired brooch. Crafted from sterling silver, sand, marble dust, and pigment, this jewelry piece is the work of South Korean artist Myung Urso (b. 1960). In a career that has had several notable iterations, Urso began as a fiber artist, which informs her preference to work with mainly organic materials, such as natural paper and textiles. But this dramatic brooch belongs to a fifteen-year period when she turned her attention to creating unexpected art jewelry. Urso works organically, choosing not to sketch her pieces first but to create them directly as she works with her array of materials. She has attributed her hands-on practice to her upbringing, both helping her mother with everyday crafts in the home and being educated in the Korean school system, which championed tactile arts, such as calligraphy and sewing. Having studied both fiber art and biology at university, her contemporary one-off pieces reflect her interest in the often-overlooked elements of the natural world. When Urso moved to the United States in 2006, she settled in Rochester, New York, where she found constant inspiration in her new surroundings. Combining elements of the innate sculptural qualities of flora and mushrooms with their imperfections resulted in each piece being wholly unique. The beauty of imperfection is an important concept in Korean culture, and Urso has applied it consistently to her art, whether fiber artworks, jewelry, or her more recent paintings.

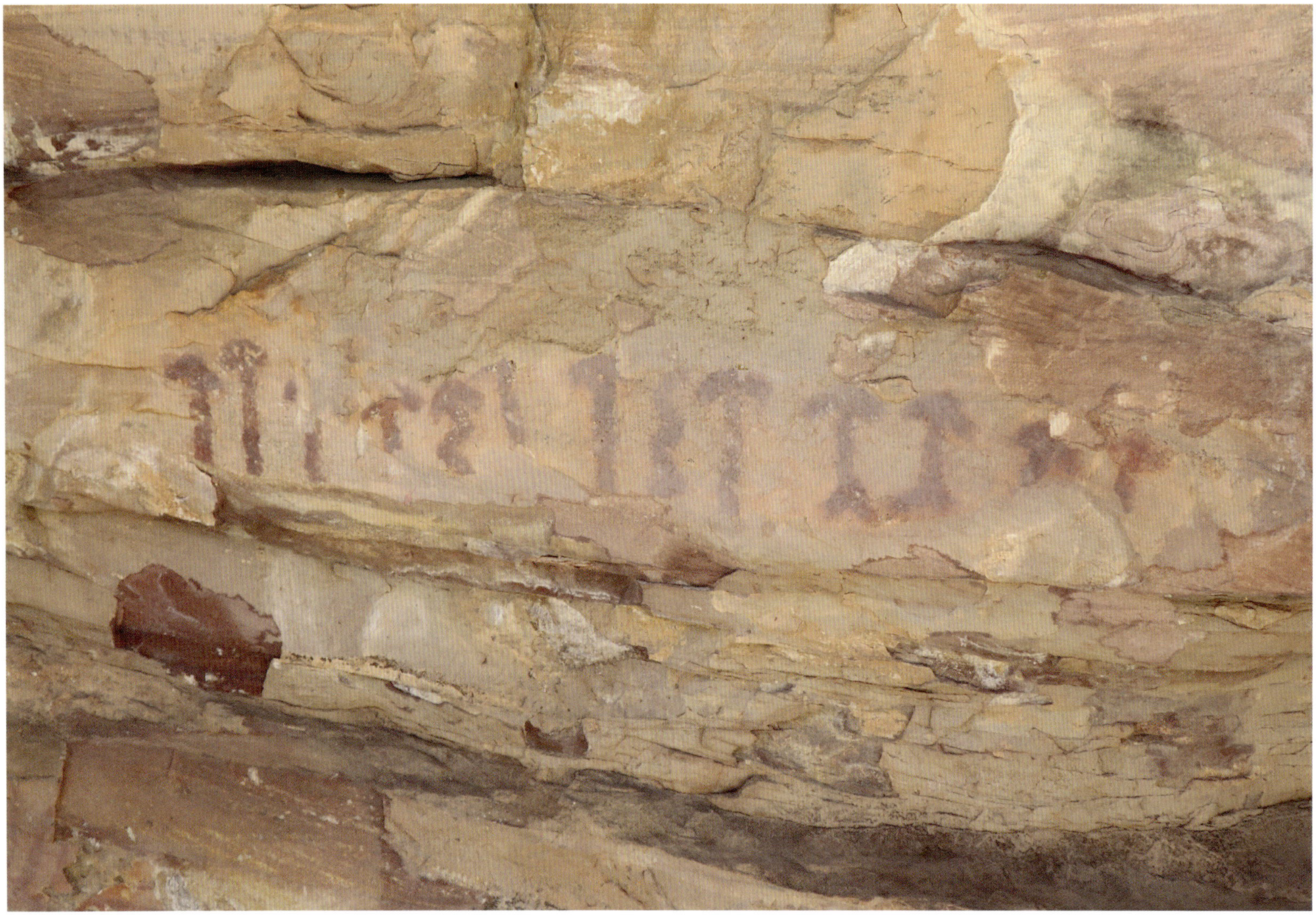

ANONYMOUS

Rock painting featuring mushrooms,
c. 6000–4000 BCE
Pigment on rock wall, entire shelter
10 ft. 6 in. × 40 ft. / 3.2 × 12.2 m
Selva Pascuala, Cuenca, Spain

The Selva Pascuala rock shelter is a landscape formation in Cuenca, Spain, rediscovered by archaeologist Eduardo Hernández Pacheco in 1918. It is host to two panels of prehistoric painted rock art, including the one here featuring a bull and the earliest known visual depiction of mushrooms. The site is similar to a couple dozen others in Spain's Mediterranean basin that also have Mesolithic-Neolithic rock art. Researchers and scholars have different opinions about the meaning of the mural, but the most widely held belief, and the one for which there is the most evidence, is that the Indigenous people who made the paintings knew about the psychedelic properties of certain varieties of mushrooms, especially those containing the organic compound psilocybin. Psilocybin mushrooms—also known as "magic mushrooms" in modern parlance—produce what we think of as "psychedelic" effects when consumed, including hallucinations, altered sensory perceptions, and euphoria. But for the Indigenous people who made this painting, it is highly possible that these mushrooms played an important role in various rituals and customs, and were not eaten recreationally. Scholars point out that the mushrooms in the painting resemble the species *Psilocybe hispanica*, which would have grown in the area; this species also happens to be coprophilic—meaning it thrives on dung—which would explain the close proximity of the bull in the mural.

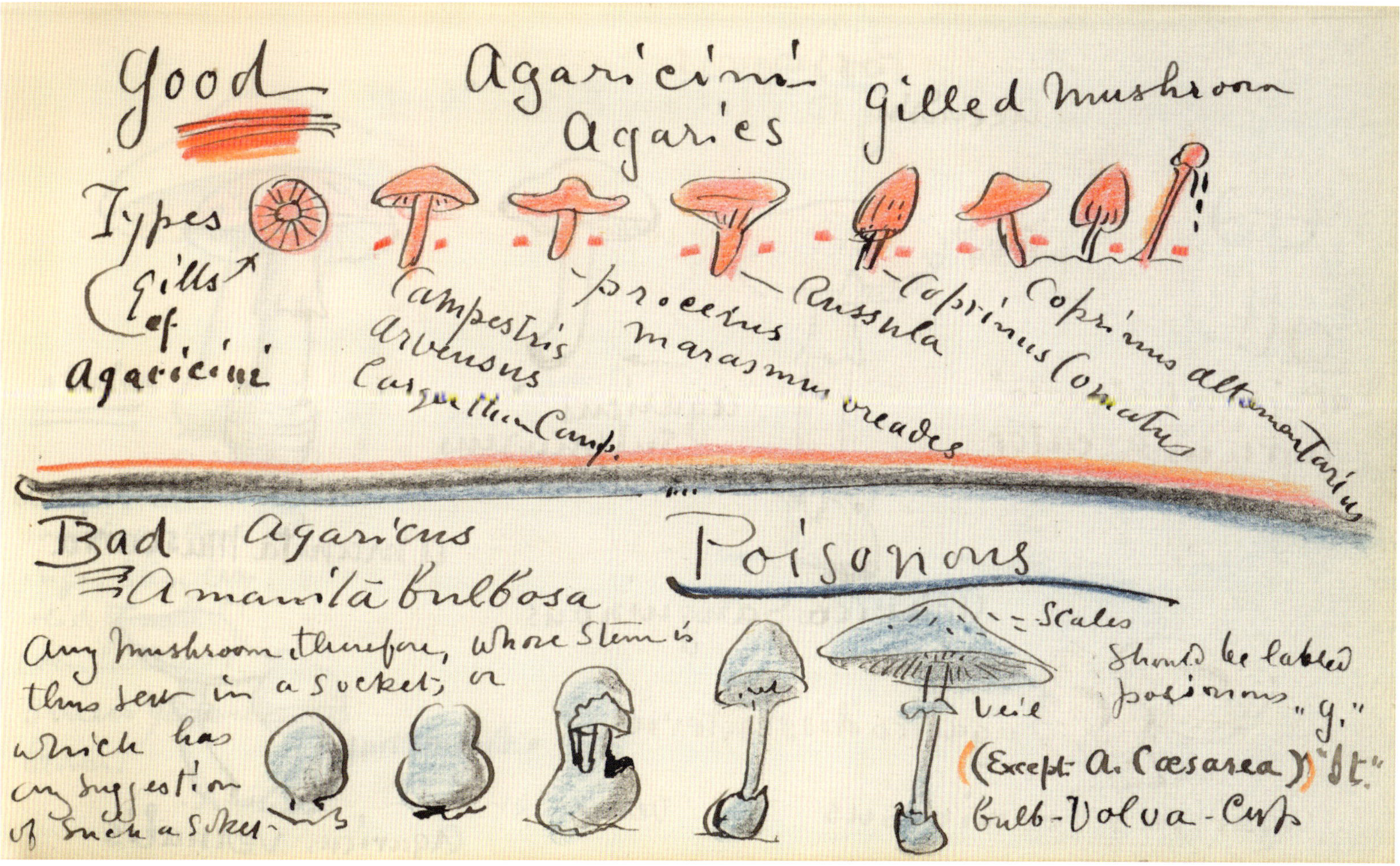

ELIHU VEDDER

Disegui Funghi (Mushrooms), c. 1906
Ink, crayon, and graphite on paper,
3½ × 5¾ in. / 8.9 × 14.6 cm
Huntington Library, Art Museum, and
Botanical Gardens, San Marino, California

Prolific American artist Elihu Vedder (1836–1923) is perhaps best known for his mural paintings lining the hallway of the Reading Room at the Library of Congress in Washington, DC. However, the symbolic murals were just one part of a career that spanned more than six decades. Influenced by English poet William Blake and the pre-Raphaelites, among others, Vedder had a varied output, including designing glassware for Tiffany (see p.35) and illustrating Omar Khayyam's *Rubaiyat* (1884). These sketches of mushrooms from the genus *Agaricus*, which includes both poisonous and edible species, come from one of the artist's personal sketchbooks. Bound in green leather, the sketchbook was gifted to him, blank, by American expatriate William Rinehart. The book's pages became home to illustrations of fungi that were possibly intended for inclusion in a project Vedder worked on from the late 1890s until 1910, *The Book of Mushrooms*, before ultimately leaving it unfinished. Vedder divided the page featured here in two: in the top half he has labeled various species of "good" mushrooms, and on the lower half, he provides instructions on how to recognize a poisonous mushroom, alongside an illustration of the growth of a "bad" mushroom, the *Amanita bulbosa*, named for its bulbous stem. The mushrooms were most likely drawn after Vedder moved permanently to Italy in 1906, where the landscape, flora, and fauna remained a continual inspiration, and where he lived until his death in 1923.

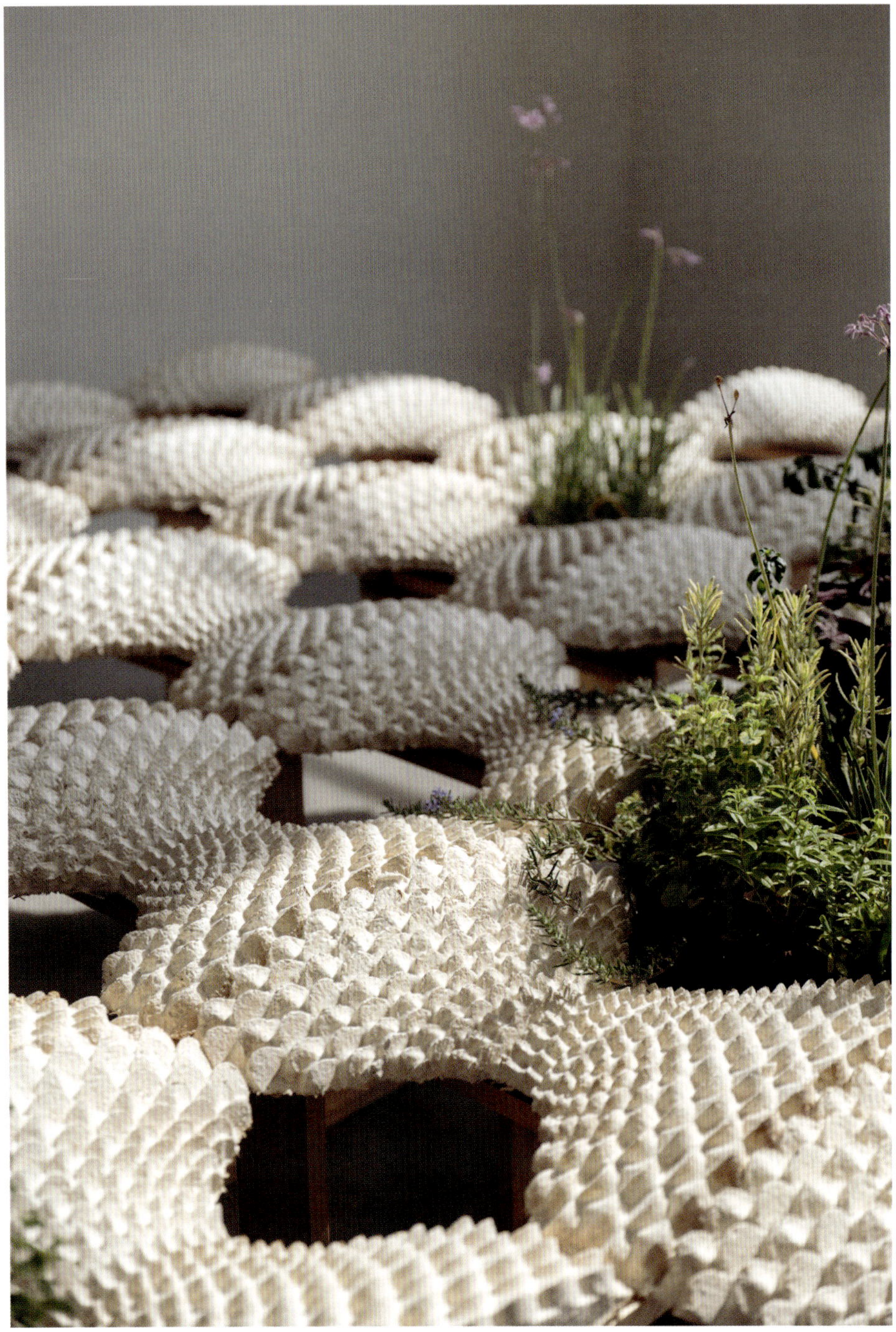

MAE-LING LOKKO

Healing Meadow, 2021
Mycelium, hemp, and local herbs,
26 ft. 3 in. × 10 ft. 6 in. / 8 × 3.2 m
Z33 House for Contemporary Arts, Design, and Architecture, Hasselt, Belgium

Filling the floor of an exhibition space, the aptly titled *Healing Meadow* is formed from a grid of architectural textured panels made from eco-friendly mycelium, interspersed with living plants. This installation is from a series of functional mushroom-based prototypes created by building-material technologist, architectural scientist, designer, educator, and historian Dr. Mae-ling Lokko (b. 1987). Originally from Ghana and the Philippines, and based in New York, Lokko creates work that centers on the upcycling of agricultural waste and biopolymers into building materials, including bricks. She is part of a wave of architects, engineers, and designers looking for sustainable, mycelium-based building materials that could transform one of the world's most polluting industries. Lokko is an assistant professor at Yale School of Architecture, where she teaches environmental design and directs doctoral research at the Yale Center for Ecosystems in Architecture. She is also the founder of Willow Technologies, a Ghanaian sustainable materials company that works with mushroom-based packaging. Her own work has been exhibited internationally, including at the Venice Architecture Biennale, MoMA in New York, Stockholm's Nobel Prize Museum, and Stedelijk Museum Amsterdam, among other respected institutions. Driven by a focus on human and planetary ecological health and social justice, Lokko's aesthetically ambitious work raises big questions about contemporary technology, the environment, politics, and culture. This is work that explores design, logistics, and sustainability as much as ideas of the home and our relationship to security, comfort, and nature.

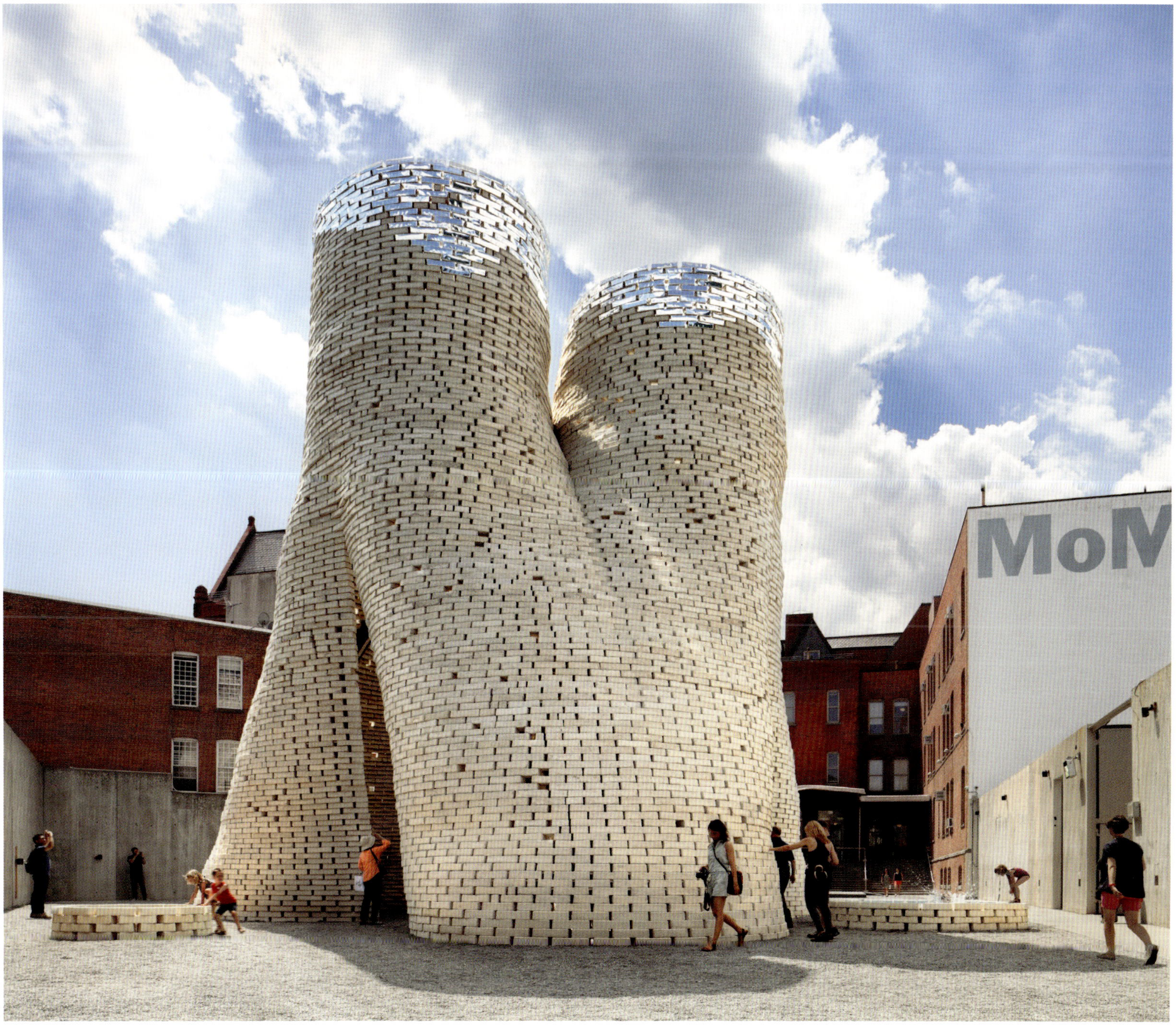

THE LIVING AND DAVID BENJAMIN

Hy-Fi, 2014
Carbon negative, compostable mycelium and corn stover bricks, 40 × 30 × 30 ft. / 12.2 × 9.1 × 9.1 m
MoMA PS1, Queens, New York

Standing at 40 feet (12.2 m) tall and looking like a giant chimney stack, this temporary structure was constructed using ten thousand bricks made from fungi. Its undulating, organic shapes and circular towers, which form a striking contrast with the surrounding buildings, were designed by architect David Benjamin (b. 1974), founding principal of The Living design studio, whose projects employ living organisms to create alternative kinds of architecture. The innovative bricks used here are the invention of biomaterial company Ecovative, which harnesses the many unique properties of fungi to develop new and innovative materials. Each one is made from shredded corn stalks and other agricultural by-products mixed with mycelium. The raw mixture is packed into molds, and as the mycelium grows, it binds the material into a dense, solid brick. An innovation in sustainable architecture, the process uses no energy and generates no waste. Tests at Columbia University in New York demonstrated that each lightweight brick could carry weights equivalent to fifty cars but were two hundred thousand times more flexible than steel. Titled *Hy-Fi*, this environmentally friendly pavilion was installed at MoMA PS1 in New York, where it provided welcome shade and a cool microclimate throughout the hot summer months, while offering an optimistic vision for the future of sustainable architecture. After three months, it was dismantled and the mushroom bricks were composted, enriching the soil of local community gardens. *Hy-Fi* showed how fungi could be a significant building material for the future, offering a low-carbon alternative to concrete.

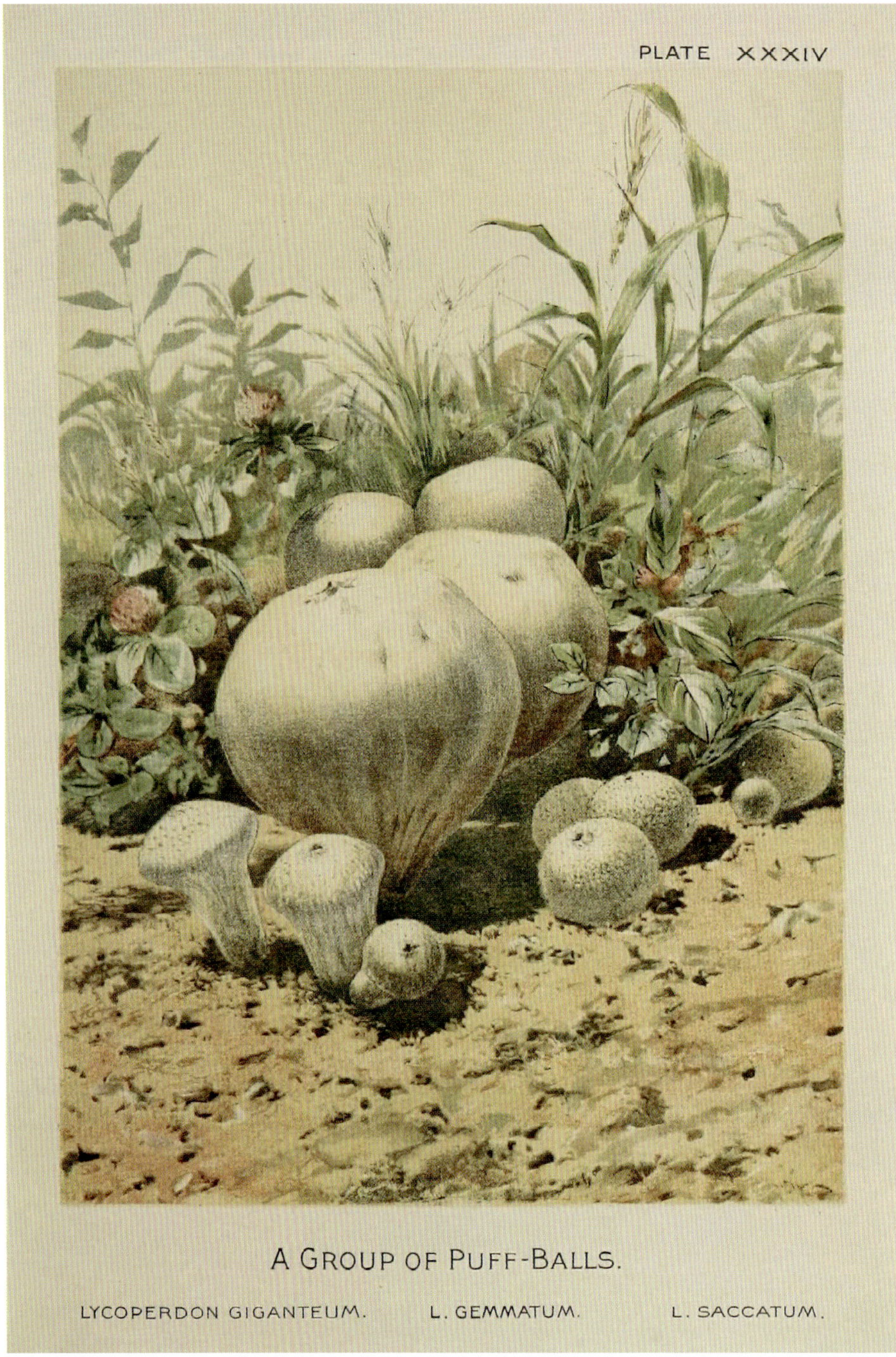

WILLIAM HAMILTON GIBSON

A Group of Puffballs, from *Our Edible Toadstools and Mushrooms and How to Distinguish Them*, 1895
Chromolithograph, H. 9½ in. / 24 cm
Cornell University Library, Ithaca, New York

Arranged in an artful composition, a cluster of puffballs sits against an idyllic backdrop of the local countryside in which it is growing. A carpet of beige-brown leaf detritus forms its base, while a tangle of pale green grasses and leaves, with a hint of pink flowers, forms the backdrop. Accurately detailed, this illustration is by American artist, author, and naturalist William Hamilton Gibson (1850–1896), created for his book *Our Edible Toadstools and Mushrooms and How to Distinguish Them*, published in 1895. The biggest of the group is the giant puffball (*Calvatia gigantea*, formerly *Lycoperdon giganteum*), a large edible fungus shaped like a white soccer ball, which is safe to eat as long as the flesh inside is white and firm. Alongside are *Lycoperdon perlatum* (formerly known as *Lycoperdon gemmatum*), the common puffball, also edible, and the pestle puffball (*Lycoperdon excipuliforme*, previously known as *Lycoperdon saccatum*), which is edible but only when young. In his book, Gibson was focused on explaining which mushrooms are edible and how best to prepare and cook them. He does, however, warn the reader of the dangers of mistaking the poisonous "globular condition of the *Amanita* or other fungi of the same deadly group" for harmless and delicious puffballs. Born in Connecticut, Gibson studied at the Brooklyn Polytechnic Institute before turning to natural history illustration, and his images appeared in *Harper's Weekly*, *Scribner's Monthly*, and *The Century Magazine*. Gibson was also an expert wood engraver, lithographer, and photographer, and his drawings display an almost photographic and microscopic accuracy of detail.

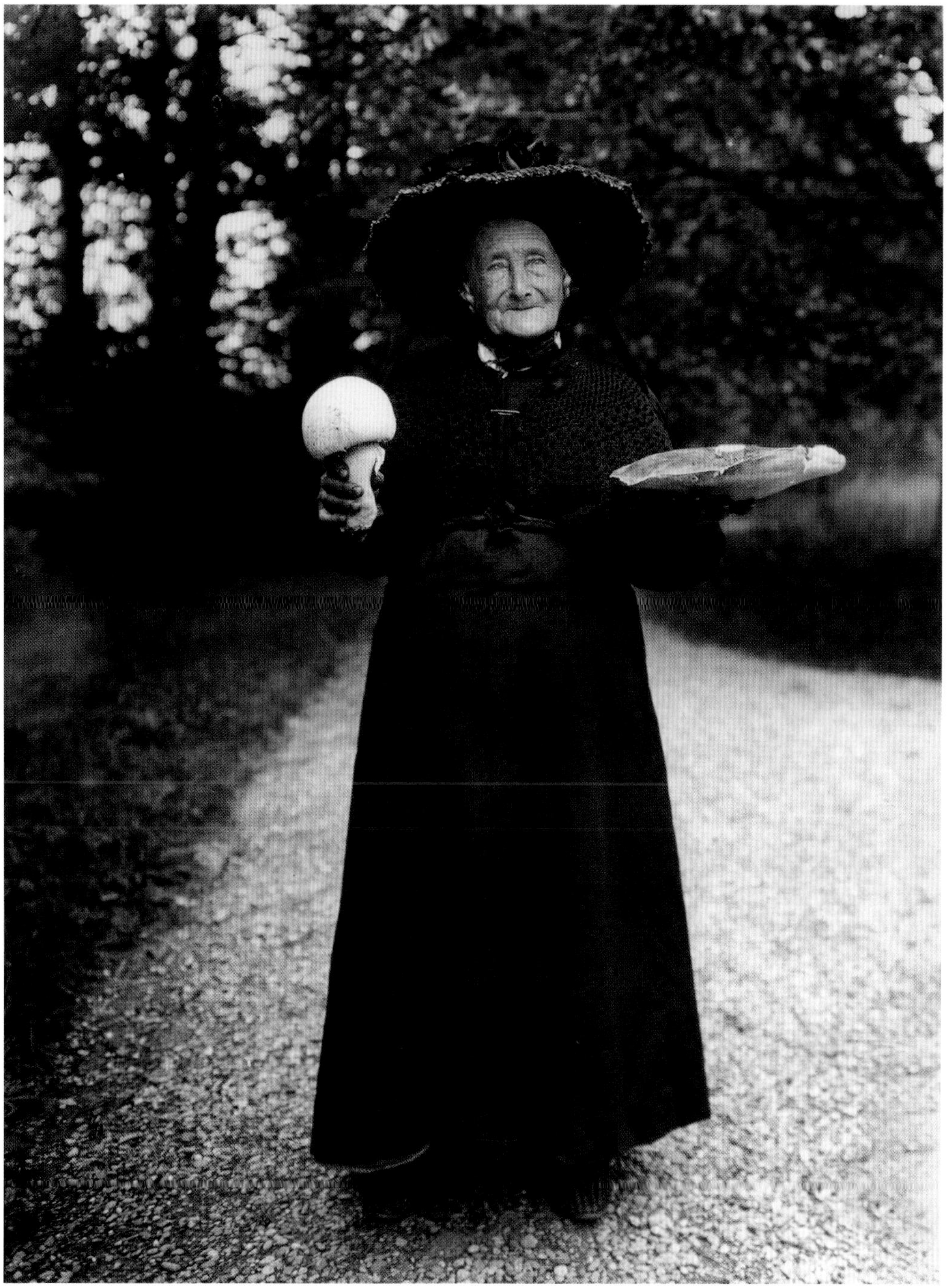

ERNEST BLOCH

Old Woman with Mushrooms, Satigny, Switzerland, 1912
Gelatin silver print,
14 × 11 in. / 35.6 × 27.9 cm

Until a decade after his death, no one knew Ernest Bloch (1880–1959) was a photographer outside of his family. Bloch was born in Geneva, Switzerland, to Jewish parents, and he was a musical prodigy; he began playing violin at nine years old and composing shortly after. Bloch traveled around Europe for years, studying violin under various teachers and at several conservatories before moving to the United States in 1916. He held numerous teaching appointments there, the last of which was at the University of California, Berkeley, from 1941 to 1952, at which point he retired, and died of cancer seven years later. In 1969 Eric Johnson, an undergraduate student at the University of Oregon, was encouraged to look into Bloch's photography by a professor who had heard about it from a mutual friend of the Bloch family. What Johnson found was a trove of more than six thousand negatives spanning Bloch's life, starting when he was a teenager. Bloch's photos—most of which were shot using a Leica camera—show a darkness that innately lends the images a sense of drama. *Old Woman with Mushrooms, Satigny, Switzerland* is a perfect exemplar. The subject seems almost otherworldly, swathed in black, as if she has stepped out of a fairy tale and is brandishing mushrooms that will soon be added to a potion. Thousands of Bloch's negatives and prints (many made by Johnson) are now housed in the Ernest Bloch Archive at the Center for Creative Photography at the University of Arizona in Tucson.

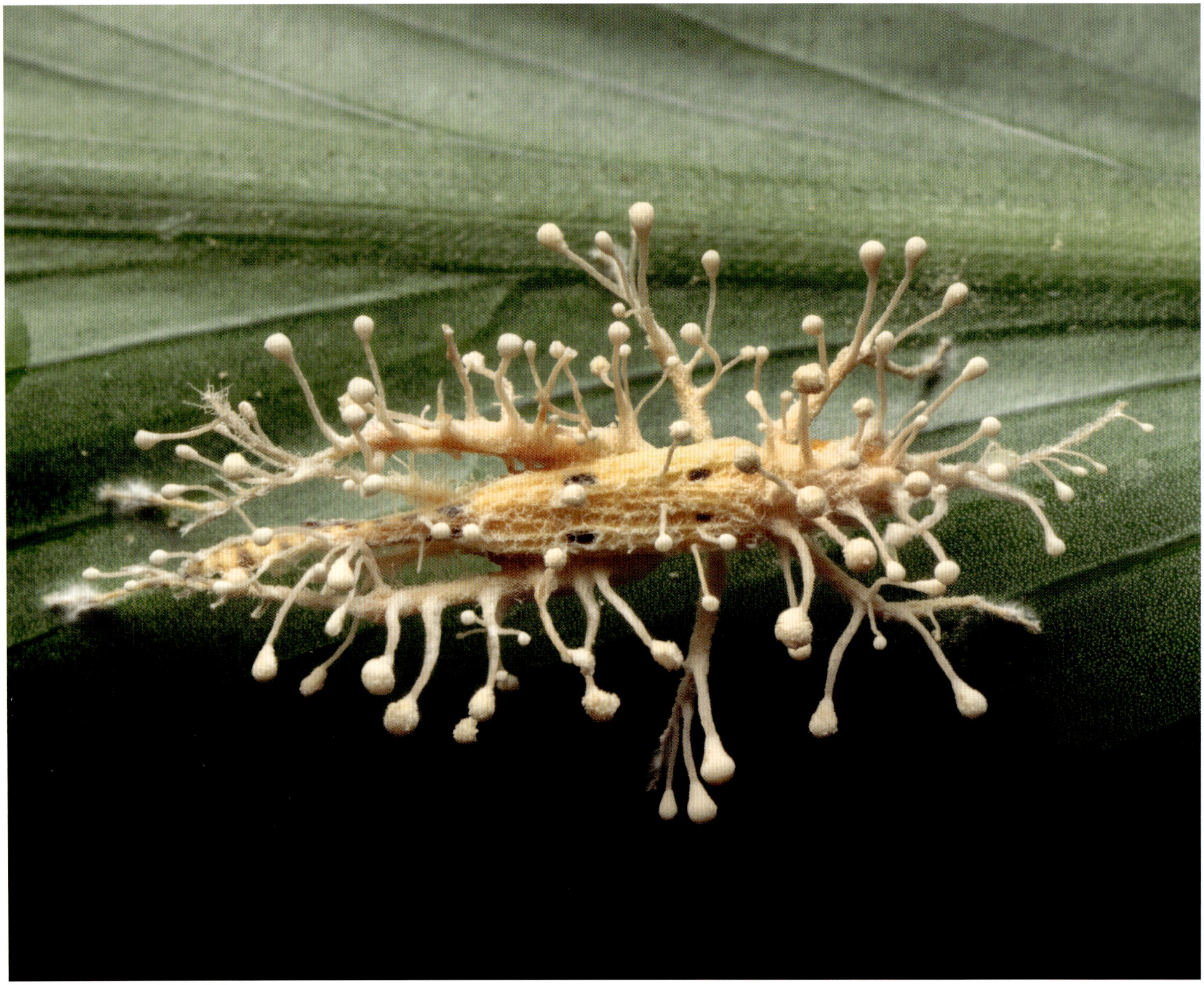

ALEX HYDE

Cordyceps Fungus Erupting from Its Host, 2024
Photograph, dimensions variable

Deep within the Chocó cloud forest along the Andes in northwestern Ecuador, photographer Alex Hyde (b. 1983) captured this beautifully macabre scene of a *Cordyceps* fungus erupting from the body of a cricket. Known colloquially as a "zombie fungus"—and recently popularized by the 2023 television adaptation of the video game *The Last of Us*—the *Cordyceps* seen here, along with other entomopathogenic fungi, is parasitic, infecting and penetrating the bodies of insects. The mycelium takes over, replaces the insect's tissue, and eventually kills its host. While the process may seem slightly gruesome, entomopathogenic fungi are ecologically crucial for controlling insect populations, infecting more than half of the known insect orders. There are approximately 260 species of *Cordyceps* fungi worldwide, many located in Asia, where they have been used in traditional Chinese medicine for more than 1,500 years. The species photographed by Hyde, however, was found in one of the most biodiverse yet least mycologically explored places on Earth: the Pacific Forest of Ecuador. Within this region lies the Chocó cloud forest, a spectacular type of rainforest found at high altitudes and characterized by persistent cloud cover. The humidity generated by the clouds creates a highly biodiverse environment where many undocumented fungal species thrive. It was named a UNESCO Biosphere Reserve, which aims to support local communities while protecting the unique ecosystems of the region, where resource extraction and agricultural development pose a significant threat.

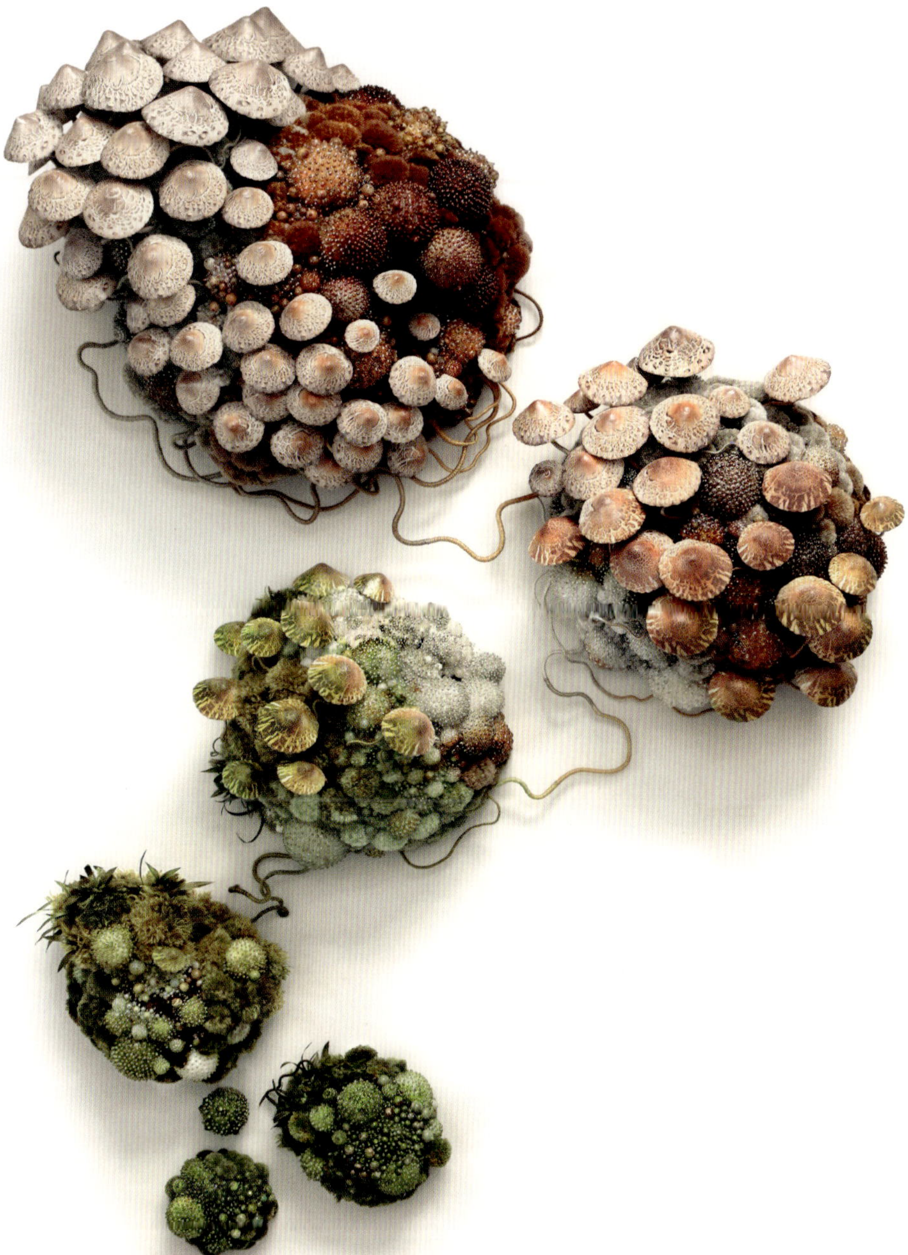

AMY GROSS

Mycorrhiza, 2018
Paper, beads, yarn, thread, ribbon, and wire,
47 × 28 × 8 in. / 119.3 × 71 × 20.3 cm
Fazz Collection, Hendersonville, North Carolina

For Florida-based fiber and mixed-media artist Amy Gross (b. 1965), her embroidered and beaded sculptures start with synthetic and inorganic materials. For *Mycorrhiza* (the word comes from the Latin word for "fungi," *myco*, and the Greek word for "root," *rhiza*), she created several clusters of linked microcosms that diminish in size from top to bottom. Each cluster is covered in mushrooms that are connected by mycelia, the subterranean fungal network that, in nature, is largely hidden from view. Gross's aim is to freeze a particular aspect of nature's transience in a moment of time. She wants to create organisms that endure forever, never aging or changing, because they are made from synthetic yarn, beads, wire, paper, and fabric. The inspiration for her work came following a move in 2000 from New York to Florida, where the subtropical environment causes growth and decay to happen at an accelerated speed. In this sculpture, the linking strands of mycelia physically embody what is called a mycorrhizal, or mycelial, network, comprising a mutually beneficial relationship that exists between plant roots and fungi. It is a phenomenon that emerged as far back as five hundred million years ago: As fungi colonize a plant root system, they enlarge it, thereby helping the plant absorb more water and nutrients from the soil. In return, the plant passes molecules, typically sugar, to the fungi. It is estimated that 90 percent of all plants have mycorrhizal fungi associated with their roots. Gross sees her sculptures as representations of "not only what can be seen but also what cannot: the early alterations of time."

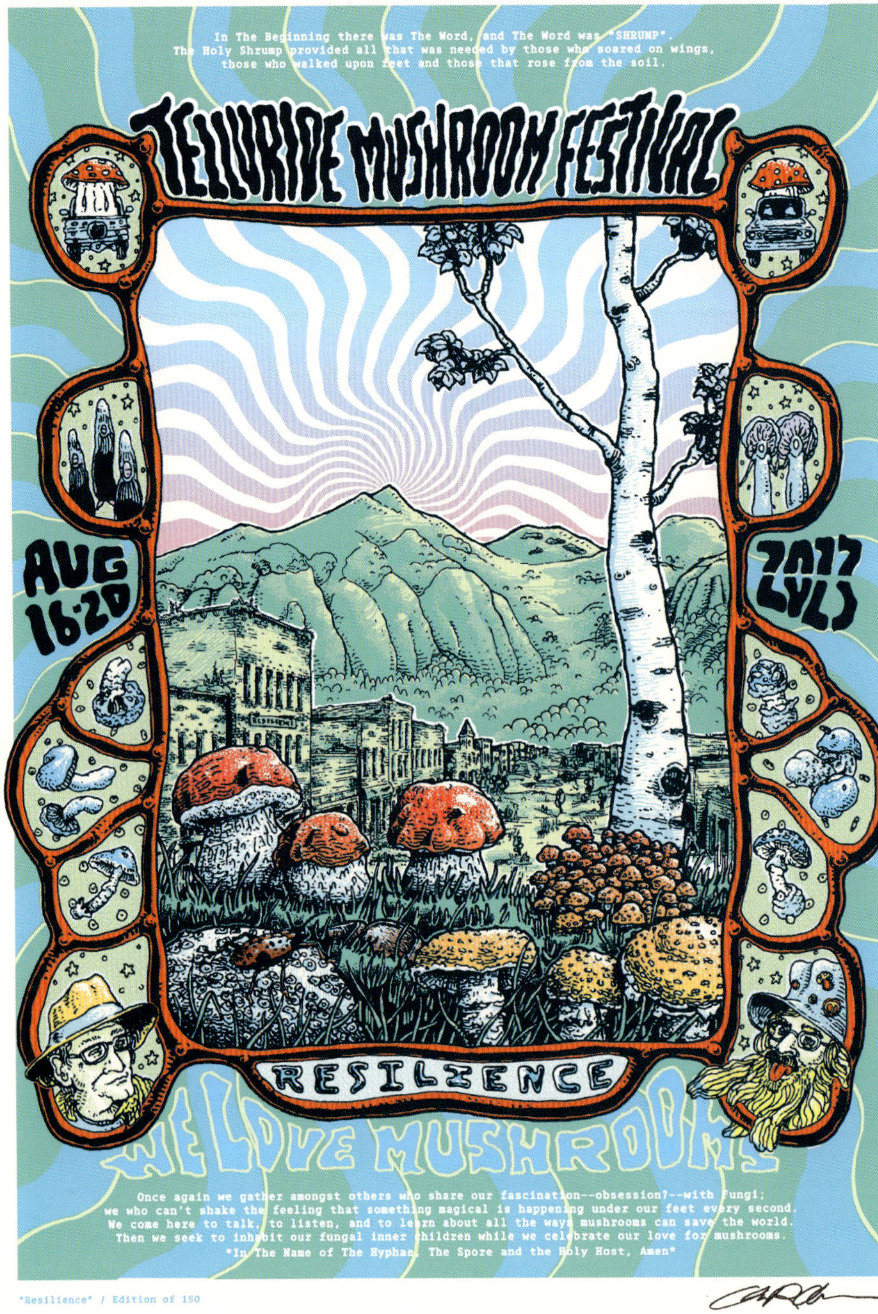

CHRIS ADAMS

Telluride Mushroom Festival Poster, 2023
Pen and ink on paper, digitized,
18 × 12 in. / 45.7 × 30.5 cm
Private collection, Corvallis, Oregon

The Telluride Mushroom Festival, held each third weekend of August in Telluride, Colorado, is the oldest and longest-running wild mushroom festival in the United States. Since 1981 the festival has brought together scientists, psychedelic seekers, foragers, foodies, artists, and people who are simply *really* into mushrooms for several days to "focus on science, culture, and community." Through workshops, presentations, and forays into the surrounding mountains to hunt mushrooms, the festival provides a place for fungophiles to hang out as a community and find in-*spore*-ation. But it's not all scholarly pursuits and foraging—there is also live music, the Mycolicious Mycoluscious Mycological Poetry Show, the Puff Ball costume dance party, and the legendary Mushroom Parade. In other words, it is a place for people to learn *and* have a blast while they do it. The Telluride Mushroom Festival now attracts thousands of attendees, and as it has grown, so has the collectability of its annual posters. The one shown here, for the 2023 festival, was designed and printed by Oregon-based artist Chris Adams (b. 1983) in a hand-signed and numbered limited edition of 150 copies. Surrounded by Adams's psychedelic artwork, the poster boldly proclaims, *WE LOVE MUSHROOMS*. Adams is involved with the fest on a number of levels, and if any attendees are interested in designing their own Telluride Mushroom Festival poster, as recently as the 2025 event Adams was teaching a class called How to Draw a Mushroom.

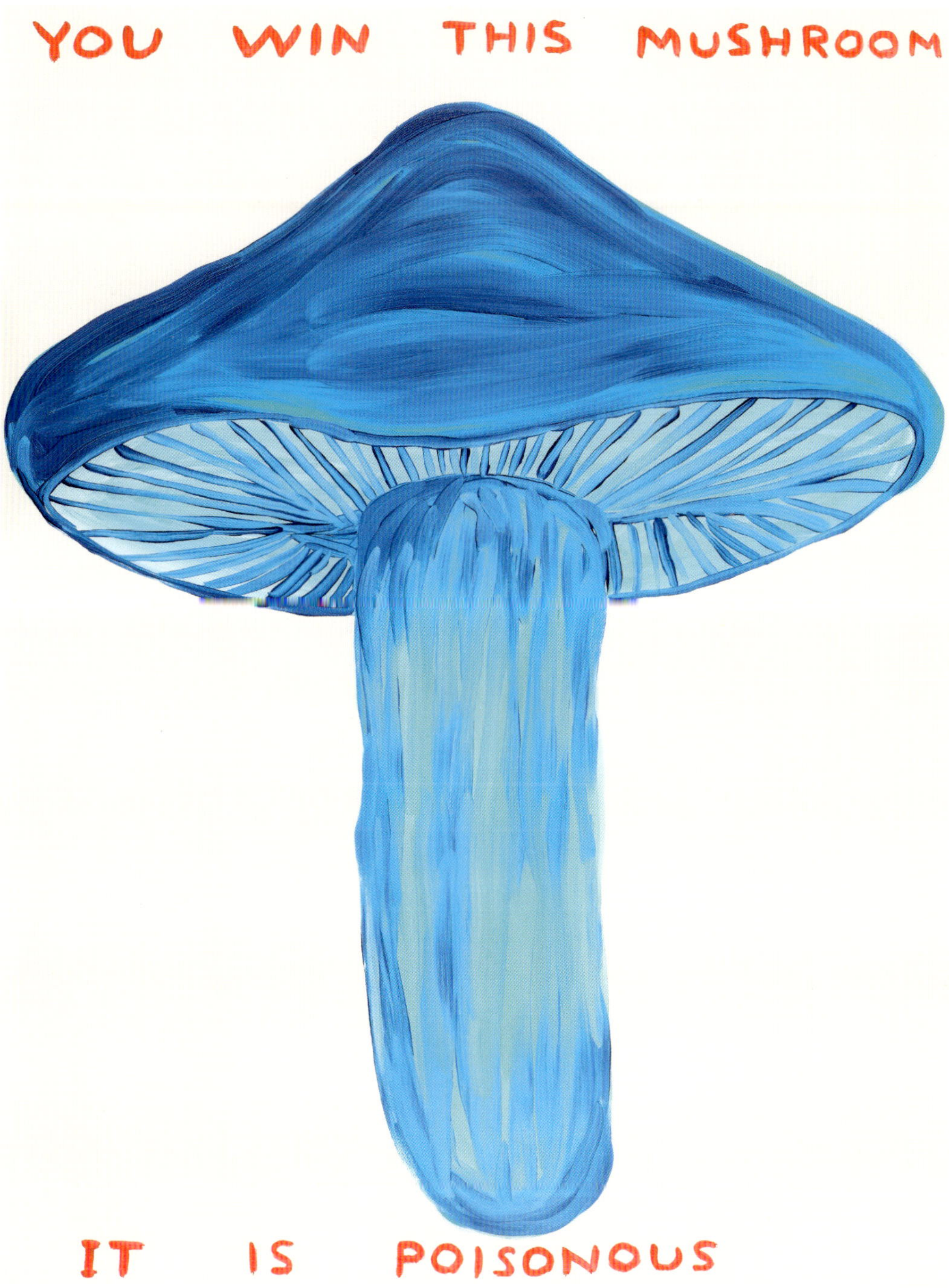

DAVID SHRIGLEY

You Win This Mushroom, 2020
Acrylic on paper, 30 × 22 in. / 76 × 56 cm
Private collection

As the diverse imagery of this book reveals, mushrooms come in many different colors. However, the vivid blue in this drawing by prolific British artist David Shrigley (b. 1968) is a shade not often seen in natural foodstuffs, so perhaps even without the "IT IS POISONOUS" words of warning, a recipient might be wary of consuming it. The work is typical of the artist's deadpan approach, in which words and images are brought together in simple compositions that communicate his reflections on life's realities and absurdities. Without further information to go on, the viewer may ponder what kind of competition would offer such a prize, and indeed whether its poisonousness is something to be alarmed by or something to cherish. While the risk of poisoning from an unknown fungus is a real and well-known fact, the toxicity of certain species has been prized by some cultures for their psychoactive and hallucinogenic effects. Although Shrigley has carefully rendered the stem, gills, and campanulate cap of his specimen, it does not appear to be any identifiable variety but instead a generically recognizable representation of a mushroom. Though, among the tens of thousands of named species of mushrooms, there do exist a number that are blue, such as *Lactarius indigo* (indigo milk mushroom), *Stropharia caerulea* (blue roundhead), and *Entoloma hochstetteri* (blue pinkgill)—also known by its Māori name, *wewerewere-kōkako*—which has featured on the reverse of New Zealand's fifty-dollar bill since 1999.

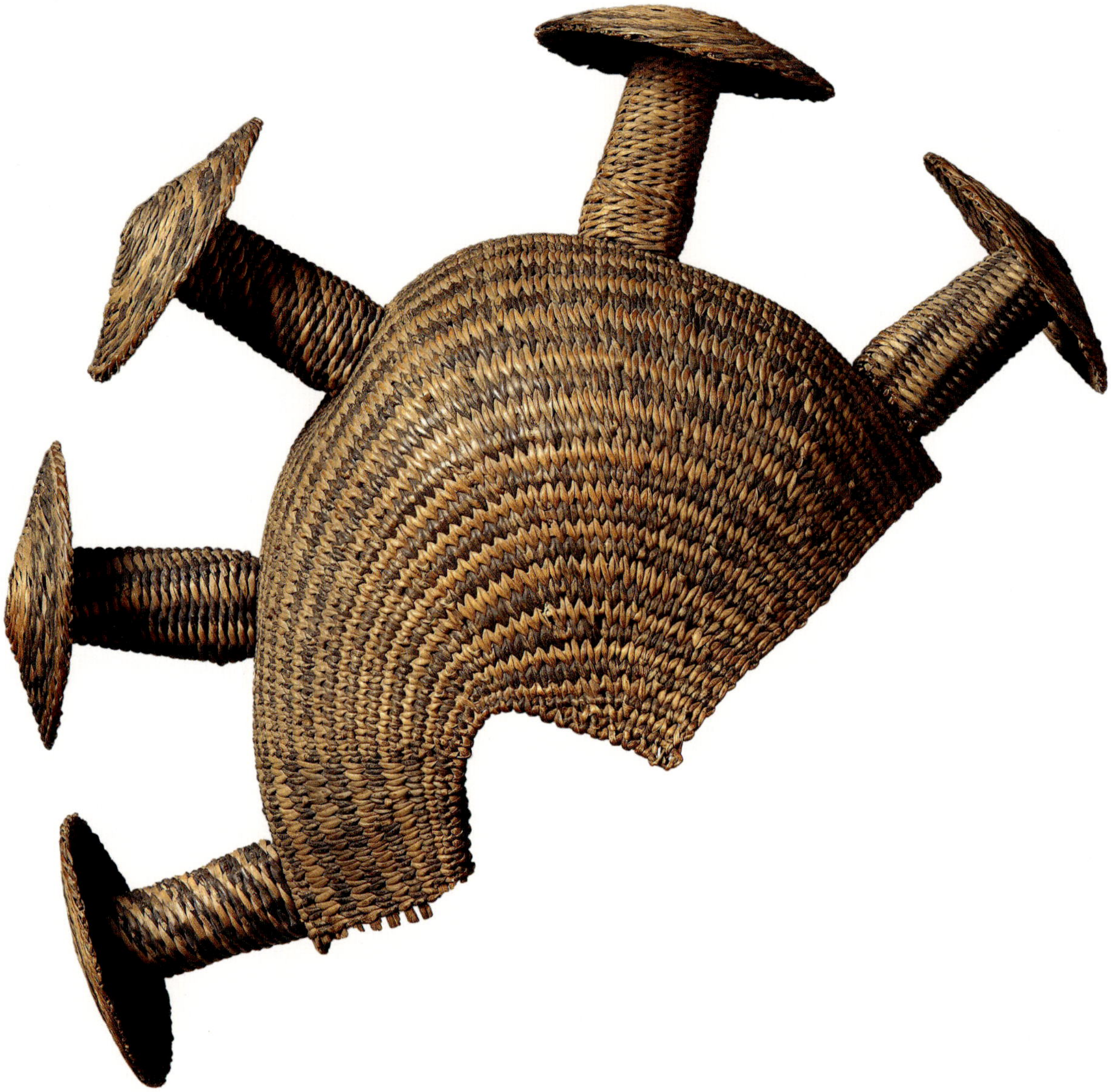

ANONYMOUS

Mahiole (helmet), c. 1819
Woven plant fibers,
12¼ × 13⅜ × 11½ in. / 31 × 34 × 29 cm
Musée du quai Branly – Jacques Chirac, Paris

This helmet, known as a *mahiole*, was made by an unknown Hawaiian artist in the early nineteenth century. Highly prized, such helmets were worn by high-ranking Hawaiian men and chiefs during ceremonies or in battle, and were typically made from braided basketwork topped with red and yellow bird feathers, two colors considered sacred. Here, the upper part of the *mahiole* is decorated with what appear to be five mushroom-shaped growths, complete with stems and large caps, made from the same basketwork as the helmet itself. The Hawaiian archipelago is home to a wide variety of mushrooms, with more than 640 known species. Tropical and saprotrophic—living off dead organic matter they break down—mushrooms grow widely in the damp, volcanic landscape, including endemic species that are found nowhere else, thanks to the islands' geographic isolation. Among them is the rare, endangered pink rose in the mist (*Hygrocybe noelokelani*), which is found in only three specific locations of tropical montane cloud forest. While the eventual owner of this helmet, French explorer Joseph Paul Gaimard inventoried it as "a Sandwich Island helmet whose crest is formed of mushrooms," the species and significance of the mushrooms being used on the *mahiole* is unclear. Gaimard purchased the helmet on a visit to Hawaii in 1819 with fellow maritime voyager Louis-Claude de Saulces de Freycinet, and took it back to Europe, where it remains today at the Musée du quai Branly – Jacques Chirac in Paris.

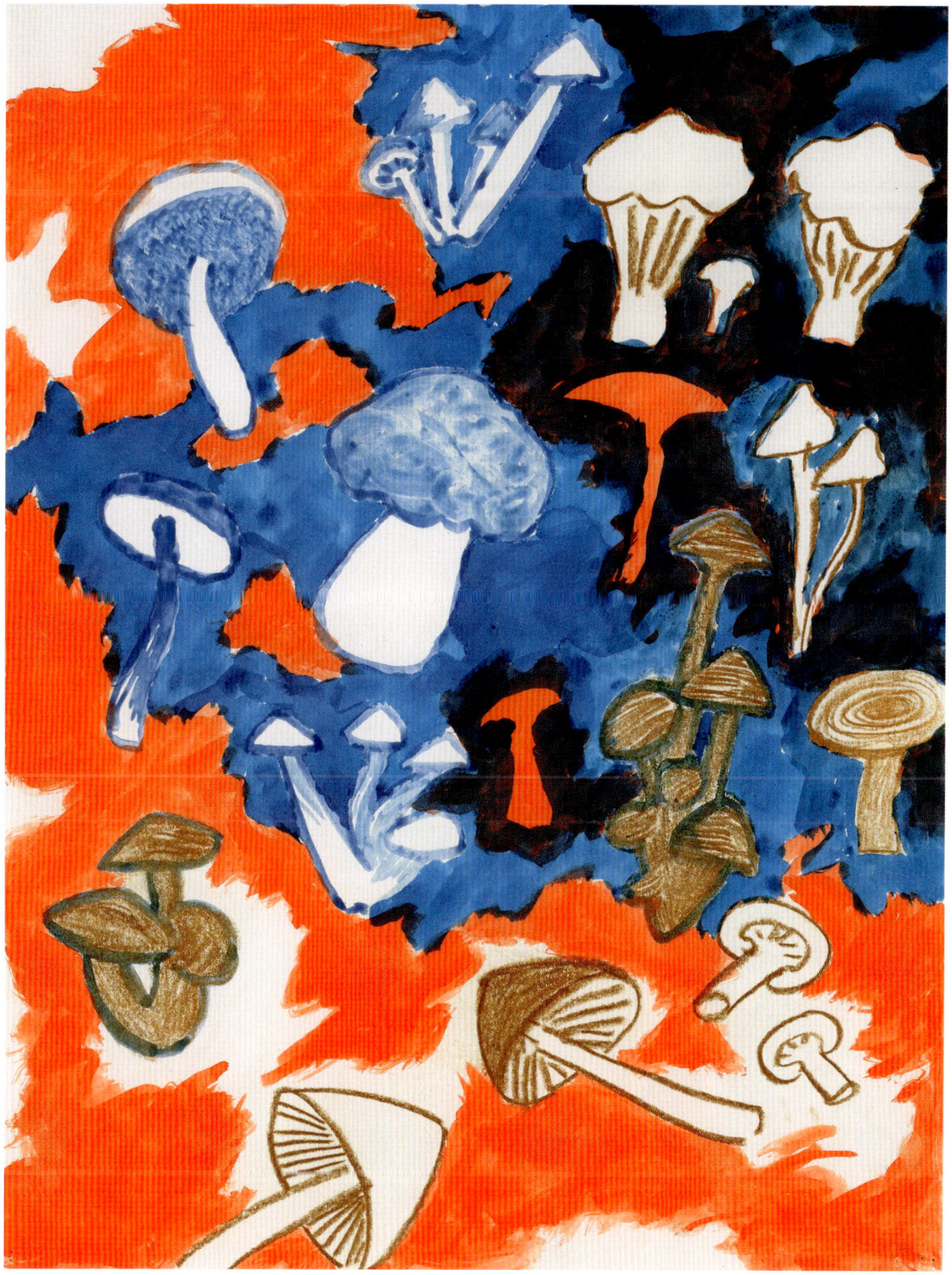

INGEBORG GABRIEL

Untitled (mushrooms), 1994
Watercolor and crayon on paper,
15¾ × 11¾ in. / 40 × 29.8 cm
Museum of Modern Art, New York

Rendered with simple lines, a variety of mushroom forms are scattered across the canvas, some taking their shape from the negative space in between the blocks of vibrant blue and orange. Created late in her career, *Untitled (mushrooms)* is the work of German artist Ingeborg Gabriel (1951–1996). Gabriel drew on the familiar and pastoral in her work, and she preferred to draw that which she saw around her home in the countryside outside Cologne, Germany: insects, birds, flowers, frogs. When she alighted on a subject, she examined it over and over, producing countless variations that allowed her to explore the dimension of each object. She experimented with media—paints, inks, crayons—and she varied the paper stocks she used, from roughly textured parchments to smooth and luminous bright whites. Some forms she sketched in silhouette, while she captured others in intricate detail. Through repetition, she was able to create innovation through experimentation. In a massive career retrospective that was mounted with the artist but opened shortly after her death, Cologne's Galerie Christian Nagel presented the first major showing of her work under the banner *Ohne Titel* (*Untitled*). The exhibition featured dozens of mushroom studies, done in a range of colorways and expressions. Hung together, they underscored Gabriel's exacting attention to detail as well as her freedom of artistic interpretation. She was not bound to a particular style or convention; she was defined by the doing itself.

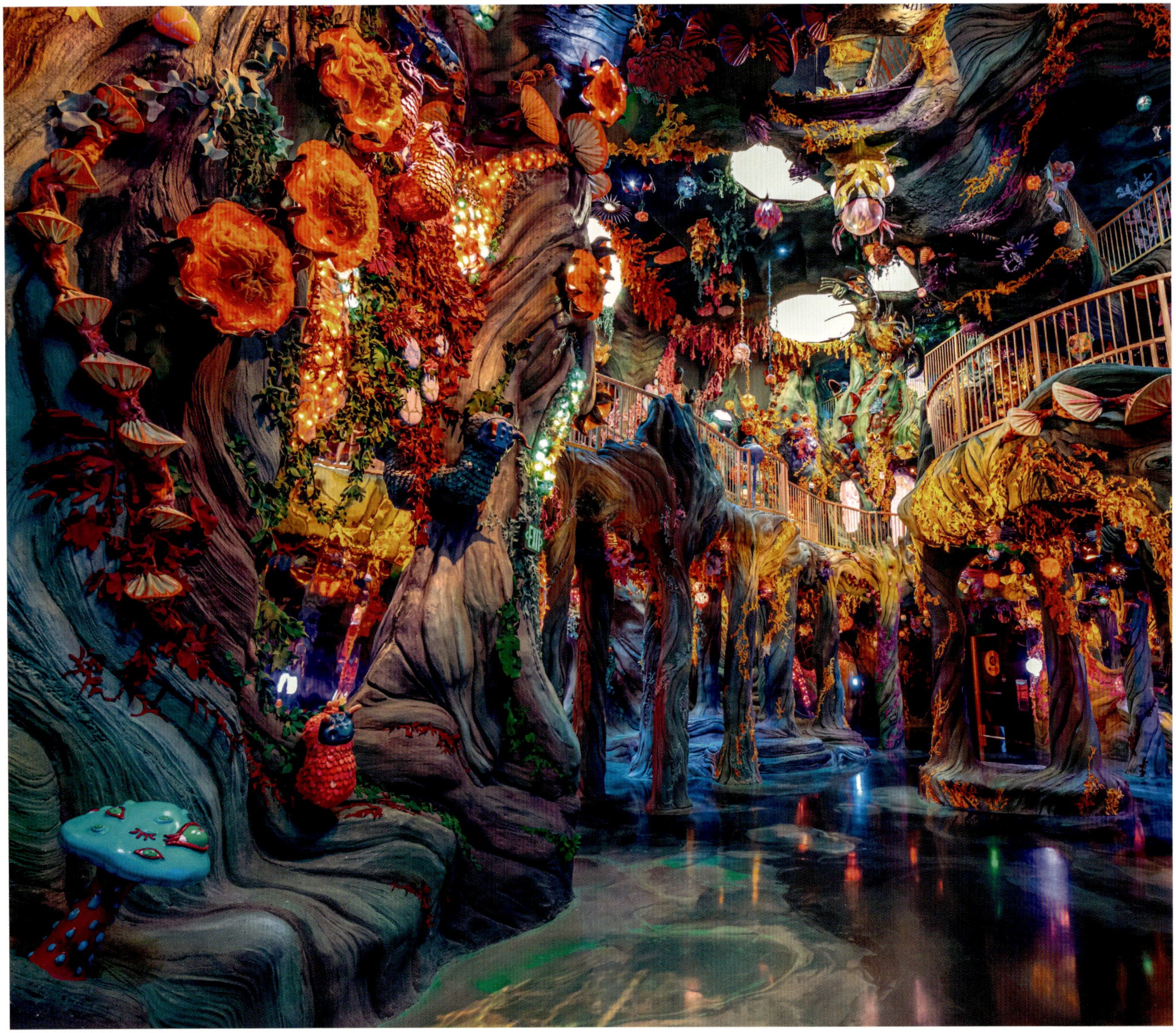

MEOW WOLF

Numina, in *Convergence Station*, 2021
Mixed-media immersive installation, dimensions variable
Denver, Colorado

Numina, one of the anchor worlds of Meow Wolf's immersive experience in Denver, Colorado, titled *Convergence Station*, feels like entering a living organism rather than an art installation. Designed as a walk-through narrative environment, saturated phosphorescent colors surround the visitor, mushroom-like sculptures tower overhead, and every surface and nook teems with fanciful life. Beyond the visual elements, sound, light, and texture combine to create an ecosystem that encourages movement around, over, under, and through the space, melding storytelling with purely sensory explorations. Founded in 2008 in Sante Fe, New Mexico, Meow Wolf is a large collective of artists and fabricators specializing in dynamic interactive experiences. At *Convergence Station*, among the installations encountered are *Foliage*, which spills across walls and ceilings in glowing greens and violets, with textures ranging from soft and delicate to spiked and sculptural, and even pulsating with glowing lights, suggesting otherworldly origins. *Rainbow Lichen* clings to the walls in thick ribbons of saturated color and tactile ornamentation. Tucked-away chambers, such as the *Swamp Pocket*, feature miniature fungi and fauna that cluster and creep like slime molds, while overhead the porcelain and wool *Parasites* and *Epiphytes* hang like a ghostly white canopy of growths crawling with frogs—a nod to deadly chytrid fungal pathogens that infect amphibian populations worldwide. Throughout, the hidden networks of the fungal world are mirrored in imaginative and interconnected ways, allowing *Numina* to turn ideas of interdependence, transformation, and symbiosis into an immersive space itself.

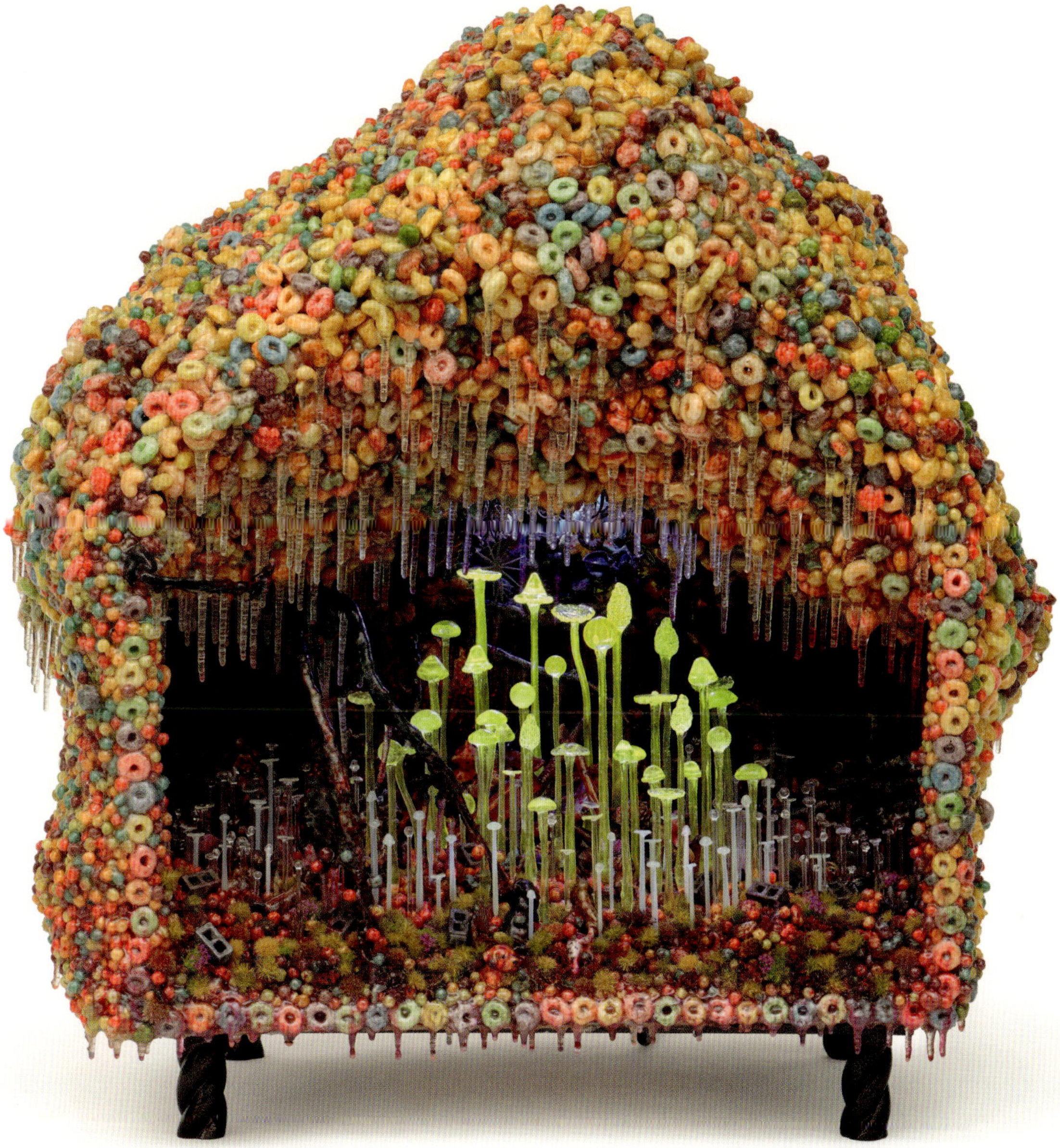

MAX HOOPER SCHNEIDER

Cereal Cave, 2023
Modeled plywood box, UV blown-glass mushrooms, epoxy resin, assorted matter, toy dogs, cereals, UV LED light, and powder-coated aluminum table, 23¼ × 19¾ × 15 in. / 59 × 50.2 × 38.1 cm
Private collection

A table is transformed into a sculptural diorama made from acid-colored breakfast cereal and containing a miniature cave of glowing, unreal fungi. This elaborate and vibrant work is typical for Los Angeles–based artist Max Hooper Schneider (b. 1982). His hyper Pop artworks mix biology and philosophy into uncanny, excessive objects that feel both artificial and natural. Hooper Schneider studied landscape architecture at Harvard Graduate School of Design and urban design and biology at New York University, before shifting into contemporary art. His tableaus and cavity-filled sculptures reflect the influence of pop culture, trash TV, and plastic Americana, as much as deep research on habitats, natural and artificial systems, and the strangeness of nature that will outlive humanity. Mushrooms—often created in plasticky neon colors; glowing like LED lights; or in translucent pinks, yellows, and purples—are a recurring motif in his work. Here, Cooper Schneider crafts the minuscule clear and milky white fungi and the luminous green mushrooms with their spindly stems and bulbous caps from UV-reactive blown glass. Fungi appear as a reminder about the ideas of symbiosis and metamorphosis in the face of moral, economic, and social decay. All of the many elements of the diorama are painstakingly positioned to create a scene that appears at once completely organic and obviously unnatural. This is a work that questions human dominance over ecology and how we are destroying the world we inhabit.

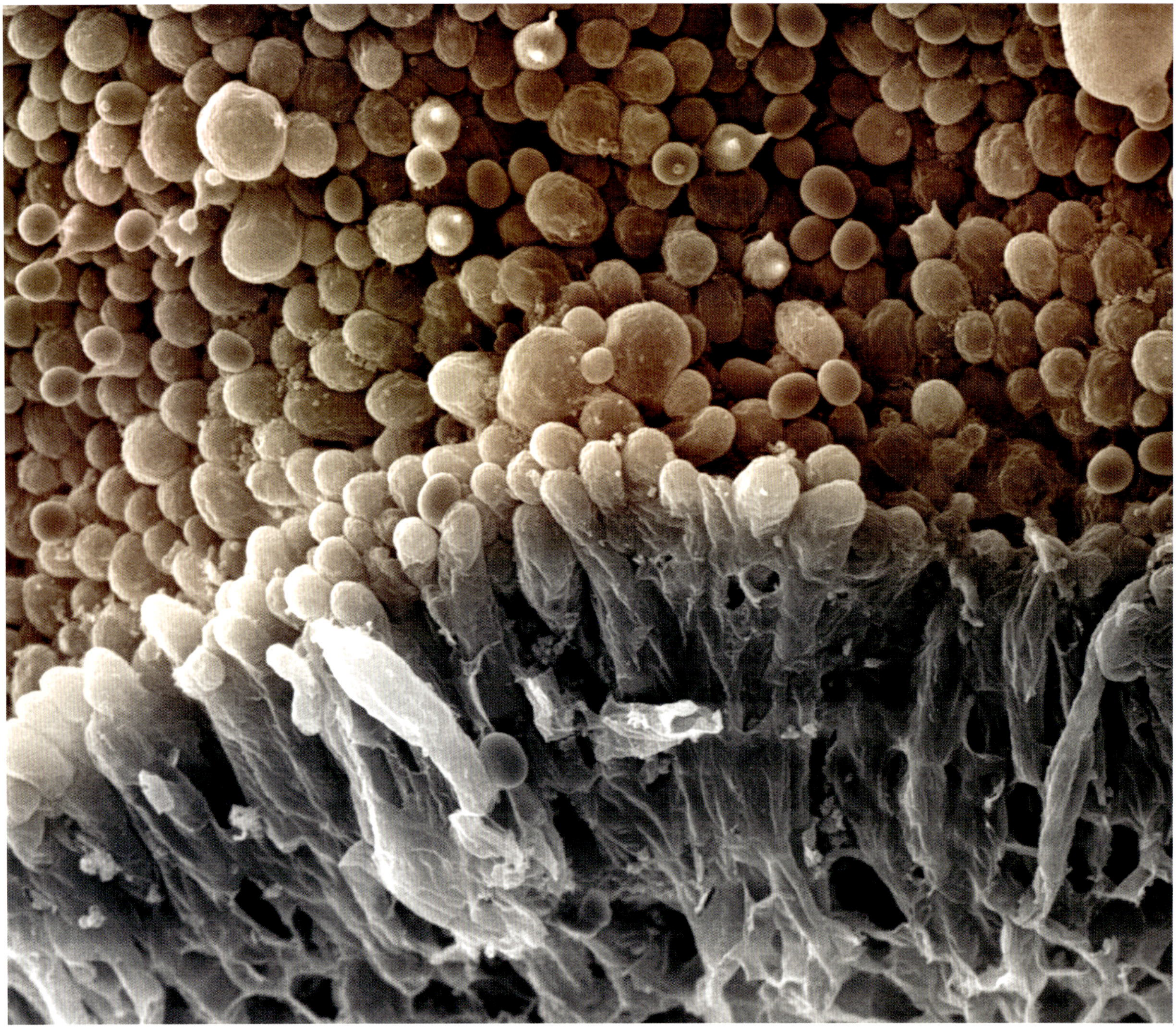

DARTMOUTH ELECTRON MICROSCOPE FACILITY

Mushroom spores (Agaricus bisporus), 2007
SEM photograph, dimensions variable

When thinking of fungi, we usually conjure an image of the mature fruiting bodies aboveground, often forgetting about the sometimes vast underground network of mycelia as well as other elements essential to mushrooms. Pictured here in a colored scanning electron microscope (SEM) photograph is one such element: spores. Each spore is typically a single cell with a nucleus surrounded by the spore wall. Vital to reproduction and dispersal, mushrooms release their spores beneath their cap, where they can easily be scattered by the surrounding air currents. While most mushrooms produce their spores in groups of four, in the species depicted here, the spores develop in pairs, hence the specific epithet in the scientific name *Agaricus bisporus*, more commonly known as portobello (or in its younger forms, button mushroom or crimini). This detailed image shows the structure of part of the gills of a mushroom with the spores developed at the tips of stalks known as sterigmata. When mature, the spores break free to drift away. Although *Agaricus bisporus* is not particularly common in the wild, it is the most widely cultivated species, especially in Europe and North America. Most commercial cultivation takes place in sheds, free from pests and diseases, using a specially formulated compost, with temperature and humidity carefully controlled. Although spores are the natural method of reproduction, in cultivation the rich compost is “seeded” with a culture of sterilized grain inoculated with mushroom mycelium.

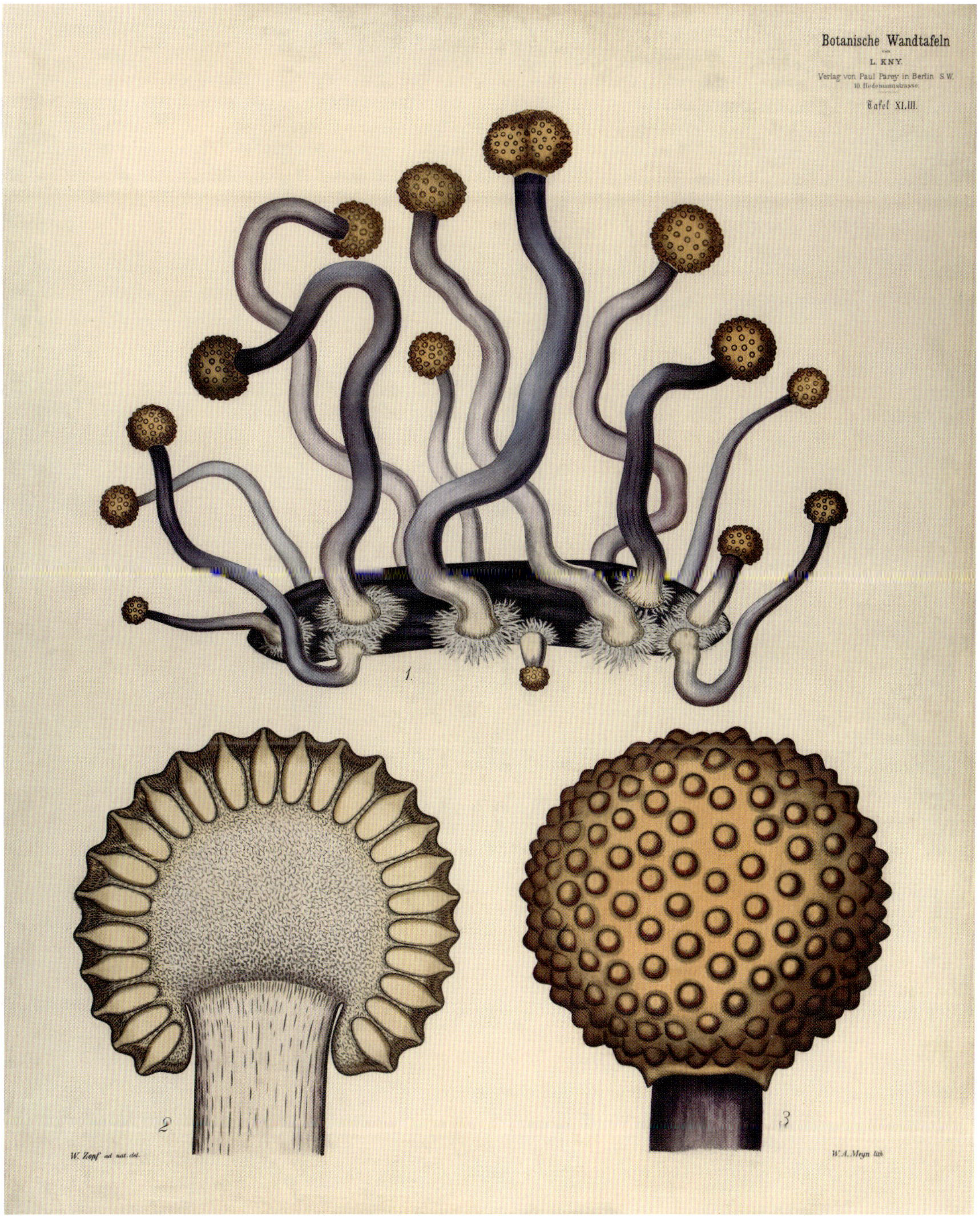

FRIEDRICH WILHELM ZOPF

Fungal Development in Claviceps purpurea, from *Botanische Wandtafeln* (*Botanical Wall Charts*) by Leopold Kny, 1874–1911
Color lithograph, 32 × 26 in. / 81.3 × 66 cm
Hunt Institute for Botanical Documentation, Carnegie Mellon University, Pittsburgh

Rising from the center of the page, more than a dozen round-headed tendrils emerge in a spindly tangle, while below, a cross section provides a closer inspection of the alien-looking organism. Despite its otherworldly appearance, this plate—from a collection of educational wall charts by German botanist and mycologist Friedrich Wilhelm Zopf (1846–1909) published between 1874 and 1911—shows the fruiting stage of the rather common ergot fungus *Claviceps purpurea*. A troublesome species, both medically and commercially, it causes the eponymous ergot disease in several cereal crops, notably rye, and also infects several grasses in the wild. Ergot infection reduces crop yield, and if the affected hay is fed to livestock, it may cause the disease ergotism. If infected cereal is consumed by people over a period, it can lead to them developing convulsions, hallucinations, or even gangrene. If a spore (technically an ascospore) of the fungus infects the developing flower of a grass or cereal, it grows into the ovary of the floret, which it destroys, and is then nourished by the flower as if it were itself the developing seed. It may then later develop into a hard, black, dry stage known as a sclerotium, which eventually falls to the ground. From the sclerotium, one or more fruiting bodies resembling tiny mushrooms emerge on snaking stems, as shown at the top of this illustration. Below are two depictions of the fruiting head of the mature fungus containing threadlike spores, which are released to begin the life cycle again.

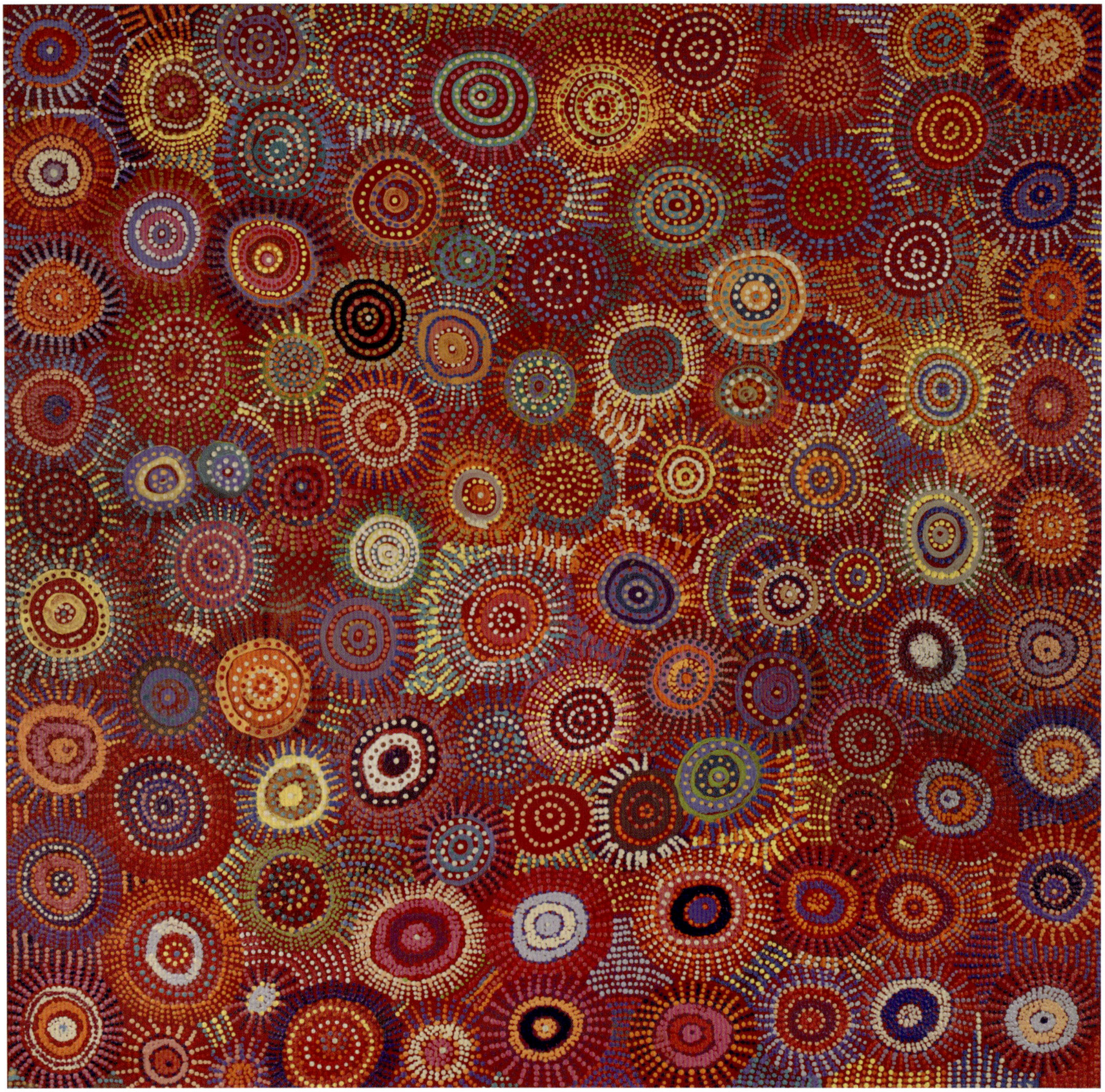

TINA NAPANGARDI MARTIN ROBERTSON

Jintiparnta Jukurrpa (Desert Truffle Dreaming) – Mina Mina, 2025
Acrylic on Belgian linen, 48 × 48 in. / 122 × 122 cm

Desert truffle (*Elderia arenivaga*) is one of a number of edible fungi that can be found in the arid Australian outback. Genetically different from European truffles, which belong to the genus *Tuber*, the native antipodal version evolved independently yet shares similar traits, such as growing and fruiting underground and spreading its spores via animal consumption. In common with Middle Eastern varieties (genus *Terfezia* and *Tirmania*), the Australian desert truffle is also notable for appearing rapidly after heavy rain. The mature, spherical fruiting bodies, ranging from around 1½ to 3 inches (4 to 8 cm) in diameter, provide bush food for Indigenous populations and resulted in one species being named *Choiromyces aboriginum* by European settlers. In native communities, it and other edible fungi are known as *Jinti-parnta* ("edible fungi") and, in addition to being essential foodstuffs, they feature in folklore and ceremonies. This colorful composition of the truffles is by Tina Napangardi Martin Robertson (b. 1960), a member since 1996 of the Warlukurlangu Artists community of Yuendumu, a town 180 miles (290 km) northwest of Alice Springs. The artist often draws on her father's *Jukurrpa* (dreamtime) stories, as exemplified in this work. Desert truffles were collected by the ancestral Napanangka and Napangardi women as sustenance on their long walks across the land from west to east. At the sacred site of Mina Mina Jukurrpa, in the remote Tanami Desert to the west of Yuendumu, the women danced and digging sticks (*karlangu*) rose up out of the ground—useful tools for foraging desert truffles on their journey ahead.

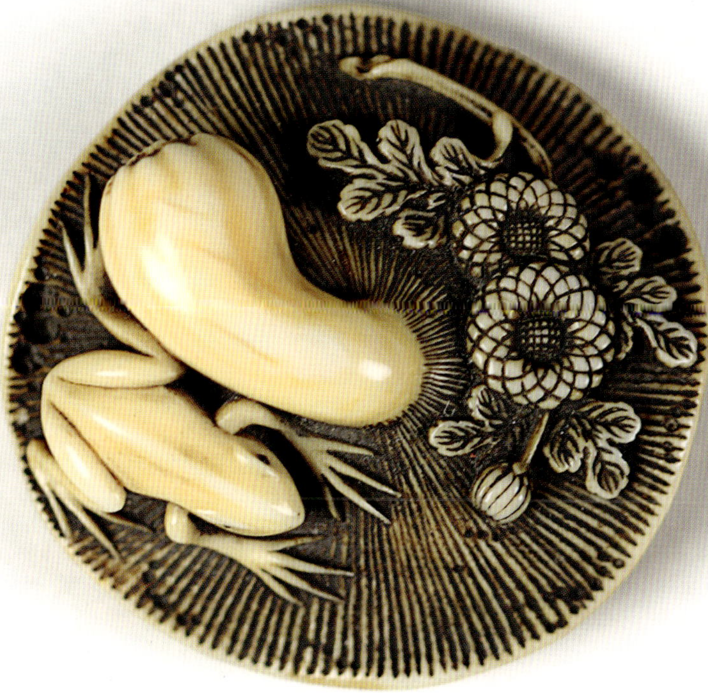

ANONYMOUS

Netsuke of a mushroom and frog, 19th century
Ivory, 1⅝ × 1⅝ × ⅜ in. / 4.1 × 4.1 × 1 cm
Metropolitan Museum of Art, New York

This miniature three-dimensional composition in ivory features a frog and the flowering and fruiting sprig of a plant, probably a poppy, resting upon the gills of an upturned mushroom. The stem of the mushroom lies twisted alongside the frog, which it resembles in both color and texture. Carefully wrought, each element fits snugly and smoothly within the rim of the circular mushroom, in a harmony of cream and brown. Small enough to fit in the palm of the hand, this is a Japanese netsuke, a small carved sculpture commonly made of wood or ivory, as here, but sometimes from antlers or coral. Netsuke gradually developed in Japan from the seventeenth century and served both functional and aesthetic purposes. The traditional form of Japanese dress, the kimono, had no pockets. Women would tuck small personal items into their sleeves, but men suspended their *sagemono*—hanging objects, such as tobacco pouches, pipes, purses, writing implements, and other items of daily use—on a silk cord passed behind their sash, or obi. The netsuke functioned as a toggle or fastener, attached to the other end of the cord, preventing it from slipping through the obi. Evolving over three centuries from simple toggles to complex art forms, netsuke often depict animals, mythical figures, people, and natural objects, serving as subtle displays of wealth and taste, especially among the merchants of the Edo period. Today, they are collectible art forms and are highly prized as fine examples of Japanese craftsmanship.

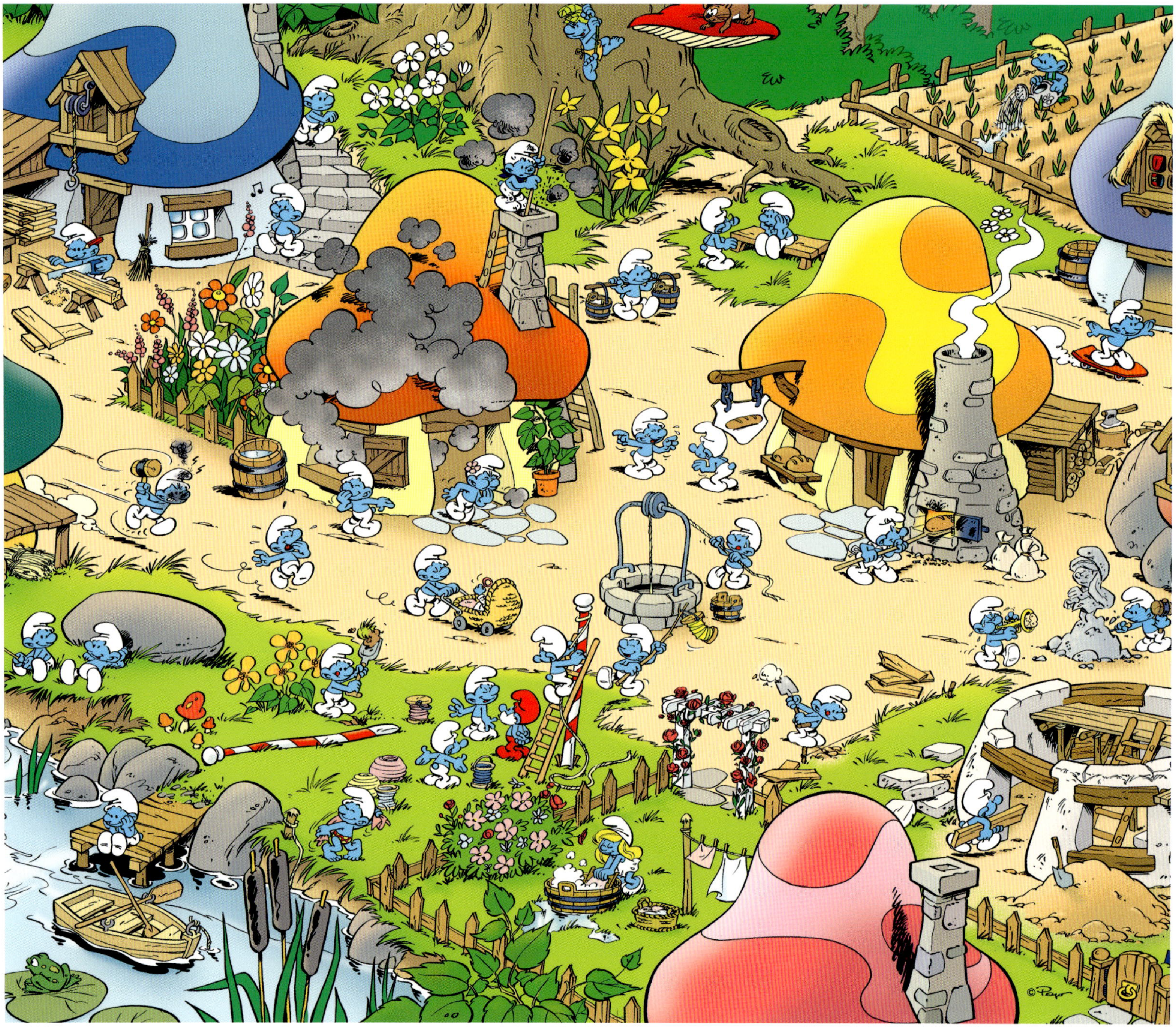

PEYO (PIERRE CULLIFORD)

The Smurfs, published by Bastei, 1998
Digital image, dimensions variable

The Smurfs were the brainchild of Belgian comic artist Pierre "Peyo" Culliford (1928–1992), who first introduced the series of small blue creatures living in a forest in 1958 in *Spirou* magazine. Originally named *Les Schtroumpfs* (after their creator forgot the word for *salt* during a meal, creating an alternative nonsense word on the spur of the moment), they inhabited a natural world Culliford brought vividly to life over the course of three decades. A cast of much-loved characters, including Papa Smurf, Smurfette, Hefty, and Grouchy, lived in villages of red or brown mushroom houses, as the mushroom cup was a convenient size for a Smurf who, as Culliford noted, is only three apples tall. Symbolically, the mushroom houses are integral to the Smurfs' fantastical world, which is rich in references to fairy tales, from Victorian classics to the Brothers Grimm. The intimate, folklore traditions the mushrooms recall also speak to the Smurfs' warm ecological philosophy. Smurfs live harmoniously with nature, foraging without destroying, striking a delicate balance that is epitomized in the vulnerable yet resilient mushroom, easily destroyed and easily rebuilt. This harmony with nature underpins the Smurfs' physical neighborhood: each member resides in an individual mushroom house, but all the mushrooms are part of a single cohesive village, much like the Smurfs themselves, who each have a personal identity yet also belong to a unified community.

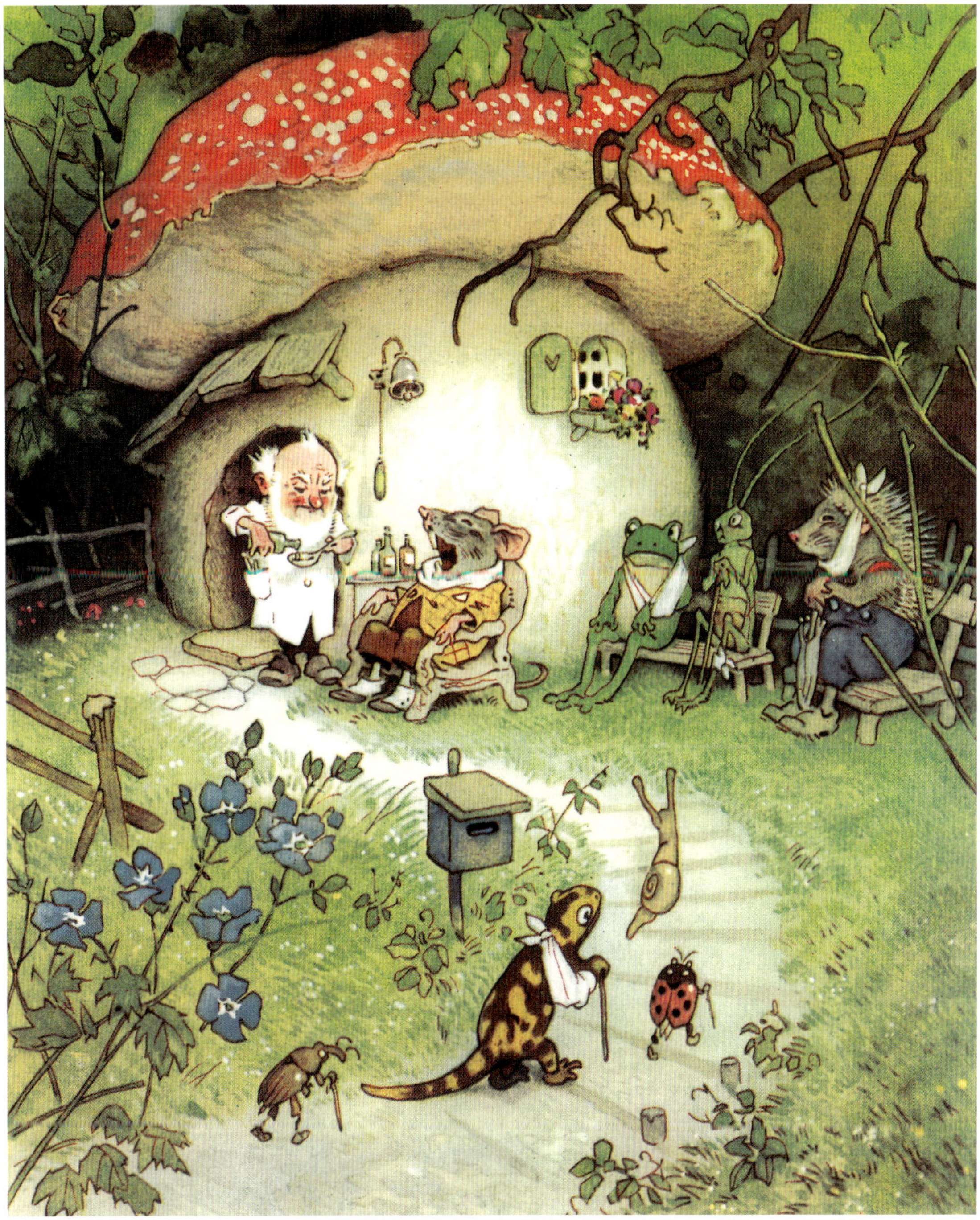

FRITZ BAUMGARTEN

Illustration from *Wichtelgeschichten* (*Elf Stories*) by Erich Heinemann, first published 1946
Color print, dimensions variable

In this charming illustration, a fairy-tale doctor stands at the door of his office administering a medicinal potion to a seated rodent. Awaiting treatment on adjacent seats are a frog with a bandaged arm, a grasshopper with a leg problem, and a hedgehog who seems to be suffering from a toothache. More animal patients are arriving up the garden path: a snail, a ladybug, a salamander, and a weevil. The dominant feature is the doctor's dwelling: a bulbous fly agaric (*Amanita muscaria*), complete with a door and a window with a flower box. The lively scene is one of many created by German painter and illustrator Friedrich "Fritz" Baumgarten (1883–1966) for the 1946 fairy tale *Wichtelgeschichten* (*Elf Stories*) by German author Erich Heinemann. The world he created to accompany Heinemann's story was one of fantastic communities of woodland elves and gnomes in a natural, but imaginary, forest setting. To the delight of his readers, his elves lived happily alongside birds, mammals, and insects, interacting with them in a believable, naturalistic fashion. The fact that the animals and plants are depicted with such accuracy and are therefore identifiable lends credibility to the involvement of his tiny humanlike characters, bringing realism to his wonderful fantasy world. Baumgarten illustrated hundreds of books between 1920 and 1960, including famous stories by the Brothers Grimm, such as *Little Red Riding Hood* and *Hänsel and Gretel*.

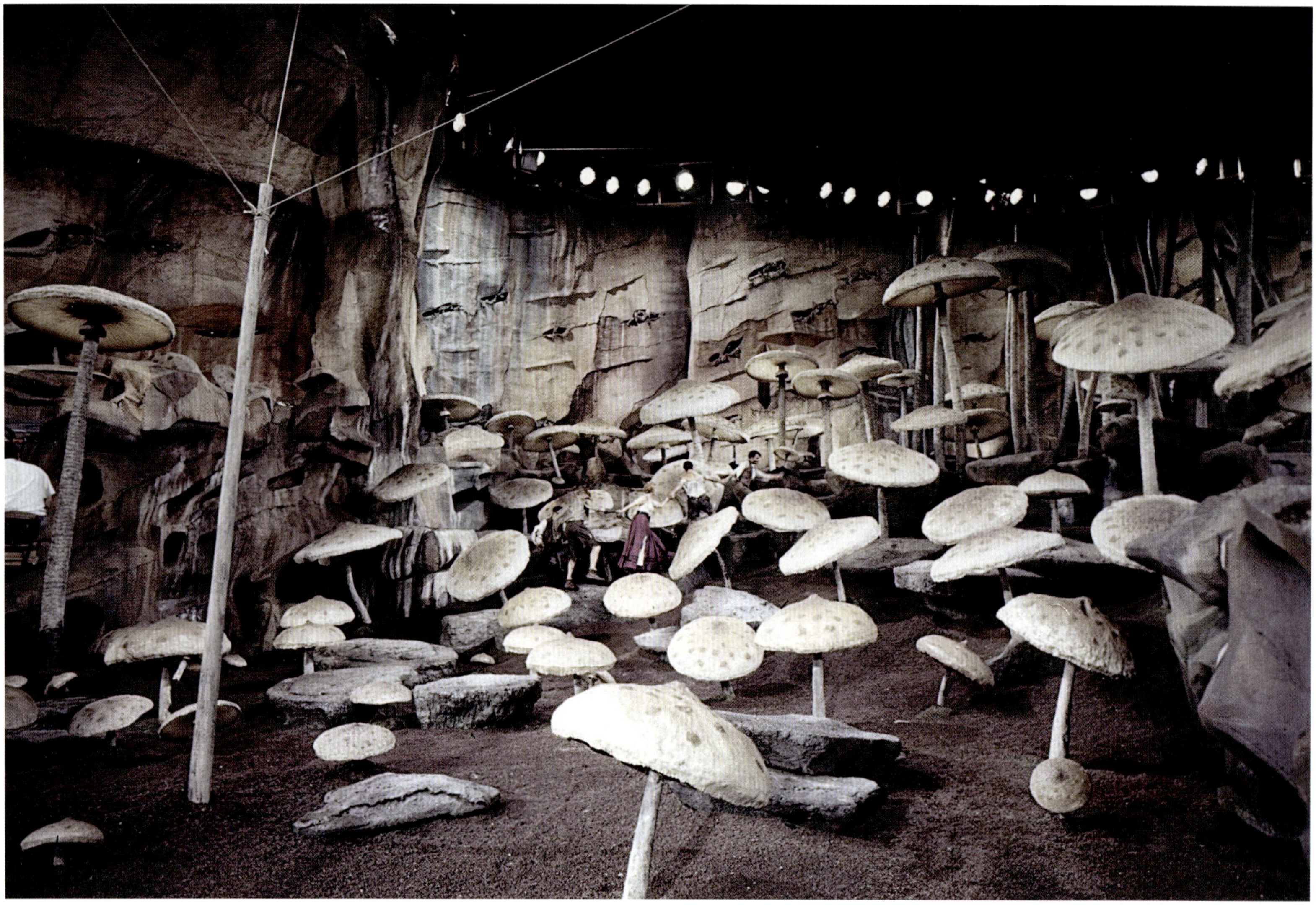

HENRY LEVIN FOR 20TH CENTURY FOX

Set of *Journey to the Center of the Earth*, 1959
Photograph, dimensions variable

French novelist Jules Verne (1828–1905) is the visionary mind behind some of science fiction's most provocative ideas. Some of them—the submarine, the helicopter, electric vehicles, space travel, and even video conferencing—have since come to pass. Others—time travel, for example—seem less likely. Verne's fertile mind made him one of the most popular writers of his day. Now widely considered to be the father of science fiction, he is the author of some of the genre's most enduring classics, including *Twenty Thousand Leagues Under the Sea, Around the World in 80 Days*, and *From the Earth to the Moon*, all of which have inspired film adaptations. His 1864 novel *Journey to the Center of the Earth* has seen numerous on-screen iterations, including the 1959 film directed by Henry Levin (1909–1980) and starring James Mason and Pat Boone.

The plot follows the quest of adventurous geology professor Otto Lidenbrock (which in the 1959 film was changed to Oliver Lindenbrook), who discovers a medieval manuscript illuminating a pathway through the mouth of a volcano and leading to the center of Earth itself. Traveling with his nephew and a guide, the trio descends the volcano's crater to alight on a vast subterranean sea. Setting sail, they travel for two months, encountering numerous creatures long presumed extinct, from dinosaurs to mastodons, and some truly astonishing flora. In one memorable scene, the group comes upon a vast forest of thousands of colossal mushrooms, each reaching 30 to 40 feet (9 to 12 m) high, where, as Verne wrote, "No light could penetrate between their huge cones, and complete darkness reigned beneath those giants."

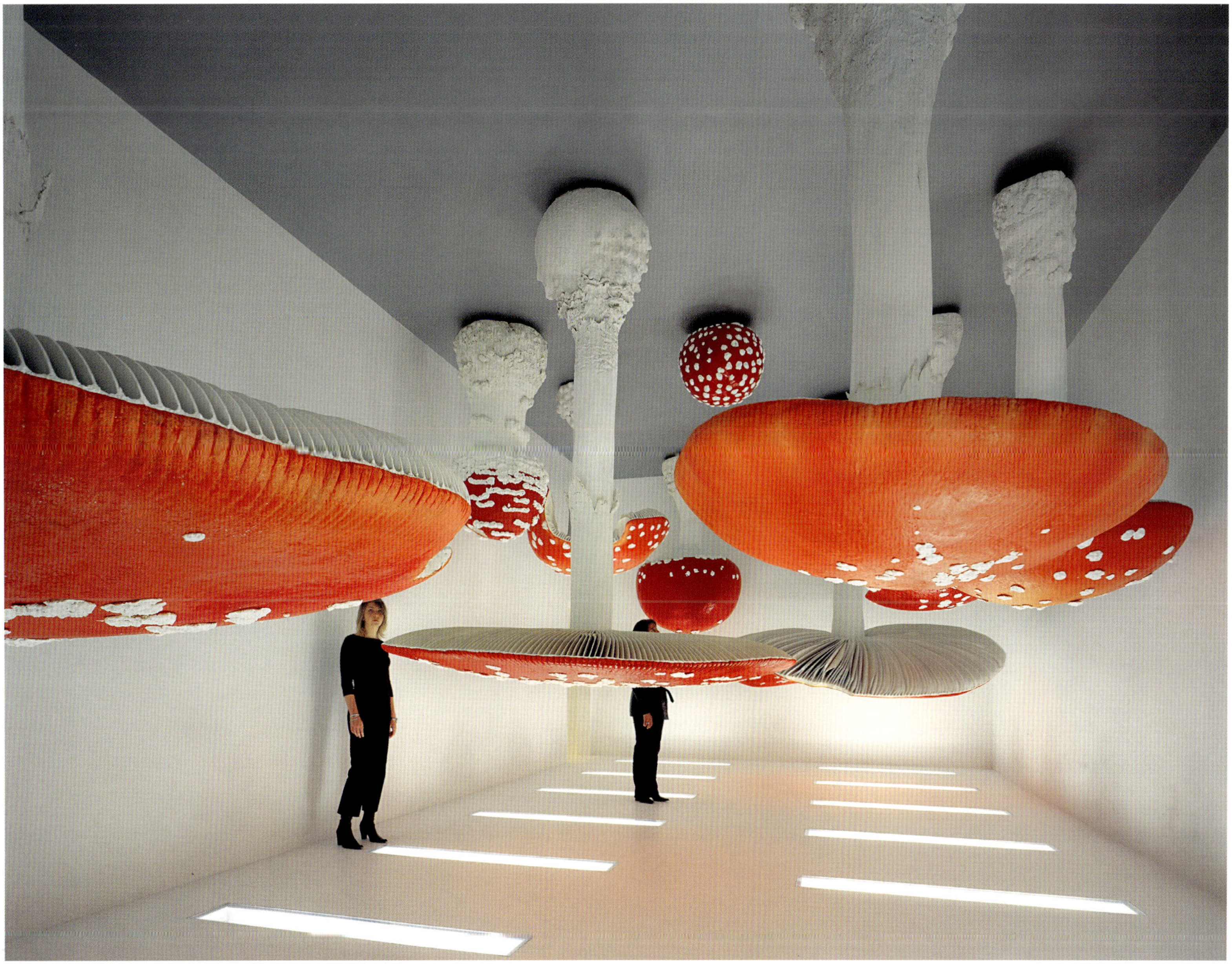

CARSTEN HÖLLER

Upside Down Mushroom Room, 2000
Installation, dimensions variable
Fondazione Prada, Milan

The installations and sculptures by German artist Carsten Höller (b. 1961) are spiritedly disorienting experiments into altered states of consciousness. Höller, who holds a doctorate in agricultural science, combines scientific methods with immersive artwork to create what he describes as a "laboratory of doubt" in work that encourages viewers to question their apparent reality. In past installations, he has examined the consequences of sensory deprivation and considered new ways of moving through the world, the latter epitomized in *Test Site* (2006–7), his installation of five spiraling, tubular slides in the Turbine Hall of the Tate Modern in London, connecting the ground level to the upper floors. The artist has also turned his scientific eye to the possibilities of perception and is particularly drawn to mushrooms, which feature frequently in his work, for their hallucinogenic properties. Poised on the boundaries of the familiar and unknown, they hold the potential to transport humans to another world, and their unexpected shapes and colors present a jarring contrast to mundane reality. In 2000 Höller dove into these surreal fungal possibilities in his exhibition *Synchro System* at Milan's Fondazione Prada, forcefully shifting viewers' impressions by hanging a crop of psychedelic mushrooms upside down from the ceiling. Invited to walk underneath them, viewers journeyed through a multisensory mushroom maze, accompanied by lights switching on and off and an occasionally bewildering darkness. Höller's aim—to estrange viewers from what they know to be real while "synchronizing" them with the artwork—encourages a new and unfamiliar discombobulating physical and mental state.

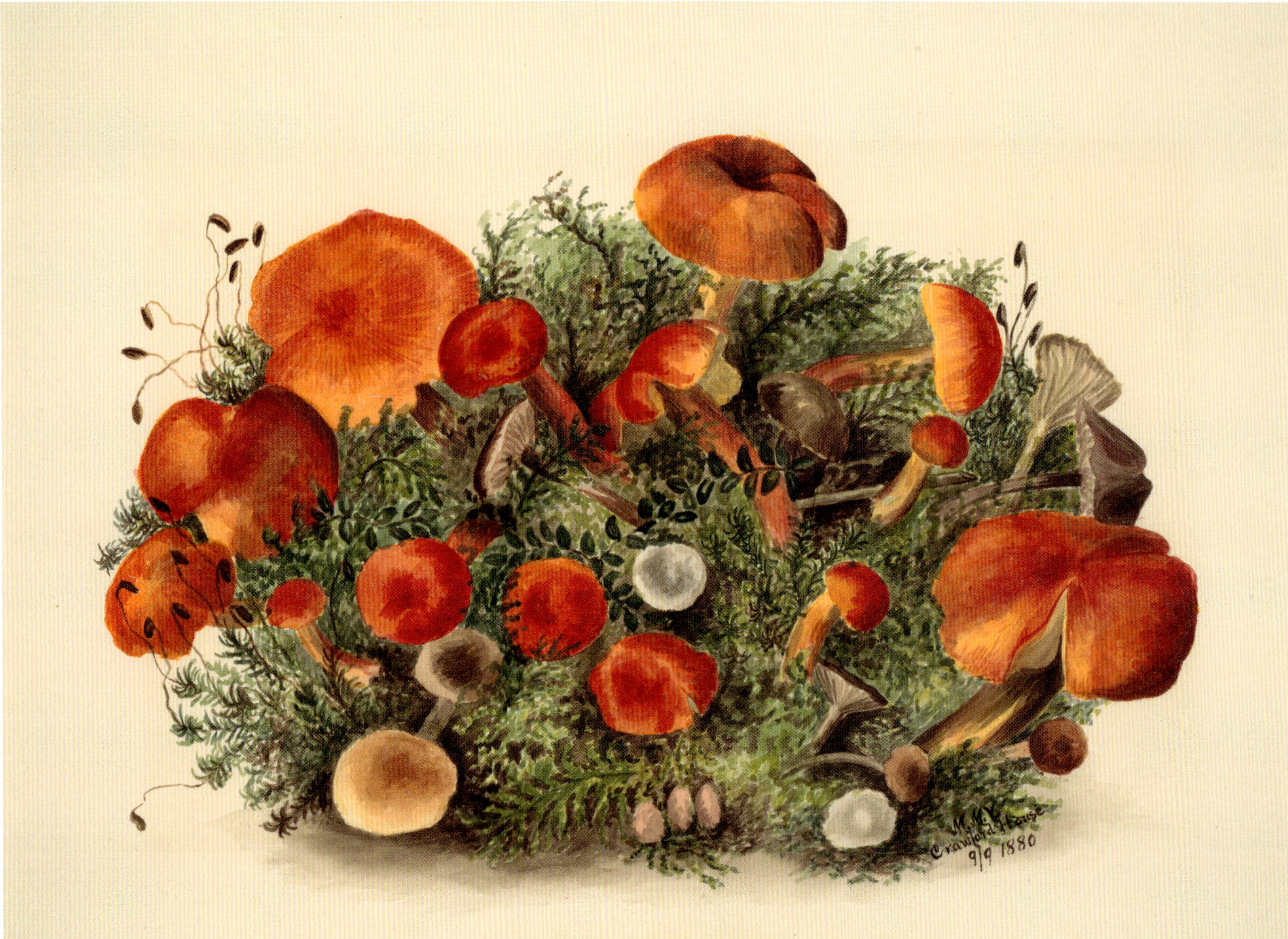

MARY VAUX WALCOTT

Untitled—Study of Mushrooms, 1880
Watercolor on paper, $9\frac{5}{8} \times 12\frac{5}{8}$ in. / 24.6 × 32.2 cm
Smithsonian American Art Museum, Washington, DC

This untitled study leaps off the page with vivid colors and a dense composition, capturing a cluster of mushrooms enveloped in the lush foliage of the forest floor. The work depicts several species, with their sturdy stems and soft caps rendered in bold shades of red and orange, muted browns, and creamy whites. The image gently suggests shadows, with variations in texture and tone created through subtle washes, and occasional striations lend the work a sketchlike quality. The tangle of overlapping greens is imprecise but evocative, enhancing the vibrancy of the reds and oranges and gently moving the eye through the composition to areas of respite among the soft creams and browns. The artist, Mary Vaux Walcott (1860–1940), is an American naturalist and painter best known for her botanical watercolors, including more than four hundred focused on North American wildflowers, published as a five-volume set in 1925 by the Smithsonian Institution. She traveled extensively throughout her life, documenting flora, minerals, glaciers, and other natural history subjects through photography and watercolor, and recently several specimens in the Smithsonian collections have been attributed to her collecting after years of being identified as belonging to her husband. This tender study—created very early in her career and when she was just beginning her life as an explorer of the far reaches of North America—is not a fanciful depiction but rather a bridge between art and science. We see a young artist carefully examining the scene and rendering it with a blend of curiosity, admiration, and a desire to document the natural world with fidelity.

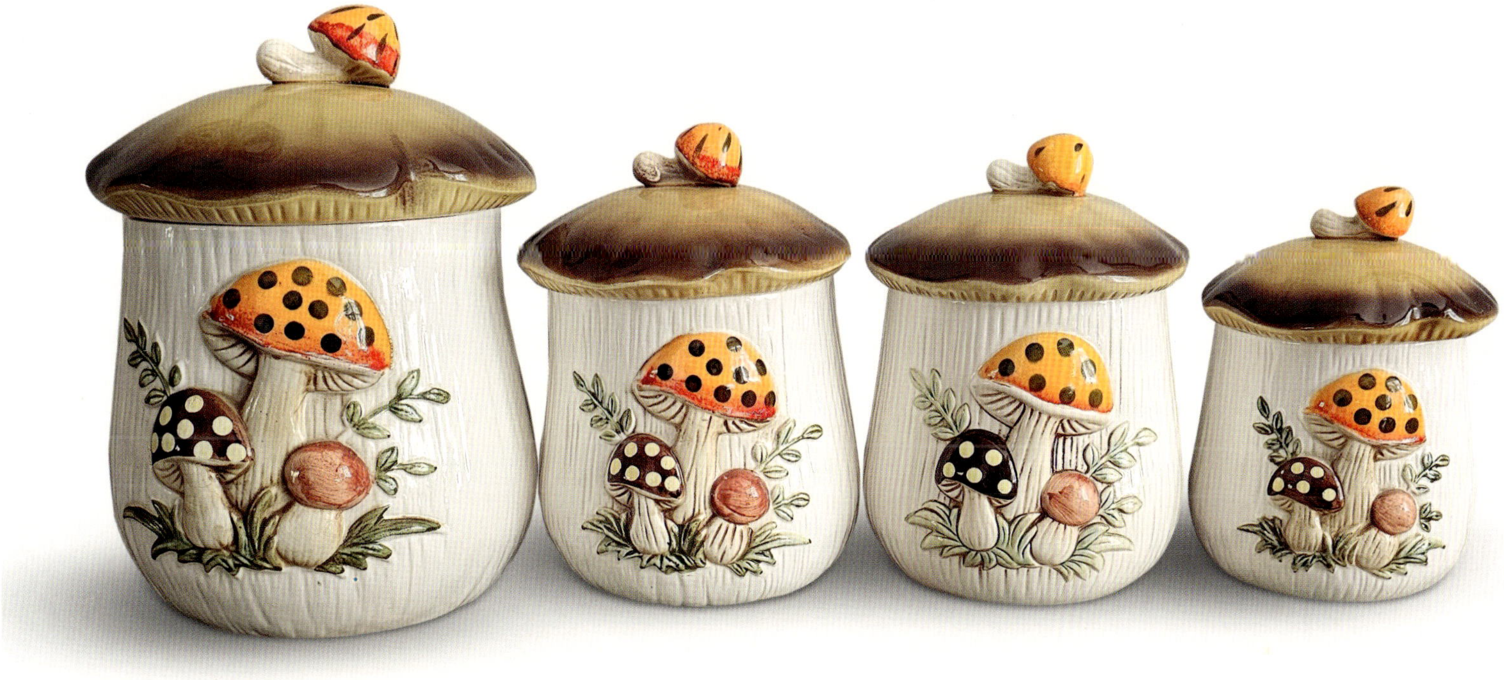

SEARS, ROEBUCK & CO.

Merry Mushrooms Canister Set, 1978
Embossed, hand-decorated ceramicware, left to right:
11 × 6¼ in. / 27.9 × 15.9 cm; 9 × 5¼ in. / 22.9 × 13.3 cm;
8½ × 5 in. / 21.6 × 12.7 cm; 7½ × 4¼ in. / 19 × 10.8 cm
Private collection

When thinking of 1970s American kitsch, one conjures the burnt oranges, avocado greens, chocolate browns, and mustard yellows that dominated domestic interiors during the decade. And perhaps no household decor is more emblematic of that groovy '70s style than Merry Mushrooms Kitchen Coordinates. Sold between 1970 and 1987 by the retail giant Sears, Roebuck & Co., the line encompassed everything from the canisters seen here to cookware, dishes, planters, cutlery, clocks, and beyond. All told, the retailer offered more than 250 products in its Merry Mushrooms line over the course of its overwhelmingly popular run. The ceramic pieces, all made in Japan, were embossed, hand-painted, and kiln-fired. The most common motif depicts a trio of cartoonlike fungi set against some greenery: one larger mushroom orange-capped and speckled brown, flanked by a spotted brown mushroom on one side—both recalling the fly agaric (*Amanita muscaria*)—and a lighter brown version resembling those most commonly found in an American diet. And although the number of fungi and colors varied over time, the orange/brown/gold/green color scheme remained a constant. Merry Mushrooms have been spotlighted in pop culture, fixtures of the Bradford family's kitchen on TV's *Eight Is Enough* (1977–81) and the Torrance family's Boulder apartment in Stanley Kubrick's *The Shining* (1980), among others. They're also highly sought-after by collectors today, whether waxing nostalgic or simply looking for a hint of cottagecore to add to their home.

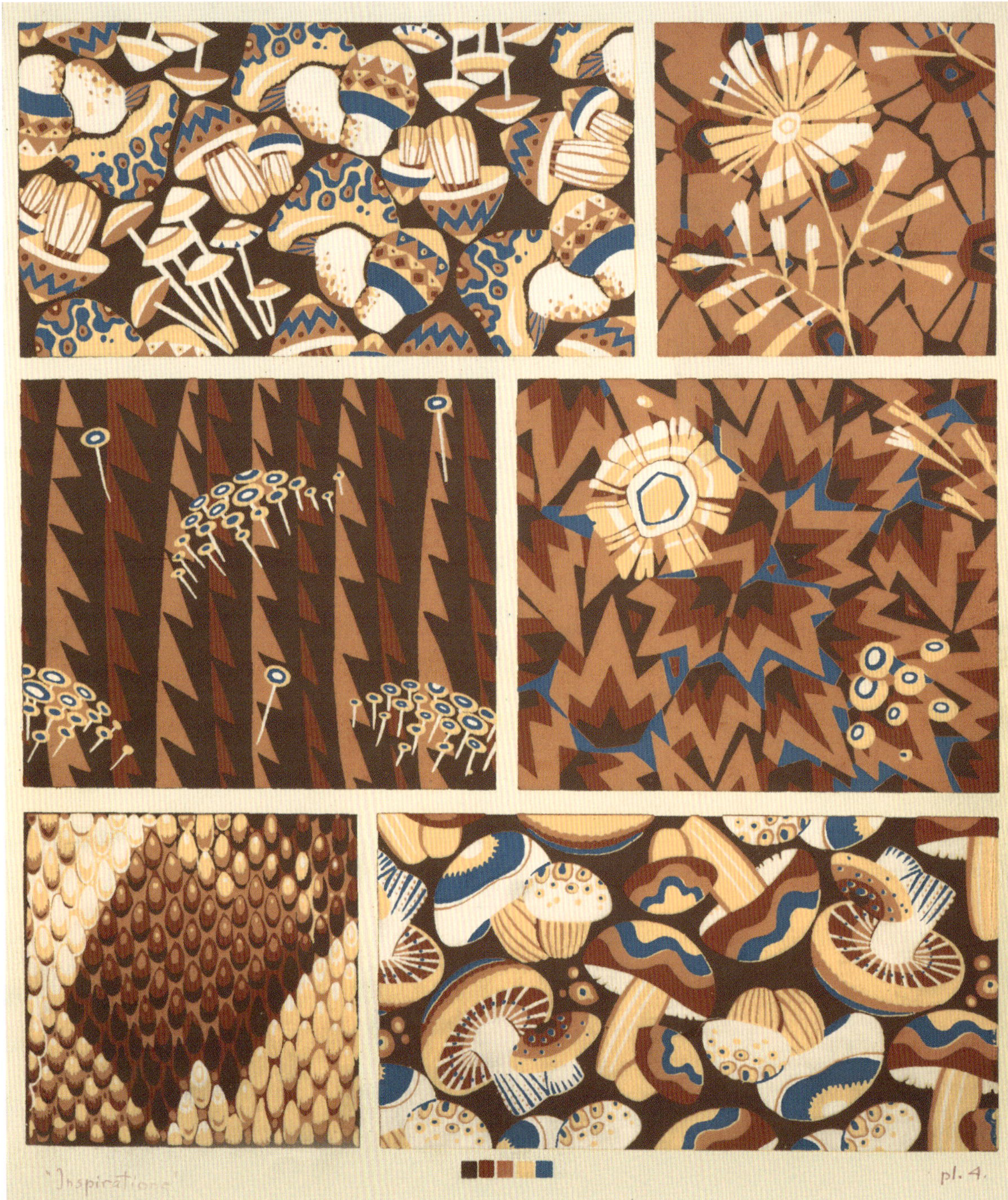

ANDRÉ DURENCEAU

Plate 4, from *Inspirations: Twenty-four Plates, with 128 Compositions*, 1928
Screenprint, $17\frac{3}{8} \times 14\frac{1}{8}$ in. / 44 × 36 cm
New York Public Library

Stylized mushrooms and plants, rendered in a riot of blue, brown, and yellow patterns, pop from the page in this 1920s design by French-born artist André Durenceau (1904–1985). Durenceau studied at the École des Beaux-Arts in Paris before immigrating to the United States in 1922 at the age of eighteen. He found his way to Hollywood, then in its golden age, where he worked as a colorist at Technicolor. Through the social connections of his wife and fellow artist Kay Kaplan, he became a muralist and decorator to the rich and famous, including Joan Crawford and Lionel Barrymore. Durenceau continued that work after moving to New York in 1933, hired by high-profile clients and taking commissions, including the 1939 New York World's Fair. Among his numerous illustrations for books and magazines was his 1928 publication *Inspirations: Twenty-four Plates, with 128 Compositions,* from which this plate originates. A gorgeous example of early twentieth-century screenprinting, the intent of the volume was to revolutionize modern industrial design by presenting appealing, graphic prints with maximum effect and minimum process, using at times just three colors and no more than six in each composition, thereby simplifying the application process. Gone were the days of color blending and half-tones, announced the publisher. Flat colors and simplified forms were the weapons in the modernist's arsenal, meant to arrest the attention of the viewer, whether in graphic design, textile art, or otherwise. Durenceau's kinetic, geometric plates in this book present a snapshot of a confident Art Deco artist at the top of his form.

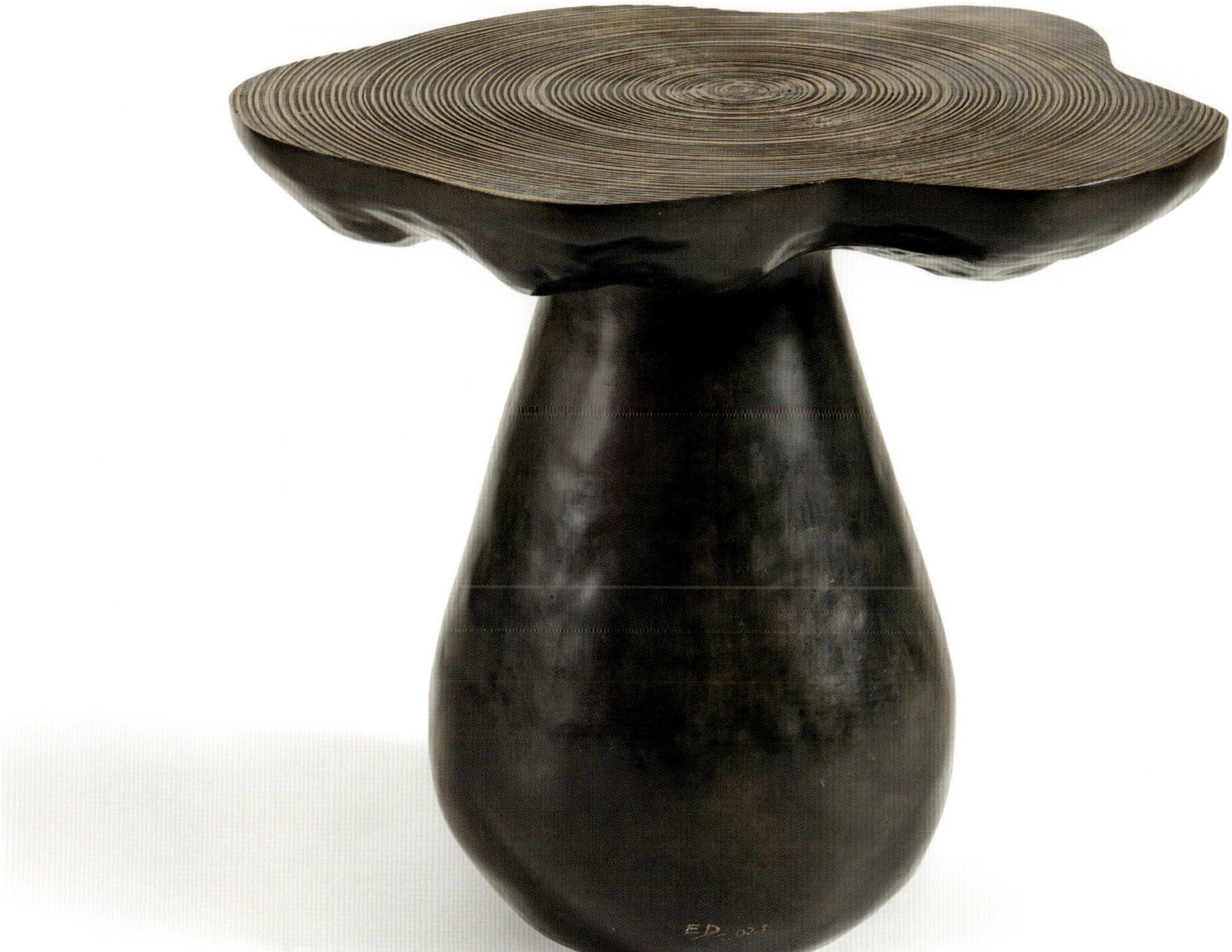

EMMA DONNERSBERG

Organika Spiral Top Table, 2022
Bronze, 24 × 26 × 22 in. / 61 × 66 × 55.9 cm
Private collection

Combining functionality with fine-art sculpture, designer Emma Donnersberg (b. 1984) reimagines the mushroom form in this bronze table, with its bulbous stem and spiral-etched, wavy cap. Donnersberg left her native Montmorency, France, for the United States after high school, first earning her bachelor's degree from Tulane University in New Orleans and then attending the New School's Parsons School of Design in New York. She landed a job at the auction powerhouse Christie's, splitting her time between their New York and Paris offices while cultivating her own design practice. She founded her company, E Donnersberg Interiors, in 2008, for which she maintains offices in Paris and New York, working with residential and commercial clients in Europe and North America. Informed by her worldwide travels, and often borrowing forms found in nature—clouds, canyons, and, as here, mushrooms—Donnersberg's pieces exist simultaneously in the worlds of art and design, pairing whimsical dynamism with functional luxury. She has said that her ultimate goal is to create movement with her pieces: chairs and couches resembling (and named after) clouds seem to float, offering pillow-soft comfort; multipiece furniture made of wood or travertine offers the option of joining or separating; and even here, in the stationary form of a mushroom, we see concentric circles decorating the top, as if a pebble has just disturbed the surface of a still pond.

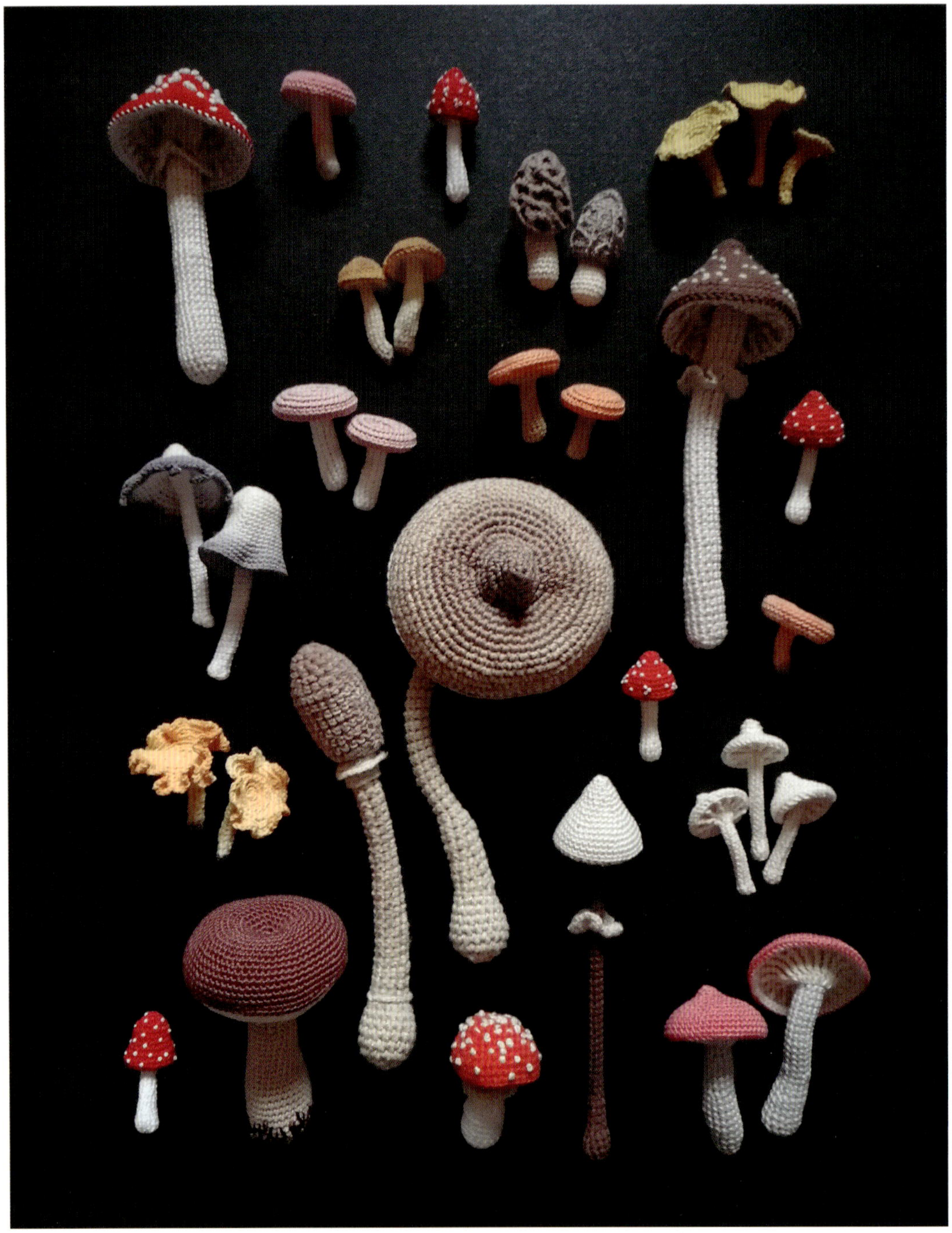

MARIANNE SEIMAN

Mushrooms, 2022
Crocheted yarn,
dimensions variable

Set against a solid black background, dozens of intricately crocheted fungi are arranged like herbarium specimens. These plush creations come from the hand of Estonian crochet artist Marianne Seiman (b. 1975), who was first inspired to craft mushrooms as part of the 100 Day Project in 2019, making a single mushroom each day for more than three months straight. The 100 Day Project was created by venerable designer and Yale professor Michael Bierut, who issued the challenge to his undergraduate students to engage in a single, repeatable design action once daily for a hundred straight days, recording the results and tracking their progress as a practitioner of graphic design. The concept proliferated when creators, including author Elle Luna, introduced the concept to social media in 2013. Today, it is a worldwide phenomenon, and #The100DayProject celebrates makers of all stripes: illustrators, painters, sculptors, quilters, photographers, crafters, and more. Seiman loved the endeavor so much that she continues to add to her collection of crocheted fungi. Some of them are inspired by nature and rendered in painstaking detail to look like their real-life counterparts, such as the golden chanterelles (*Cantharellus cibarius*), with their delicately scalloped edges and blazing yellow flesh, and the morels (*Morchella*), with their unmistakable craggy cap. Others in the series are vivid creations of the artist's imagination, stitched in a kaleidoscopic spectrum of colorways. Seiman's sense of whimsy extends beyond her textile fungarium to crochet subjects from fishbones and ashtrays with spent cigarettes to heaping plates of green salad and fluorescent pink flamingos.

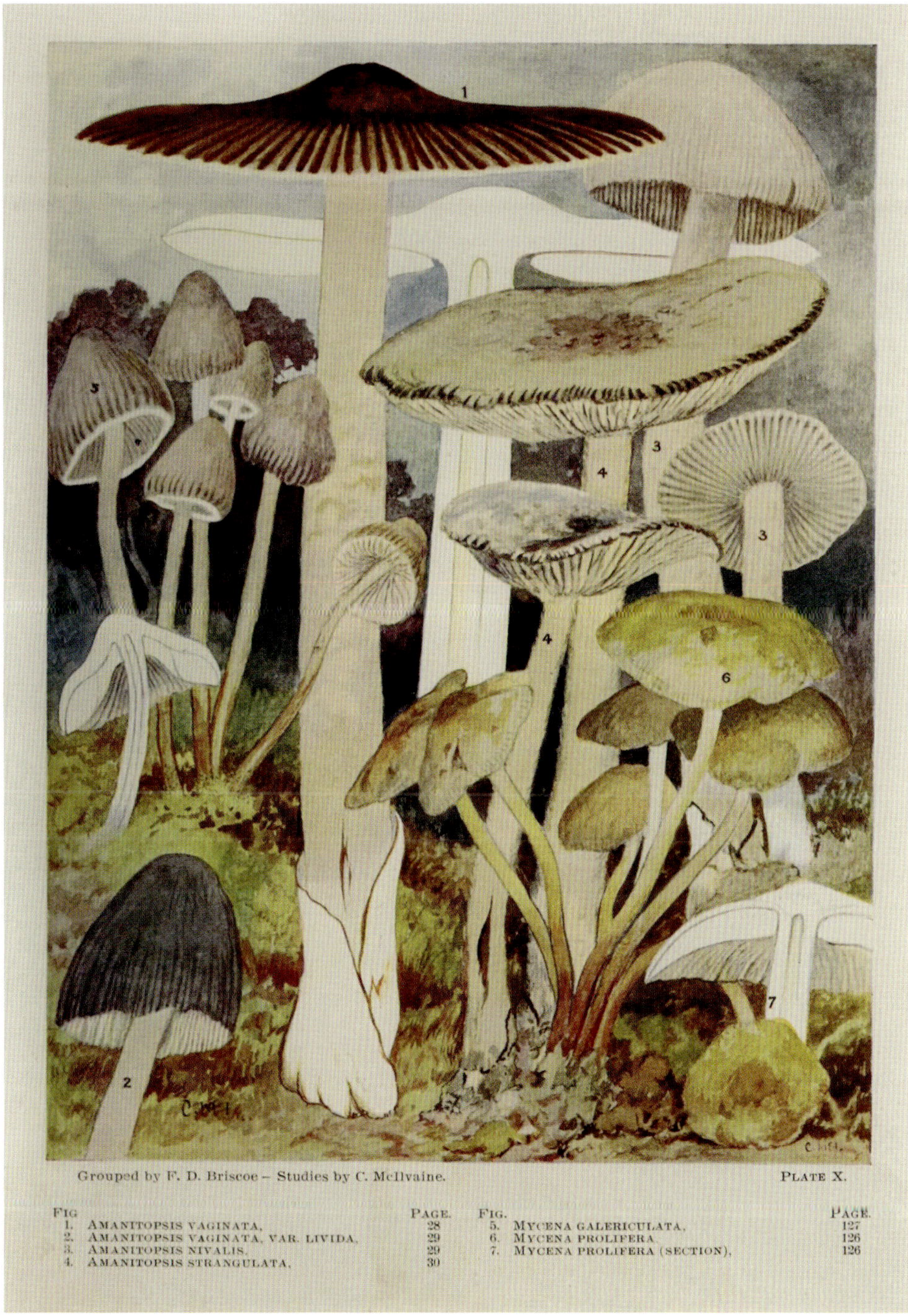

CHARLES McILVAINE

Plate X, from *One Thousand American Fungi*, 1900
Chromolithograph, H. 11 in. / 28 cm
Wellcome Collection, London

"For twenty years my little friends—the toadstools—have been my constant companions," wrote pioneering mycologist Charles McIlvaine (1840–1909) in the introduction to *One Thousand American Fungi*. The detailed illustrated guide to the edible and poisonous fungi of America, in which this hand-drawn illustration depicting various species of the *Amanita* and *Mycena* genera appears, was first published in 1900 and significantly revised in 1902. A seminal work of American mycology, it is based on years of personal research and experimentation. McIlvaine believed in getting his hands dirty, and he took it upon himself to sample hundreds of species, including several—such as *Hypholoma fasciculare* and *Russula emetica*—that are highly toxic, earning him the nickname Old Iron Guts. His interest in mycology was sparked by encounters with myriad mushrooms in the West Virginia mountains, with his book stemming from his frustration at finding no standardized text on American fungi. Remedying this, alongside precise descriptions of different species, his main text contains a wealth of miscellaneous mushroom information, including guidelines for the cultivation and preparation of edible species. As an authority on American fungi, McIlvaine presided over the Philadelphia Mycological Center, which published a regular bulletin of his findings. Furthermore, the North American Mycological Association's biannual journal *McIlvainea* is named in his honor. McIlvaine's book found a new audience of mushroom enthusiasts when it was republished in the 1970s, and it remains a comprehensive resource, useful to mycologists, amateur foragers, and culinary enthusiasts alike.

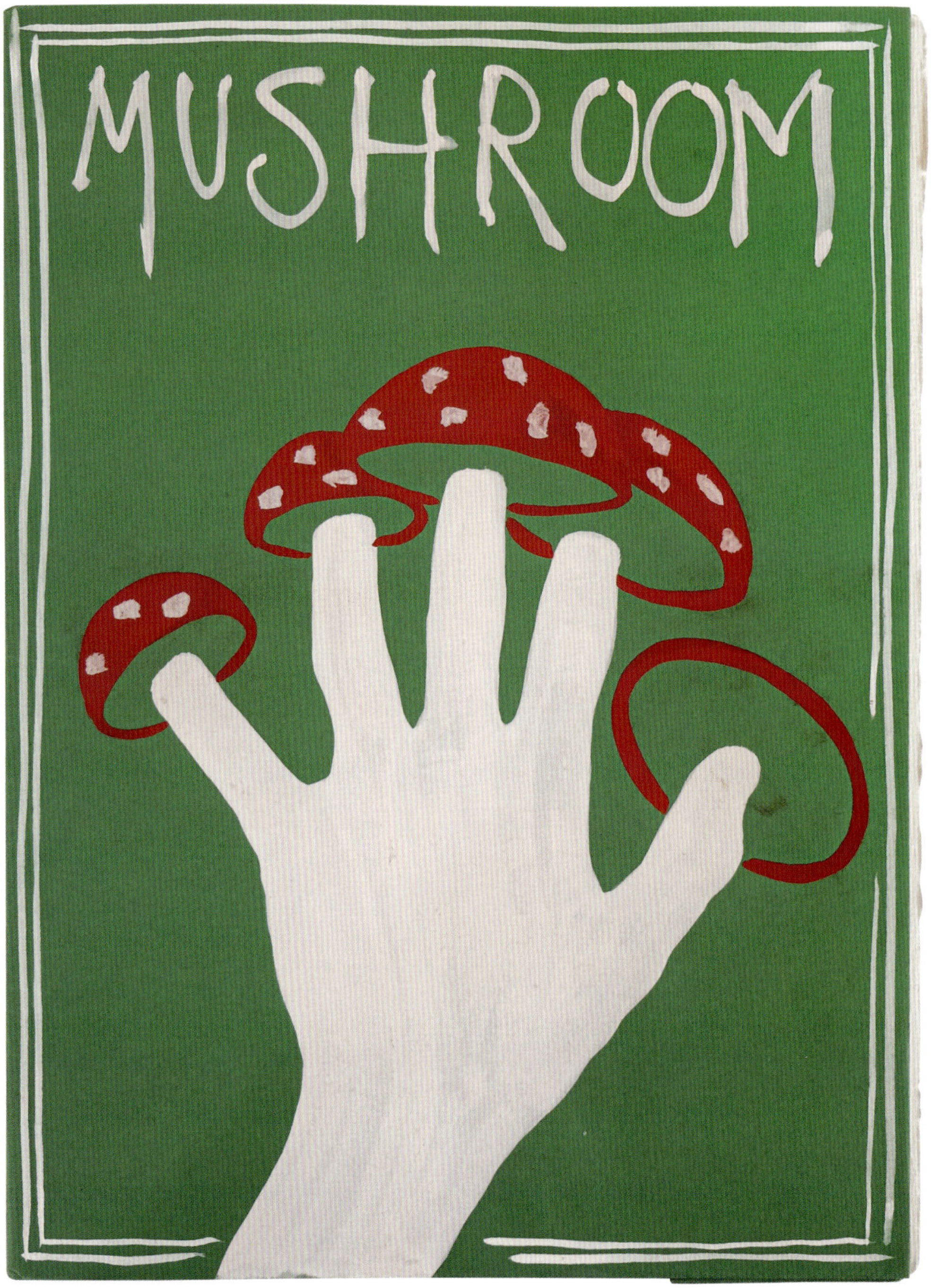

SIMON POPPER

Mushroom Fingers, 2010
Handmade book with watercolor,
14 × 9⅞ in. / 35.5 × 25 cm
Private collection, London

Set against a solid green background, five white-spotted, red mushroom caps balance on the fingertips of a hand rendered starkly in white. This boldly graphic cover was painted by eclectic British artist Simon Popper (b. 1977) for his handmade book *Mushroom Fingers*, published in a limited edition of three. Beneath the striking cover, the book contains handwritten recipes for mushroom dishes, filling each page in a beautiful, oversize script painted in watercolor. The instructional texts are interspersed with pictures of mushrooms, echoing the graphic style of contemporary artists such as Enrico David and Silke Otto-Knapp. Like both David and Otto-Knapp, Popper has developed a fluid practice that juggles aesthetics and a playful grounding in big ideas around structure, value, and meaning. After completing *Mushroom Fingers* in 2010, Popper collaborated with British chef Jeremy Lee on a similar project about potatoes in 2018. *Mushroom Fingers* is not the only fungi-inspired artwork created by Popper. For more than ten years, beginning in 2006, Popper collected international postage stamps that depict mushrooms. The hundreds of stamps were first displayed as a large table-based installation, *Mycology Philately*, at Flowers Gallery and later on the wall in the 2020 exhibition *Mushrooms* at Somerset House in London. Both works reflect Popper's wider interest in language and industry, which are an essential part of his artistic practice, alongside his often playful aesthetic and bold color palettes.

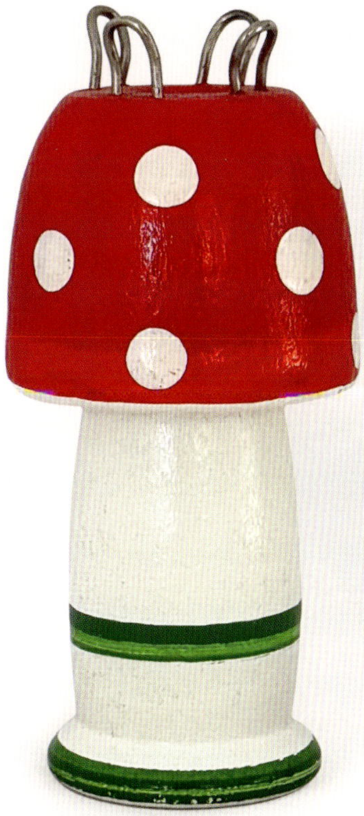

WHITMAN PUBLISHING CO.

Polly Ann Knitting Knob, 1939
Wood and metal, 3 × 1⅛ in. / 7.6 × 2.9 cm
Private collection

A tool of many names—knitting mushroom, French knitter, knitting nancy, knitting bobbin, knitting spool, knitting jenny, just to name a few—this simple but effective device has a singular purpose: to simplify knitting. The nifty device is an essential aid for fledgling knitters who wish to learn the craft. Popularized in the early twentieth century and marketed to children, these knitting helpers often take whimsical and cheerful forms. There are myriad variations, but the most common are small dolls and mushrooms, frequently resembling the red-and-white-capped fly agaric (*Amanita muscaria*), such as here. Indeed, the mushroom's stem and cap provide the perfect ergonomic balance in the knitting process. Typically carved from wood, these devices feature four pegs or loops at the top to hold the yarn, with a handle below. To start the process, one "casts on," wrapping the yarn once around each loop, working counterclockwise, until two full revolutions have been completed. Then, using a pick or by hand, the knitter pulls the bottom loop of yarn over the top loop and off each peg before restringing each with a fresh loop of yarn. The knitting mushroom has a hollow center, providing a channel through which the knitted length can feed as it increases in size. The process of tying on and looping bottom over top repeats over and over until the intended outcome is reached. The result is a neatly and uniformly knitted tube called an I-cord, which can then be twisted or joined together to form any number of knitted creations.

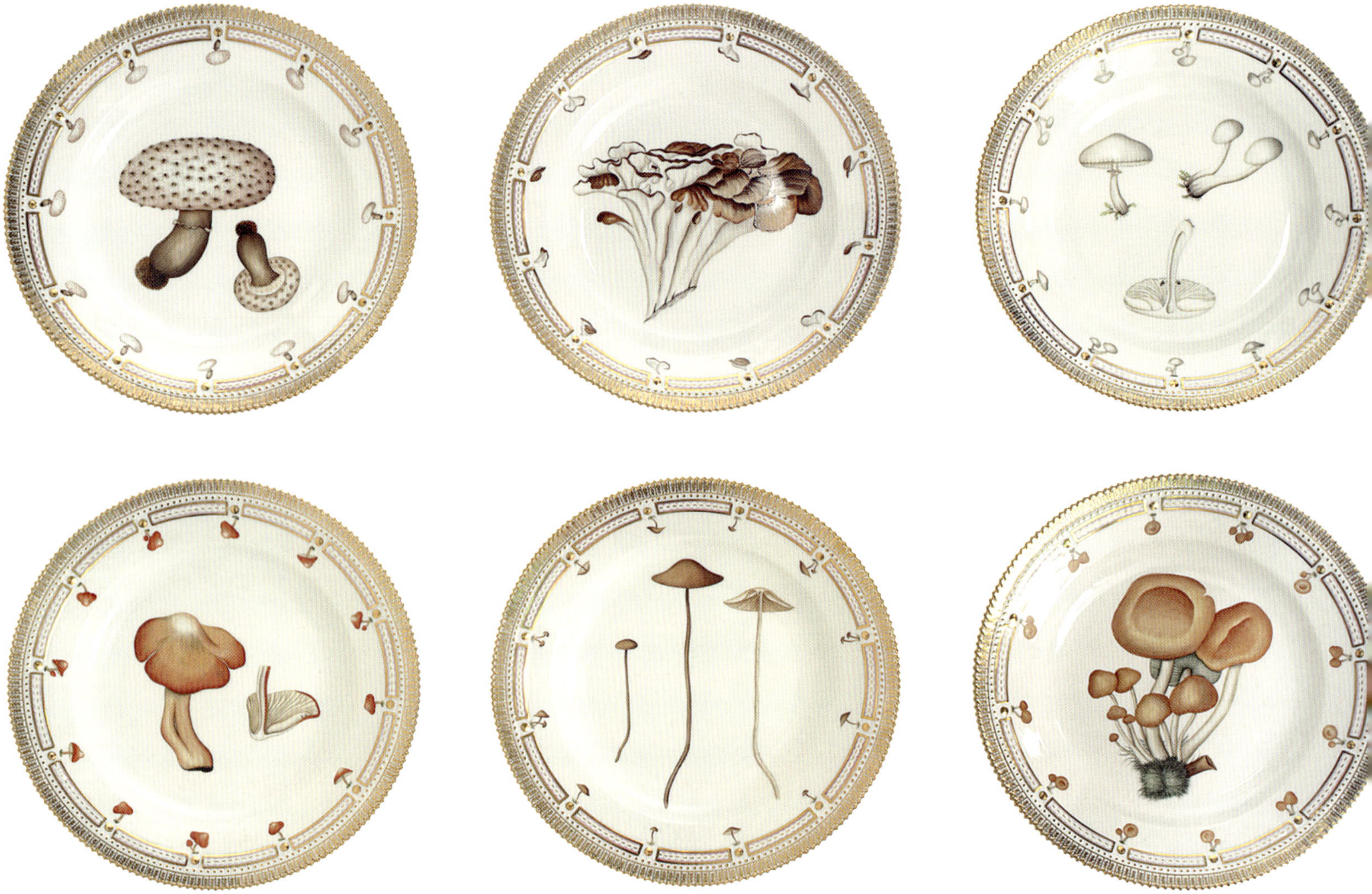

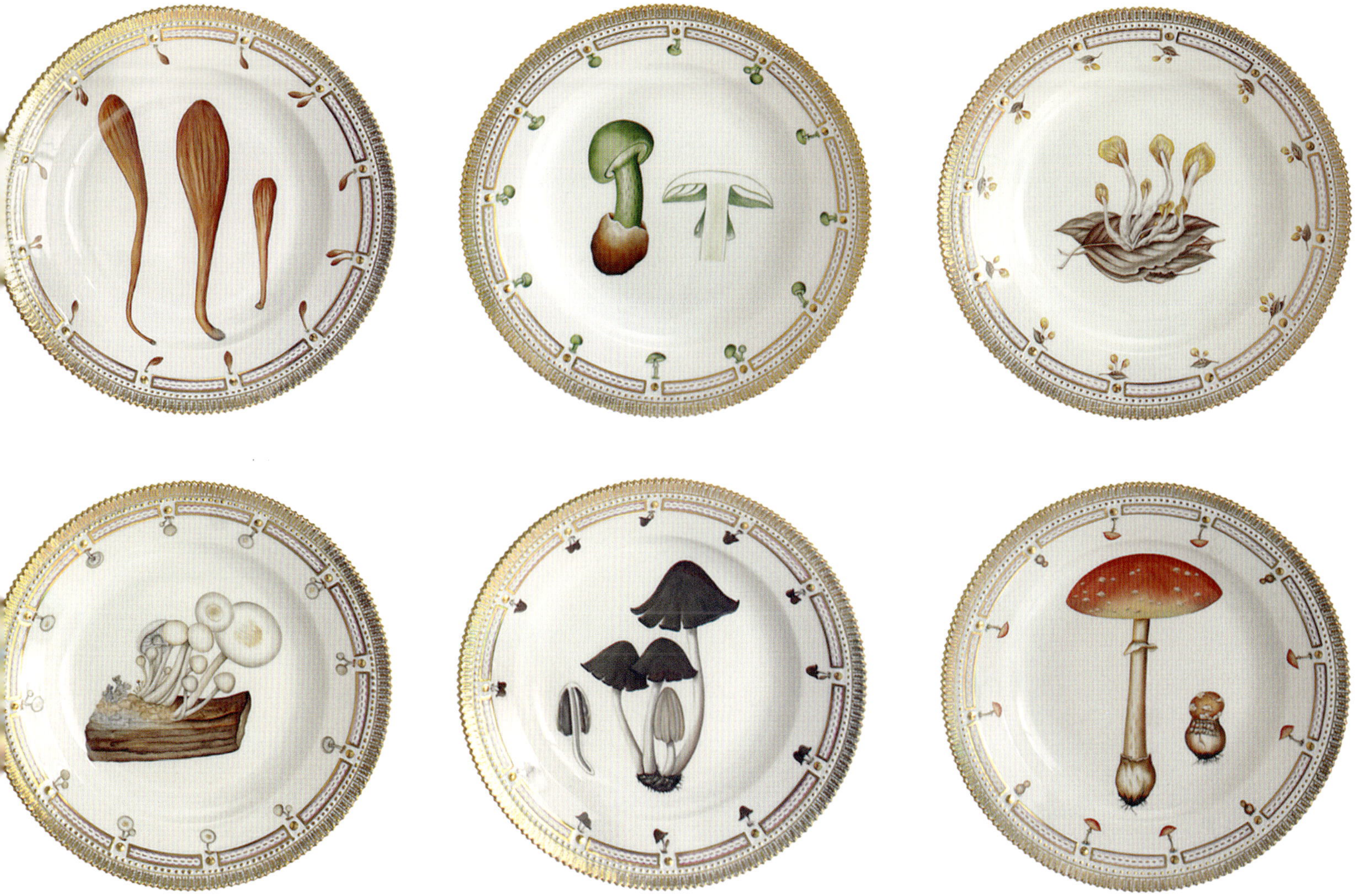

ROYAL COPENHAGEN

Set of mushroom plates, undated
Hand-painted porcelain, Diam. 10 in. / 25.4 cm each
Private collection

At the center of each of these porcelain dinner plates, enclosed within a serrated gilt-edged border, sits a hand-painted illustration of a different mushroom specimen. The plates were made by the Danish ceramics firm Royal Copenhagen, and belong to its esteemed *Flora Danica* collection. The illustrations are based on drawings from the *Flora Danica*, an ambitious undertaking begun during the Age of Enlightenment to record in book form all the flora of the Danish kingdom, which at that time included Norway and Greenland. In 1752 German botanist Georg Christian Oeder was commissioned by the prince regent, Crown Prince Frederick of Denmark, on behalf of his father, King Christian VII, to catalog and illustrate the kingdom's entire plant array of life. The encyclopedia ultimately took 122 years to complete and includes 3,240 flowers and plants depicted with a high level of botanical accuracy. The mushroom plates were first painted in 1790 as part of a gift for the Empress Catherine the Great of Russia, but it was not until 2007 that a new collection of mushroom porcelain was created. For that collection, twenty-six fungi were drawn using a color palette of grays, browns, and bright reds. The mushroom theme continues around the edge of the plates, where the same motif is repeated in miniature. On the reverse side of each piece, the Latin name of the fungus it depicts is written. With their extraordinary level of detail and accuracy, the mushroom plates have become highly sought-after by collectors and have found their way into many of the leading global auction houses.

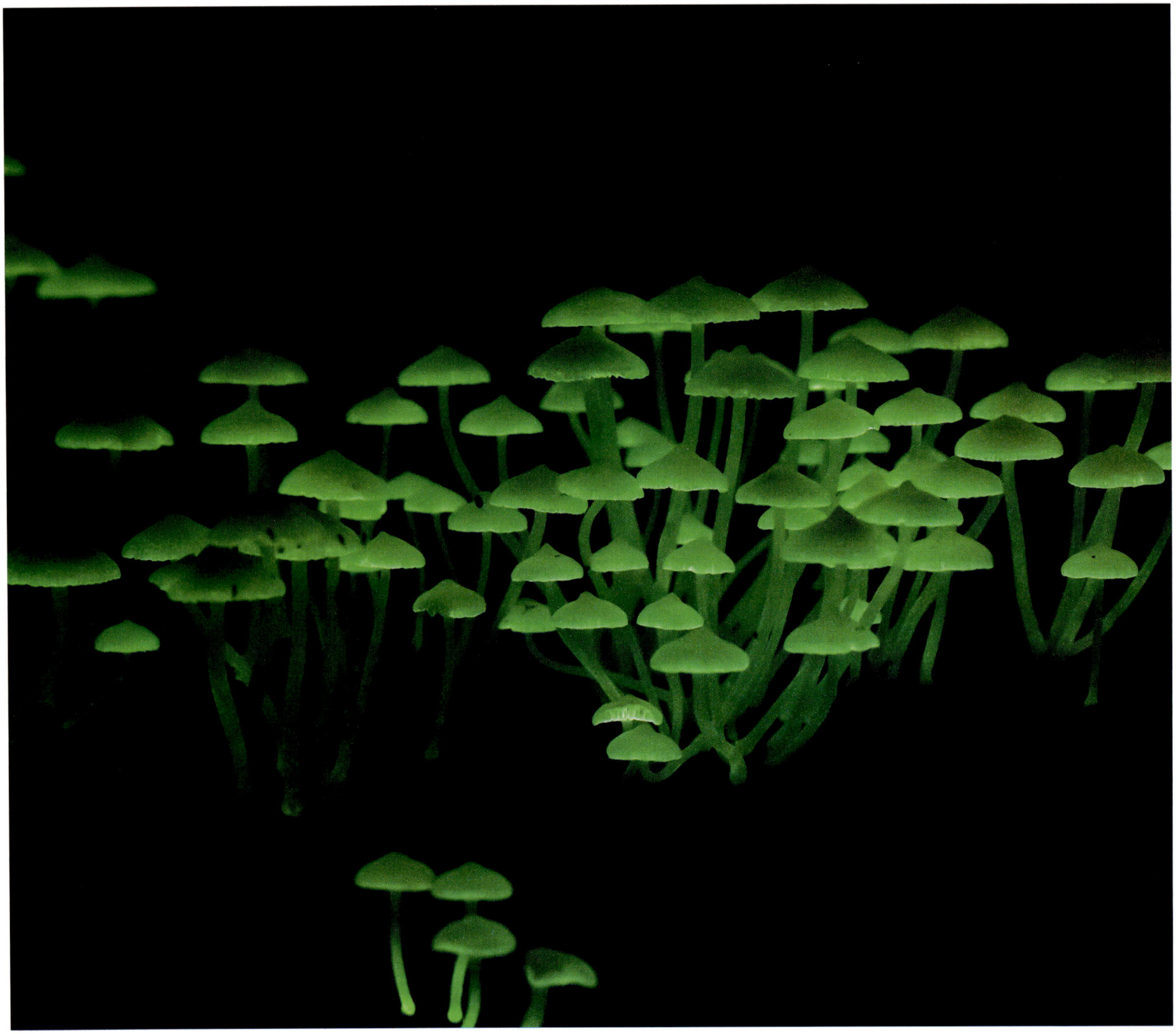

MELVYN YEO

Bioluminescent Mushrooms, 2016
Photograph, dimensions variable

These ethereal, seemingly fantastical clusters of glowing mushrooms were captured at night in a Singapore forest by photographer Melvyn Yeo, just as the exact conditions aligned for one of nature's most spectacular shows: bioluminescence. There are more than one hundred distinct species of fungi with bioluminescent abilities. Most bioluminescent fungi grow in warm, tropical climates from Brazil to Southeast Asia, but there are in excess of two dozen species native to the United States, spanning from the West Coast to New England. The majority of bioluminescent fungi belong to the order Mycena, such as the species of agaric fungus seen here, *Filoboletus manipularis*, which grows in forests throughout Australasia, Malaysia, and the Pacific Islands. These mushrooms contain the chemical luciferin (present in all bioluminescent species across plants, animals, and fungi alike), which combines with certain enzymes and oxygen to create oxyluciferin. This process causes the emission of light—a soft, greenish glow that can be seen by the naked eye under the right conditions. The fungus's bioluminescence is self-serving: its radiance attracts insects and arthropods, which then dislodge and distribute the mushroom's spores, thereby increasing its range of proliferation. While bioluminescence contributes to the mushroom's survival, it has also made them a ripe source for speculation throughout history. In Japanese folklore, the eerie green glow was thought to have been caused by *yōkai*, or supernatural creatures, and as far back as 382 BCE, the Greek philosopher Aristotle referred to it as "cold fire."

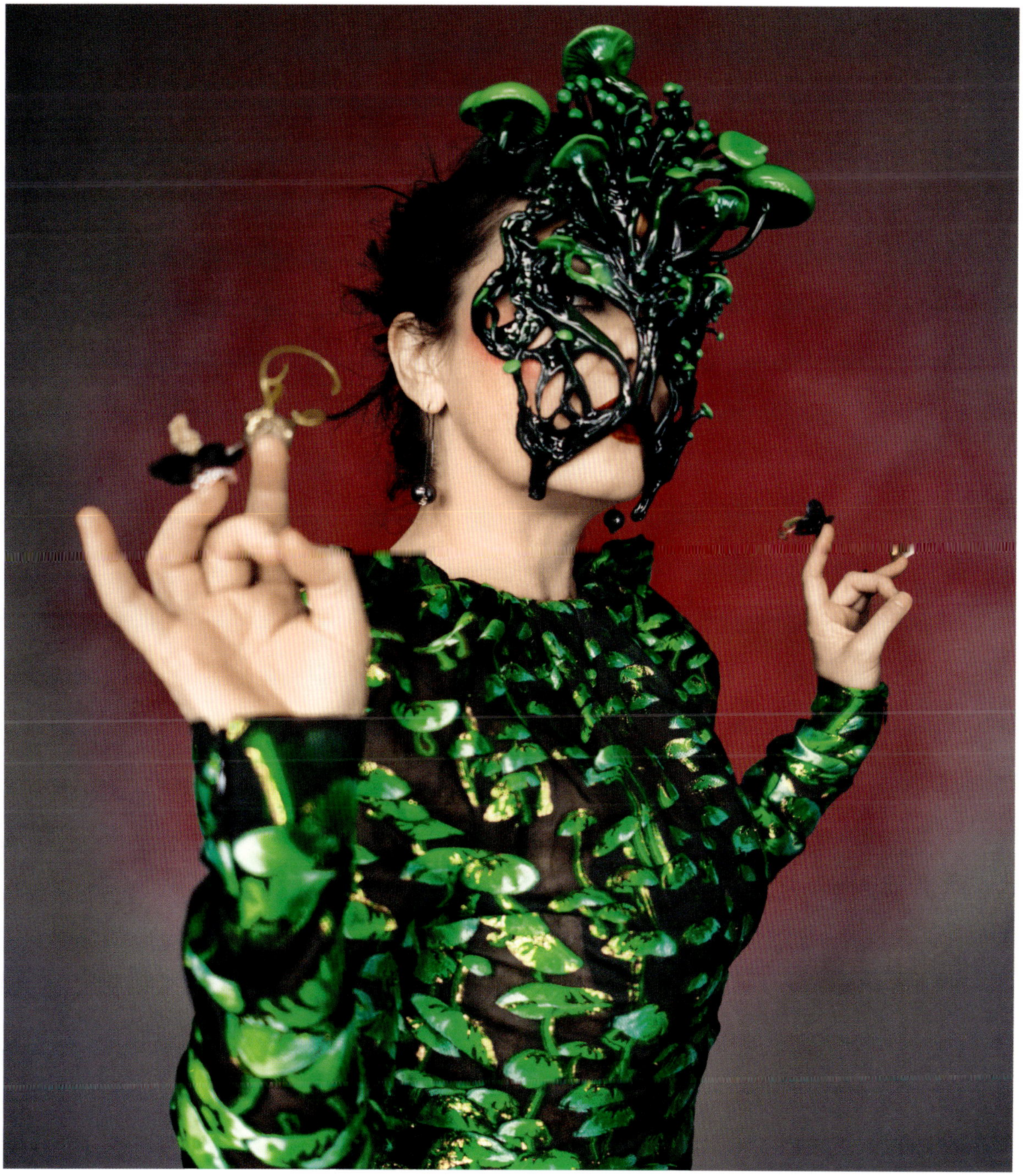

DANIEL DEL CORE IN COLLABORATION WITH DAVID ÅBERG

Mushroom Mask, 2021
Stereolithography, resin, air-brush, and lacquer,
10⅝ × 8⅞ × 7⅜ in. / 27 × 22.5 × 18.7 cm

As a musical artist, Björk is a master of invention and reinvention, moving deftly between genres and styles, constantly pushing the envelope in the pursuit of transcendence. And in the annals of fashion risk-takers, she is similarly fearless—and peerless. This fantastic fungi ensemble fits in beautifully with the artist's audacious oeuvre. Worn for the musician's launch of the album *Fossora* (itself fueled by the artist's interest in mushrooms), the dress and mask were featured at the Nobel Prize Museum in Stockholm in *Fungi—in Art and Science*, an exhibition celebrating mushrooms' wide-ranging influences in art and culture. Daniel Del Core (b. 1988), who designed for Gucci before establishing his own eponymous brand at just age thirty-two, created a stunning debut collection inspired by fungi and other wonders of the natural world. The full-length black-and-green couture gown pictured here was sparked by the idea of a field of bioluminescent mushrooms at night. The materials, Lurex and transparent nylon, further reinforce the otherworldly greenish glow. To complete the look, Del Core enlisted Swedish artist David Åberg (b. 1977), who crafted several pieces for the collection, including the mushroom mask seen here. Combining technology with traditional sculptural techniques, Åberg uses 3D-printing technology to create works that, in the artist's words, blur the line between dream and reality, something this surrealistic, face-hugging creation most elegantly exemplifies.

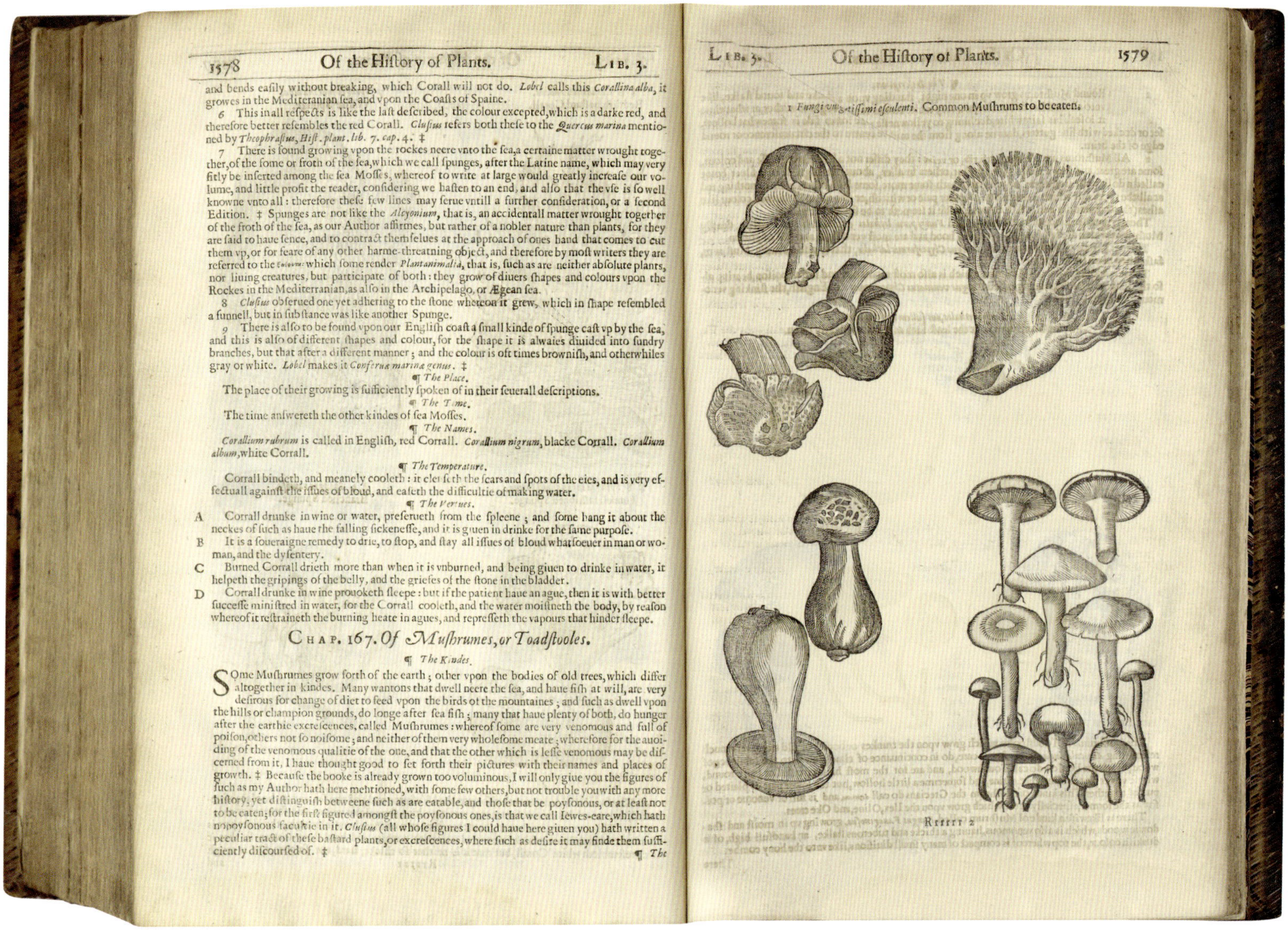
1578 Of the Hiſtory of Plants. Lib. 3.

and bends eaſily without breaking, which Corall will not do. *Lobel* calls this *Corallina alba*, it growes in the Mediteranian ſea, and vpon the Coaſts of Spaine.

6 This in all reſpects is like the laſt deſcribed, the colour excepted, which is a darke red, and therefore better reſembles the red Corall. *Cluſius* refers both theſe to the *Quercus marina* mentioned by *Theophraſtus, Hiſt. plant. lib. 7. cap. 4.* ‡

7 There is found growing vpon the rockes neere vnto the ſea, a certaine matter wrought together, of the fome or froth of the ſea, which we call ſpunges, after the Latine name, which may very fitly be inſerted among the ſea Moſſes, whereof to write at large would greatly increaſe our volume, and little profit the reader, conſidering we haſten to an end, and alſo that the vſe is ſo well knowne vnto all: therefore theſe few lines may ſerue vntill a further conſideration, or a ſecond Edition. ‡ Spunges are not like the *Alcyonium*, that is, an accidentall matter wrought together of the froth of the ſea, as our Author affirmes, but rather of a nobler nature than plants, for they are ſaid to haue ſence, and to contract themſelues at the approach of ones hand that comes to cut them vp, or for feare of any other harme-threatning object, and therefore by moſt writers they are referred to the Ζωόφυτα which ſome render *Plantanimalia*, that is, ſuch as are neither abſolute plants, nor liuing creatures, but participate of both: they grow of diuers ſhapes and colours vpon the Rockes in the Mediterranian, as alſo in the Archipelago, or Ægean ſea.

8 *Cluſius* obſerued one yet adhering to the ſtone whereon it grew, which in ſhape reſembled a funnell, but in ſubſtance was like another Spunge.

9 There is alſo to be found vpon our Engliſh coaſt a ſmall kinde of ſpunge caſt vp by the ſea, and this is alſo of different ſhapes and colour, for the ſhape it is alwaies diuided into ſundry branches, but that after a different manner; and the colour is oft times browniſh, and otherwhiles gray or white. *Lobel* makes it *Conferuæ marinæ genus.* ‡

¶ *The Place.*

The place of their growing is ſufficiently ſpoken of in their ſeuerall deſcriptions.

¶ *The Time.*

The time anſwereth the other kindes of ſea Moſſes.

¶ *The Names.*

Corallium rubrum is called in Engliſh, red Corrall. *Corallium nigrum*, blacke Corrall. *Corallium album*, white Corrall.

¶ *The Temperature.*

Corrall bindeth, and meanely cooleth: it cleanſeth the ſcars and ſpots of the eies, and is very effectuall againſt the iſſues of bloud, and eaſeth the difficultie of making water.

¶ *The Vertues.*

A Corrall drunke in wine or water, preſerueth from the ſpleene; and ſome hang it about the neckes of ſuch as haue the falling ſickeneſſe, and it is giuen in drinke for the ſame purpoſe.

B It is a ſoueraigne remedy to drie, to ſtop, and ſtay all iſſues of bloud whatſoeuer in man or woman, and the dyſentery.

C Burned Corrall drieth more than when it is vnburned, and being giuen to drinke in water, it helpeth the gripings of the belly, and the griefes of the ſtone in the bladder.

D Corrall drunke in wine prouoketh ſleepe: but if the patient haue an ague, then it is with better ſucceſſe miniſtred in water, for the Corrall cooleth, and the water moiſtneth the body, by reaſon whereof it reſtraineth the burning heate in agues, and repreſſeth the vapours that hinder ſleepe.

Chap. 167. *Of Muſhrumes, or Toadſtooles.*

¶ *The Kindes.*

SOme Muſhrumes grow forth of the earth; other vpon the bodies of old trees, which differ altogether in kindes. Many wantons that dwell neere the ſea, and haue fiſh at will, are very deſirous for change of diet to feed vpon the birds of the mountaines; and ſuch as dwell vpon the hills or champion grounds, do longe after ſea fiſh; many that haue plenty of both, do hunger after the earthie excreſcences, called Muſhrumes: whereof ſome are very venomous and full of poiſon, others not ſo noiſome; and neither of them very wholeſome meate; wherefore for the auoiding of the venomous qualitie of the one, and that the other which is leſſe venomous may be diſcerned from it, I haue thought good to ſet forth their pictures with their names and places of growth. ‡ Becauſe the booke is already grown too voluminous, I will only giue you the figures of ſuch as my Author hath here mentioned, with ſome few others, but not trouble you with any more hiſtory, yet diſtinguiſh betweene ſuch as are eatable, and thoſe that be poyſonous, or at leaſt not to be eaten; for the firſt figured amongſt the poyſonous ones, is that we call Iewes-eare, which hath no poyſonous facultie in it. *Cluſius* (all whoſe figures I could haue here giuen you) hath written a peculiar tract of theſe baſtard plants, or excreſcences, where ſuch as deſire it may finde them ſufficiently diſcourſed of. ‡

¶ The

Lib. 3. Of the Hiſtory of Plants. 1579

1 *Fungi vulgatiſſimi eſculenti.* Common Muſhrums to be eaten.

Rrrrrr 2

JOHN GERARD

The Herball, or, Generall historie of plantes, 1633
Printed book with woodcut illustrations,
each page approx. 13¾ × 9½ in. / 35 × 24 cm
Getty Library, Los Angeles

The work of English barber-surgeon turned herbalist John Gerard (1545–1612), *The Herball, or, Generall historie of plantes*, first published in 1597, was a vast undertaking of three volumes with more than eight hundred chapters—and a total of more than 1,600 pages—of botanical information. In addition to more than a thousand plant species that were then known, Gerard included information on mushrooms, mosses, and corals. Known as the Father of Botany, Gerard, who worked as the curator of the physic garden at the Royal College of Physicians in London, included the plants' common and botanical names alongside details of their medicinal properties and the habitats where they could be found. Gerard was not a botanical scientist, and his book relied on texts and illustrations from many other published sources. Its unique selling point was its breadth, but inevitably it contained many errors, which were subsequently revised in the 1633 edition. Errors notwithstanding, *The Herball* was the most widely circulated botany book in the seventeenth century. The section on mushrooms and toadstools, which appeared alongside mosses and corals in the third volume, included not only information about their appearance, habitat, and medicinal properties but also anecdotes about poisonous mushrooms. In a precursor to later mycology volumes, the book contained comparative illustrations of poisonous versus edible mushrooms—as shown in this plate of "common mushrooms to be eaten"—to guide the reader to choose correctly.

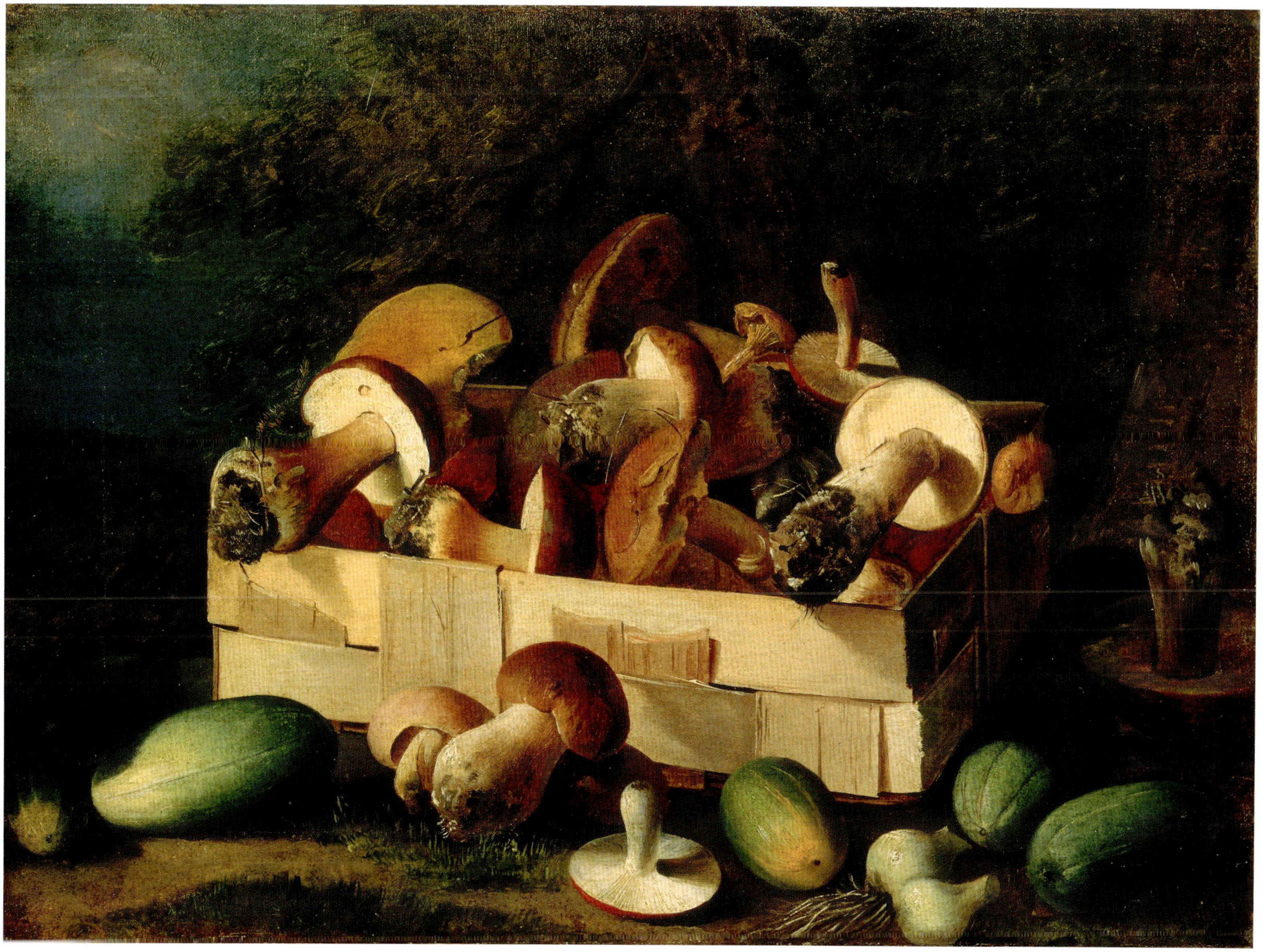

JAN CHRUCKI

Still Life with Mushrooms, 1842
Oil on canvas, 15¾ × 20⅝ in. / 40 × 52.5 cm
National Museum of Russian Art, Kyiv, Ukraine

A rectangular wooden trug overflowing with assorted fungi sits at the center of this still-life painting by Jan Chrucki (1810–1885), also known as Ivan Khrutsky. A Polish-Russian of Belarusian descent, Chrucki studied art at the Imperial Academy of Arts in Saint Petersburg. While he is best known for his detailed *nature morte* ("dead nature") genre of still-life paintings, Chrucki also painted portraits and interiors, as well as churches and other religious subjects. As part of his training, he made copies of seventeenth-century Dutch paintings, and this clearly influenced his style. After the death of his father in 1840, he often traveled to Vilnius in Lithuania, where he was commissioned to paint mainly portraits and religious themes, eventually returning to the family estate in the rural Polotsk region of Russia (now Belarus). The edible fungi depicted so faithfully here have presumably been recently gathered from their woodland habitat, hinted at by the roots and foliage that form the backdrop to the scene, and by the remnants of soil and twigs clinging to the bottom of several of the specimens. Most of the fungi shown here are boletes, with their characteristically domed caps and pores instead of gills. Several species of boletes are much prized for their edibility and fine flavor, and the culinary destination of the ones in this work is hinted at by the inclusion of onions and squashes in the foreground.

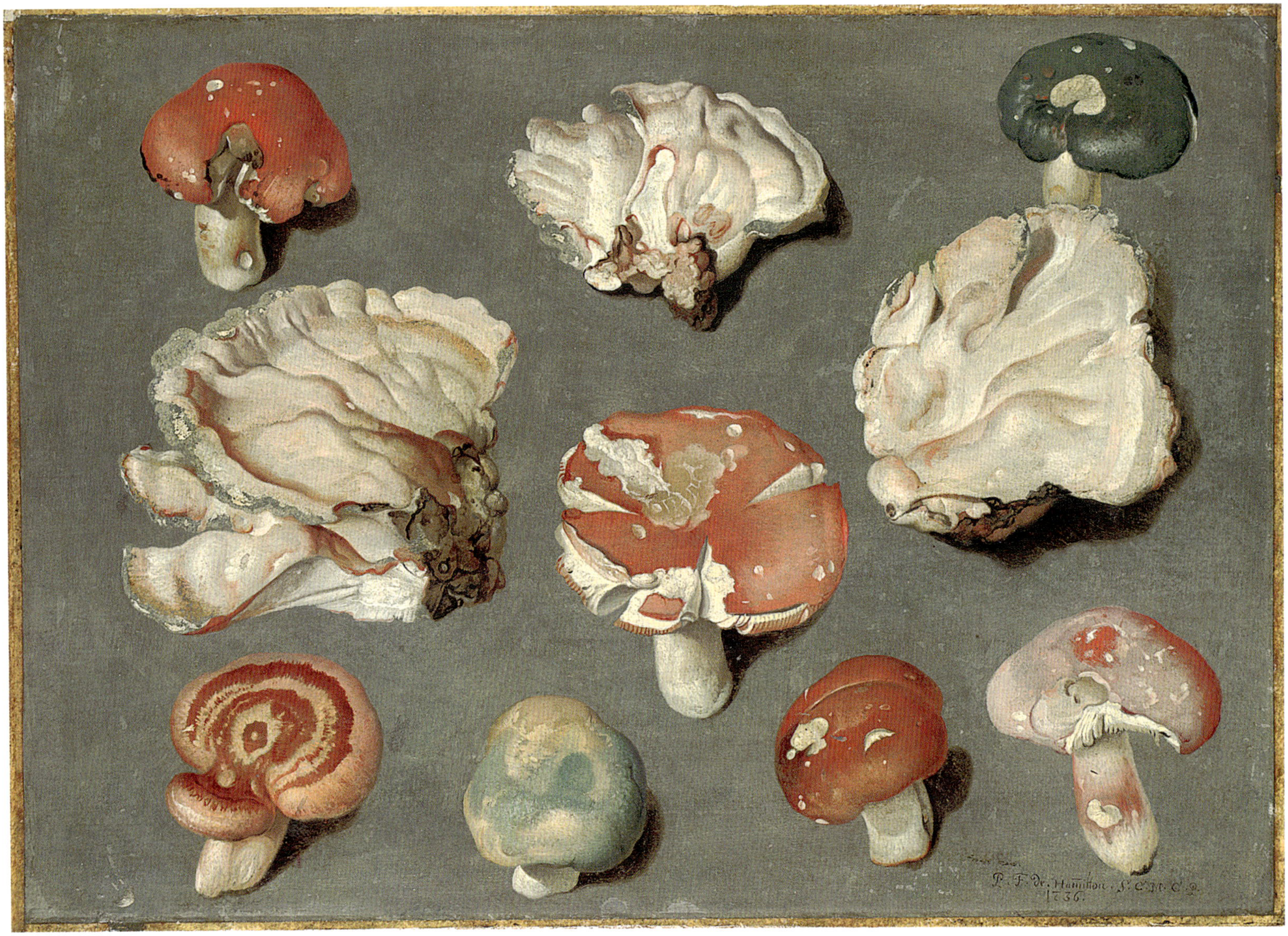

PHILIPP FERDINAND DE HAMILTON

Studies of Fungi, 1736
Oil on paper laid on canvas,
9⅞ × 13¾ in. / 25.1 × 34.9 cm
Private collection

A surprising selection of diverse fungi appears in this botanical study by Belgian painter Philipp Ferdinand de Hamilton (1664–1750). The natural subject matter is unique for Hamilton, who is best known as the court painter for a succession of Holy Roman rulers in Austria, a position he held from 1705 until his death. At court, he was commissioned to paint large hunting scenes, which typically featured a variety of birds, game, and dogs set against dark landscapes. The species of fungi here include three specimens of pale pink polypores, or bracket fungi, so called for their strong, projecting structures. Known to be inedible, the fan-shaped brackets grow in tiers on both living and dead trees and can reach a diameter of nearly 8 inches (20 cm). The edible emerald mushroom at top right, shrimp russula or crab brittlegill (*Russula xerampelina*), is usually more pinky red, but it can become green if either its stem or flesh come into contact with iron salts. At the canvas's lower left, one can spot the bull's-eye top of the poisonous yellowdrop milkcap (*Lactarius chryssorrheus*), and the equally inedible bloody brittlegill (*Russula sanguinaria*) is present at lower right. Although the reasons Hamilton painted these mushrooms is unclear, these are fungi that would be common to the temperate woods of Central Europe, making this a potential preparatory study for one of his larger hunting scenes. Even so, he painted with a high degree of accuracy, leaving the mushrooms' bruising and damage clearly visible.

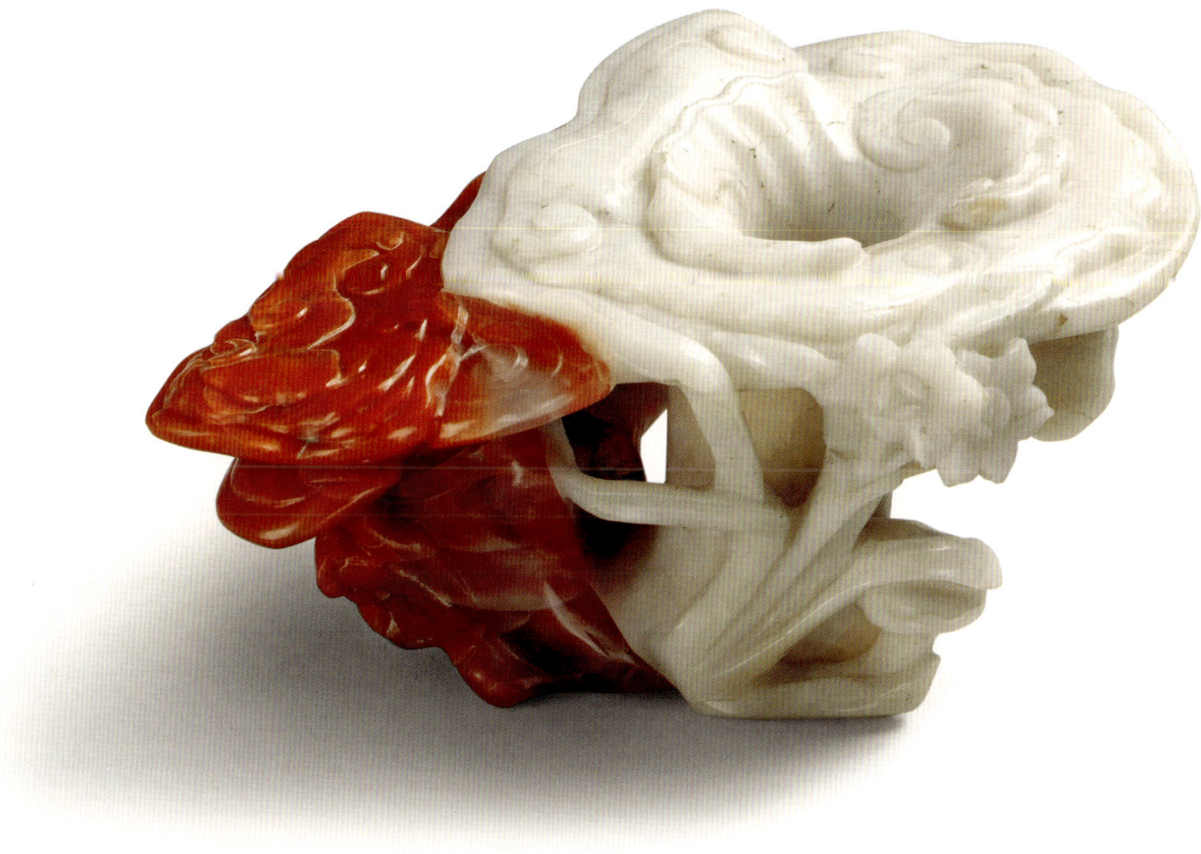

ANONYMOUS

Water cup for moistening ink, c. 1600s
Milky white chalcedony and carnelian,
$2\frac{3}{4} \times 6\frac{1}{4} \times 4\frac{3}{4}$ in. / 7 × 16 × 12 cm
Musée Guimet, Paris

The lingzhi (or reishi) mushroom (*Ganoderma lingzhi*), referred to as the "mushroom of immortality," has for centuries been valued for its use in traditional East Asian—especially Chinese—medicine, both physical and spiritual. The *Shennong bencao jing* (*Divine Farmer's Classic of Pharmaceutics*) from 200 to 250 CE claims it is beneficial for the liver, heart, spleen, lungs, kidneys, and "essence." Numerous other Chinese writings from as far back as two thousand years ago suggest it is highly advantageous for the heart and circulatory system. Taoist temples made a concentrated hallucinogenic drink from the mushroom that purportedly allowed followers the ability to see spirits, or even become them. Still today, many people throughout the world ingest the mushroom in various forms (most commonly teas and capsules) to provide such health benefits as boosting the immune system and fighting cancer. The lingzhi has also been widely viewed as a talisman of good luck and prosperity in East Asian countries for centuries. Given all of these positive attributes, it is no surprise that this seventeenth-century Chinese Qing dynasty water cup was made in the form of two joined lingzhi mushrooms. Chinese ink would have come in solid form, composed of soot, animal glue, and other additives. In order to write with the ink, a chunk of the solid would have been mixed with water in a cup such as this to yield a usable liquid form. Because most Chinese were illiterate during the Qing Dynasty, this ink cup would almost certainly have been used by a scholar or member of the wealthy elite.

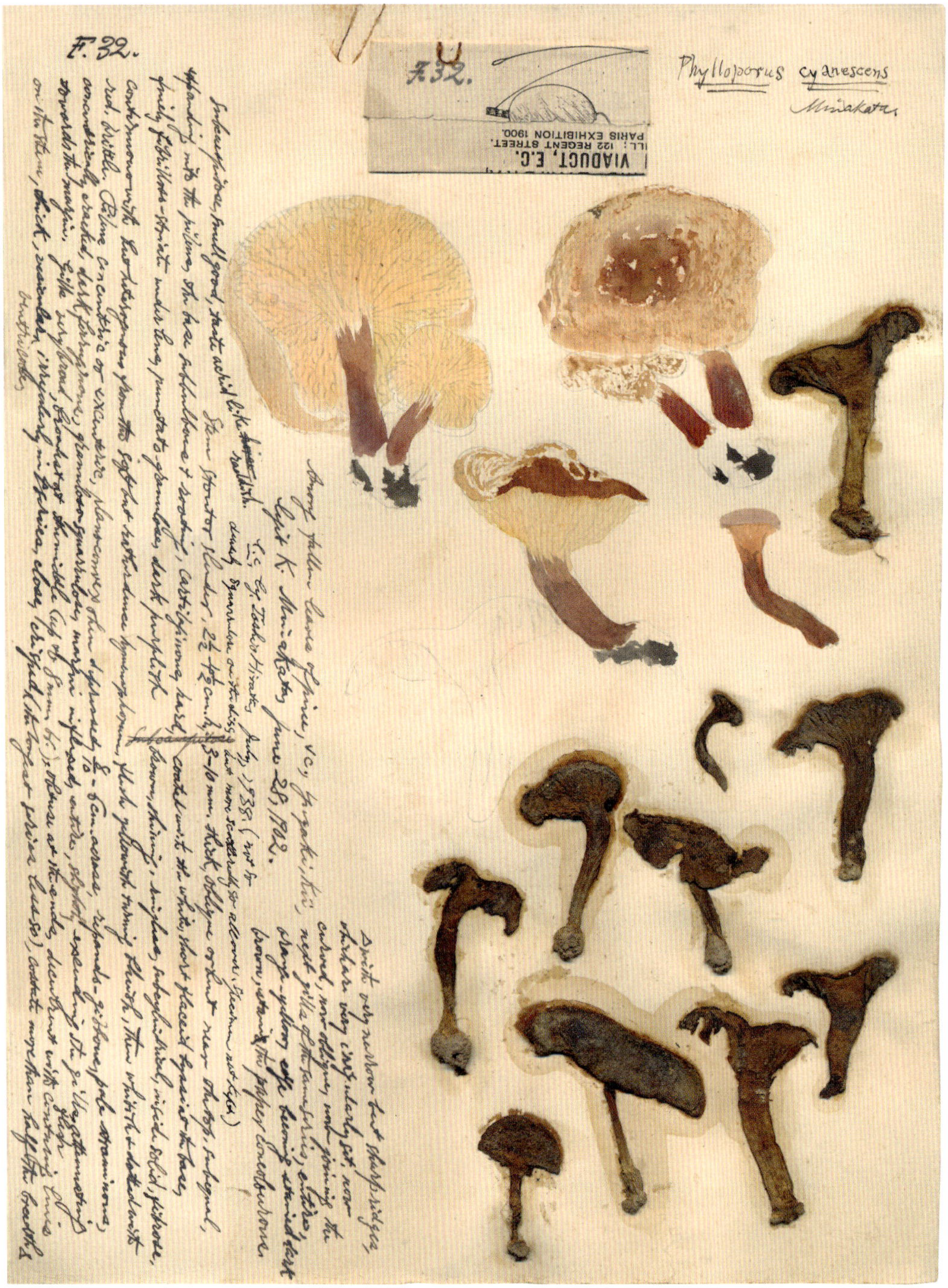

MINAKATA KUMAGUSU

Phylloporus cyanescens, 1902
Watercolor, ink, and dried specimens on paper, National Museum of Nature and Science, Tokyo

A delicate watercolor of *Phylloporus cyanescens* mushrooms sits alongside sprawling handwritten notes and pressed specimens in this page by Japanese polymath Minakata Kumagusu (1867–1941). Multilingual, Minakata studied in the United States and spent fourteen years overseas, many in London, where he was a regular at the British Museum. On his return to Japan in 1900, he moved to Wakayama, where he spent his days collecting mushroom specimens in the mountain forests. Those specimens formed the basis of his magnum opus, *The Illustrated Book of Bionomics of Japanese Fungi*, which examines his scientific discoveries about the biological interactions (bionomics) among fungi, their environment, and human culture, making it a founding text of mycology in Japan. Published in 1937, the book contains some 4,500 species and 15,000 pictures, and, although it has 500 fewer entries than Minakata originally planned, it covers fungi, slime molds, and algae. While Minakata produced thousands of studies of mushrooms, his primary interest was, in fact, slime molds. Known as false fungi, the microscopic organisms were historically classified as fungi, as they share many similarities, such as spore production, growing on decaying material, and a moldlike appearance. In addition to his essential contributions to mycology and biology, Minakata also conducted anthropological research and studies on folklore—a true Renaissance scholar whose career has drawn parallels with those of famed German polymaths Alexander von Humboldt and Gottfried Wilhelm Leibniz.

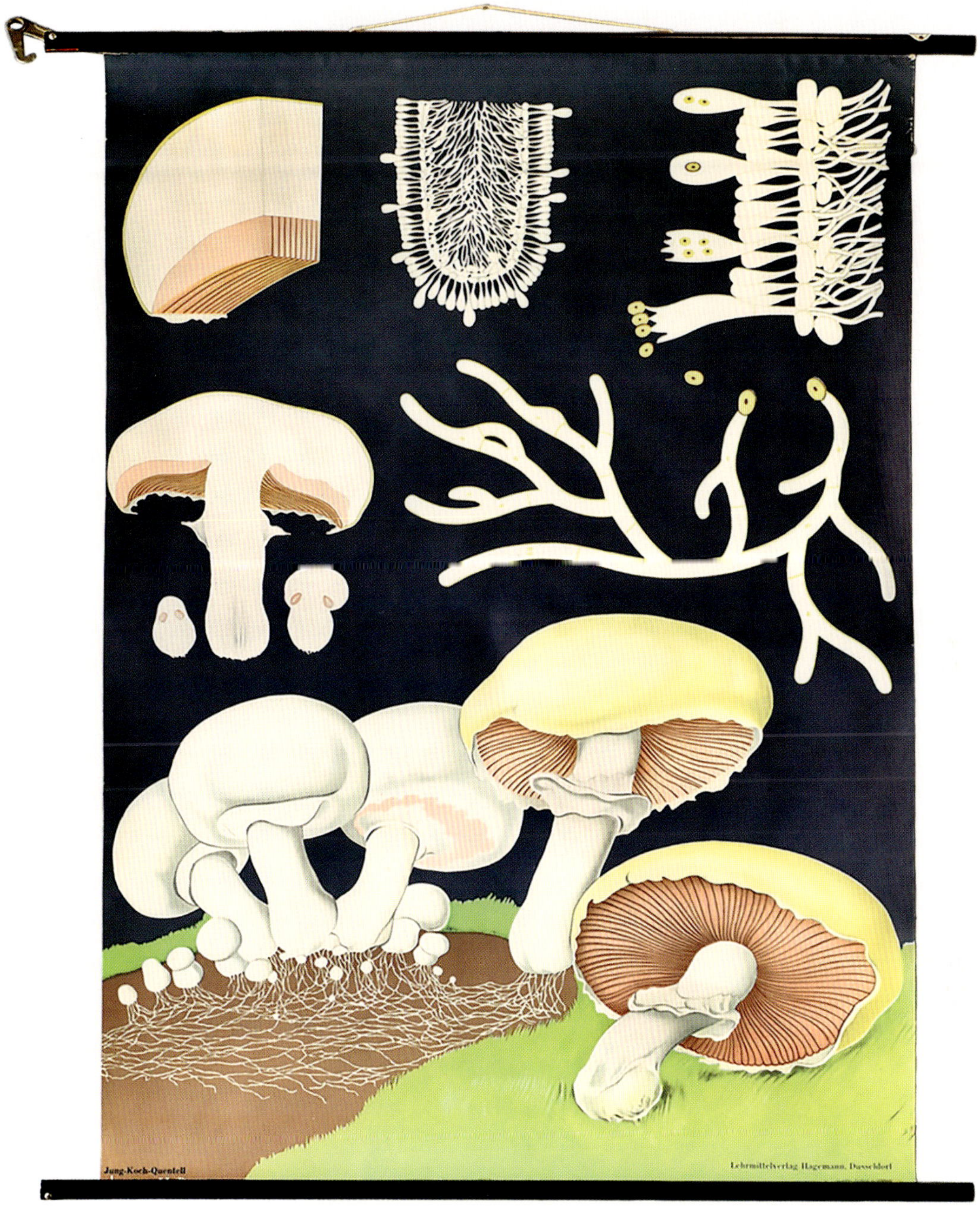

HEINRICH JUNG, GOTTLIEB VON KOCH, AND FRIEDRICH QUENTELL

Wall Chart of the Mushroom, 1966
Print, 47¼ × 39⅜ in. / 120 × 100 cm
Private collection

The vivid illustrations of naturalists and educators Heinrich Jung, Gottlieb von Koch, and Friedrich Quentell were first gathered in the 1902 volume *Neuen Wandtafeln für den Unterricht in der Naturgeschichte* (*New Wall Charts for Teaching Natural History*). Though converted into large-scale wall charts at the time, they later gained popularity when reprinted after World War II. One hallmark of these stunning wall charts is that they bear no text. This meant that they could be produced for broad audiences, no matter what language they spoke, and also enabled the instructors who used them to customize their own lessons. As seen here in a strikingly modern representation of various mushroom characteristics, each wall chart depicts a single species. The images are presented from numerous angles, as well as showing cross sections and details of organisms ranging from gross morphology—what can been seen with the naked eye—down to the cellular level. Specimens are also presented in an idealized form, free of the imperfections and variations that would be found in the wild—a characteristic also found in the works of famed scientific illustrator Ernst Haeckel (see p.141), under whom von Koch apprenticed. The beautifully rendered figures are saturated with color and texture. Set against a stark black background, they draw the viewer into the composition. The unique earmarks, superior print quality, and unassailable artistry of Jung-Koch-Quentell wall charts make them sought after today by both collectors of scientific and educational art and interior designers alike.

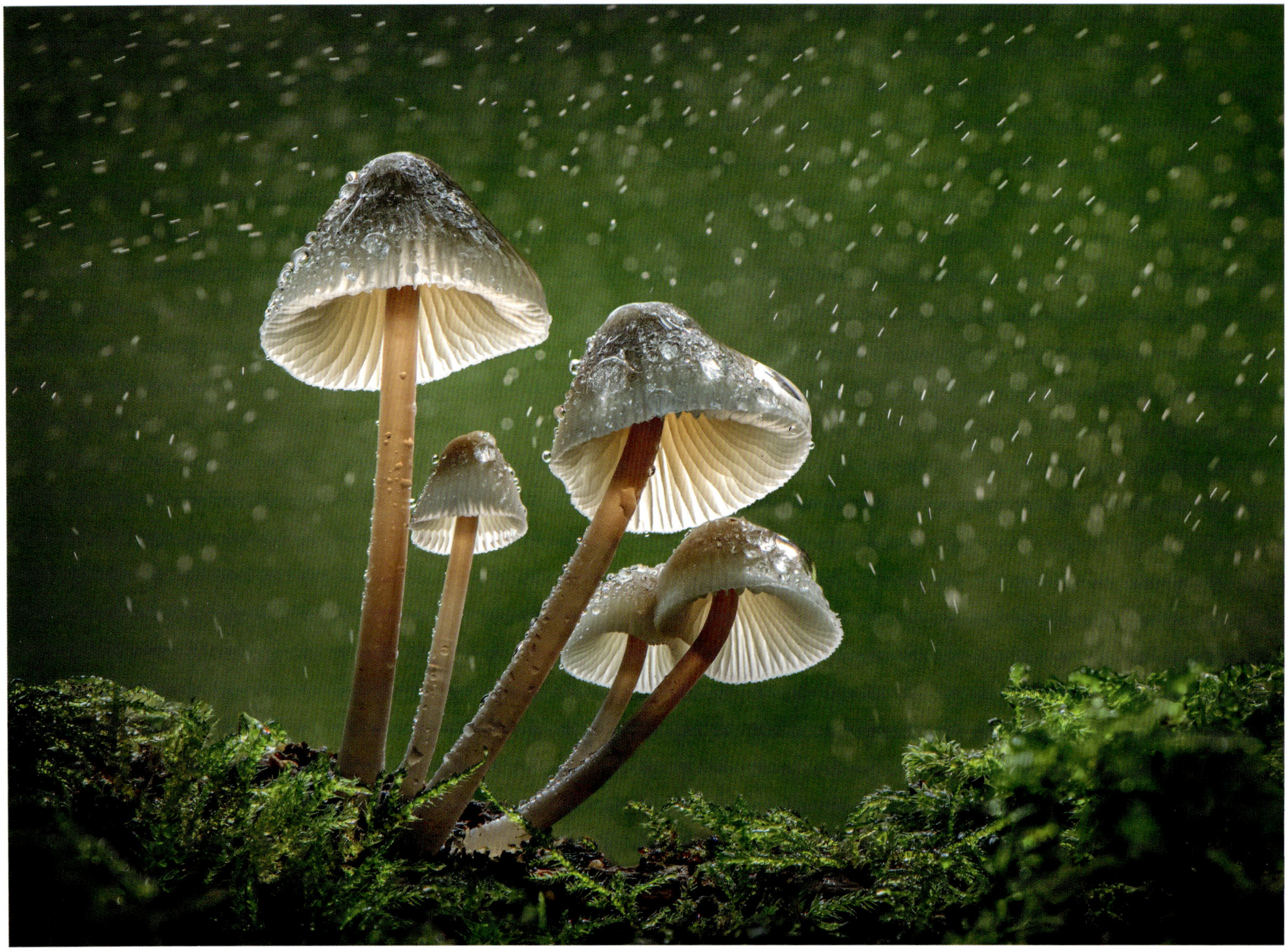

TONY NORTH

Milking Bonnets, 2023
Photograph, dimensions variable

Five frilly white "bonnets" reach up from their soft, green, mossy carpet, growing close together on an old log. They are supported by slender gray-brown stems of different heights and appear to be swaying slightly in the soft rainfall, raindrops splashing like luminous baubles onto the bell-shaped mushroom caps. Each cap is slightly grooved, and beneath it the carefully pleated gills spread out from the center, like a crinoline skirt. Captured by British photographer Tony North (b. 1967), these *Mycena galopus*, commonly known as milking bonnets or milk-drop mycena, are an inedible species of mushroom of the order Agaricales. Found in Europe and North America, they have grayish-brown, grooved caps, each up to nearly 1 inch (2.5 cm) wide. The fruiting bodies appear during summer and autumn on leaf litter in coniferous and deciduous woodland habitats. North found this group growing on a log in Marbury Country Park in Cheshire, England. His Nikon photograph was artfully adjusted, backlit by LED, with the rain effect created by a water spray. Based in Manchester, England, North trained in psychology and now has a career teaching history and science at the primary level, and he also runs courses on macrophotography. He focuses mainly on magnified natural subjects—insects, flowers, and fungi—as well as landscape photography and astrophotography. An award-winning photographer, North took first place in the World of Fungi section of the 2022 International Garden Photographer of the Year competition and was declared the overall winner of the competition in 2023.

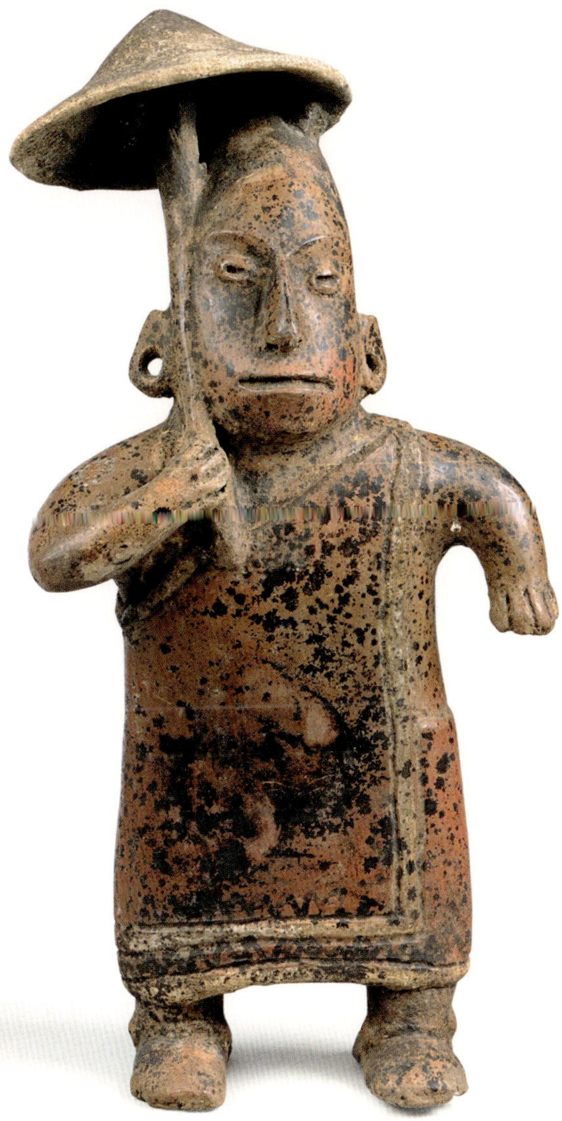

ANONYMOUS

Standing figure holding a mushroom, 250 BCE–250 CE
Ceramic and pigment, 21⅝ × 11¾ in. / 55 × 30 cm
The Israel Museum, Jerusalem

In its right hand, this standing ceramic figure holds the long stem of a mushroom to form what appears to be an umbrella or parasol to shield their head. Made by an unknown artisan from the Comala region of Colima state in western Mexico, the statue reflects the influence of one of the earliest Mesoamerican civilizations, the Olmec, who are best known for their huge sculptures of helmeted heads carved to honor their rulers and to communicate with their gods. This figure dates from sometime between 250 BCE and 250 CE and was probably intended to be placed in the shaft tomb—so called because bodies were buried in underground shafts and tunnels—of a member of the local elite. The Comala style of pottery is known for its distinctive effigy vessels, which include human figures, animals, and vegetables, such as squashes. The ceramics, made from a reddish clay, are often decorated with geometric designs, and many of the effigies were finished with a pigment to create a mottled effect, as seen here. The inclusion of the mushroom as an ersatz umbrella points to the important role mushrooms played in early Mesoamerican life. Psychoactive mushrooms—most likely those containing psilocybin—were used by religious leaders and rulers for divination and to communicate with deities and ancestors.

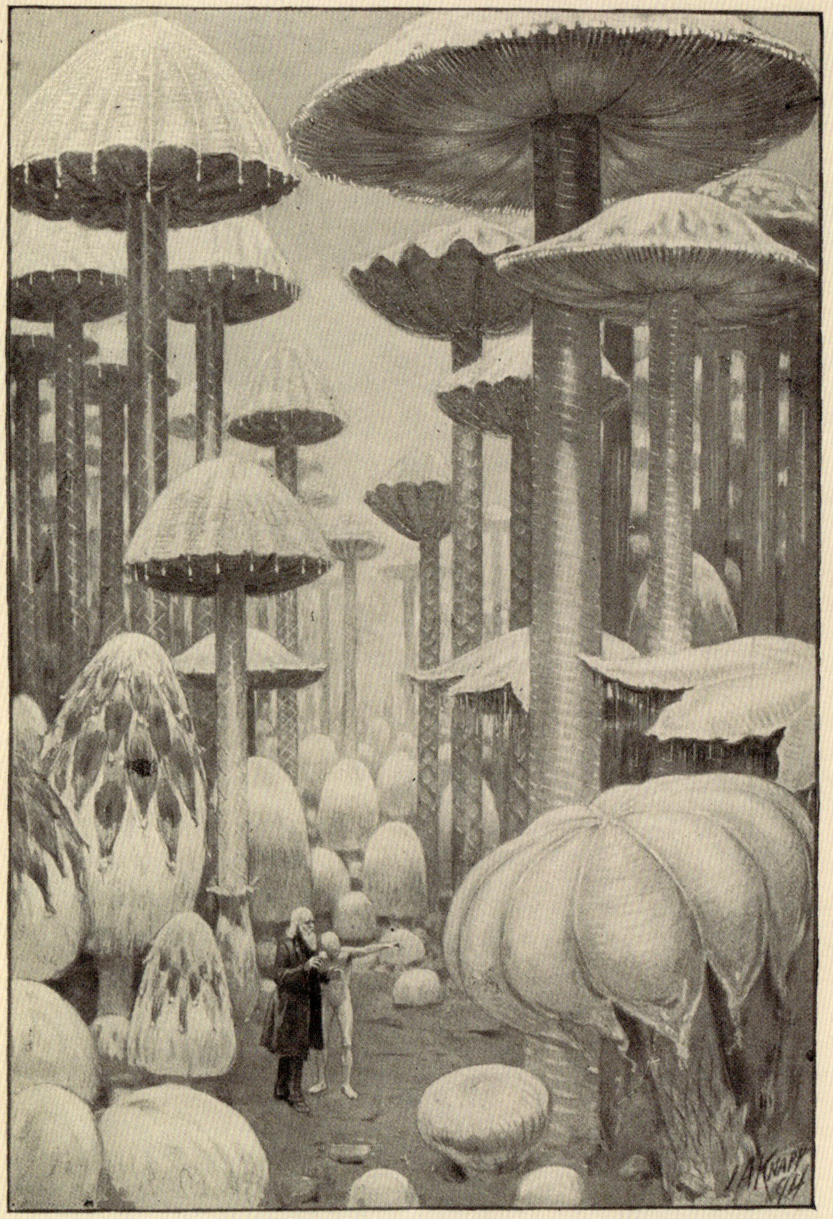

"I WAS IN A FOREST OF COLOSSAL FUNGI."

JOHN AUGUSTUS KNAPP

I Was in a Forest of Colossal Fungi, from *Etidorhpa, or, The End of the Earth* by John Uri Lloyd, 1897
Printed book, H. approx. 9½ in. / 24 cm
New York Public Library

In a glimpse of a science-fiction landscape, a strange, faceless alien-like character leads an old, bearded man through a forest of towering, thin-stemmed mushrooms. This is a scene from the book *Etidorhpa, or, The End of the Earth,* written by John Uri Lloyd and illustrated by American artist John Augustus Knapp (1853–1938). Born in Ohio, Knapp trained at the McMicken School of Design in Cincinnati in 1871 and by 1877 had established a studio there, producing theater bills, circus posters, and, later, book illustrations. In 1895 Lloyd—who was also previously Knapp's neighbor—hired Knapp to illustrate his first novel, *Etidorhpa*. The book's narrative describes an adventure through a cave in Kentucky into the core of the earth, exploring ideas of alchemy, secret Masonic orders, the Hollow Earth theory, and physical transcendence along the way. A mixture of science fiction, occultism, and early psychedelia, the book was a popular and critical success and established Knapp's career. It was also reflective of the artist's own interests in theosophy and Freemasonry. Knapp continued to illustrate for Lloyd, but he is perhaps best known for his collaborations with occultist Manly P. Hall, who wrote the book *The Secret Teachings of All Ages* (1928), whom he met during a period in Los Angeles. Knapp returned to fungi in the 1910s, painting a group of forty-two watercolors of mushrooms for John Lloyd's mycologist brother, Curtis Lloyd. However, it was his eerie, fantastical early take on science fiction and spirituality that was his greatest success.

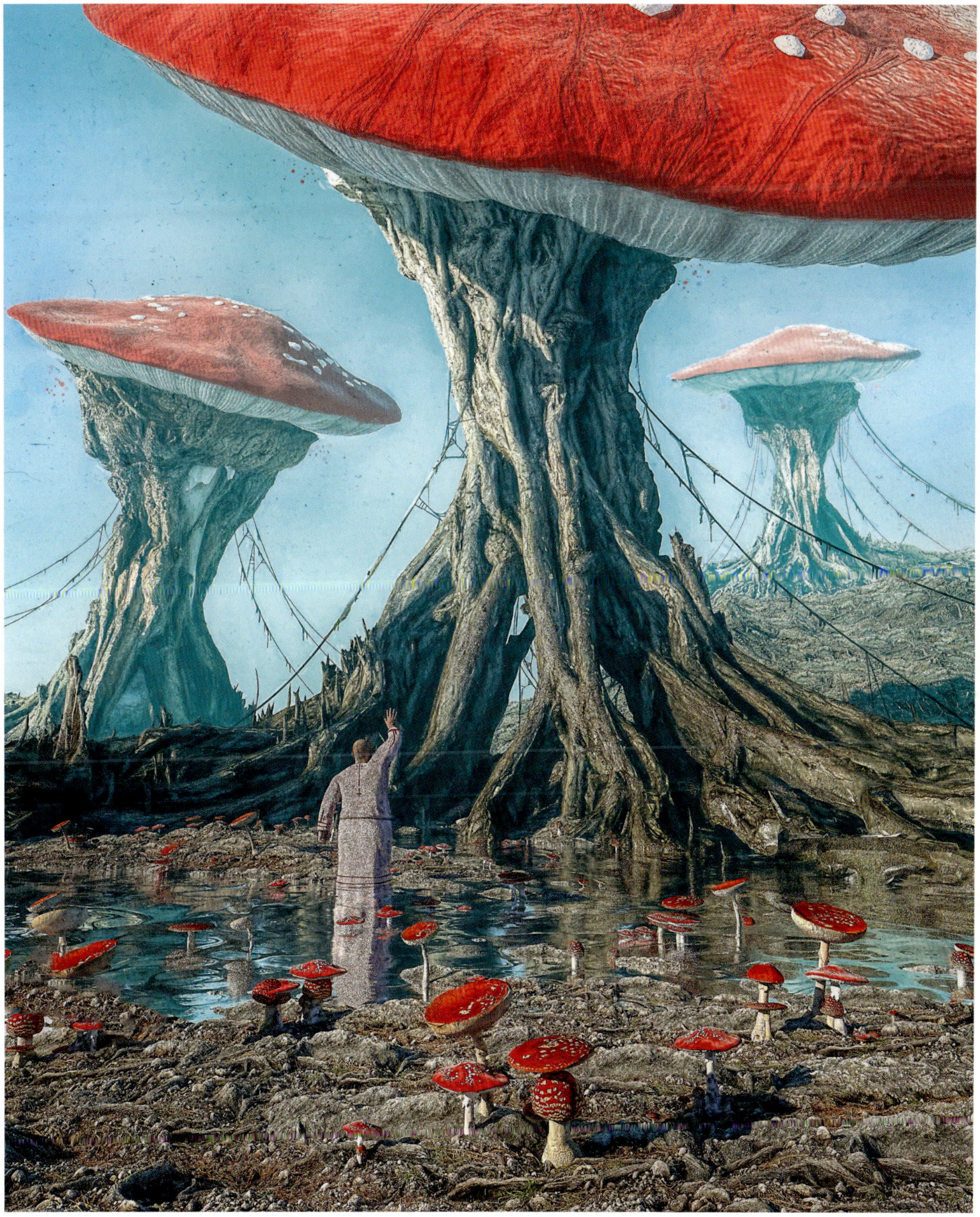

MIKE WINKELMANN (BEEPLE)

Pilgrimage, 2021
NFT, single-channel video,
00:00:52 minutes, 1920 × 1080 pixels
Private collection

Impossibly huge fly agaric (*Amanita muscaria*) mushrooms rise from a swampy landscape in this hallucinatory digital video, echoing the visionary dreamscapes of Surrealist art. Surrounded by regular-sized mushrooms, a monklike figure gestures toward the gargantuan toadstools, the stems of which resemble gnarled and ancient tree trunks. These curious plant-fungi hybrids are the creation of Mike Winkelmann (b. 1981), a trailblazing American digital artist known professionally as Beeple. Filled with phantasmagoric and dystopian visions, his images and animations are often tinged with provocative political and social commentary. This piece forms part of his landmark *Everydays* series, for which he creates and posts online a new digital artwork daily. The first five thousand images in the series, which began on May 1, 2007, were collaged into a single work and minted as an NFT (non-fungible token)—a unique digital asset that uses blockchain technology to prove ownership of digital items. It was sold for a staggering $69.3 million in 2021, the most expensive single-piece NFT sale to date. *Pilgrimage* is image number 5,132 in the ongoing series, created on May 24, 2021, with the software programs Cinema 4D, Octane Render, and Photoshop. On June 28, 2022, in a sale at Christie's, *Pilgrimage* fetched $252,000. Notorious for its hallucinogenic properties, the red-and-white spotted fly agaric is instantly recognizable, encountered across popular culture, fairy tales, art, and film. In an apparent reference to the mushroom's mind-altering effects, Winkelmann's artwork is accompanied by an artist's note stating, "It's been a long strange trip so far, but the journey has just begun."

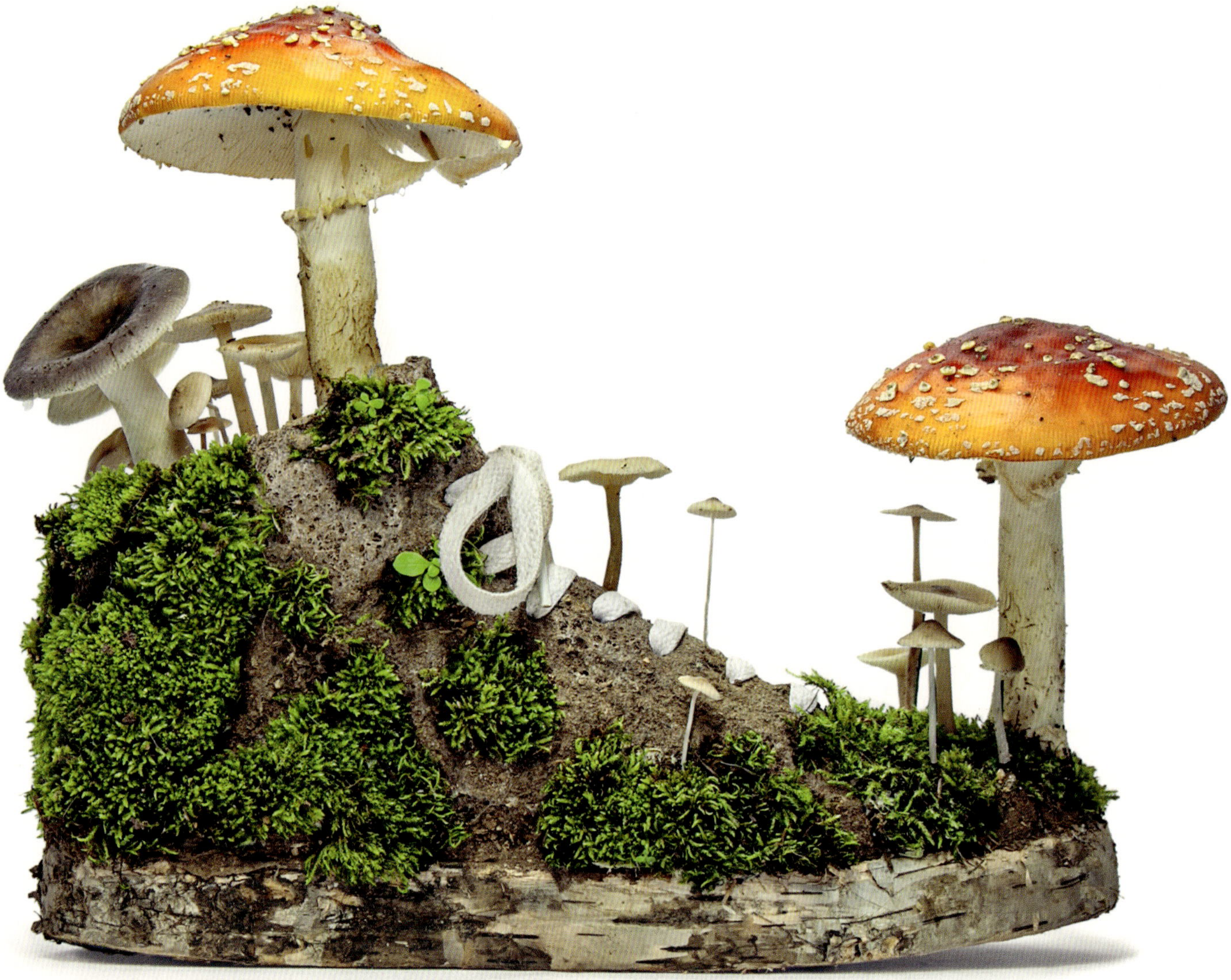

CHRISTOPHE GUINET (MONSIEUR PLANT)

Mushroom Sneaker – Amanita, from the *Mushroom Party* series, 2022
Vegetal sculpture with bark, moss, and mushrooms, W. 11¾ in. / 30 cm
Private collection, Brittany, France

Paris-born artist Christophe Guinet (aka Monsieur Plant, b. 1977) has been making art out of nature for years. Transforming everyday objects into the extraordinary, Guinet combines natural elements with man-made materials to reimagine how we use (and discard) the things around us. Echoing his previous series of flower shoes and *Sneakarium*, in which he imagined nature reclaiming discarded shoes, he created the *Mushroom Party* series, incorporating foraged materials from around his countryside home in Breton, France, including moss, grass, leaves, tree bark, and fungi to create compositions teeming with life. Everything seen in the sneaker here, except the shoelaces, is completely natural. To create the work, Guinet painstakingly assembled the scraps piece by piece to make a verdant mosaic that appears completely organic, as if the sneaker just happened to emerge from the woods themselves. Taking center stage are two fly agarics (*Amanita muscaria*), with their distinctive white-spotted red caps. In the artist's perspective, in the eternal battle forged between man and nature, the latter will be the ultimate victor. His work demonstrates the natural world's ability to adapt to circumstance, whether it's a tree contorting its shape—as in his *TWIST* series—or greenery reclaiming an abandoned surface, as in *Think Nature*. A powerful indictment of consumer culture in the twenty-first century and a reminder to appreciate our native environment, his works are as serious as they are playful. Whether sculpture, installation, or otherwise, Guinet puts the natural world at the forefront, where it rightly belongs.

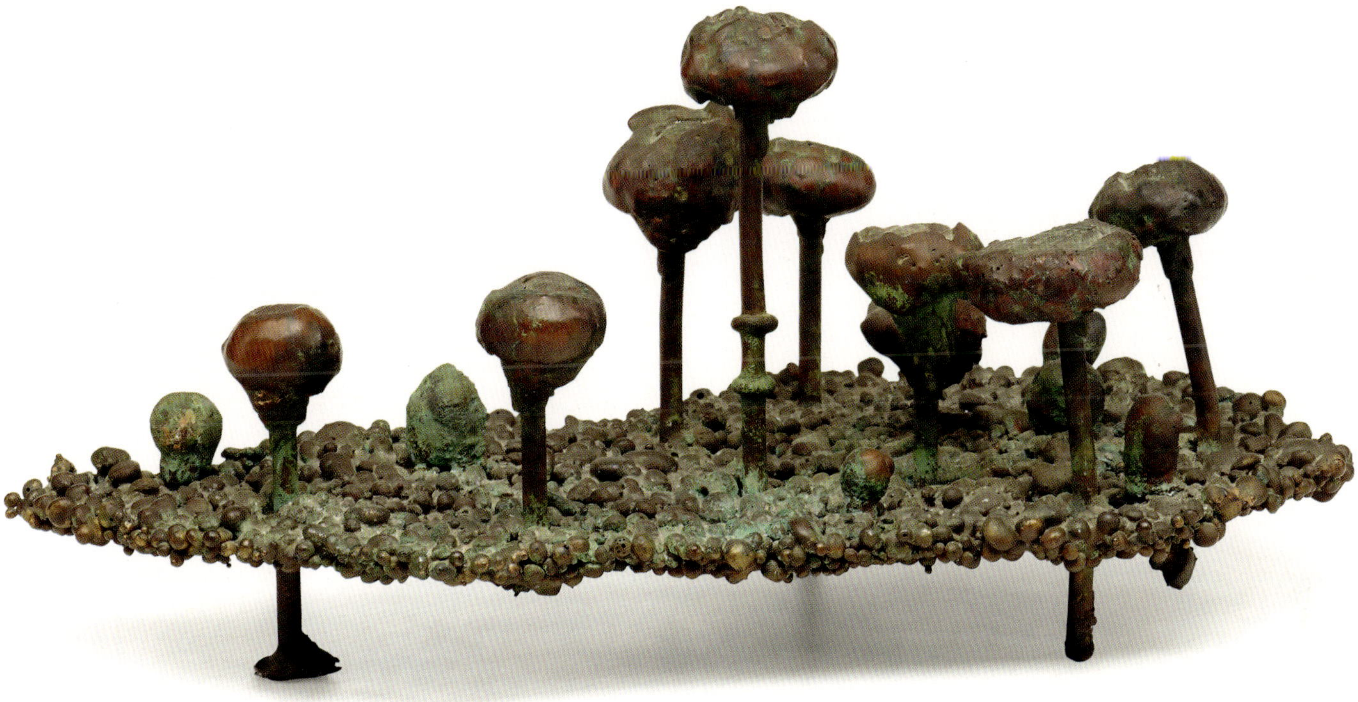

HARRY BERTOIA

Mushroom Landscape, c. 1965
Welded bronze,
6½ × 14¾ × 9¾ in. / 17 × 37.5 × 24.8 cm

Despite a career that spanned more than four decades and a plethora of media, artist and designer Harry Bertoia (1915–1978) is most closely identified with the collection of furniture he designed for the American company Knoll between 1950 and 1952. Knoll introduced Bertoia's wire furniture collection—which includes some of the most iconic and important pieces of furniture in the pantheon of twentieth-century design—in 1952 and has been selling it continuously ever since. But still, Bertoia was at Knoll for only two years, and his time designing furniture represents but a blip on his artistic timeline. Bertoia was truly a Renaissance man and produced an estimated ten thousand or more pieces of jewelry, works on paper, sculptures, installations, and assorted other objects in his lifetime. Born in Italy in 1915, Bertoia traveled to the United States in 1930 and enrolled at Cass Technical High School in Detroit to study jewelry-making. He spent the 1930s and into the '40s primarily making jewelry, including at the Cranbrook Academy of Art in Bloomfield Hills, Michigan—where he met Florence Knoll—starting in 1937. The later success of his collection for Knoll in 1952 gave him the financial freedom to pursue his artistic career as he saw fit, and he largely made various types of sculpture until his death in 1978. *Mushroom Landscape* is from this later era, produced around 1965, and is a beautiful example of Bertoia's static sculptures from this time. Clearly on display here is Bertoia's core approach to sculpture: his belief that appropriate space should be left between forms to allow light to enter and "dance."

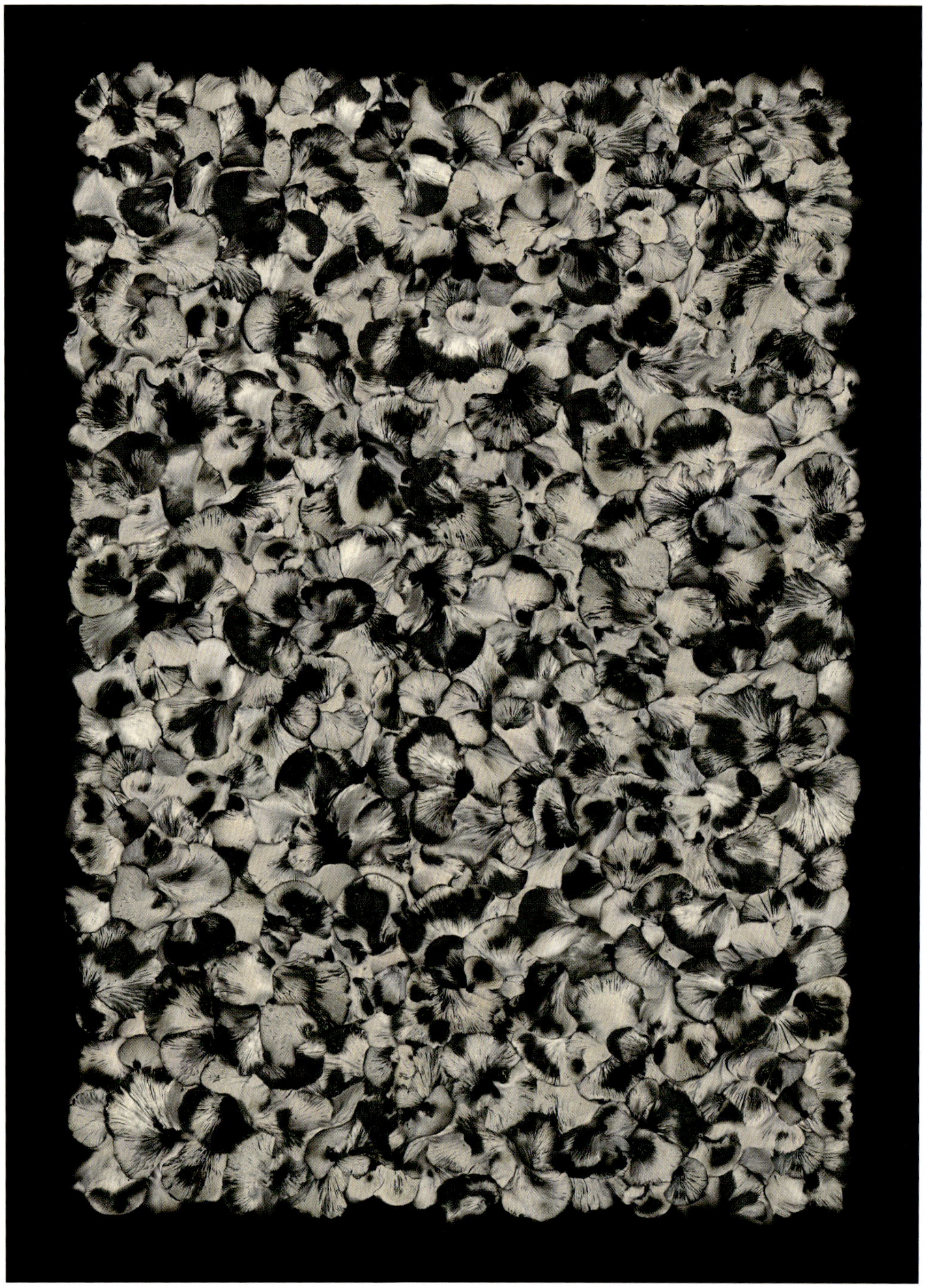

ADAM FUSS

Untitled, 2016
Pigment print,
83½ × 60 in. (212.1 × 152.4 cm)
Private collection

A large, black-framed monochrome print is crowded with the images of hundreds of mushrooms, filling the page like flower petals. Based in New York, British artist Adam Fuss (b. 1961) has been creating images of the natural world, such as this untitled work, since the 1980s. His process is intentionally simple, utilizing only his subjects, light, and light-sensitive material. In fact, no camera is used in the production of his work at all. Fuss places his subjects on photosensitive paper, and that is precisely the image that we see in the final print. In his fungi works, Fuss positions the mushrooms gills down on the paper, which is then covered and pierced to affect the light and airflow. On close examination, the powdery spores of the mushrooms create a delicate pattern on the final images, which are preserved with fixing spray. There is a spiritual element, as well as an informed reference to photographic history and techniques in Fuss's photograms, daguerreotypes, and pinhole images. Other motifs in his oeuvre include birds, insects, flowers, smoke, babies, butterflies, and snakes. Although rooted in reality, there is something abstract in Fuss's compositions—echoing the paintings of Jackson Pollock or Lee Krasner—as well as something ghostly, particularly in his conceptual work. Fuss's prints are held in prestigious collections around the world, including those at the Victoria and Albert Museum in London and the Metropolitan Museum of Art in New York, among many others.

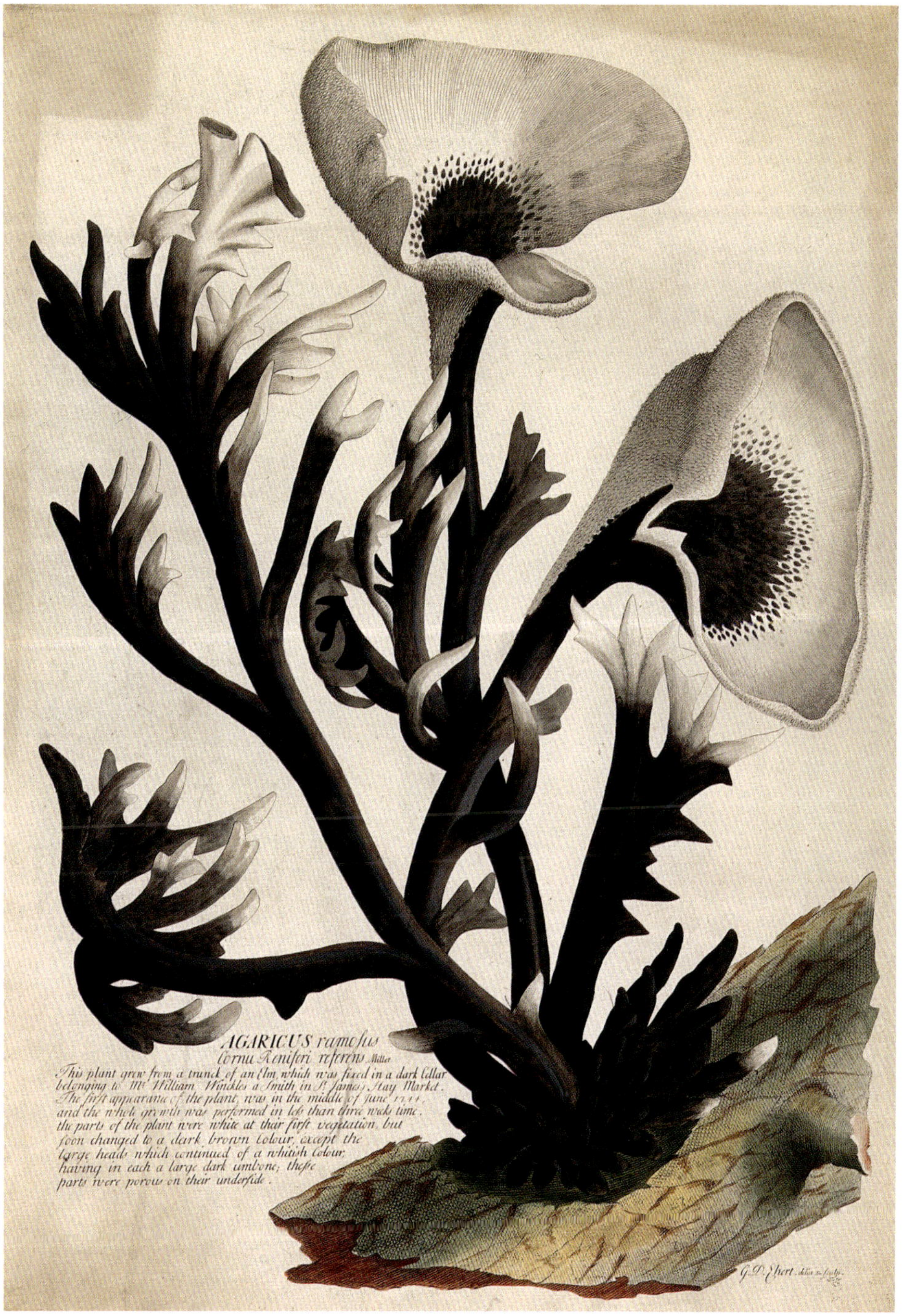

GEORG DIONYSIUS EHRET

Agaricus ramosus Cornu Reniferi referens, c. 1744
Etching with watercolor, 21⅜ × 14¾ in. / 54.3 × 37.6 cm
Wellcome Collection, London

Rendered in bold black lines and shades of gray, a branching fungus with two large flowerlike forms emerges from a trunk of rotting wood in this eighteenth-century botanical etching by German painter Georg Dionysius Ehret (1708–1770). An early Latin name, now obsolete, refers to this species as *Agaricus ramosus Cornu Reniferi referens*, which roughly translates to "branching Agaricus resembling a reindeer's horn." Now, it is scientifically known as *Xylaria hypoxylon* but is still commonly called the stag's horn, or sometimes candlesnuff, fungus. It was first described in the *Clavaria* genus by Carl Linnaeus around the time of this etching (Linnaeus identified its habitat as in the hold of a ship), and it was later transferred to the *Xylaria* genus because of its relationship to wood. This moldy-looking bioluminescent species is one of the most familiar fungi found in temperate woodlands and one of the last to colonize rotting wood in the late fall and winter. One of the most prolific botanical artists of his time, Ehret also worked with Linnaeus at the beginnings of his work on binomial classification. It is perhaps no surprise, then, that the accompanying text attempts to more scientifically classify the fungus with additional contextual information, including where it was growing: "This plant grew from a trunk of an Elm, which was fixed in a dark cellar belonging to Mr. William Winckles, a smith in St. James Hay Market."

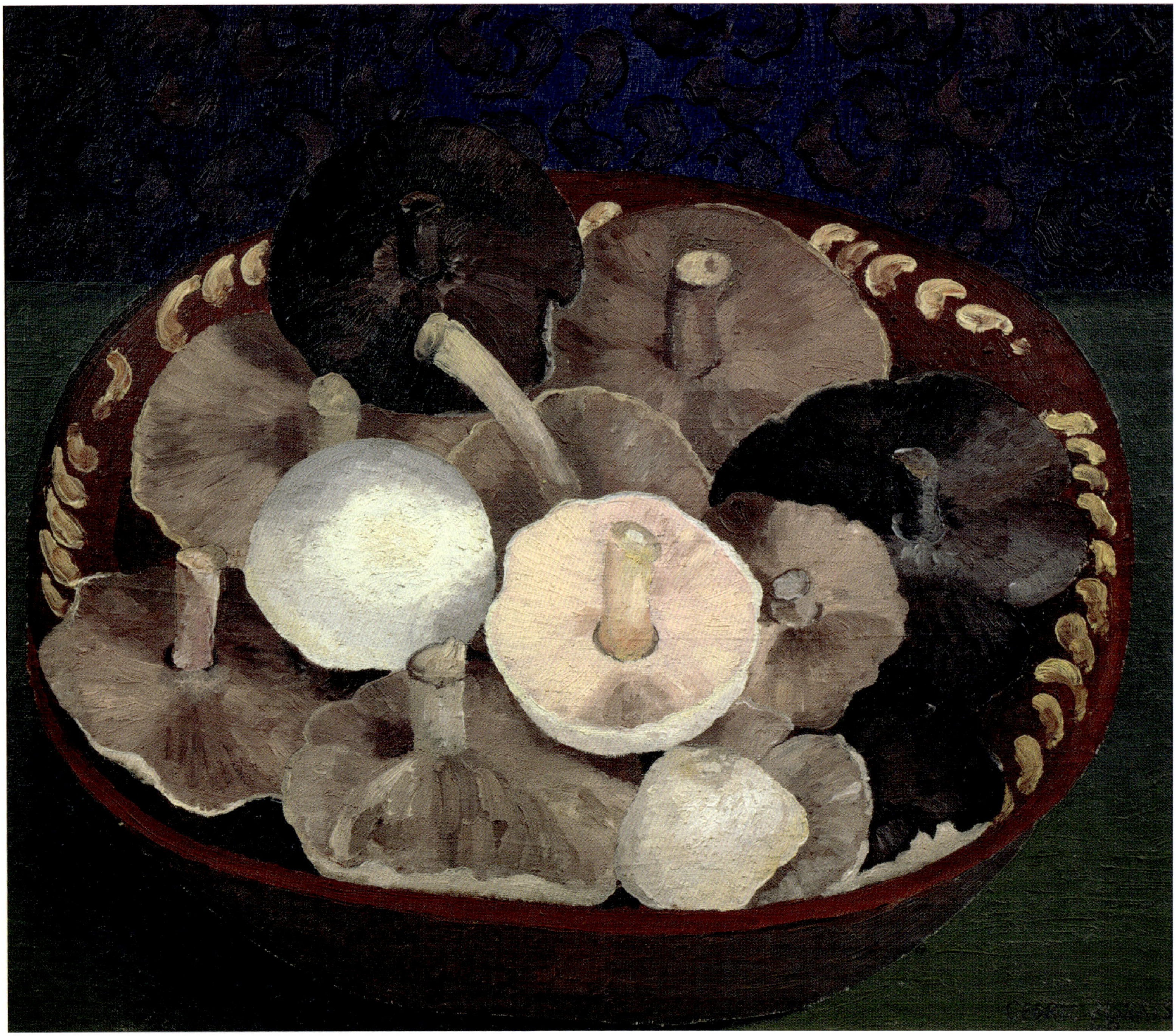

CEDRIC MORRIS

Mushrooms, 1948
Oil on canvas,
19½ × 23 in. / 49.5 × 58.4 cm
Private collection

Filling a ceramic plate, a jumble of largely upturned mushrooms gives the appearance of having been freshly gathered, their stems—some short, some long—pointing upward, rendered in a palette ranging from near black to a pale pink. The painting is the work of English artist, horticulturalist, and polymath Cedric Morris (1889–1982), who is perhaps better known for his flower paintings and for cultivating irises in his garden at Benton Edge in Suffolk. He and his partner, Arthur Lett-Haines, moved to the sixteenth-century farmhouse near Hadleigh in eastern England in 1939, founding an art school there, the East Anglian School of Painting and Drawing. An accomplished art teacher, Morris counted among his students the future stars Lucian Freud and Maggi Hambling. Morris was known for his portraits, landscapes, and, as here, still lifes, but gardening was his true passion. At Benton Edge, he and Lett-Haines transformed part of their 3-acre (1.2-hectare) plot into a well-organized vegetable garden, although growing vegetables remained a secondary passion after the iris breeding, for which Morris gained a considerable reputation. Here, Morris treats a humble plate of mushrooms with the same botanical precision he brought to his flower paintings. Morris's skill is clear to see: the oil paint gives the delicate gills of the upturned mushrooms a velvety succulence. He referred to himself as an "artist plantsman," and the description appears on his gravestone in Friars Road Cemetery in Hadleigh.

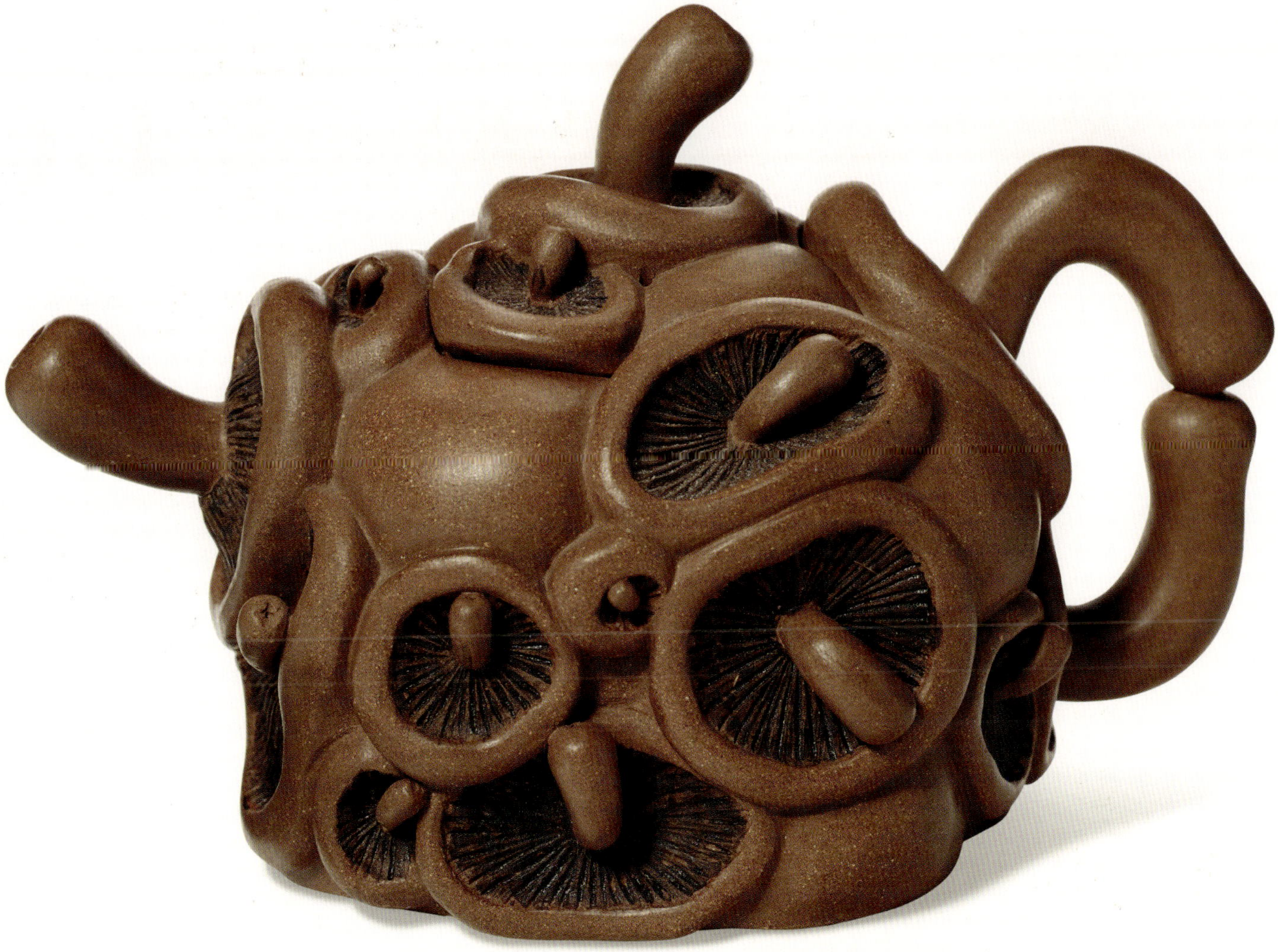

ZHOU DINGFANG

After Rain, undated
Clay, Diam. 6 in. / 15.3 cm
Private collection

A contemporary potter known for her highly creative sculptural forms, Zhou Dingfang (b. 1965) created this whimsical clay teapot composed entirely of mushroom forms. Zhou's bulbous and exaggerated textures often reflect the flora and fungi of her native China in breathtakingly realistic detail, from the craggy texture of petrified wood to the striated reediness of a bamboo shoot. The oversize mushrooms here almost seem alive, like they are overtaking the teapot's form; their caps are upturned, as if to let their gills breathe unfettered, and their stems are outstretched to form the teapot's handle and spout. Born in Yixing in Jiangsu Province, Zhou apprenticed under master sculptor Xu Xiutang, and herself achieved master status by just age thirty, in 1995. The region has been renowned for Yixing clay wares, in particular teapots, since the Song dynasty (960–1279). But in the late seventeenth century, Yixing pottery began to be exported to Europe, bringing the artisanal craft to the world stage. While Zhou was trained in traditional techniques and materials, in the late 1980s she began to experiment with other types of clay, embarking on the avant-garde sculpted works that would come to define her career. Often drawing on inspiration from the natural world—as with the exuberant mushrooms in *After Rain*—and everyday objects, Zhou has crafted teapots to resemble bamboo shoots, decaying branches, carts laden with pots, wooden boats, and leather satchels.

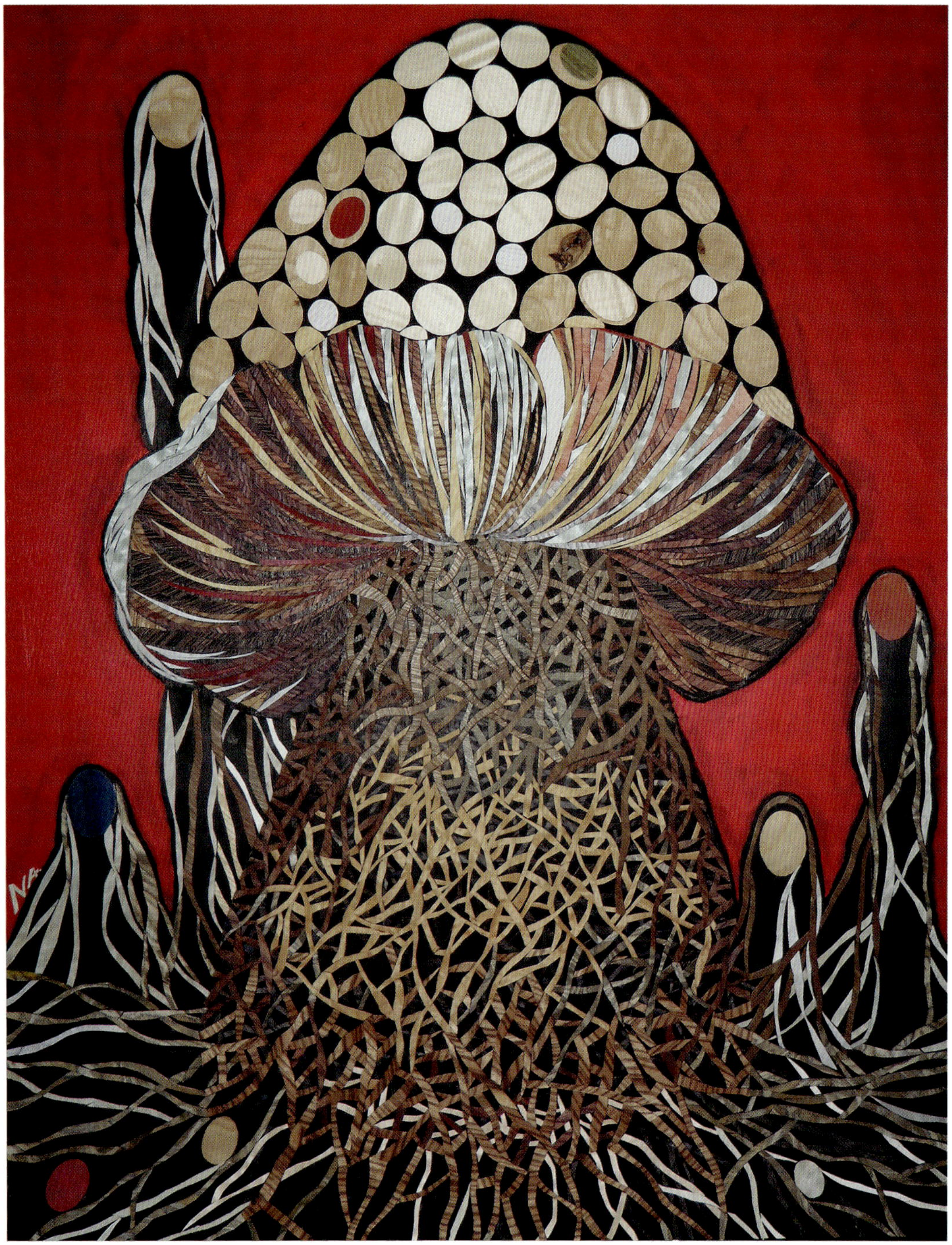

NAOKO MORISAWA

Magic Mushroom World VI Africa – Nairobi, 2015
Oil-stained wood mosaic, acrylic, oil, and sumi,
30 × 24 in. / 76.2 × 61 cm
Private collection, Japan

In an artistic interpretation of fungi from around the world, Japanese artist Naoko Morisawa (b. 1963) created fifteen works in her medium of choice: wood mosaic. Morisawa's intricate works are composed of thousands of small tiles of wood, each one hand-stained to bring out the wood's natural grain and texture. Many of them feature an exuberant display of colors and eye-catching, vibrant swirls that mesmerize the viewer with their vitality and, when viewed from afar, give the appearance of a painted canvas. *Magic Mushroom World VI Africa – Nairobi* celebrates the spectrum of wood's natural tones. The central figure of the mushroom is constructed from small wood discs that converge to form its cap. In contrast, long, thin strands of wood comprise the mushroom's gills, creating a neat curvilinear arrangement. Below the cap, those same wooden strands interweave and tangle to form the stem and surrounding environs. Set against a blood-red background, the mushroom looms large. Among the Kalenjin people of Kenya (of which Nairobi is the capital city), the mushroom has strong symbolic significance. It is a marker of ancestral connection—some believe that fungi's sudden appearance is a gift from ancestors in the afterlife—and a totem of considerable power. Other mosaics in the series draw on New York, Paris, São Paulo, Moscow, and other international locations, including Tokyo, where she was born. Based in the United States since 2004, Morisawa lives and works in Seattle, but she has said that through her *Magic Mushroom* series, she hopes her work travels like the spores of a mushroom, reaching destinations across the world.

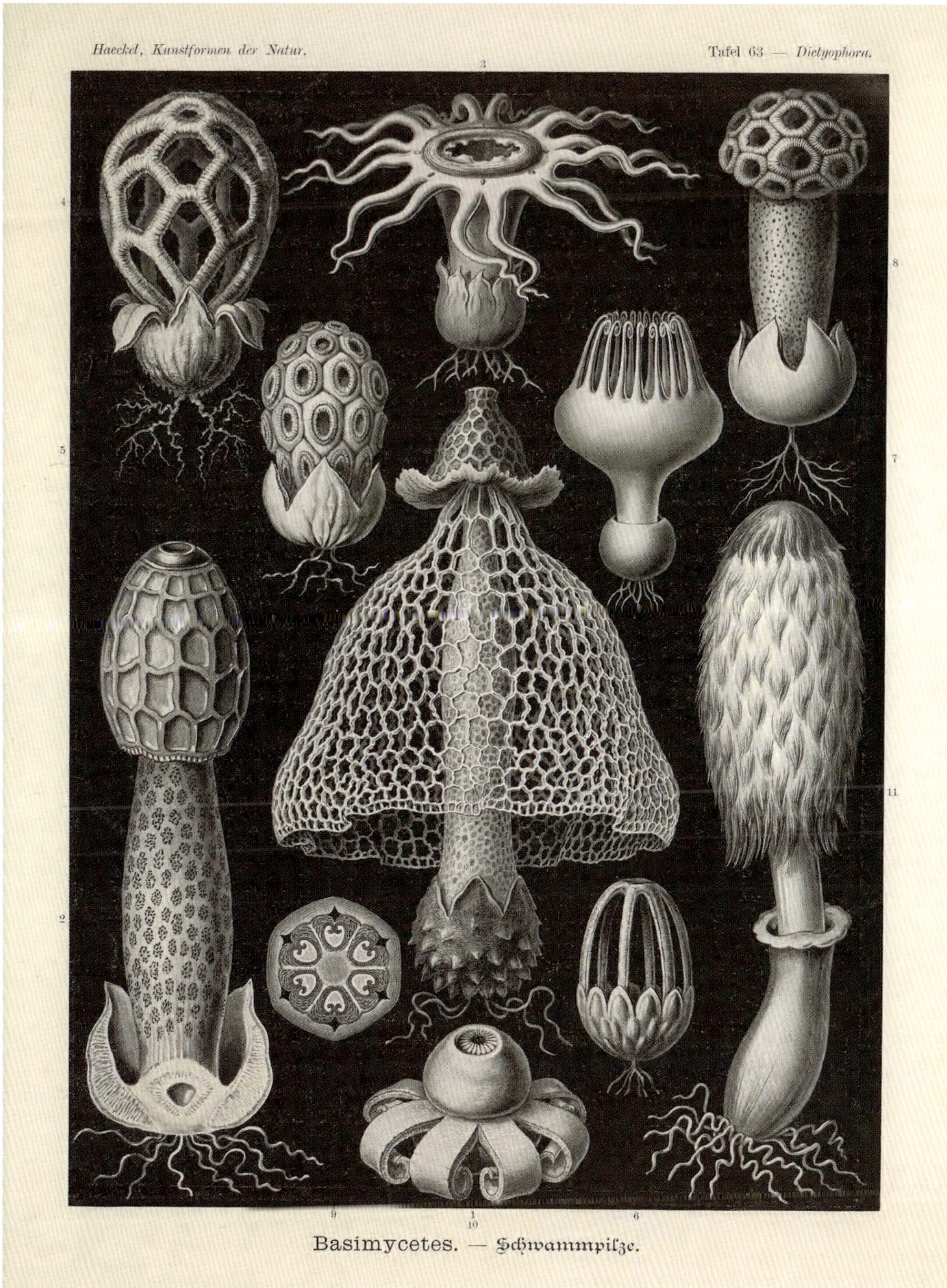

ERNST HAECKEL

Basimycetes, from *Kunstformen der Natur* (*Art Forms in Nature*), 1899–1904
Lithograph, 14½ × 11 in. / 37 × 28 cm
Smithsonian Libraries, Washington, DC

To say Ernst Haeckel (1834–1919) was a man whose life seemingly knew no bounds is not an overstatement. Born in Potsdam, Prussia (now part of Germany), Haeckel was a philosopher, zoologist, physician, professor, marine biologist, and artist—and he was a master at each. He discovered and named thousands of new species and coined numerous scientific terms, including *ecology* and *phylum*, that are integral to the field of biology. Haeckel was a staunch Darwinist, and as such, he believed in an ordered and logical development of life forms in the known world. As he studied these life forms, and discovered new ones, Haeckel made more than a thousand sketches and watercolors to capture his observations (photography was neither refined nor portable during the bulk of Haeckel's working years). Between 1899 and 1904, one hundred of Haeckel's best sketches were made into prints by master lithographer Adolf Giltsch and released ten at a time, comprising what is now known as *Kunstformen der Natur* (*Art Forms in Nature*). Many of the prints in the seminal tome focus on patterns and organization in the natural world, such as the symmetry of arachnids and arthropods, the patterns of fish scales and reptile skin, and, as seen here, the repetition found in mushrooms and other fungi. Titled *Basimycetes*, this print captures a type of fungi called basidiomycetes, which includes puffballs, agarics, jelly fungi, chanterelles, and many others. The detail found in Haeckel's various fungi artworks served to advance the science of mycology at large.

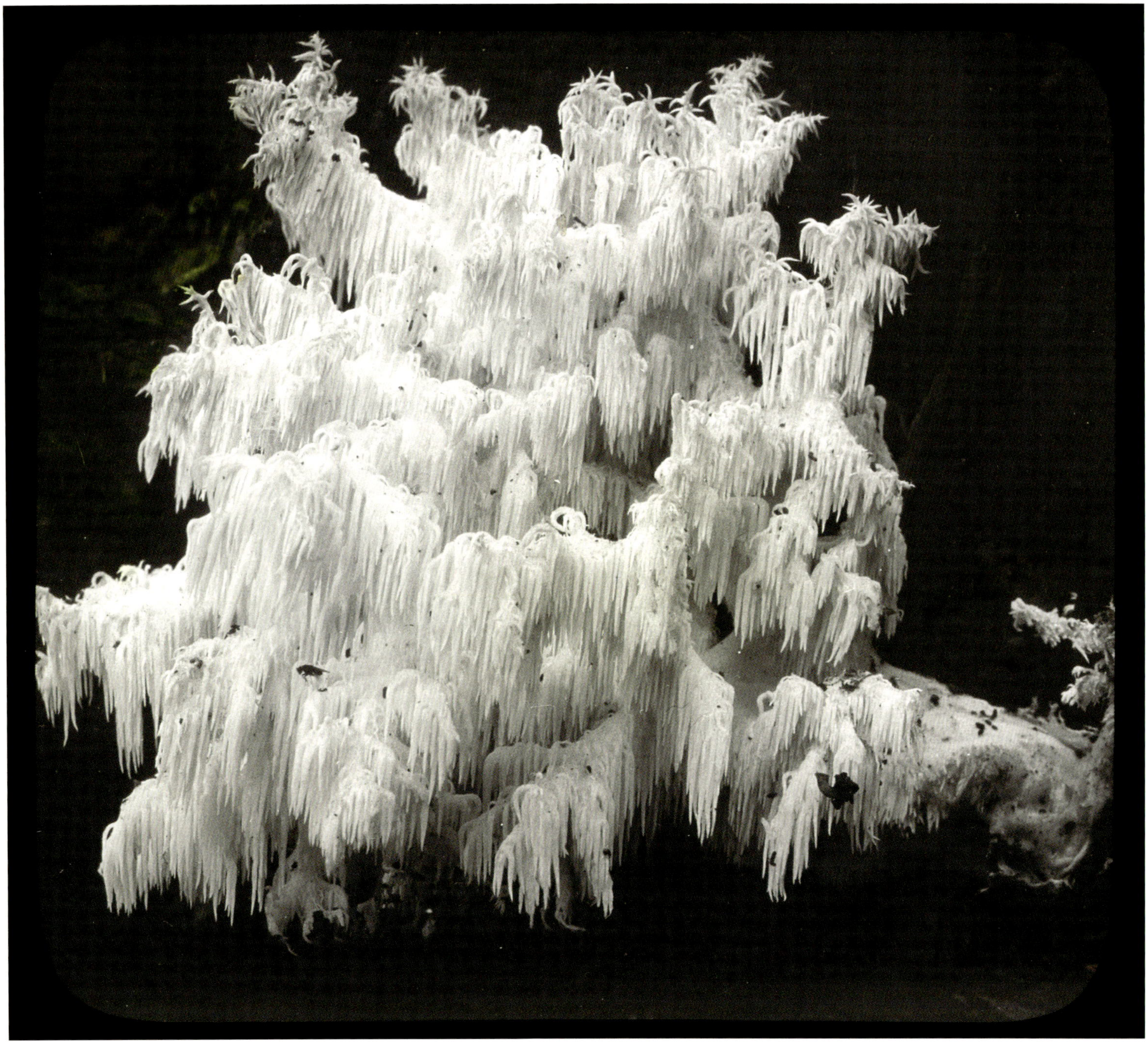

NED L. HUFF

Hydnum, 1936
Lantern slide, 3¼ × 3 in. / 8.3 × 7.6 cm
University of Minnesota Libraries,
University Archives, Minneapolis

With soft, wispy tendrils, this organism appears as if it would be more at home underwater than in the forest. Indeed, the *Hericium coralloides* (formerly *Hydnum coralloides*) is commonly known as the coral tooth fungus. Like its more popular cousin lion's mane (*Hericium erinaceus*), the species of the *Hericium* genus are edible, with a taste similar to lobster or crabmeat. Besides being a prized culinary delicacy, they are also potent comestibles that have been used for centuries in Eastern medicine. Recent scientific studies of *Hericium coralloides* in the United States have even shown its benefit in treating neurodegenerative conditions, such as Alzheimer's disease, and promoting overall immunity and digestive health. This image is one of more than five thousand slides left to the University of Minnesota by Ned L. Huff (1876–1964).

Huff was an assistant professor of botany at the university for nearly four decades, and he also taught the public—educators, scout and camp leaders, and sundry other nature enthusiasts—through an extension course he initiated on Minnesota plant life. The centerpiece of his instruction was his collection of slides depicting the state's diverse plant life and fungi. Huff spent his summers traveling the region, photographing and cataloging its natural bounty. He then undertook the meticulous work of printing his photographs on sensitized glass, hand-coloring many of them to vivify his images. While his work was relatively lost to obscurity until the university began scanning his slide collection in 2014, Huff has now been recognized for both the artfulness of his photographs and his diligence in documentation of natural history.

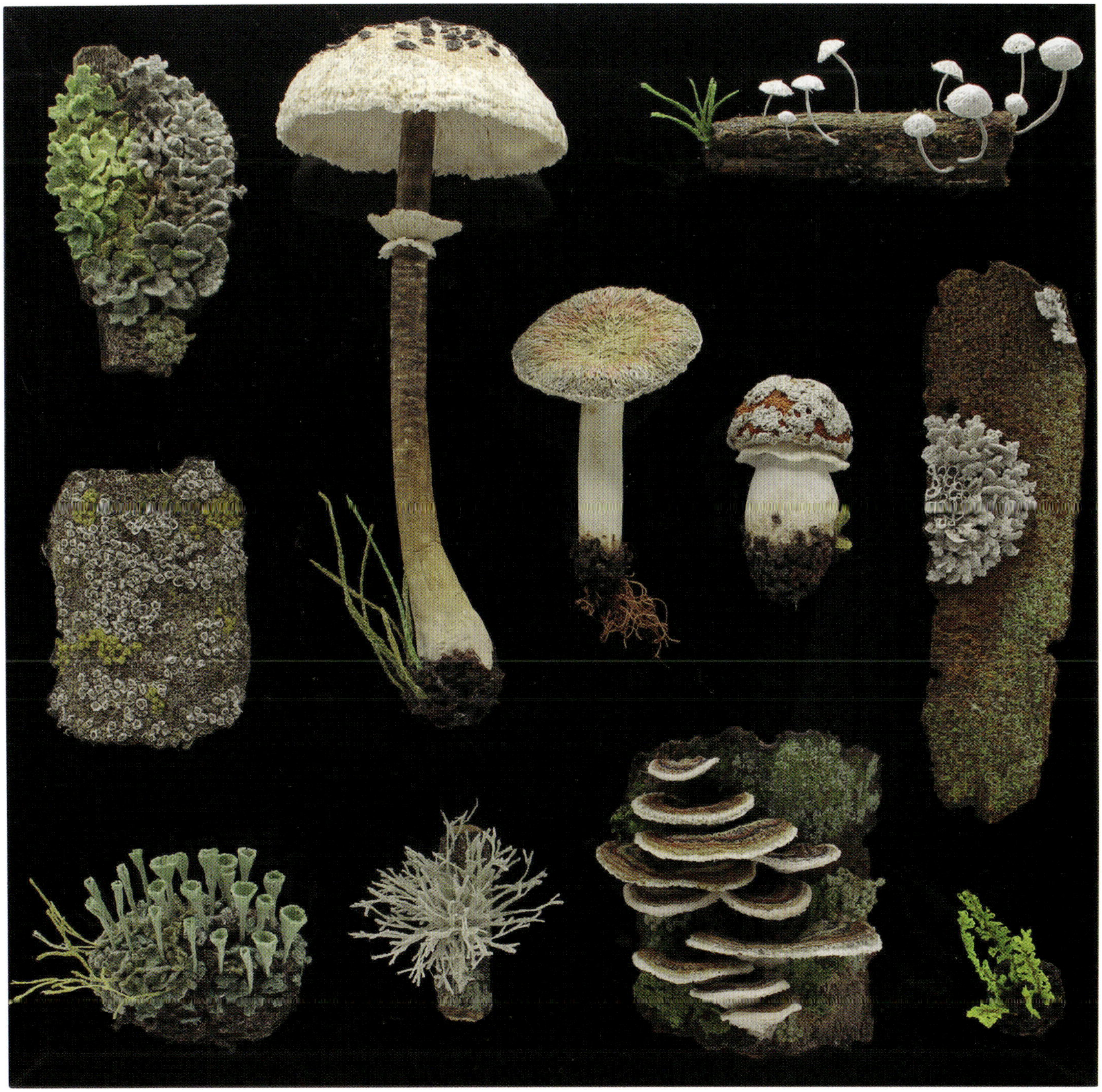

AMANDA COBBETT

Macrolepiota procera, Amanita excelsior (var. spissa), Russula ionachlora, Trametes versicolor, Mycena (on wood shard), Leconora dispersa, Physcia aipolia, Ramalina farinacea, Hypnum cupresseforme, 2023
Mixed threads and paper,
13⅜ × 13⅜ × 4 in. / 34 × 34 × 10 cm
Private collection

Deeply detailed embroidered textile facsimiles of fungi sit in a Perspex box, seemingly plucked from the forest floor and chopped from the trunks of fallen trees. Among them are a large parasol mushroom (*Macrolepiota procera*), a gray spotted Amanita (*A. excelsior* var. spissa), oilslick brittlegill (*Russula ionochlora*), turkey tail (*Trametes versicolor*), bonnets (*Mycena*) and a number of lichens. British artist Amanda Cobbett (b. 1972) is known for her hyper-realistic, three-dimensional sculptural works depicting the breadth and nuance of nature in fabric form. Her multimedia works are made by interweaving colorful rayon threads using free-motion machine embroidery, essentially "drawing" with thread. She works with dissolvable fabrics, building up her pieces in layers that she molds into her final embroidered works. Some of her pieces play with the traditional mushroom shape itself, using hand-dyed, scorched papier-mâché tubes to represent decaying stems. Her visible roots are made from hand-stitched silks, the threads hanging like hyphae, the branching filaments that collectively form a fungus's mycelium. Cobbett studied at Chelsea College of Arts in London before working in the printed textile design industry and sidestepping into her own art practice. She was inspired to depict nature during her daily dog walks in a nearby forest, where she would collect and photograph fallen debris. She then transformed this source material into her unique works through sketches and up to 130,000 stitches per day. Cobbett embeds a sensitivity to nature in all her creations, ranging from her use of soy-based inks to her choice of biodegradable packaging.

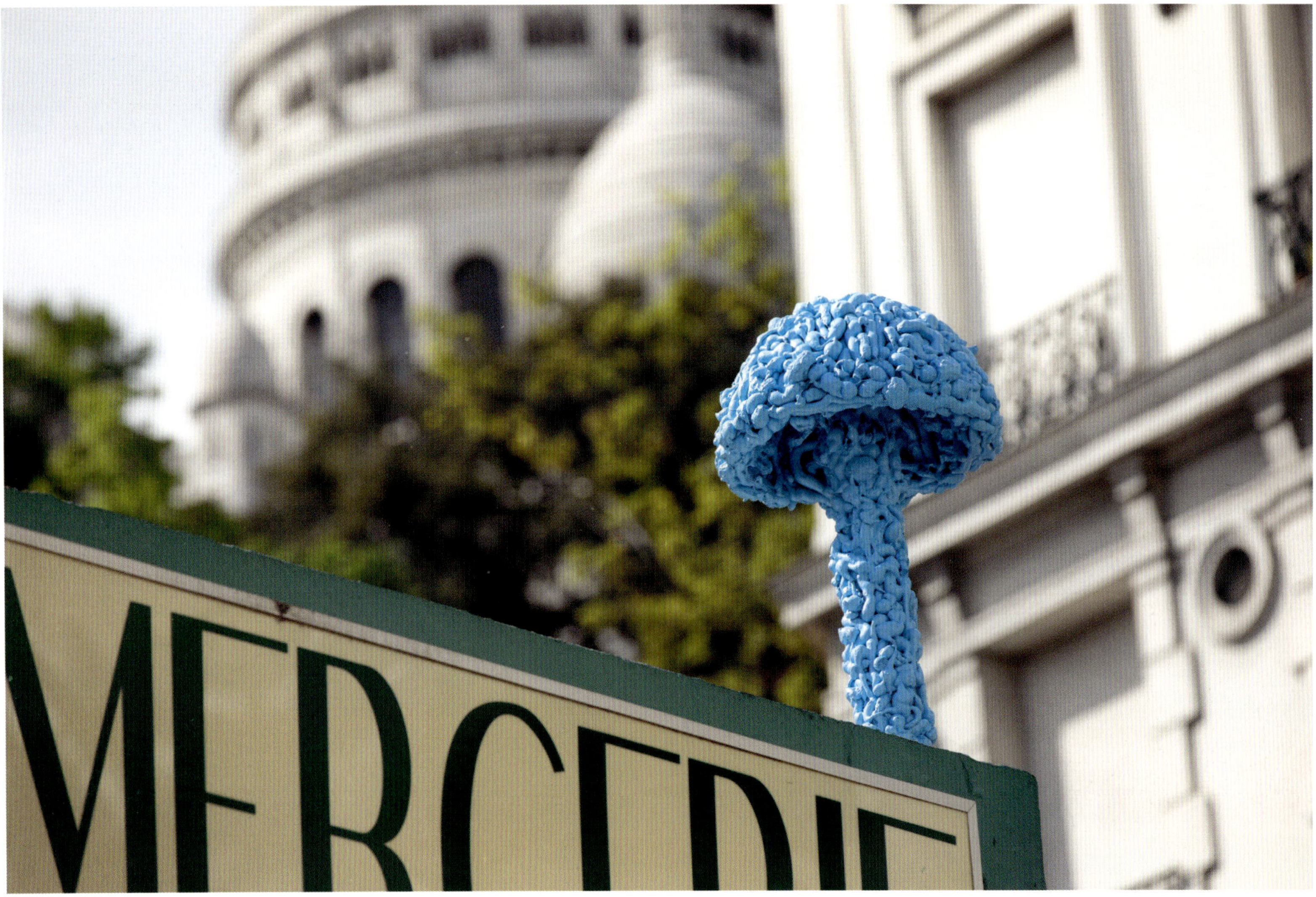

CHRISTIAAN NAGEL

Mushroom in Blue, 2019
Polyurethane, H. 39⅜ in. / 100 cm
Rue Cazotte, Paris

Noticed only by those glancing up from the street, this oversize blue mushroom—high above the heads of passersby on Rue Cazotte in Paris—delighted and perplexed in equal measure. Attached to the rooftop of a commercial property, the lightweight polyurethane sculpture was created by South African street artist Christiaan Nagel (b. 1982), who has installed his vividly colored toadstools atop buildings and other tall structures around the world, from London and Cape Town to New York and Berlin. In the early 2010s, his vibrant fungi began appearing all over East London's rooftops as the area's street art scene burgeoned primarily on the walls below. For Nagel, mushrooms are an apt metaphor for street art, popping up unexpectedly overnight with legal and illegal implications. Fungi on buildings is generally considered a problem, a sign of disrepair, damp, or rot. Yet Nagel's vibrant and uplifting mushrooms enliven their urban locations. Before turning to sculpture, Nagel was a painter, and his mushrooms, made with expanding polyurethane foam, have a distinctly painterly quality. The domed caps are made by squirting foam over an exercise ball, and while each one adopts the classic mushroom shape, their squiggly surfaces resemble worm castings, albeit colored with sunshine yellow, neon pink, electric blue, and other artificial hues. In addition to his playful street-art installations, Nagel makes bronze mushrooms for private galleries, a range of mini mushroom collectibles, and has even designed a mushroom-themed chessboard.

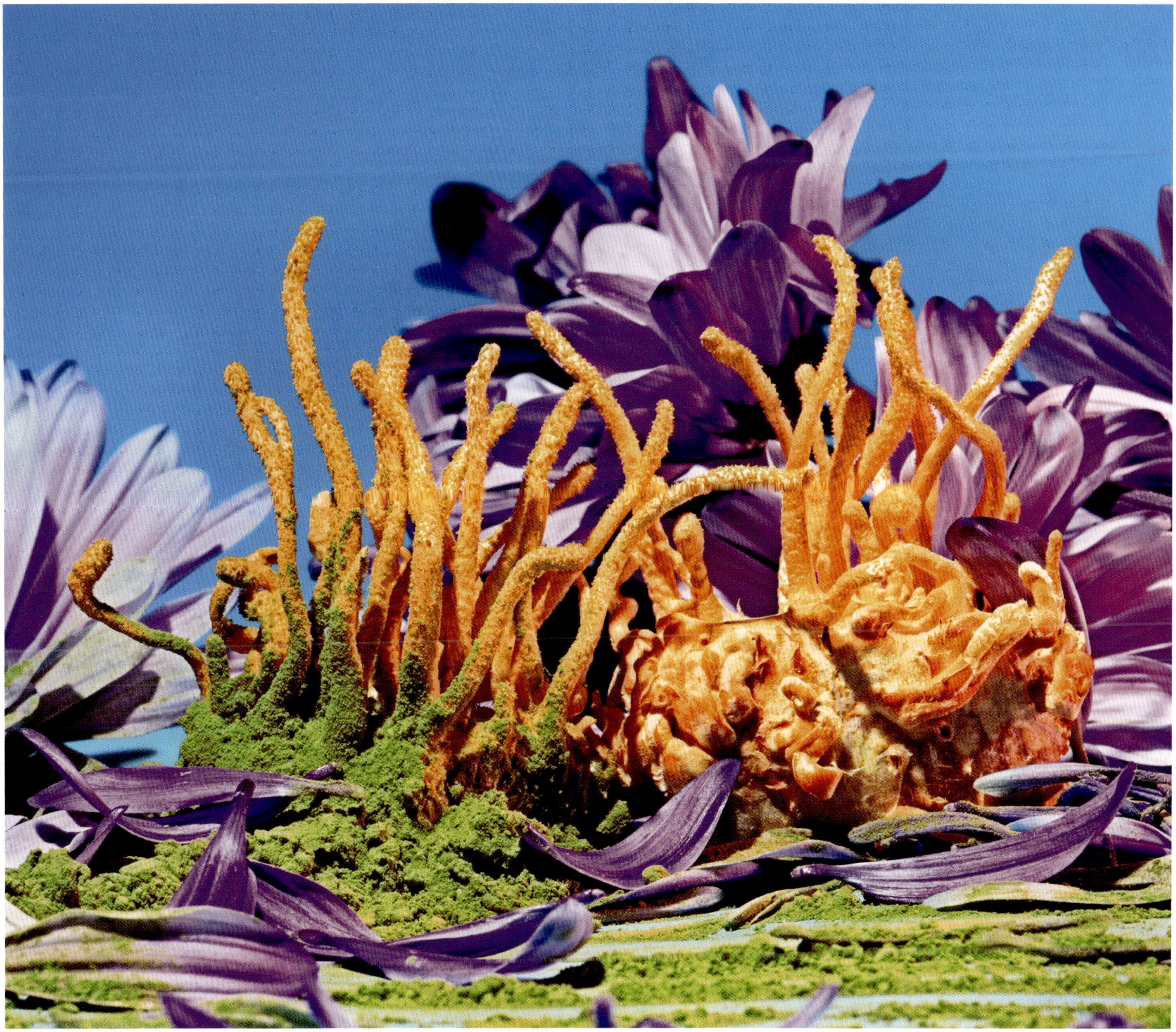

PHYLLIS MA

Cordyceps militaris, 2020
Digital photograph, dimensions variable

To Switzerland-based artist Phyllis Ma (b. 1987), every species of mushroom has its own distinct personality, each one informing the delightful photographs in her *Mushroom & Friends* series. Here, the vibrant orange *Cordyceps militaris*, or caterpillar fungus, emerges with spindly fingers from a mossy green ground, set starkly against purple flower petals. The alien-like setting is an apt atmosphere for the mushroom, which parasitizes insects, growing from dead underground pupae. Born in Guangzhou, China, Ma immigrated to the United States with her family at age eight, ultimately earning a degree in visual arts from Columbia University in New York. In 2014 she launched the Instagram project *Lazy Mom* with her fellow artist and Columbia grad Josie Keefe. Lampooning the prevailing trends in food photography—highly fussy plates of haute cuisine—they presented imagery created by an imaginary mother who preferred playing with food to cooking it. What began as a lark ultimately led to professional commissions and germinated the seeds for what would become Ma's *Mushroom & Friends* zines. Ma's fantastical still lifes have a dreamlike, almost hallucinatory quality. The luminous backgrounds of her works emphasize the unique topography and shape of their spotlighted shrooms, which stand in stark relief to their Technicolor settings. But Ma's interest in mushrooms is not purely aesthetic. An amateur mycologist, she is fascinated by their inherent function as nature's recycling agent. She echoes their sustainability superpowers when putting together her own creations: most of her images are assembled from fungi she's foraged herself, combined with on-hand household items.

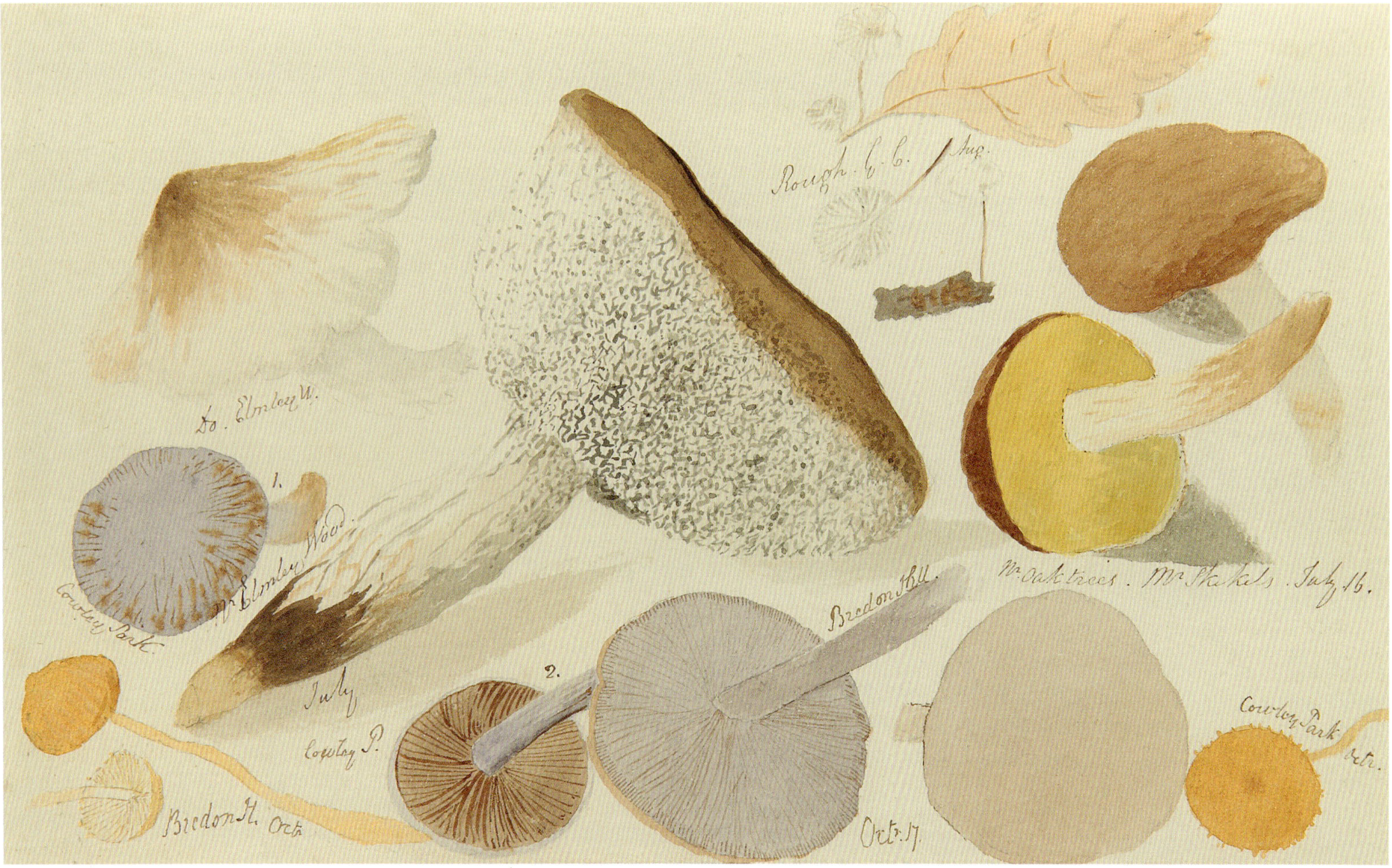

ANONYMOUS

Fungi manuscript album, c. 1850
Pen, ink, and watercolor on paper, sheet 5⅜ × 8⅝ in. / 13.5 × 22 cm
Private collection

Delicate watercolors of various mushrooms fill this sheet from a sixty-four-page album of around three hundred species of mushrooms and fungi by an unknown English artist. Each page depicts different mushrooms, painted in grays, whites, yellows, and browns, alongside handwritten details of where and when they were found. Dating from the nineteenth century, the album belonged to Sylvia Joan Thurlow of Driffield, a town in Yorkshire, England, although it is undetermined if Thurlow was the artist and amateur mycologist. There was a long tradition of female nineteenth-century mushroom hunters—many of them English—who recorded their discoveries with artistically skillful and botanically accurate drawings and illustrations. Women such as author Beatrix Potter—best known for *The Tale of Peter Rabbit* and other children's books—were dedicated mycologists (see p.6). Another woman mycologist of note was Anna Maria Hussey, whose *Illustrations of British Mycology* combined hand-colored lithographs with foraging instructions and recipes that widened the fungi audience (see p.78). At the time, the study of fungi was largely ignored by the male-dominated scientific world, as mushrooms were not considered worthy of study. Women, restricted from participating academically in the Victorian obsession with botanical study and analysis, turned to the overlooked humble mushroom instead. Their illustrations, notes, and descriptions helped grow the understanding and knowledge of fungi in what was the relatively new field of mycology.

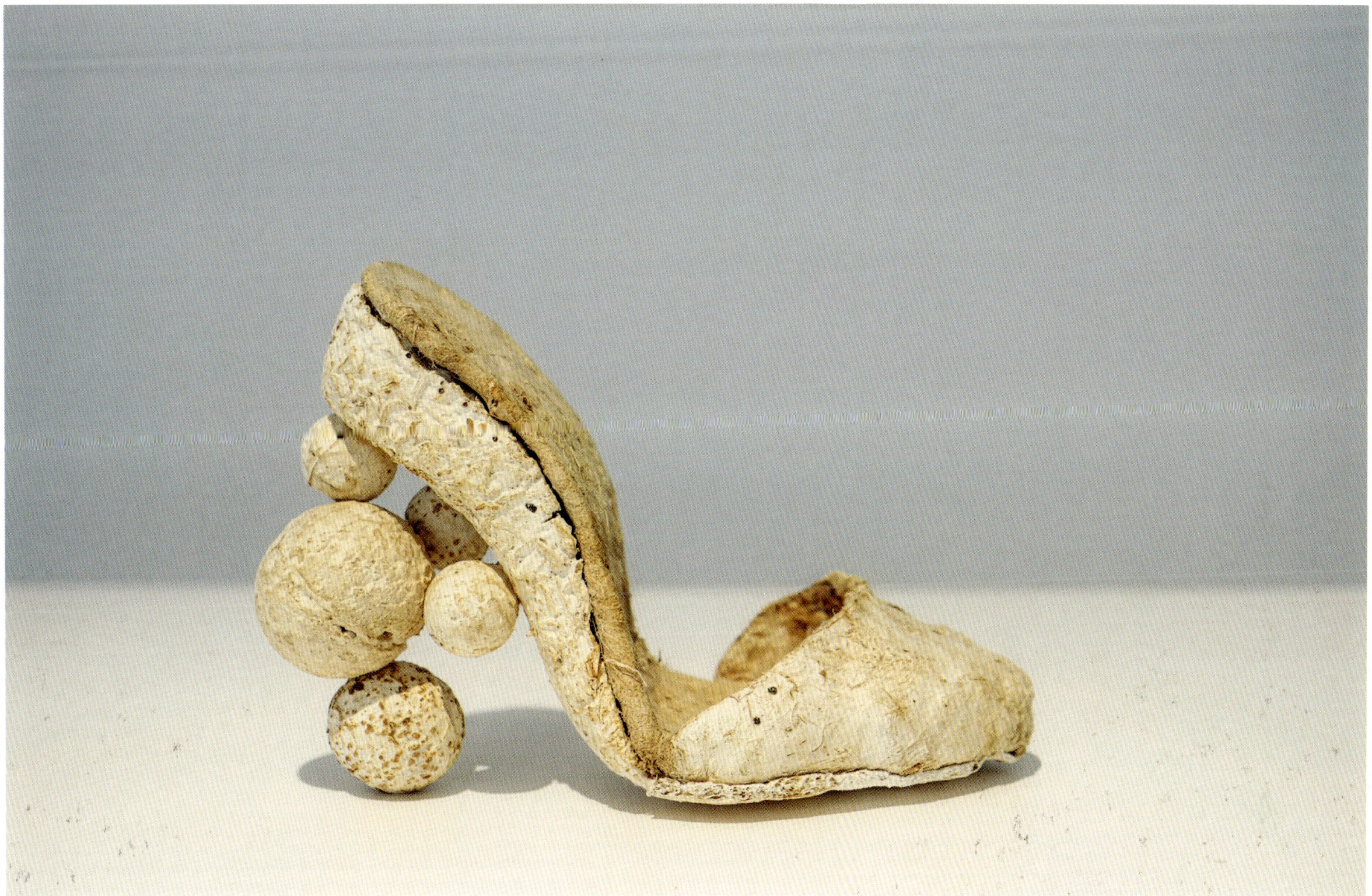

KRISTEL PETERS

Mycoschoen #1, 2018
Trametes versicolor mixed with linen, cotton, and hemp, 6¼ × 9½ × 3½ in. / 16 × 24 × 9 cm
Private collection, Graz, Austria

A single prototype stiletto made from treated mycelium balances precariously, its heel made from a series of round, mushroomlike balls. *Mycoschoen #1* is one of a series of fashion experiments by Belgian shoe designer and consultant Kristel Peters (b. 1974) created in 2018 that explored the use of mycelium as a material. The aim was to develop shoes with little to no environmental impact. Peters had previously worked with fashion houses Dries Van Noten and Bottega Veneta before concentrating on sustainable fashion design. Existing at the intersection of biotechnology and fashion, Peters's work is at the most artistic, experimental end of contemporary research in fashion for new biomaterials. With an industry progressing toward sustainable fashion, GANNI, Stella McCartney, Hermès, and Balenciaga have all used mycelium fabrics or leather (see pp.42, 64, 177). And in the years since Peters's work, mycelium-based fabric has been used in shoes by Vivobarefoot, and footwear designer Nicholas Rapagnani published a study titled *Bio Design for Footwear Innovation*, focused on growing mycelium-based components for sneakers. In 2025 the prototype of the world's first mushroom-made sneakers was developed and launched by the Korvaa Consortium, created entirely without plastics or fossil fuels. These shoes were a collaboration among Ecovative (who created the sole), Ourobio (who made the base structure), and Modern Synthesis (who developed the shoe upper out of bacterial nanocellulose), clearly demonstrating that mushrooms are at the cutting edge of sustainable footwear design.

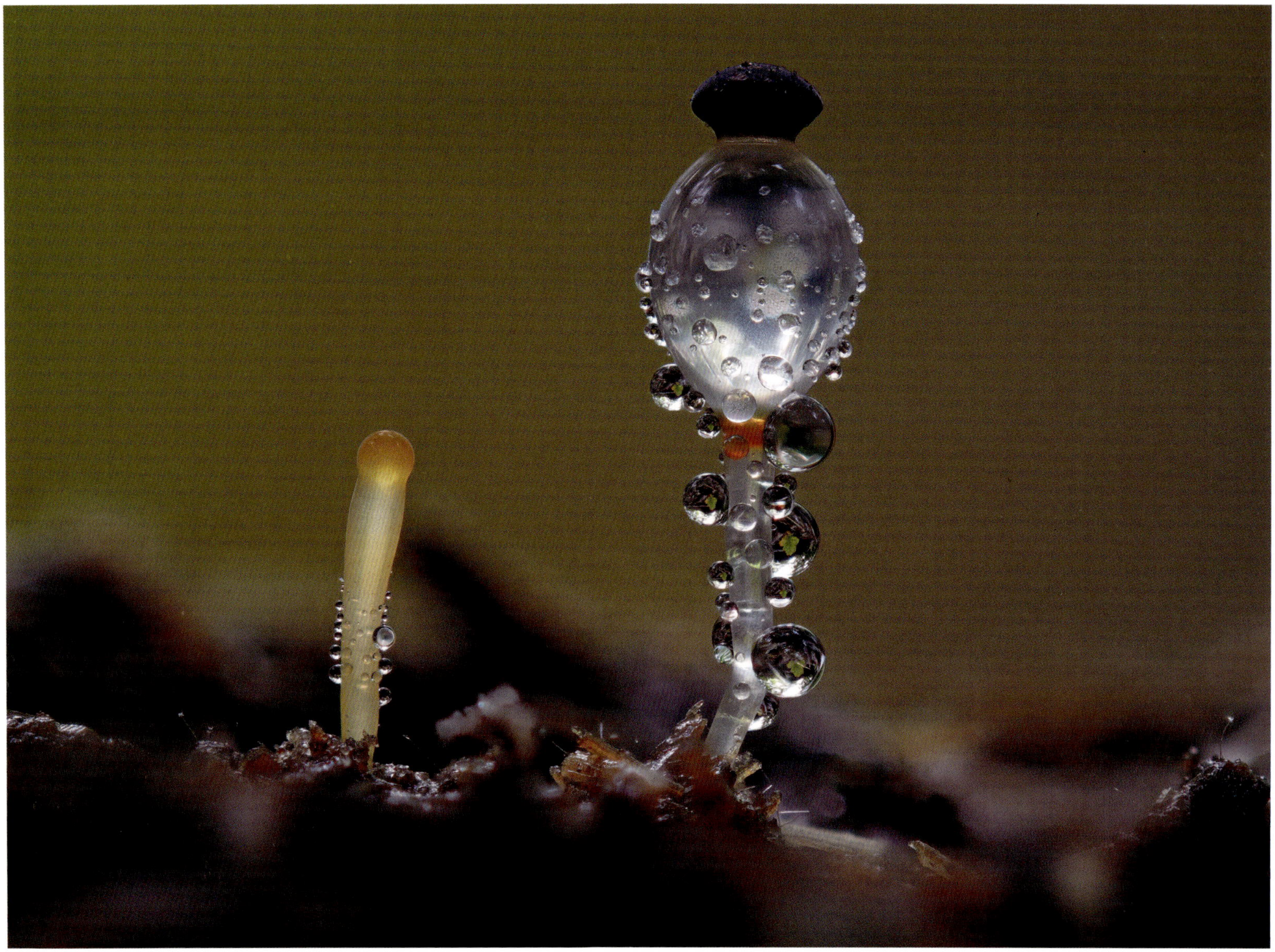

MAX MUDIE

Pilobolus sp., 2022
Digital photograph,
11 × 14 in. / 27.9 × 35.6 cm

Through his highly popular Instagram posts, British macrophotographer Max Mudie (b. 1985) has done much to promote the remarkable multicolored world of fungi and slime mold that is usually hidden from view. As he says, "When you find out how integral fungi are to our existence, it makes everything else seem insignificant." Mudie's passion for fungi was reignited by a move to the East Sussex countryside, where he often forages in the local woods and forests, particularly enjoying discovering undocumented species. Many of the fungi he finds are so minuscule that the only way to record them is by using a macro lens to photograph them. This image is of *Pilobolus crystallinus,* more commonly known as the dung cannon or hat thrower. It rarely reaches more than a quarter of an inch (5 mm) in height, but what it lacks in stature it more than makes up for with the speed with which it releases its spores. The spores are stored in the fluid-filled bladder at the top of the stem, and to reproduce, the bladder expands until it bursts, sending spores at speeds of 56 miles per hour (90 km/h) as far as 656 feet (200 m) across the grass for grazing mammals to eat and then disperse. For Mudie, the thrill of fungi comes from their sheer diversity in terms of shapes, colors, and forms. Such is his passion that he cofounded the Sussex-based festival All Things Fungi, which has become an annual three-day event for myco-enthusiasts.

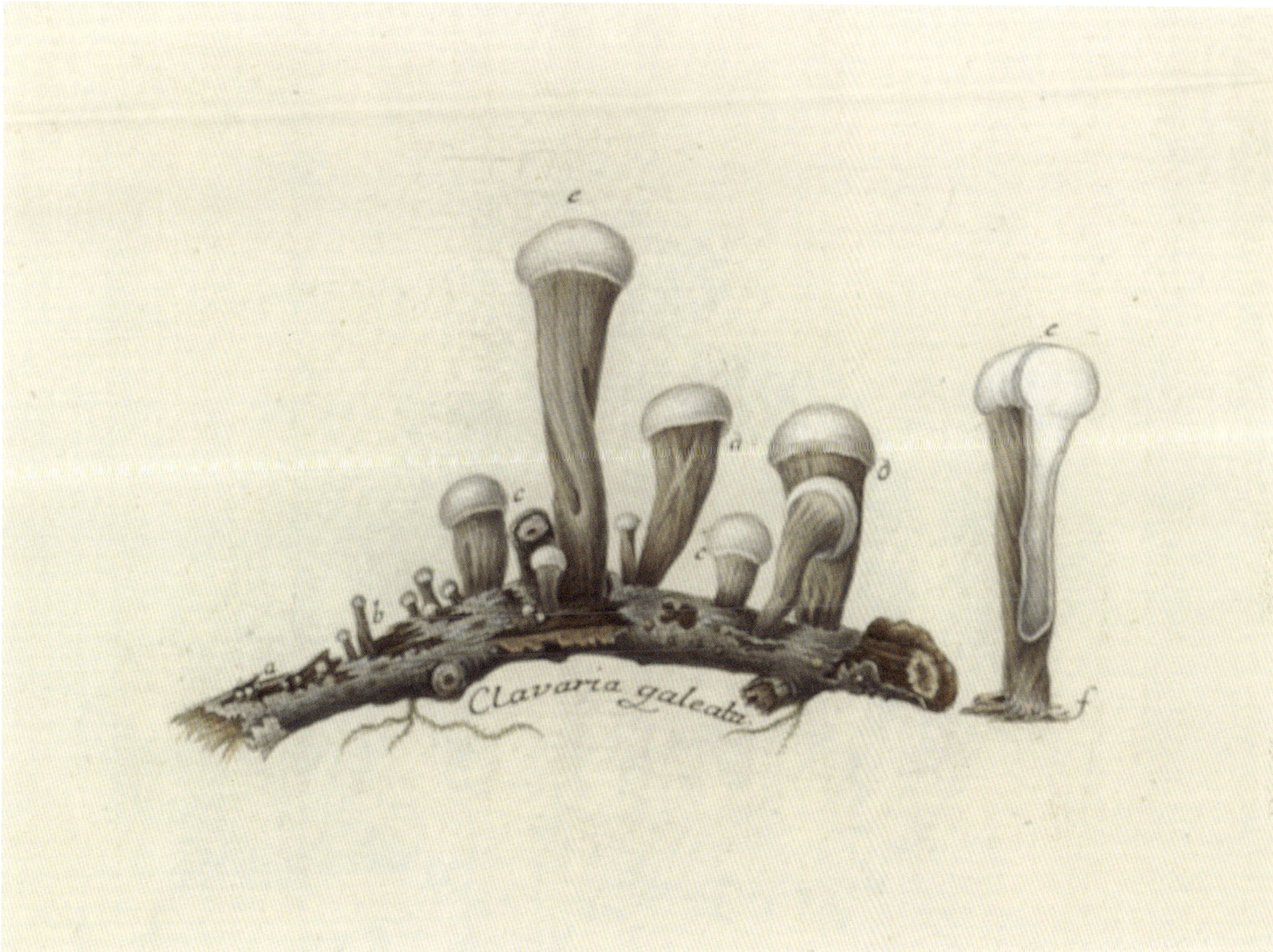

JOHANN NEANDER

Clavaria galeata, from *Beata ruris otia fungis Danicis* (*Happy Resting Periods in the Country Studying Danish Fungi*) by Johan Theodor Holmskjold, 1799
Copper engraving, H. 15 in. / 38 cm
University of Michigan, Ann Arbor

This illustration of the slender and delicate fruiting bodies of the fungus *Clavaria galeata* (now known as *Ascocoryne sarcoides*), which typically grows in clusters on dead tree trunks or branches, is by Danish painter and engraver Johann Adolf Neander. It first appeared in the two-volume work on Danish fungi *Beata ruris otia fungis Danicis* (*Happy Resting Periods in the Country Studying Danish Fungi*), a groundbreaking addition to the field of mycology published in 1790 with an initial printing of forty-three copies. *Beata* is the work of Danish botanist Johan Theodor Holmskjold (1731–1793), who commissioned Neander to make botanically accurate, life-size drawings of the fungi he had discovered over a two-year period living near the Danish seaport of Aarhus. The result of Holmskjold's discoveries was game-changing: the book includes fifty-seven then-new fungi he named—five new combinations and fifty-two new taxa. Holmskjold was the first person to realize that *Cordyceps militaris* (scarlet caterpillarclub) grows on dead insects found in the soil rather than, as was previously believed, transforming into fungus from the insects. That the book was ever published is just as remarkable. Holmskjold retired in 1765, at the age of just thirty-four, from the Academy of Sorø, where he was a professor of natural history, and embarked on his fungi-research sabbatical. But in 1767 he was appointed general director of the Danish postal service, a position that led to him dedicating the rest of his life to serving the Danish monarchy, leaving him little time for anything else.

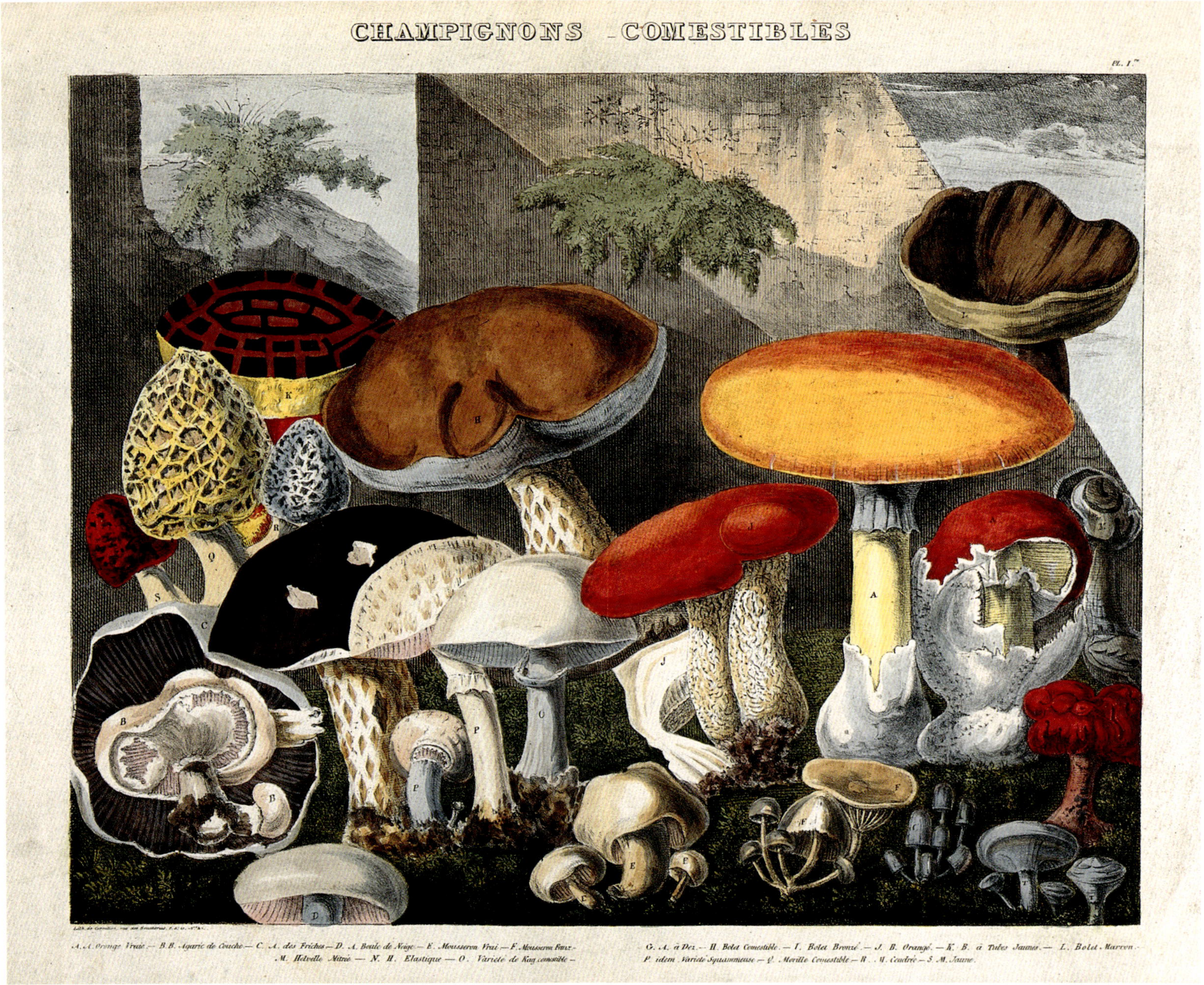

A. CORNILLON AFTER PRIEUR

Champignons Comestibles, from *Atlas des Champignons: Comestibles, suspects et vénéneux* (*Atlas of Mushrooms: Edible, Suspicious, and Poisonous*) by Michel Étienne Descourtilz, 1827
Lithograph with watercolor,
approx. 15¾ × 19⅝ in. / 40 × 50 cm
Wellcome Collection, London

Set against a rather dramatic background of stone ruins among a mountainous landscape, this plate depicts nineteen species of edible fungi. Created by French lithographer A. Cornillon after an original illustration by an artist only identified as Prieur, the vibrant tableau was included in the 1827 publication *Atlas des Champignons: Comestibles, suspects et vénéneux* (*Atlas of Mushrooms: Edible, Suspicious, and Poisonous*) by Michel Étienne Descourtilz (1775–1835/6). Descourtilz was a physician and botanist who led a somewhat unexpected life. Born in France, he studied medicine and trained as a surgeon before traveling in 1799 to Haiti, where his family owned several plantations. The country was in the midst of a revolution, and Descourtilz was promptly arrested and served for four years, against his will, as physician to the revolutionary forces, narrowly avoiding death. Upon his return to France, he seemingly put the events behind him and returned to medicine and a renewed interest in mycology. The aim of his *Atlas* was to educate the public, especially the poor, about the uses and the dangers connected with foraging for wild fungi. Descourtilz offered his guide "out of a devotion to the alleviation of human suffering." The work includes ten plates by Cornillon, with each species keyed by letters to French names at the foot of the plates. Among the edible fungi shown here are the common cultivated mushroom at the lower left (B, *Agaricus bisporus*), with a trio of morels above, including the common morel (Q, *Morchella esculenta*), and at center the highly prized penny bun, or cep (H, *Boletus edulis*).

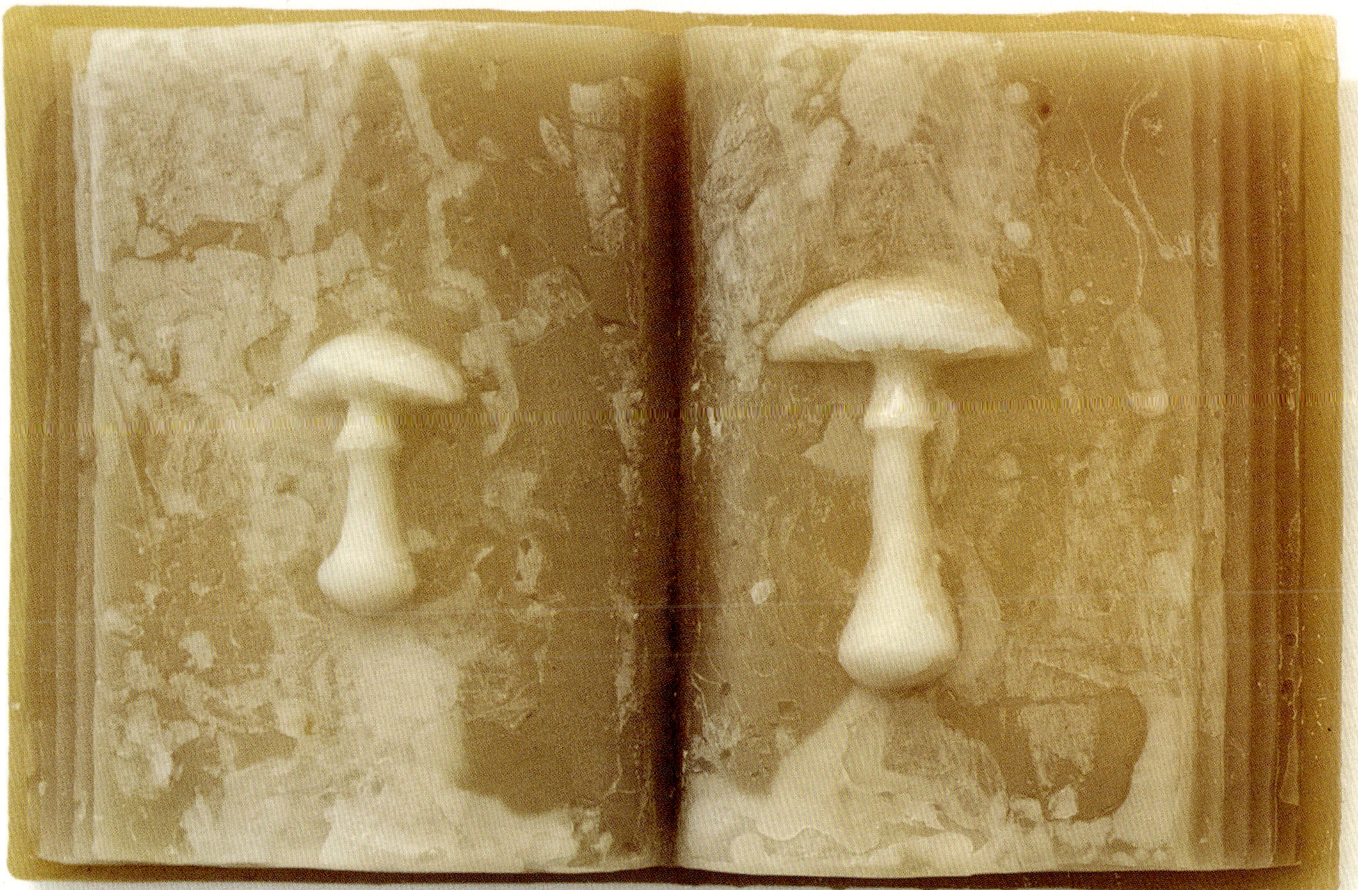

STEPHAN DOITSCHINOFF

Sculpture from *Remotis Visum*, 2017
Paraffin, wick, and pigment,
13¾ × 9 ⅞ × 2¾ in. / 35 × 25 × 7 cm
Private collection, São Paulo, Brazil

A paraffin wax sculpture in the form of a book lies open, its pages replaced with the relief imprint in the shape of two mushrooms. This is one of a series of wax wall pieces, including hands that function as candles, with mushroom images pressed into their palms, and other open books showcasing various flora. Created by Brazilian artist Stephan Doitschinoff (b. 1977), the series is just one facet of his practice, which includes painting, sculpture, videos, installation, and performance. Much of Doitschinoff's work expands on a fictional cosmology—a fusion of Afro-Brazilian traditions, folkloric symbolism, and diverse religious visual systems. Mushrooms have also appeared in his graphic paintings and works on paper, and he is notably interested in fungi and psychoactive plants for their role in spiritual practices as a way to connect the human and the divine. These works also touch on the tension between the culture of Indigenous rituals and colonialism innate in Brazil's history. A self-taught artist, Doitschinoff is based in São Paulo and has exhibited at international institutions such as the Irish Museum of Modern Art, the Museum of Contemporary Art San Diego, and the Fondation Cartier in Paris, as well as MAM and MASP in Brazil. In recent years, he has been working on *Cvlto do Fvtvrv*, an ongoing project structured as an invented philosophical order across multiple media. No matter the medium, symbolism and mythology are deeply central to Doitschinoff's approach.

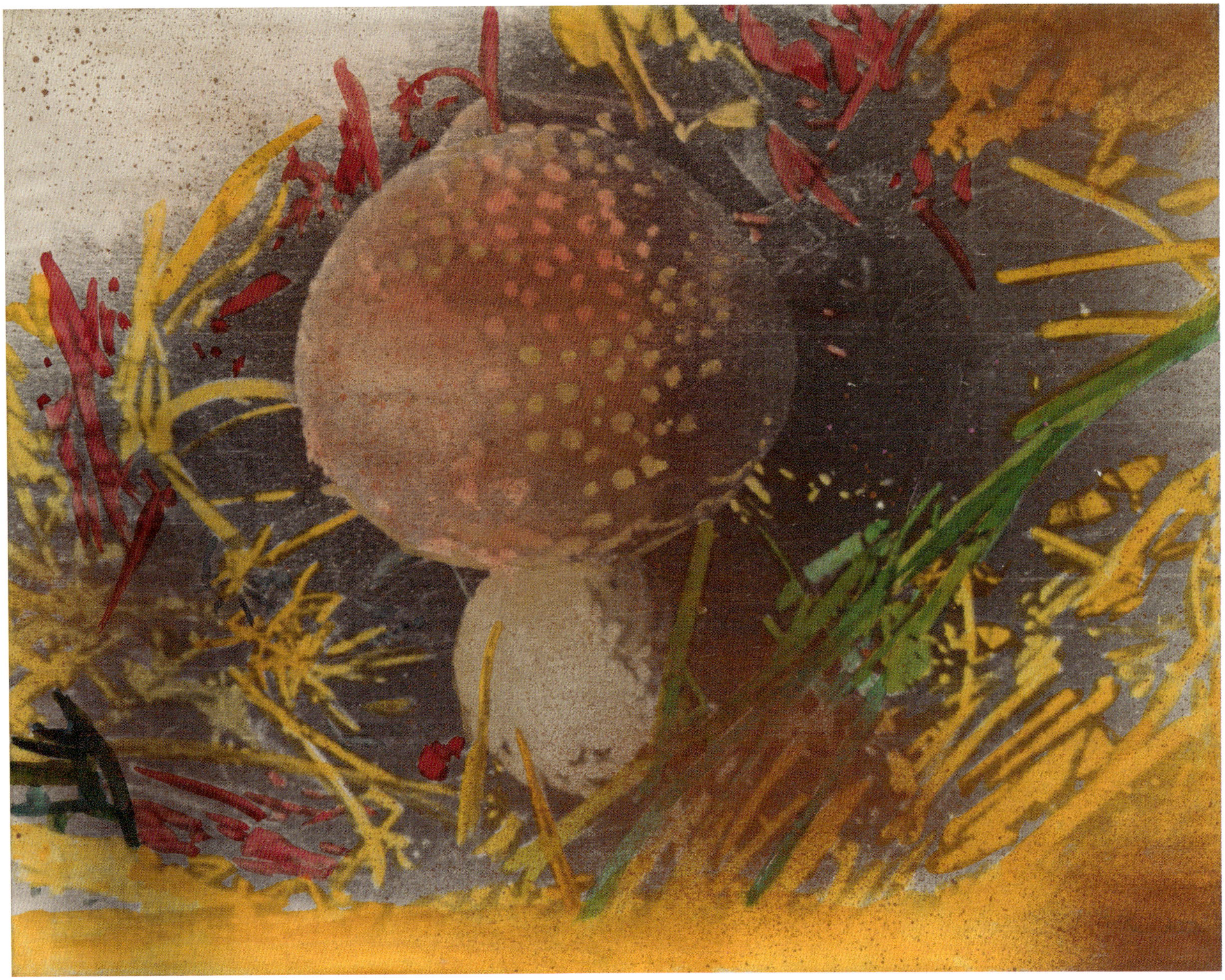

SIGMAR POLKE

Untitled (Fly Agaric), 1975
Diaphoto retouching dye and metallic spray paint on photosensitive canvas, 15¾ × 19¾ in. / 40 × 50 cm
ACT Art Collection, Germany

Clusters of expressive and energetic brushstrokes in yellow, green, and red swirl around a fly agaric (*Amanita muscaria*) mushroom, distinguished by its red-and-white spotted markings. Created by German artist Sigmar Polke (1941–2010), this colorful mixed-media painting combines photographic and painterly techniques to evoke the featured fungus's hallucinogenic properties. Belonging to a generation of rebels and nonconformists who rose to prominence in the 1960s, Polke was one of the most experimental and influential artists of the twentieth century. He spoke openly about his experiences experimenting with psychoactive substances in the 1970s, and much of his work from this period features surreal and drug-inspired imagery, some of which was actually created while under the influence of hallucinogenic mushrooms, LSD, and peyote. For Polke, mind-altering fungi and other drugs were a tool for expanding consciousness, opening up new creative possibilities that liberated him from the fetters of conventional art-making. These were fruitful years for the artist, leading to wildly experimental photomontages, prints, and mixed-media paintings that combined figurative and abstract elements, and that used an array of unusual materials, including heat- and light-sensitive chemicals, silver oxide, arsenic powder, resin, and even meteor dust. In many works, Polke employed chemical agents to disrupt and distort his imagery in a seeming attempt to portray the experience of altered consciousness. Although he produced many mushroom-inspired works on paper, this is one of the few canvases he made addressing the subject directly.

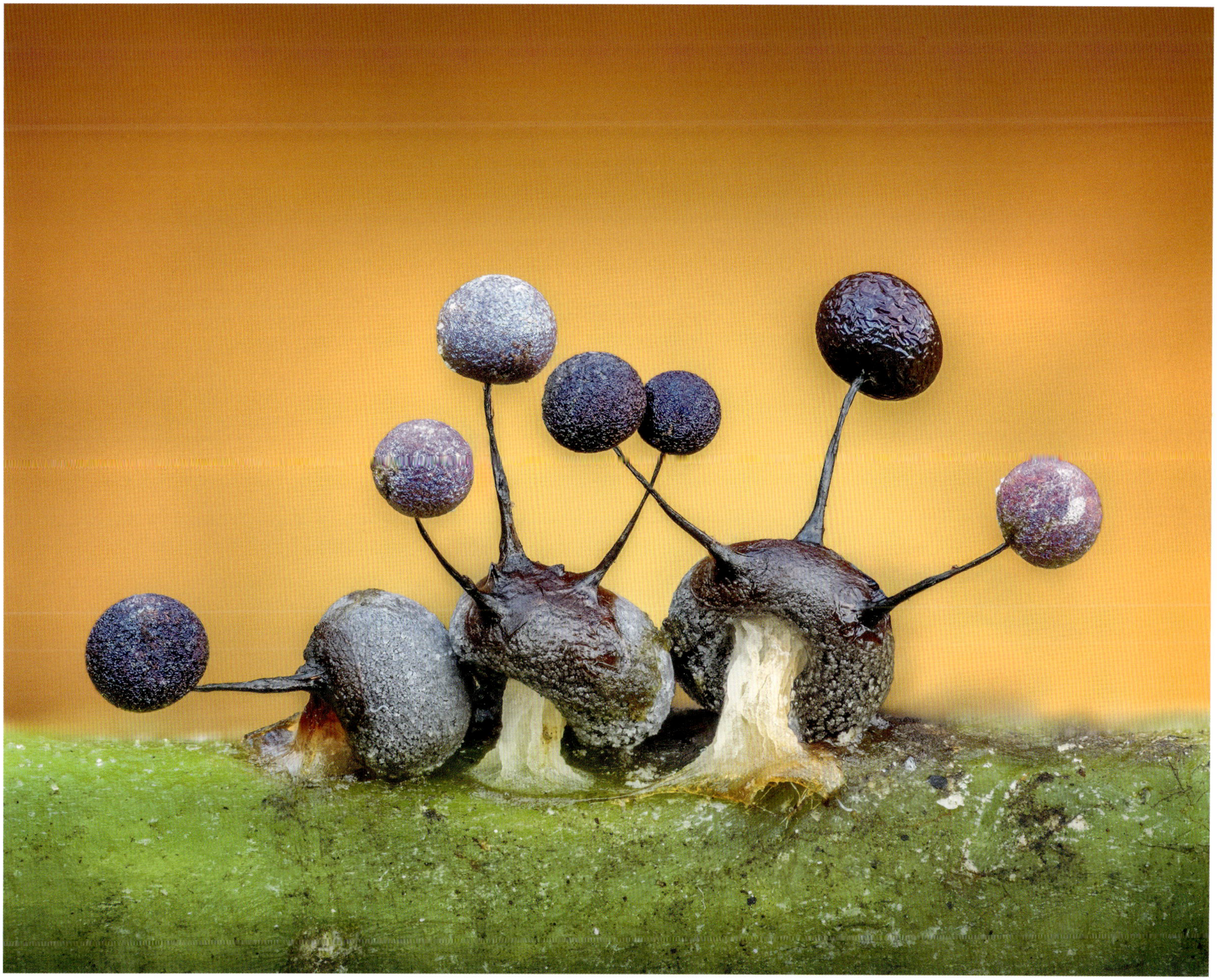

BARRY WEBB

Lamproderma scintillans on Didymium squamulosum, 2024
Digital photograph,
8 × 10⅝ in. / 20.3 × 27 cm

Despite what might at first glance look like something out of science fiction, this photograph shows not extraterrestrial beings but the fruiting bodies of slime molds, specifically those of *Lamproderma scintillans* sitting atop *Didymium squamulosum*. The scale of the photograph is also deceptive: both of these fungi species measure no more than one millimeter tall. To make such minuscule molds visible to the human eye, British photographer Barry Webb (b. 1958) spends numerous hours in an ancient woodland in southeast England, in the county of Buckinghamshire, photographing his subjects using a macro lens to magnify the tiny fungi. The final image was assembled from fifty separate shots digitally stacked together. Since 2019 Webb has spent much of his time photographing and documenting slime molds: the single-celled amoebas that thrive in damp, decaying places and contribute to the decomposition of organic material. For such tiny fungi, slime molds are remarkably versatile, including being able to climb trees and other surfaces in search of food and to gain height to increase the dispersal of their spores. *Lamproderma scintillans* is notable for its iridescent metallic peridium—the protective ball that contains the spores—which vary widely from shades of silver and bronze to blue and violet, as Webb's photograph clearly shows here. When the peridium bursts, spores are scattered across the woodland floor, where they feed on bacteria. The *Didymium* slime mold is also found on leaf litter, but, as with the *Lamproderma scintillans*, its tiny size makes it a challenge to photograph.

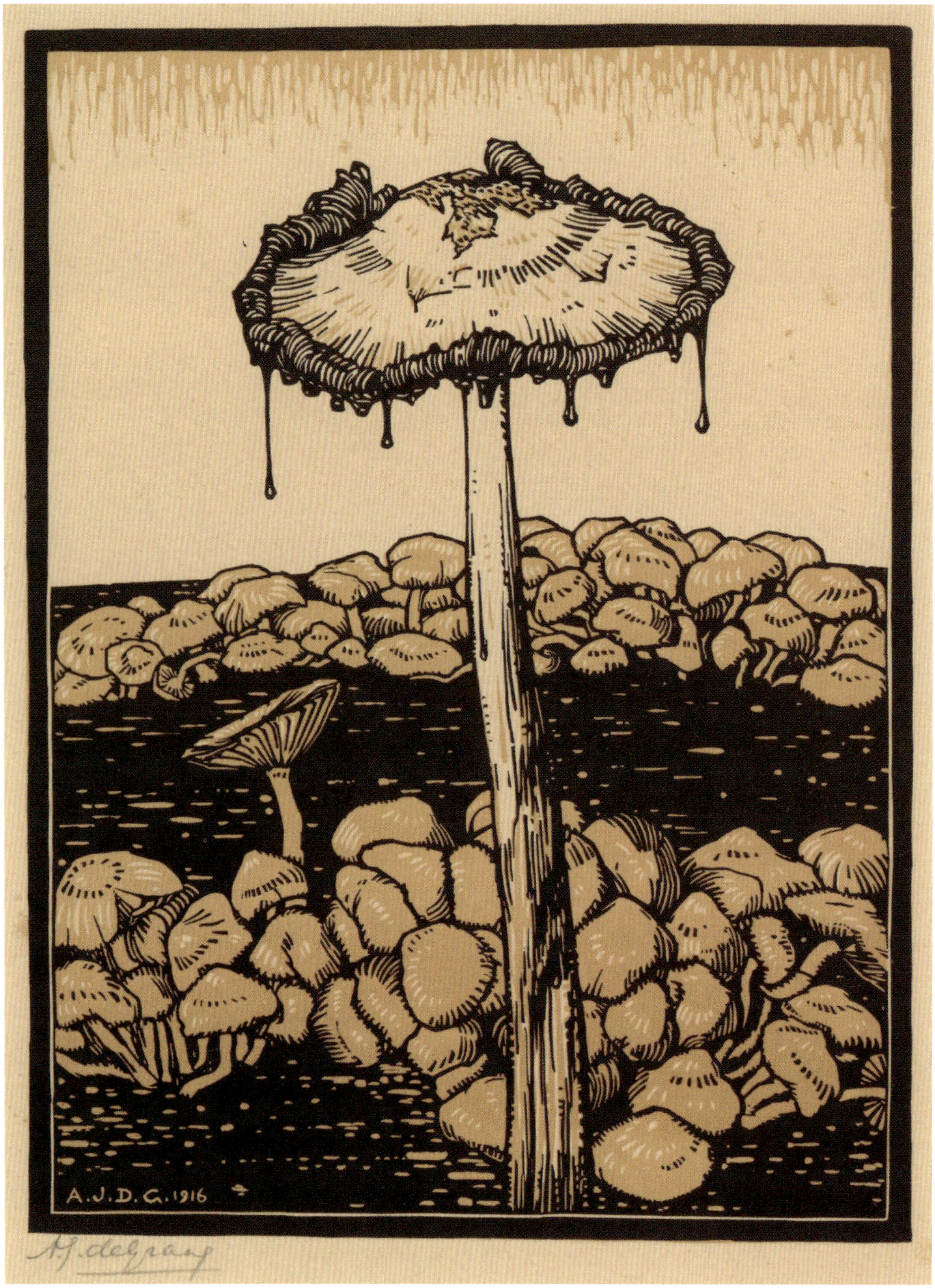

JULIE DE GRAAG

Druipende paddenstoel (*Dripping Mushroom*), 1916
Woodcut, $11\frac{3}{8}$ × 9 in. / 28.8 × 22.7 cm
Rijksmuseum, Amsterdam

A single mushroom stands tall, the withered edges of its cap turned in while moisture drips down onto the clusters of mushrooms below. This highly stylized woodcut is the work of Dutch graphic artist and painter Julie de Graag (1877–1924). Unlike her famed contemporary M. C. Escher, de Graag and her work were largely forgotten to history. Ill health during childhood prevented de Graag from interacting with the outside world. Instead, she created her own world, becoming an acute observer of nature and a self-taught expert in botany and zoology, in addition to studying at the Royal Academy of Fine Arts in the Hague, where her family lived. She brought her expertise to her printmaking, particularly after a 1901 meeting with the influential Dutch art teacher and critic Henk Bremmer, who reinforced her belief that woodcuts best served her talent. After she moved to the artistic village of Laren, North Holland, in 1904—home to artists that included Piet Mondrian—she produced woodcuts of birds, animals, flowers, shells, and mushrooms, finding beauty in everyday flora and fauna, as well as portraits. De Graag's distinctive style, which married clean lines with a monochromatic tonal contrast, was achieved by working with the end grain of the wood. It was harder to work with than the more usual longitudinal grain, but it produced finer prints. A large number of her works were lost in a fire in 1908, and after struggling with depression for most of her adult life, de Graag died by suicide at the age of just forty-six.

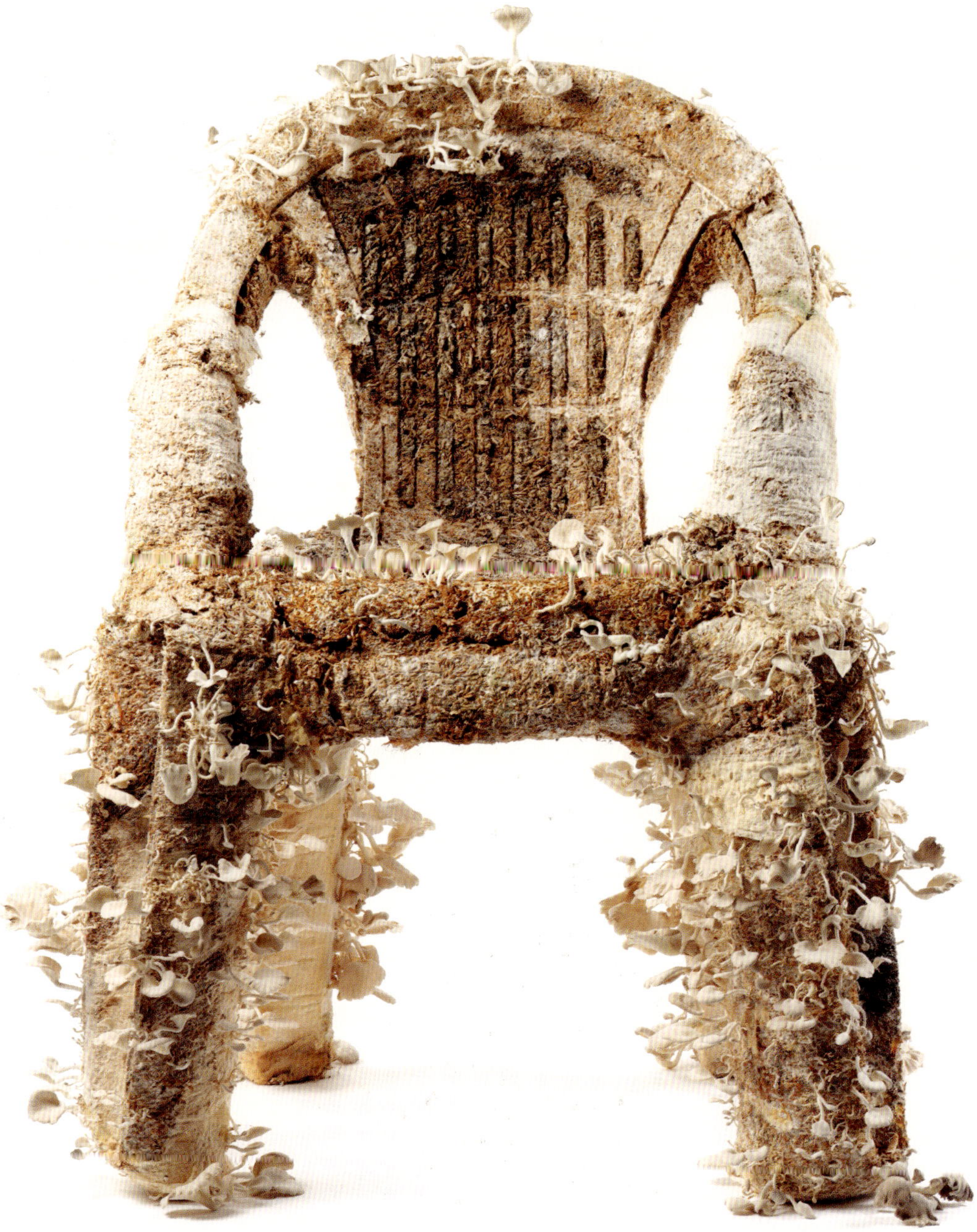

NONGZAO

Coffee Grounds Mycelium Practice, 2024
Coffee grounds, rice husks, straw and other agricultural waste, 31½ × 23⅝ × 19⅝ in. / 80 × 60 × 50 cm
Sustainable Design (Material) Museum, China

The cycle of buying and consuming cheap plastic goods is ever present in today's society, and its environmental impact is staggering. More than 430 million tons of plastic waste are produced every year, two-thirds of it comprising items meant for short-term use. Chinese design studio NONGZAO examined this crisis in its 2024 *CONSUMER* project. They remind us that plastic products, while ephemeral in our own lives, do not simply fade away, and offer us another alternative: mushroom-based plastics. The root system of mushrooms, or mycelium, forms a strong, variegated structure that can be used as a natural substitute for anything from foam and vegan leather to the kind of hard plastics used to make innumerable products. In a split-screen video contrasting consumer dependence on plastic goods with the painstaking process of nurturing and growing mycelium across those same plastic forms, NONGZAO shows us that there is a better way forward. Using coffee as a substrate material on which to grow fungi, they carefully wrapped nonbiodegradable objects—a chair, lamp, basket, step stool, and pair of sandals—in it and encouraged the fungi to proliferate. Over time, those discarded plastic forms became hosts for living, breathing ecosystems. Although each plastic item was completely covered, it was not—and never would be—broken down. Completely biodegradable and sustainable, mushroom plastics may indeed be the wave of the future. Forward-looking companies, including IKEA, are already embracing the trend, turning to biodegradable mushroom plastics in place of Styrofoam in all their packaging.

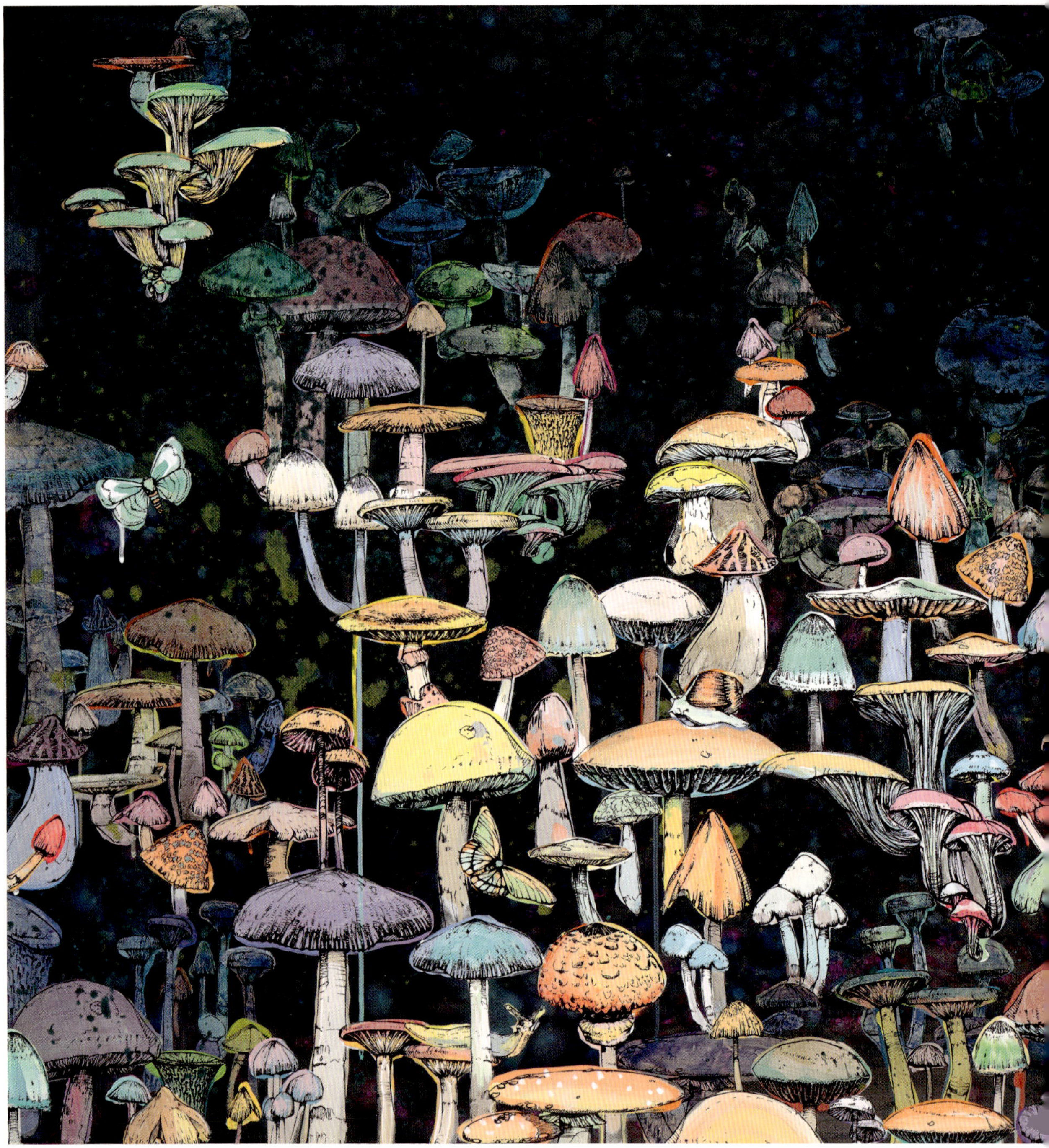

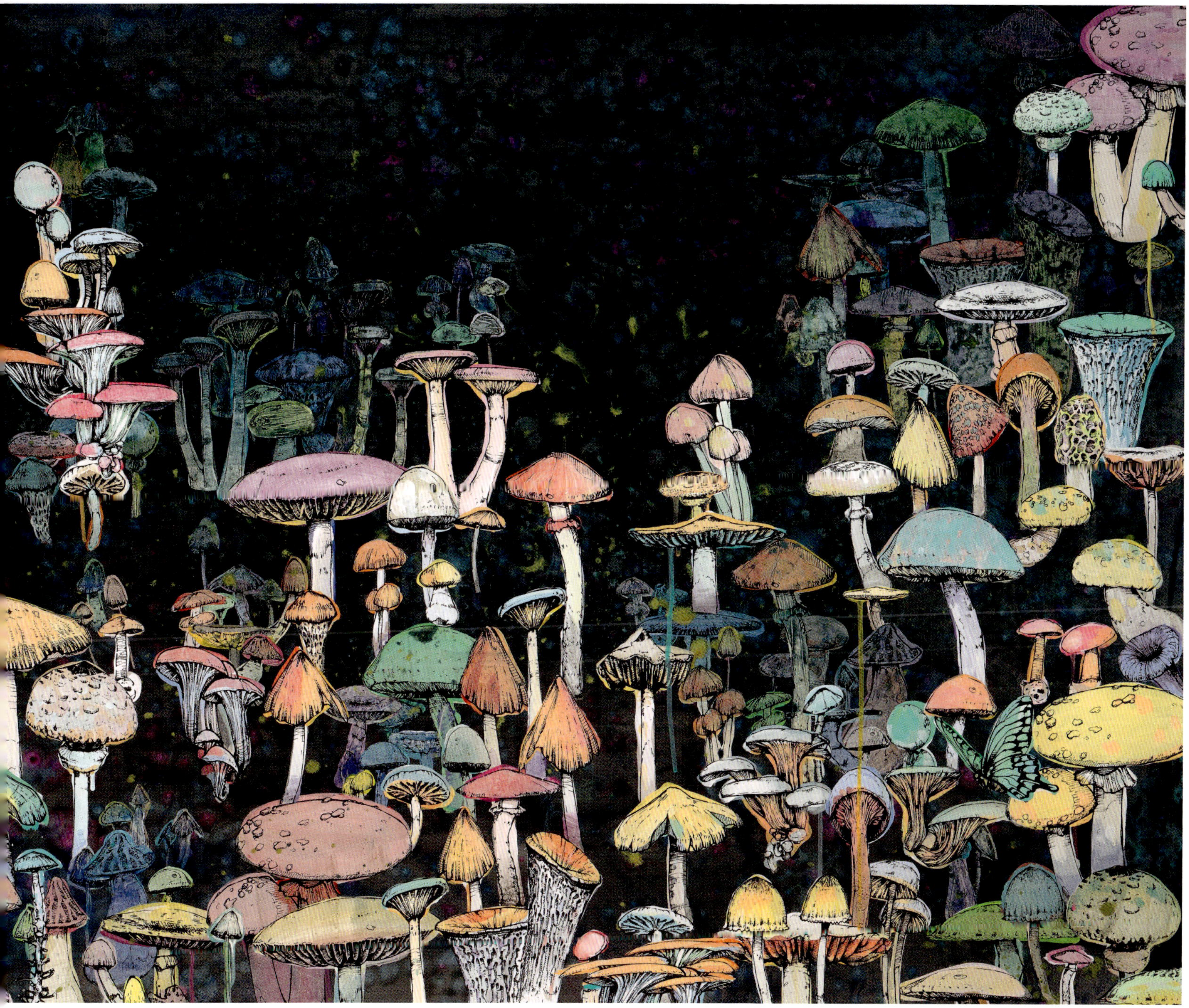

SAGE VAUGHN

Night Garden, 2016
Acrylic, ink, and vellum on canvas,
30 × 65 in. / 76.2 × 165.1 cm
Private collection

Los Angeles–based artist Sage Vaughn (b. 1976) explores many of the familiar aspects of his native terrain, from rangy coyotes prowling the cityscape to psychedelic fields of mushrooms, as well as flowers and other flora. A self-described child of two hippies, Vaughn sees the objective of his practice as capturing the point where the real meets the surreal. He examines the often bitter clash between man and nature and asks us to reconsider the way in which we view the world we inhabit. Rendered in acrylic and ink, the nocturnal scene depicted in *Night Garden* is far from restful. Butterflies take flight and mushrooms glow against the pitch-black background, almost as if they are fluorescing. The field of fungi seemingly stretches endlessly into the distance. It is a landscape that stands in stark contrast to other works in his oeuvre, where similar compositions burst with an array of pastel and candy-colored blossoms and foliage that seem to explode off the canvas. With *Night Garden*, Vaughn captures a perhaps lesser-appreciated layer of the natural world, that which is always teeming underneath but much more seldom noticed. Primarily a painter and illustrator, he has exhibited in solo and group shows across the United States and Europe, and has also created large-scale public works. An avid surfer, Vaughn connects the practices of surfing and art as two disciplines that require creativity and intuition in equal measure.

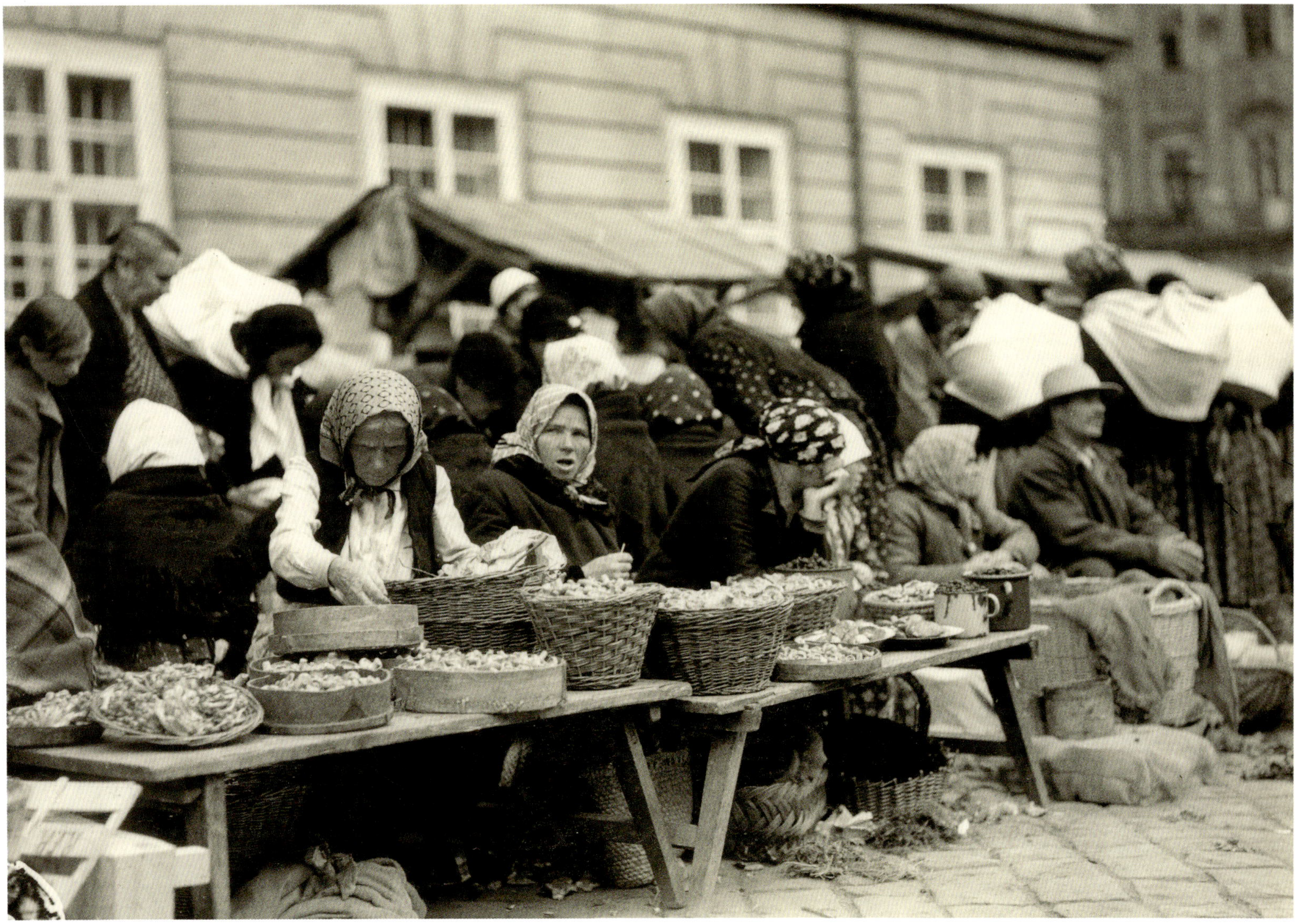

LOUISE ARNER BOYD

Poland, Mushroom Stall at Lwów Market, 1934
Photograph, 3¼ × 4½ in. / 8.3 × 11.4 cm
American Geographical Society Library
Photo Collection, University of
Wisconsin–Milwaukee Libraries

Baskets of mushrooms fill two trestle tables at a market stall in Lwów, Poland (now L'viv in present-day Ukraine), watched over by a group of babushka-clad women as they sit and wait for customers. Captured in the autumn of 1934, the black-and-white photograph was taken by American polar explorer and self-taught photographer Louise Arner Boyd (1887–1972). In August 1934 Boyd was commissioned to photograph the lives of ethnic Poles, Ukrainians, Belarusians, and Lithuanians and spent three months traveling through the Polish countryside before ending her trip in Vilnius, the Lithuanian capital. In the 1930s mushrooms were an important part of the Polish diet, particularly among poor rural communities. Foraging for varieties such as boletes (including porcini mushrooms), chanterelles, saffron milk caps, and parasol mushrooms was a family enterprise. The mushrooms were free, nutritious, and had multiple uses: parasol mushrooms could be breaded and cooked so they resembled meat cutlets, porcini were perfect for drying, and some other mushrooms could be pickled. Any additional quantities of edible mushrooms could be brought to market—Boyd noted on this photograph that each basket sold for 2.20 złoty, or about ten dollars today. Knowledge of the best forest foraging sites and of how to identify edible mushrooms was passed from generation to generation. Such was the significance of mushroom-hunting that it appears in *Pan Tadeusz*, the national epic poem written by Adam Mickiewicz in 1834 that captures the heart of Polish culture, and that every Polish child reads in school, even today.

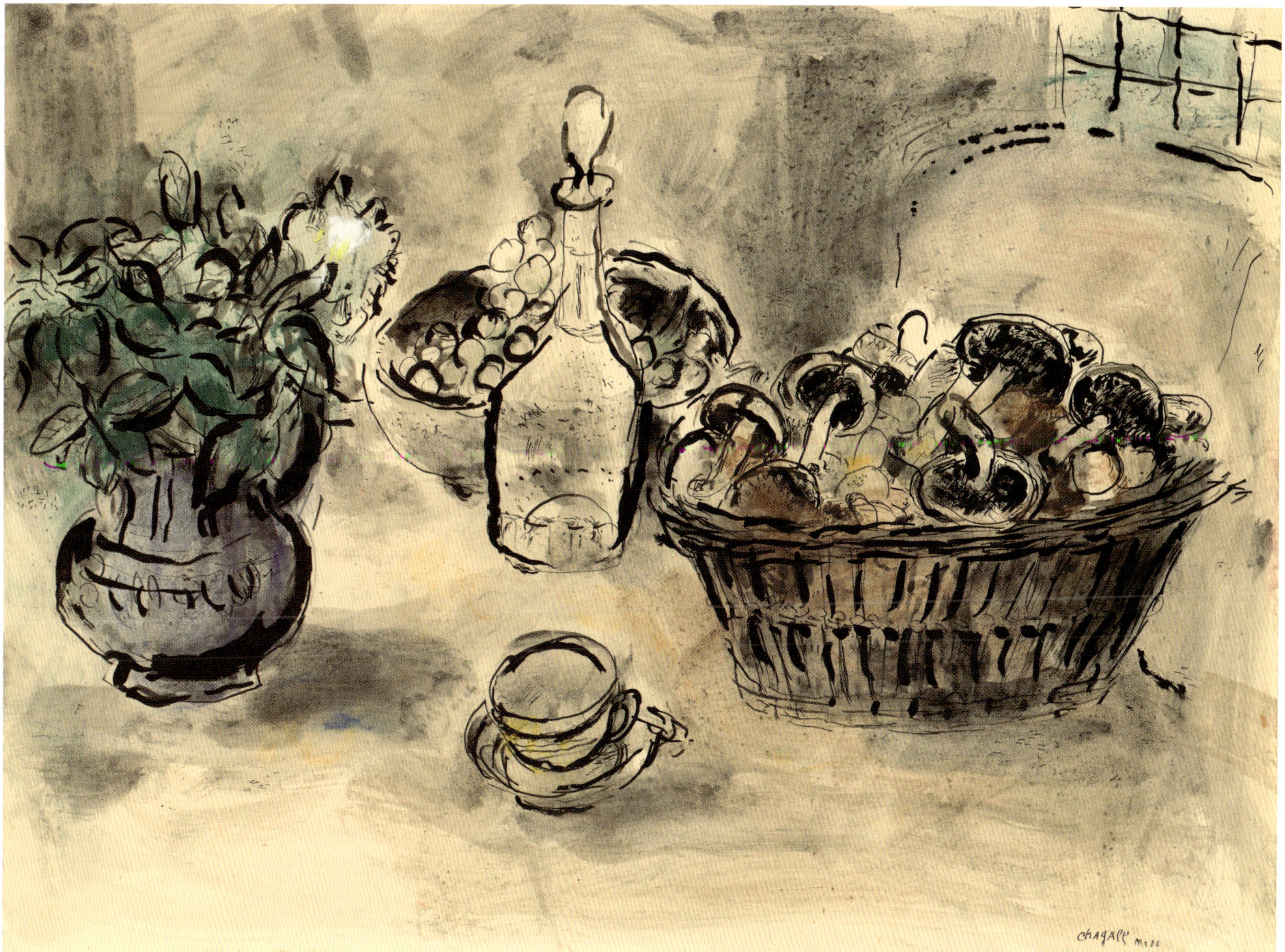

MARC CHAGALL

Basket of Mushrooms, undated
Brush and black ink, gray wash, gouache, and pastel on wove paper, 19⅜ × 25¼ in. / 49 × 64 cm
National Gallery of Art, Washington, DC

The eldest of nine children, Moyshe Segal, better known today as Marc Chagall (1887–1985), was born in the largely Jewish town of Lyozna, near Vitebsk, Russia (now Belarus). His father hauled barrels for a herring merchant, and his mother sold groceries. Chagall attended the *heder*, a traditional Jewish elementary school, where he took singing and violin lessons. After learning the elements of drawing at a local school and studying painting at Zvantseva School of Drawing and Painting in Saint Petersburg, he moved to Paris in 1910, where he changed his name and began to cultivate his artistic aesthetic, which blended elements of Cubism, Fauvism, and Surrealism. *Basket of Mushrooms*, however, is typical of the subdued palette of his earlier paintings, enlivened by changes in perspective and a sense of movement from brushstrokes and shading. Instead of using the traditional colors for the grapes, carafe, and flowers, he washed the whole painting in the neutral tones of the fungi and basket. Other than a single touch in the shadow of the flowers, hints of yellow and brown in a few mushroom stems, and blue on the glass in the window, this work does not hint at the genius for which he became renowned later in life: his unparalleled use and understanding of color. Nor does the simplicity of the composition of this still life foreshadow the development of his personal, distinctive style of dreamlike, mystical imagery often filled with Jewish tradition and Christian symbolism.

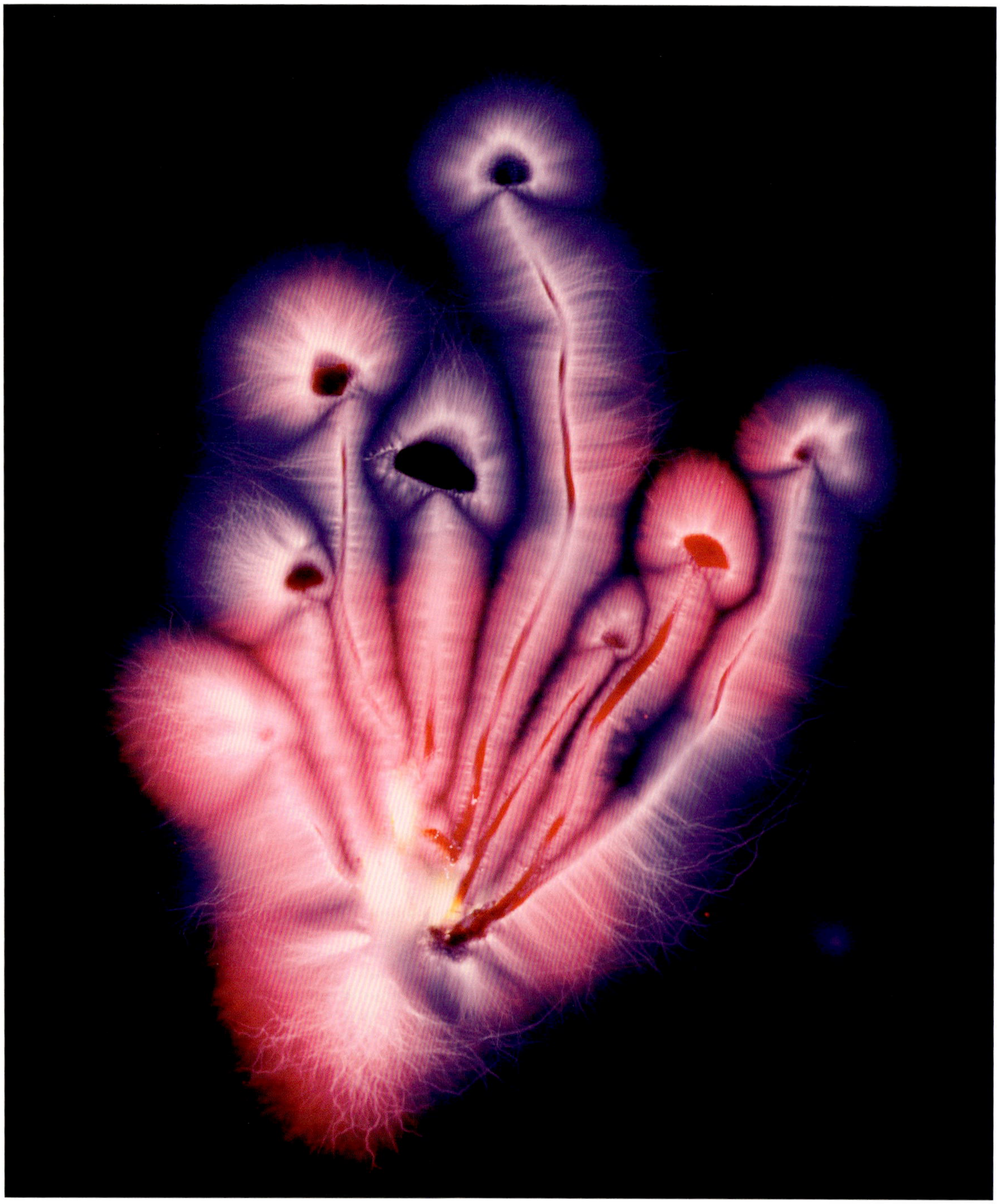

NIGEL HUTCHINGS

Liberty Cap Mushrooms, 1997
Digital photograph, dimensions variable

Mushrooms that contain the organic compound psilocybin, also known as "magic mushrooms," produce psychedelic effects when consumed, including visual and audio hallucinations; altered perceptions of time, place, and reality; and euphoria and giddiness. One of the most common (and most potent) species of these mushrooms is *Psilocybe semilanceata*, or liberty caps, which tend to grow in clusters and favor the decaying grasses and roots of moist grassland regions. This image shows an electromagnetic discharge around the edges of a cluster of liberty caps, which was captured through a process called Kirlian photography, also known as radiation field photography and electrophotography. Soviet inventor Semyon Kirlian accidentally discovered this type of photography in 1939. Explained at its most basic, a Kirlian "camera" has no lens; instead it uses a positively charged plate (cathode) and a negatively charged point (anode) and sheet film to capture an image. The film is placed directly on the cathode, and whatever is being photographed is placed upon the film, becoming the anode. The interaction between the cathode and anode in this electric sandwich is captured on the sheet film, manifesting as unique, vibrant patterns that vary from object to object. Recent developments allow for this process to be captured digitally, directly to a computer in real time, as employed here by Nigel Hutchings, a pioneer in the field of modern Kirlian photography.

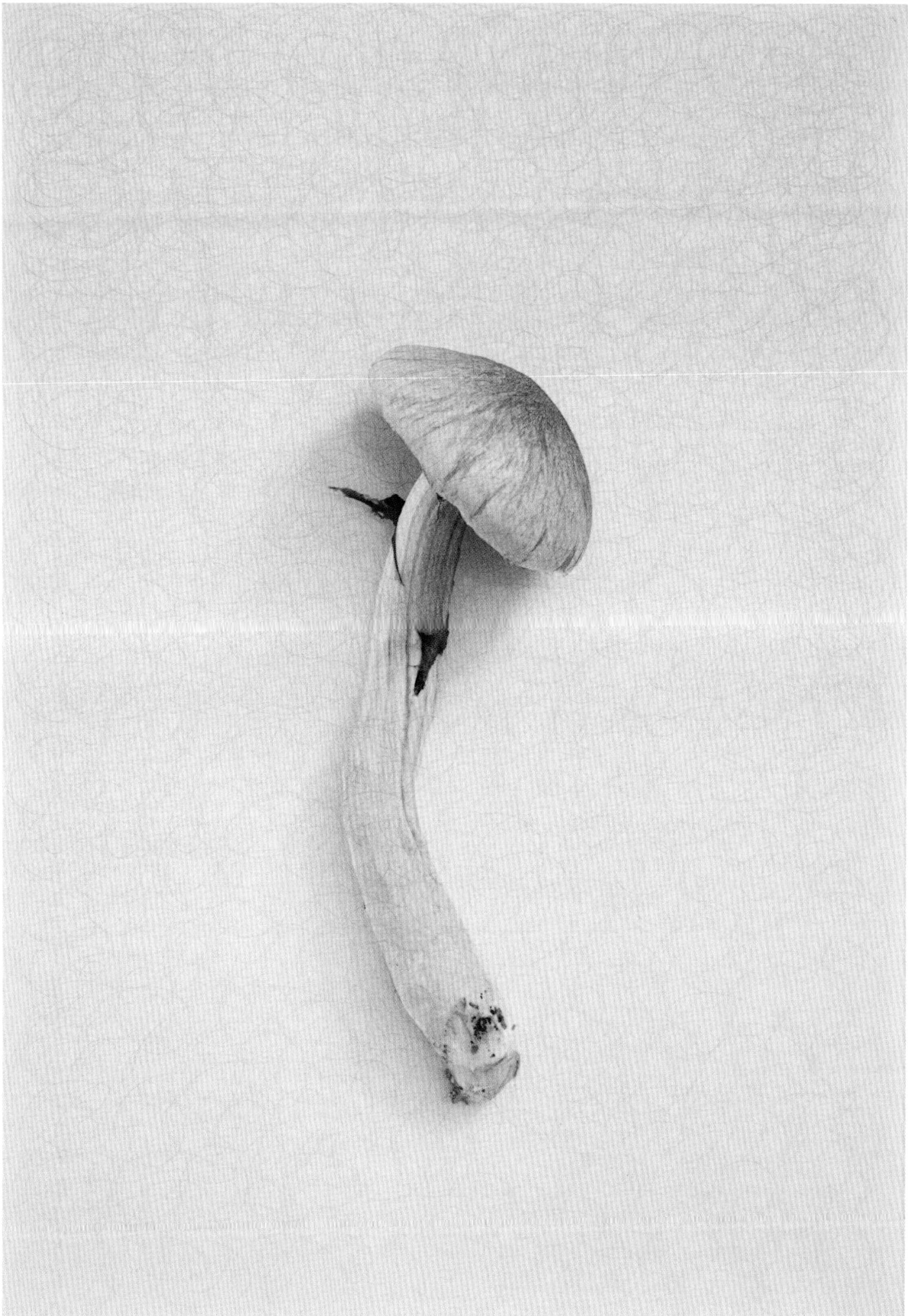

ANNIE RATTI

Psilocybin Mushroom I, 2012
Graphite on photographic print,
19 × 13 in. / 48.5 × 33 cm

A solitary psilocybin mushroom is the central subject of this photograph by London-based Italian artist Annie Ratti (b. 1956), the striations of its cap and stem subtly blending with the overlay of graphite scribbles. The swirling, slightly hypnotic pattern nods to the 144 species from the genus *Psilocybe* that contain hallucinogenic properties when ingested. Ratti works with photography, video, installation, and texts as a way to question social complexity, human conflict, and the experience of participation in art. Her work around mushrooms began as a research project in 2013 during an exhibition at Vleeshal Center for Contemporary Art in Middelburg, the Netherlands, and continued until 2020. In particular, she was drawn to exploring the ethnographic, scientific, and cultural significance of psilocybin mushrooms. Her methods of investigation varied, including filming the mushrooms growing with surveillance cameras, documenting their growth forms, and creating photographic and drawing installations that explored their formal and metaphoric possibilities. Ratti, who has exhibited at Somerset House and the Whitechapel Gallery in London and MoMA PS1 in New York City, has often used research as a methodology and aesthetic driver in her work. She expands on the cultural idea of the rhizome, a theme popular in contemporary art in the early twenty-first century and the philosophical writings of Gilles Deleuze and Félix Guattari. Comparing the structure of knowledge to the branching rhizomatic system of roots in plants, in her mushroom works Ratti parallels how the innate properties of mycelium provide a way of thinking about nonlinear, decentralized networks.

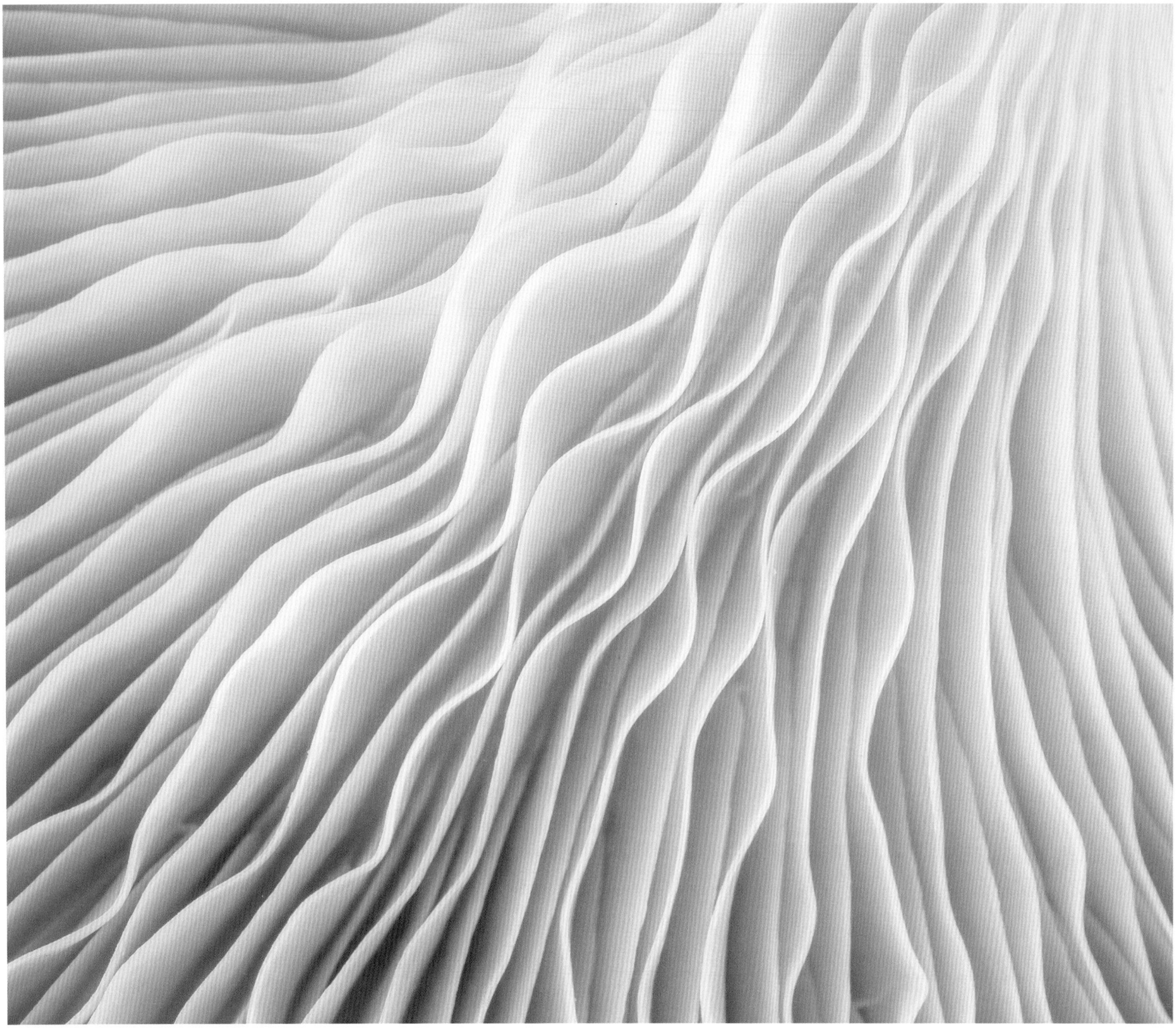

LAMYAI

Sajor-caju Mushroom, 2017
Digital macrophotograph,
dimensions variable

What at first might appear as an aerial photograph of white sand dunes or perhaps delicately pleated fabric is in actuality a macrophotograph of the underside of a funnel woodcap mushroom (*Lentinus sajor-caju*). These mushrooms reach a height of 1½ to 3 inches (4 to 8 cm) and are said to be infundibular, meaning they are conical or funnel-shaped, hence their name. The striations visible here are the mushroom's lamellae, the gills that serve to produce and disperse its spores to facilitate its reproduction. The lamellae on the funnel woodcap, and other similar species, are often referred to as decurrent gills, which means they are not contained to the underside of a defined cap, as in many other species, but run down the stem in a pronounced fashion. Funnel woodcaps are saprophytic, thriving on decaying matter, in this case usually on rotting wood. They are edible, offering a firm, meaty texture and a subtly sweet flavor, and they are rich in antioxidants, vitamins, minerals, and protein. For that reason, they are widely cultivated as a food source in India, and to a slightly lesser extent in China, Thailand, Malaysia, Tanzania, and other warm, often tropical, regions. It is worth noting that experienced foragers often warn novices to beware of the fool's funnel (*Clitocybe rivulosa*), a rather common species of poisonous mushroom often found in Europe and North America that is nearly identical to the funnel woodcap. They contain the toxin muscarine, which can lead to severe illness and, in high enough doses, even death.

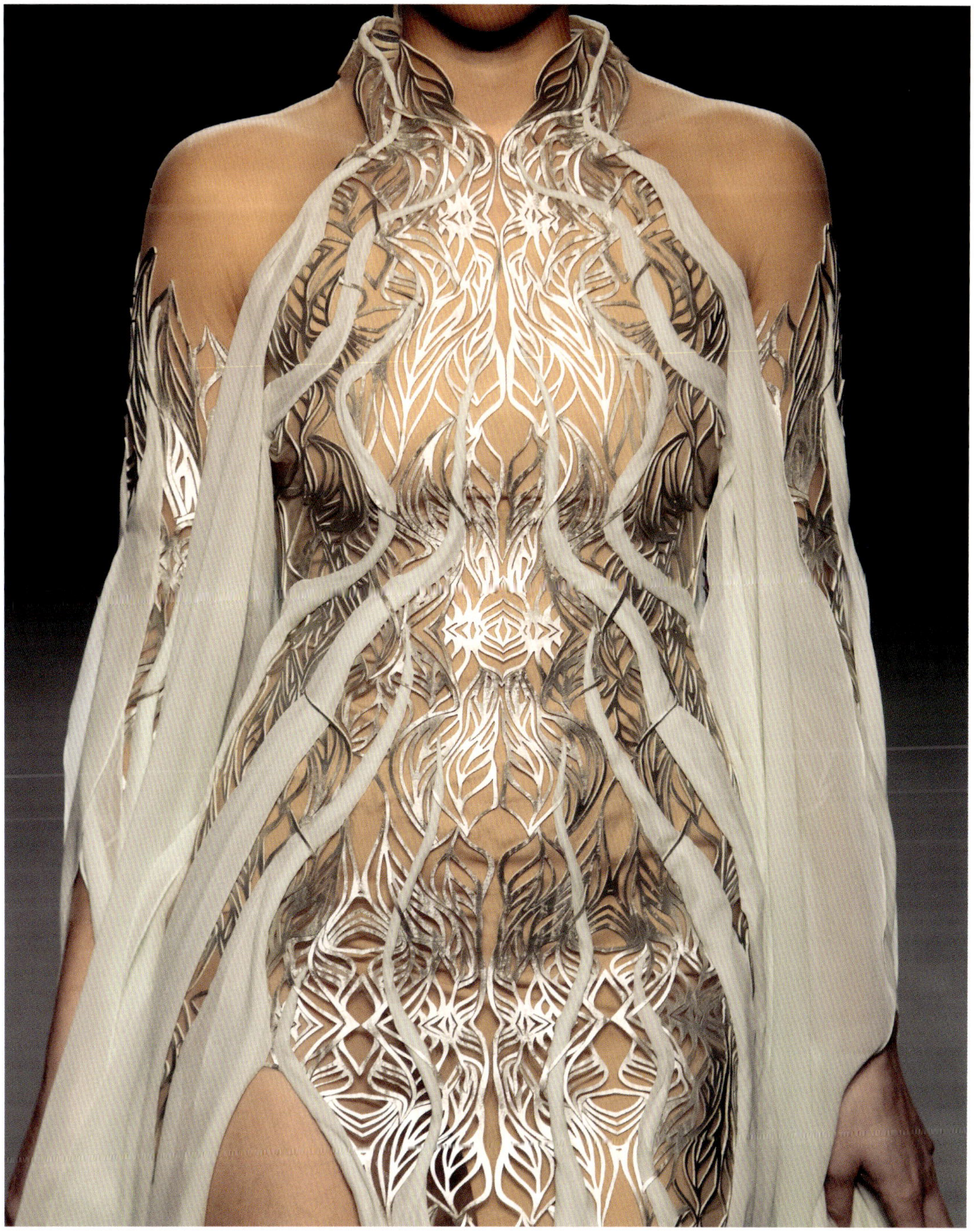

IRIS VAN HERPEN

Luminous Lichen gown (detail), from the *Roots of Rebirth* collection, Spring/Summer 2021
Sustainable laser-cut Majilite fabric and mint-dyed crepe-silk, dimensions variable

Dutch designer Iris van Herpen (b. 1984) is one of fashion's most innovative and forward-thinking designers, whose avant-garde silhouettes and materials blend futurism and technology with traditional couture techniques. She has collaborated with sculptors, architects, microbiologists, choreographers, and particle physicists, and made clothes using metal, glass, plaster, and even bones. Van Herpen honed her skills and penchant for pushing boundaries under the tutelage of British couturier Alexander McQueen before launching her eponymous label in 2007. In 2010's *Crystallization,* she became the first designer to send a 3D-printed garment down the runway, and with 2011's *Capriole,* she made her debut as a member of the Chambre Syndicale de la Haute Couture. For her Spring/Summer 2021 couture collection *Roots of Rebirth*, van Herpen looked to nature's "wood wide web"—the underground symbiotic system of tree roots and mycelium—for inspiration. In creating the collection, she referenced a quote from author and mycologist Merlin Sheldrake's *Entangled Life* (see p.223): "Fungi is the ecological connective tissue, the living seam by which much of the world is stitched into relation." In a series of twenty-one looks, the designer makes visible her interpretation of roots, spores, and mycelial tendrils in garments fashioned from hand-pleated silk and organza and 3D-embroidered and laser-cut lace, complemented with kinetic crowns. Each intricate gown is brought to life on the runway, every movement causing the fabrics to appear as if they entwine around the model, surrounded by lacy swirls of roots and fanning arrays of mycelial threads.

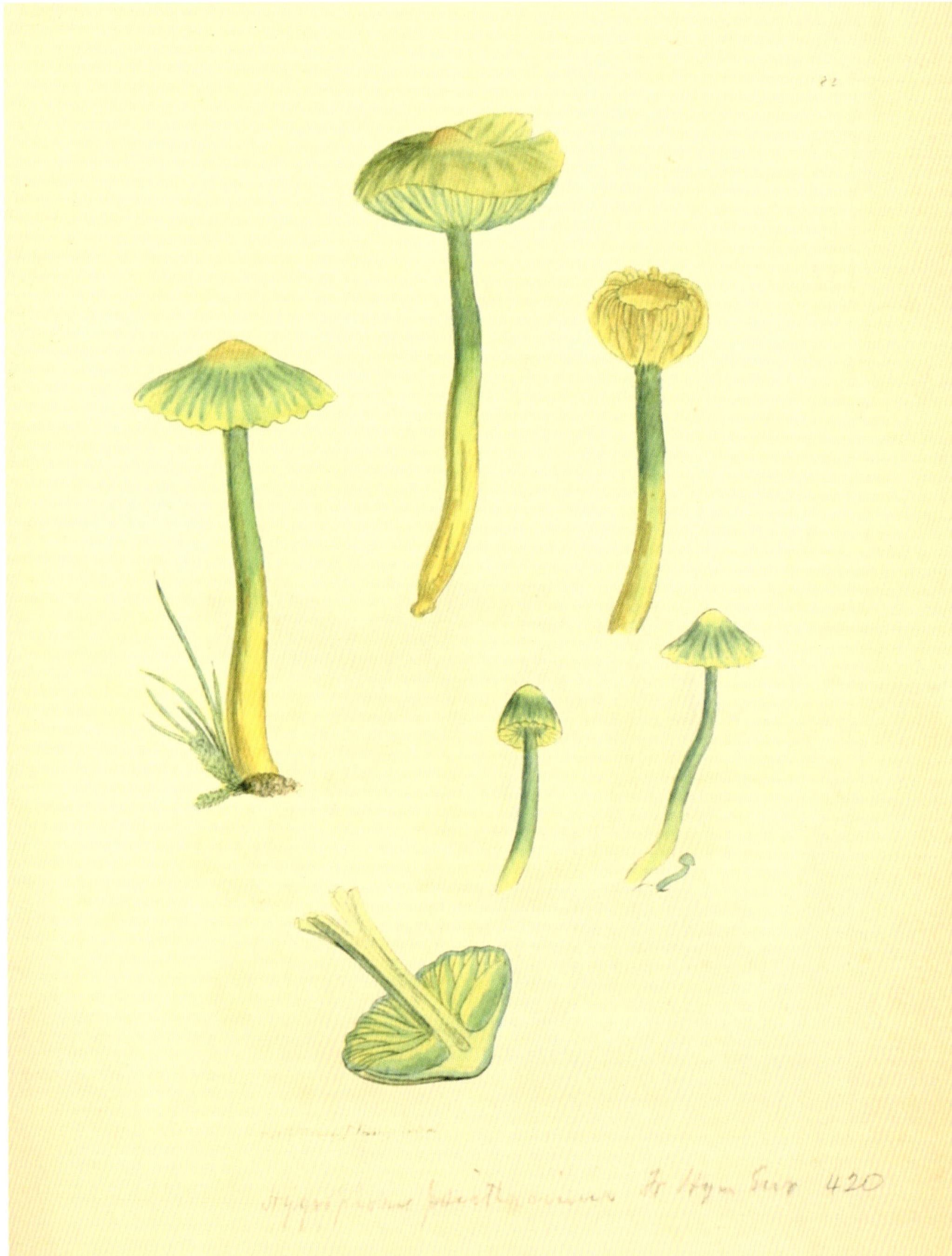

JAMES SOWERBY

Agaricus psittacinus, from *Coloured Figures of English Fungi or Mushrooms*, vol. 1, c. 1797
Hand-colored engraving, 13 × 8¼ in. / 33 × 21 cm
LuEsther T. Mertz Library, New York Botanical Garden

In this delicately hand-colored print, slender green-and-yellow mushrooms captivate with vivid hues and precise portraiture against a stark ground, inviting close inspection. The illustration emphasizes the willowy stems of parrot waxcap (*Gliophorus psittacinus*, formerly known as *Agaricus psittacinus*), shown at different ages, all leaning as if struggling to support the cap, which is, in turn, depicted as maturing from a small, closed bell into an upturned skirt, frilled yellow gills on full display. Created by James Sowerby (1757–1822), one of Britain's most accomplished natural history illustrators, *Coloured Figures of English Fungi or Mushrooms* was published in four parts from 1797 to 1809, a time when fungi were poorly understood, and it is considered the first mycological iconography from England. Throughout the book, all manner of fungi are similarly illustrated at different stages of growth, sometimes sliced open to expose internal structures, from puffballs and coral fungi to creeping molds and brackets clinging to tree trunks. Sowerby, with evident care, set out with a goal that was practical and scientific: to document English fungi in a way that would help readers distinguish edible species from poisonous, and he succeeded with artistic sensitivity and empirical rigor. In this work, he marked a shift in the history of fungi in art and science, presenting them as subjects worthy of deep study and aesthetic appreciation. The book reflects a culture increasingly interested in cataloging nature for both personal enjoyment and public benefit, part of a trend in eighteenth-century regional mycological floras, and it remains a compelling reference today.

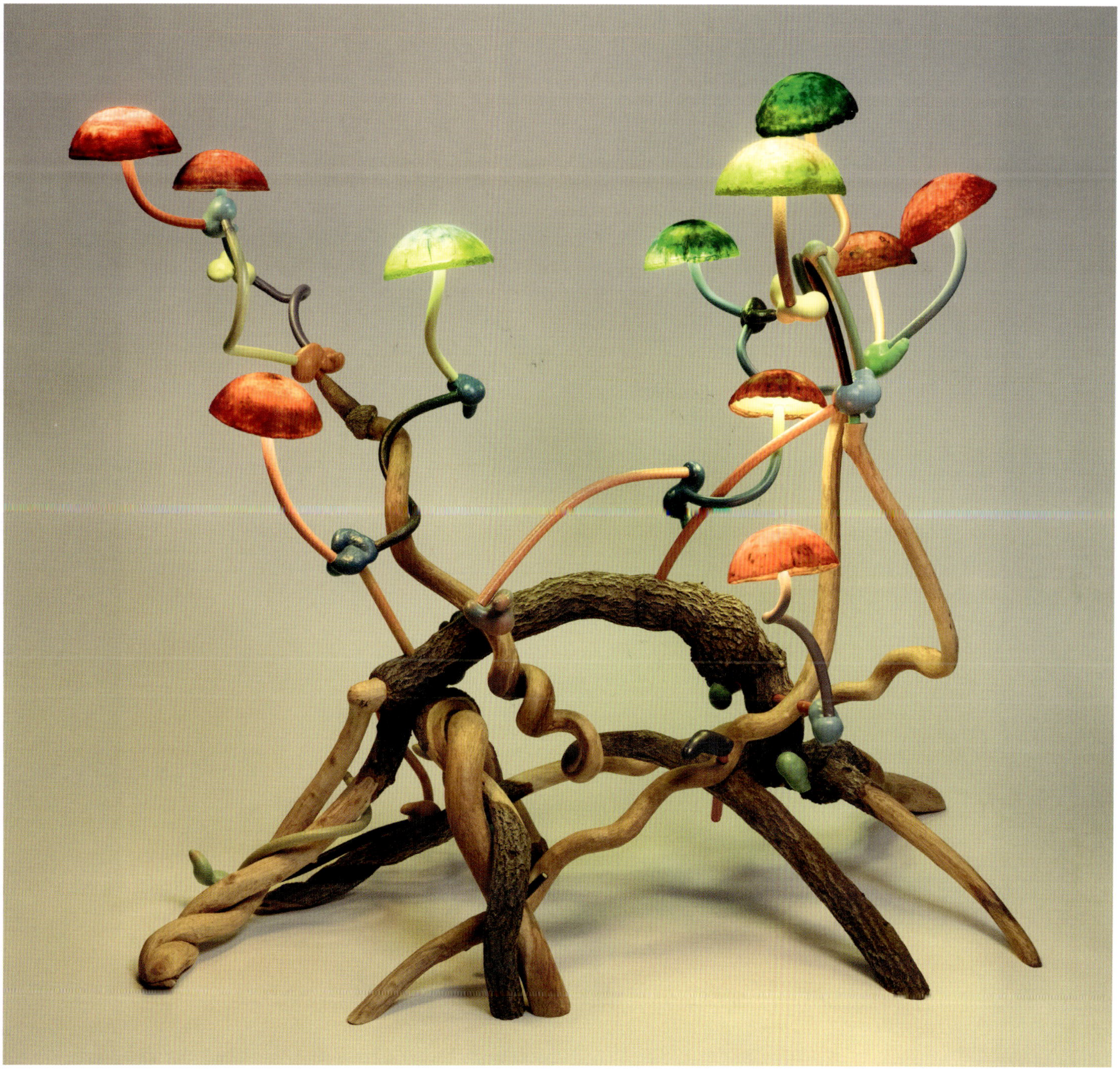

MIKA ROTTENBERG

Lampshare (P13), 2024
Milled reclaimed household plastic and bittersweet vines with lighting components, approx. 48 × 48 × 69 in. / 121.9 × 121.9 × 175.3 cm

Emerging from a tangle of woody vines, whimsical green and red mushroom caps, sitting atop sinuous stems, emit an unearthly glow. Part of an ongoing series by contemporary artist Mika Rottenberg (b. 1976) titled *Lampshares*, these sprouting pods of mushrooms are made from discarded New York City plastic waste (Rottenberg calls plastic a "contemporary natural resource") and invasive bittersweet vines. She began the series in 2023 with the idea of working within a "mini circular economy." Her team collects the plastics and melts them, forms them, and presses them with the vines, creating something spontaneous and new from what originally amounted to trash. This process is not unlike the life cycle of real mushrooms, which often spontaneously appear atop rotting wood and compost. A native of Buenos Aires, Rottenberg attended art school in Israel before moving to New York to continue earning her BFA from the School of Visual Arts in 2000 and her MFA from Columbia four years later. Rottenberg works across several media, including sculpture, film, and architectural installation. Her work often ponders the human condition, keenly aware of the simultaneous beauty and absurdity of these modern times. She is particularly interested in issues of labor, production, and value within this context, including matters of ecology and the generation and management of waste in our capitalist system.

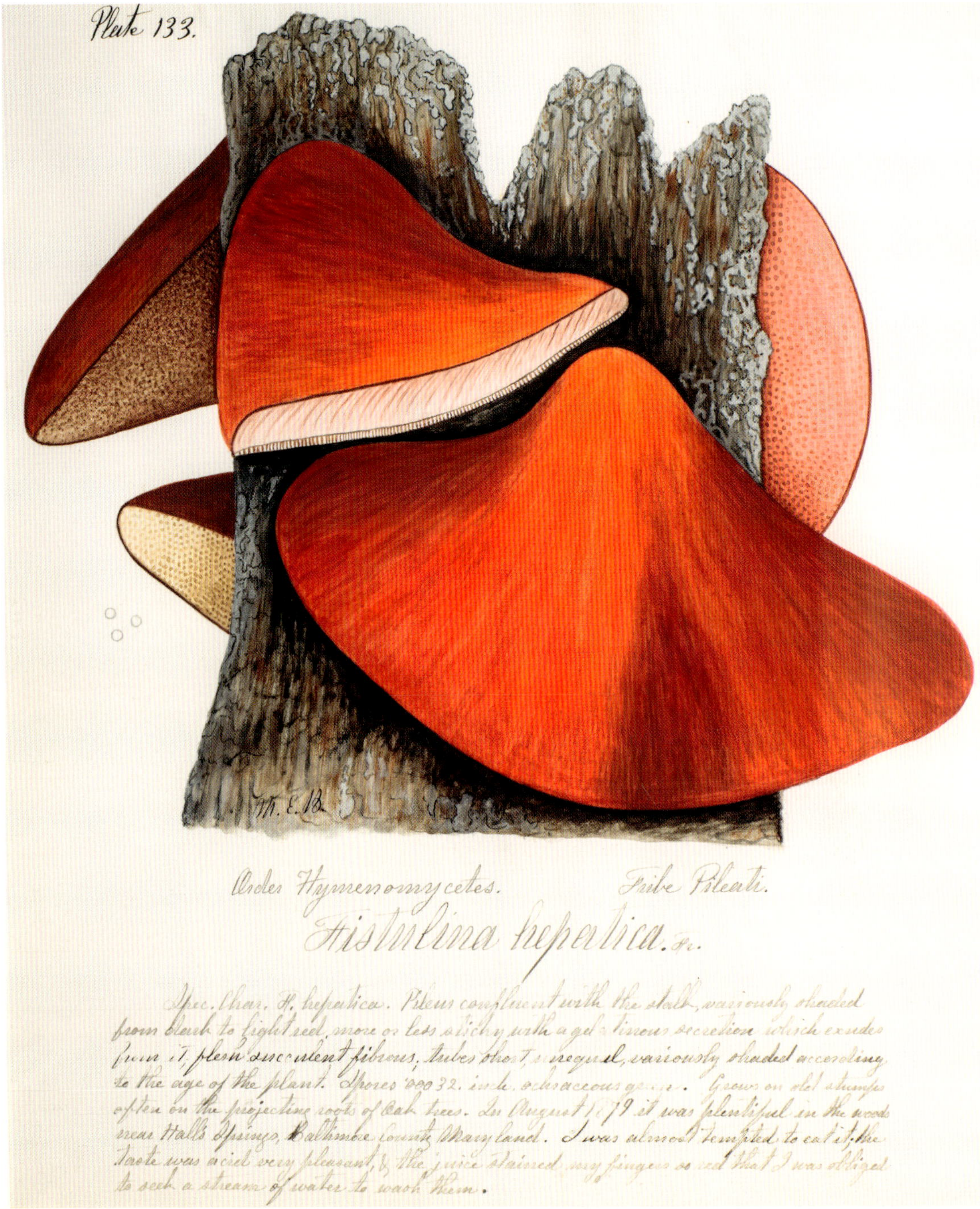

MARY ELIZABETH BANNING

Fistulina hepatica, c. 1879
Watercolor and ink on paper,
15 × 13 in. / 38.1 × 33 cm
New York State Museum, Albany

Beneath this striking and botanically accurate illustration of a beefsteak fungus (*Fistulina hepatica*) is a detailed handwritten description of the mushroom's appearance, where it grows, and when and where it was found. Part of a much larger project, it is the work of pioneering American mycologist Mary Elizabeth Banning (1822–1903), who was consistently overlooked in her lifetime but whose expertise and genius are now acknowledged. Banning dedicated her life to studying and recording fungi, and she even described twenty-three previously unrecorded species in the scientific journal *Botanical Gazette*. Referred to as the Toadstool Lady of Maryland, Banning was well-known locally for her hobby—she once traveled in a carriage with a stinking fungus that attracted flies and the disdain of her fellow travelers in equal measure. However, she lived at a time when women were excluded from the serious study of natural history. Fungi were considered unworthy of attention from male botanists, relegated to a lower taxa along with lichens, mosses, and algae, thus providing Banning and other women mycologists with a field where they could make a considerable impact. Although she lacked formal education and money, Banning dedicated her life to documenting fungi and spent twenty years writing an unpublished manuscript, *The Fungi of Maryland*, that featured her illustrations, including this one, alongside scientific descriptions and anecdotes. In 1890 she sent the manuscript to the one man who had championed her career, mycologist Charles Peck at the New York State Museum. He put it in a drawer, where it remained for ninety-one years.

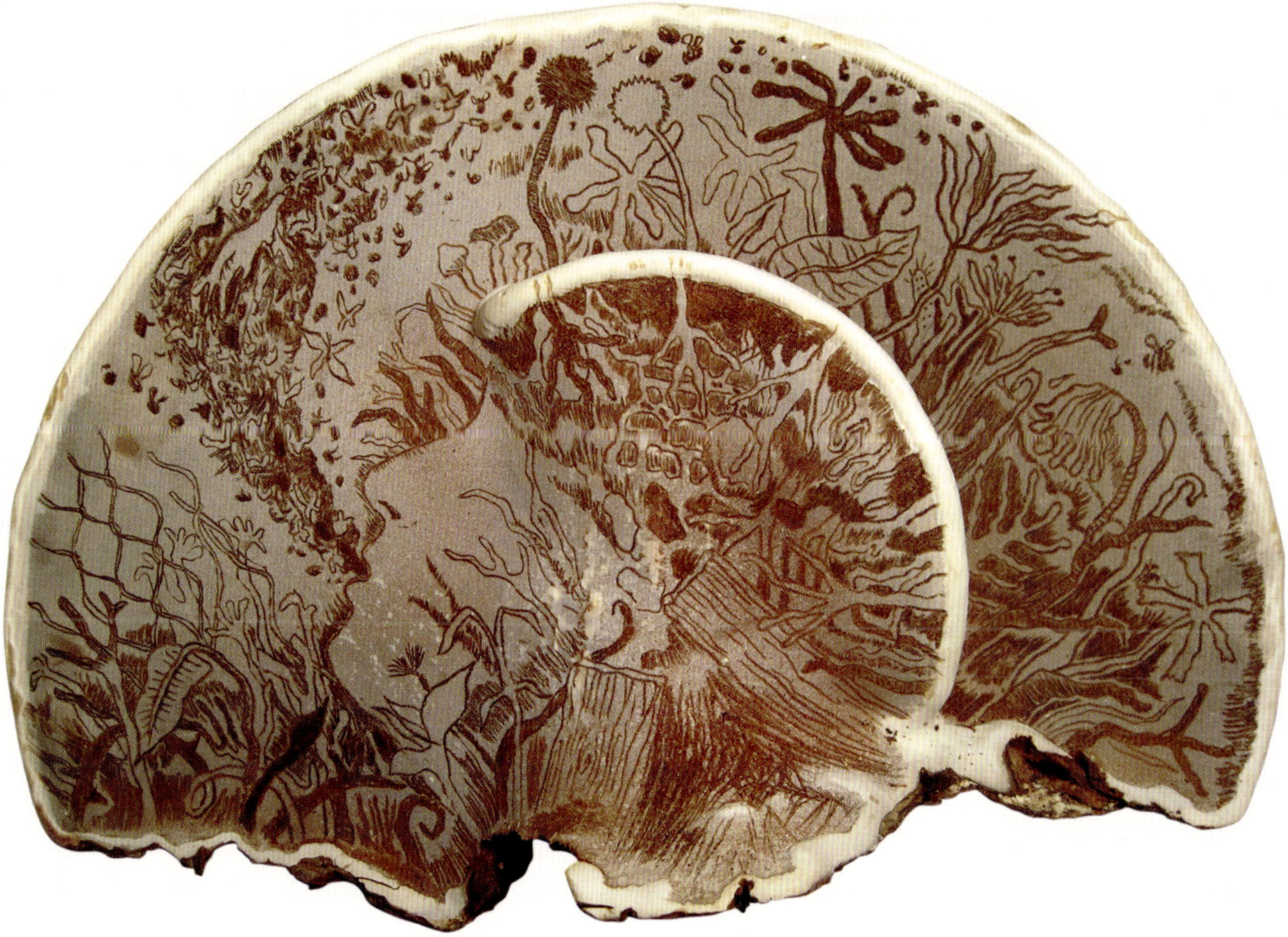

COREY CORCORAN

Spores, 2011
Engraving on artist's conk
(*Ganoderma applanatum*),
7 × 11 in. / 17.8 × 27.9 cm
Private collection

Nature's Etch A Sketch is how Boston-based multimedia artist and illustrator Corey Corcoran (b. 1985) describes the artist's conk fungus (*Ganoderma applanatum*). Also known as the artist's bracket, this tough, flat-faced mushroom stands out from the usual soft-bodied species of the fungi world. While there is a soft layer on its underside, the rest of its surface is as hard as a piece of wood. Another notable feature of the mushroom is its size: it can grow from as small as 4 inches (10 cm) to as large as 2 feet (60 cm) in diameter. Found on deciduous trees, it is the flat cap, or shelf, of the fungus that has long been used by woodsmen, particularly in North America, as a canvas for folk art or as a commemorative marker for a particular excursion, much like sailors' scrimshaw. Once Corcoran has removed a cap from its host tree, careful to avoid bruising, he sets to work immediately. Using an etching needle, rather than a nail as he used early in his practice, he draws his image in a single sitting—he only has up to around forty-eight hours before the mushroom's surface becomes too hard to engrave. Once the surface has been scratched, the mushroom changes color: the deeper the incision, the darker the color. Each design is unique—dictated by the organic shape of the mushroom—and once the mushroom dries out, it will shrink and warp, so that the final artwork takes on a slightly different appearance from the original design.

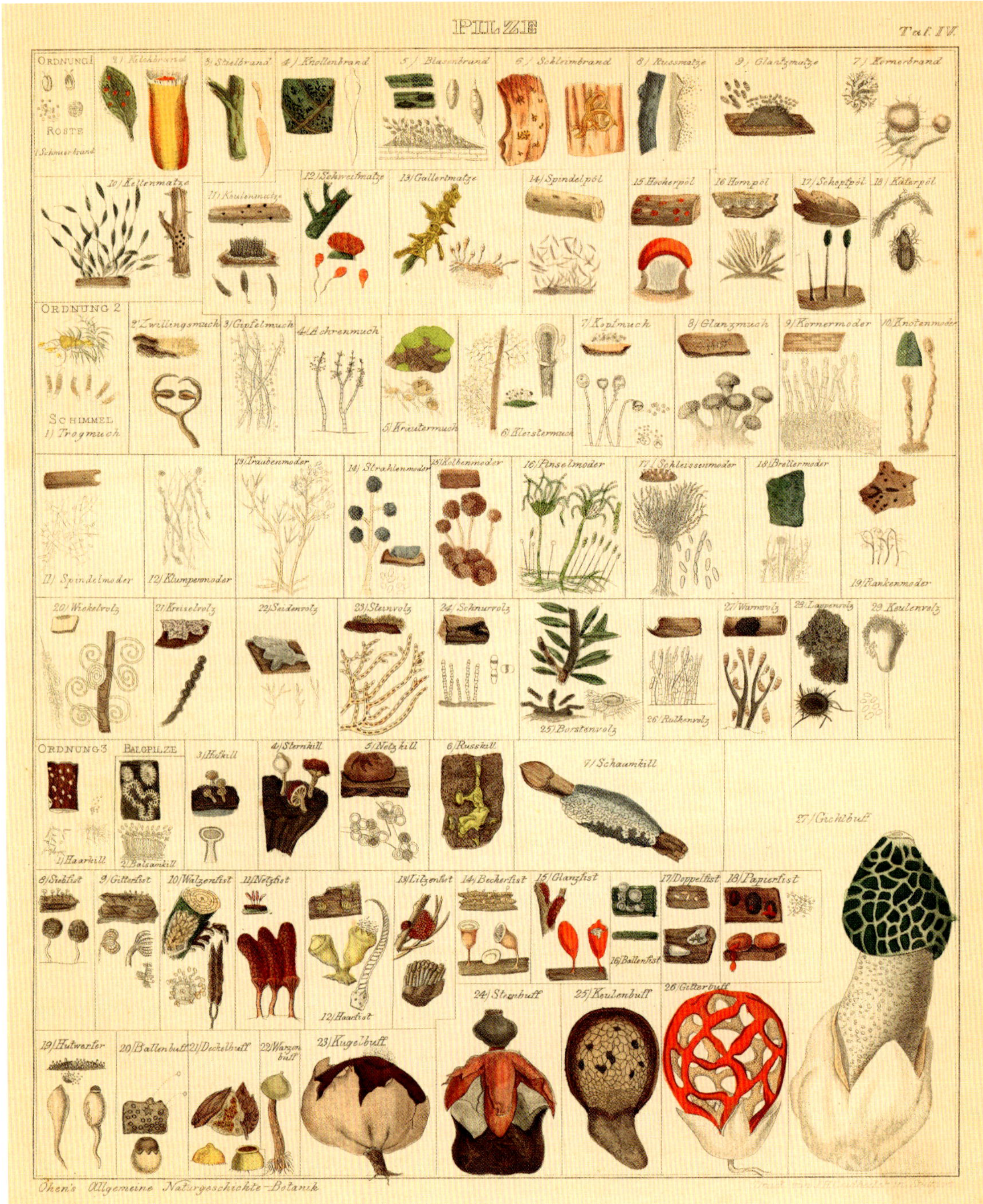

LORENZ OKEN

Pilze (Mushrooms), from *Allgemeine Naturgeschichte für alle Stände (General Natural History for All Classes)*, 1843
Lithograph, H. 13¾ in. / 35 cm
Smithsonian Libraries, Washington, DC

This kaleidoscopic array of various fungi species is from a monumental work by German biologist and philosopher Lorenz Oken (1779–1851). Oken's *Allgemeine Naturgeschichte für alle Stände (General Natural History for All Classes)* was published in thirteen volumes between 1833 and 1841, with an accompanying atlas volume of illustrations published in 1843, in which this plate was included. While the first two rows show various species of rust fungus and the next three rows a variety of molds, the final three rows depict perhaps more familiar species. They include a puffball (23, *Bovista plumbea*) with spores emerging from its surface, an earthstar (24, *Geastrum quadrifidum*), a red cage fungus (26, *Clathrus ruber*), and a stinkhorn (*Phallus impudicus*) emerging from its egglike base at bottom right. Featuring meticulously rendered and vibrantly colored-in engravings by Johann Susemihl, the immensely detailed publication became an important reference for the teaching of biology. Oken studied natural history and medicine at university, after which he became a lecturer at Germany's University of Göttingen. It was there that he began touting his *naturphilosophie* ("philosophy of nature") movement. In 1807 Oken was invited to take up a professorship of medical sciences at the University of Jena, where Johann Wolfgang von Goethe was rector, and in 1816 Oken began publishing his journal *Isis*. The publication mainly covered the topics of natural history, comparative anatomy, and physiology, but also poetry and politics, the latter of which caused *Isis*, and Oken, no shortage of legal and personal problems.

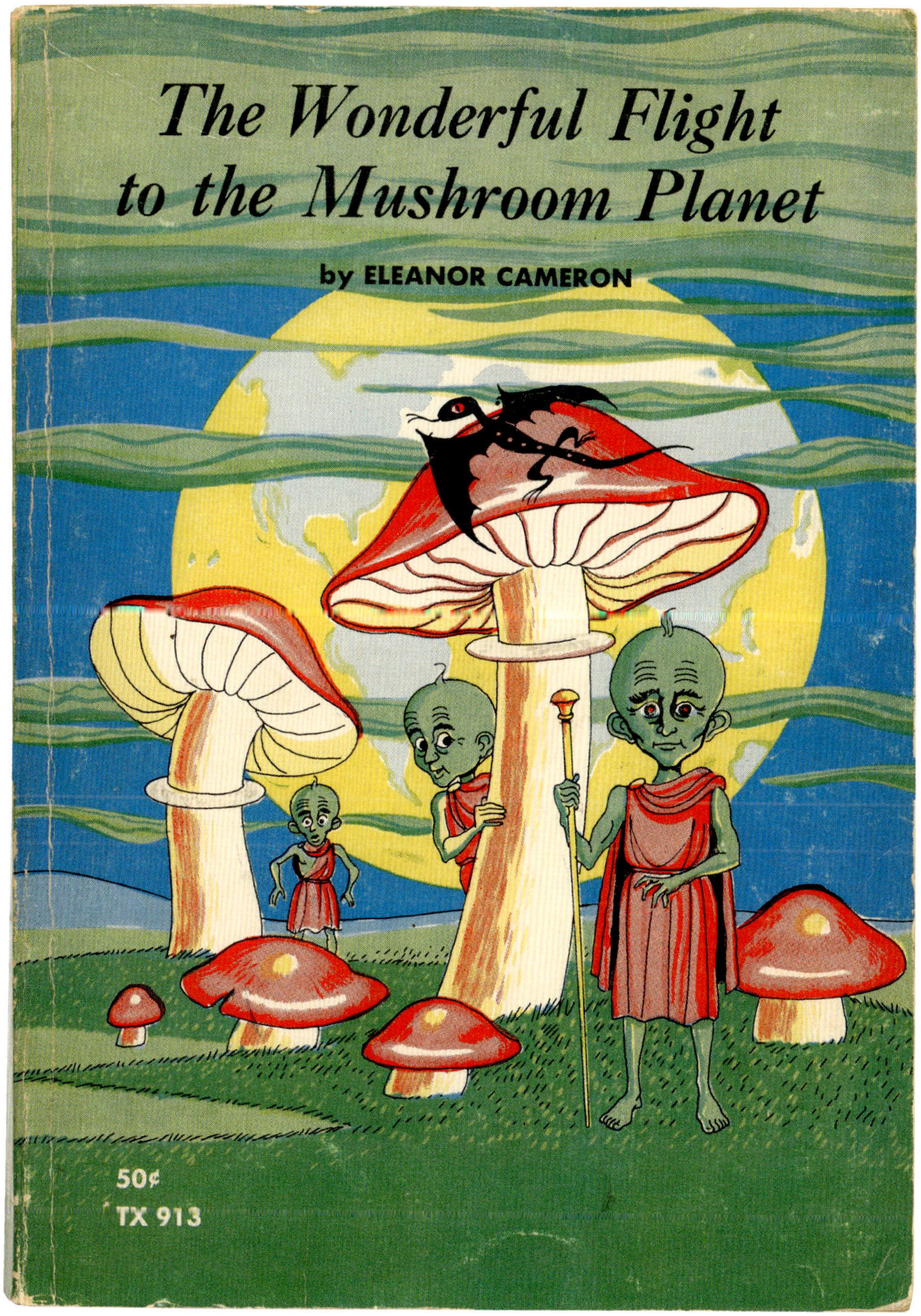

ROBERT HENNEBERGER

Cover of *The Wonderful Flight to the Mushroom Planet* by Eleanor Cameron, 1954
Printed book, 7½ × 5¼ in. / 19.1 × 13.3 cm
Private collection

Eleanor Cameron (1912–1996) was a librarian, literary critic, and prolific author, mainly of science fiction books for young readers. She won the National Book Award in 1973 for *The Court of the Stone Children*, but it was her delightful novel *The Wonderful Flight to the Mushroom Planet* that first put her on the map when it was published in 1954. The story follows the adventures of young boys David and Chuck, who are drawn in by a mysterious newspaper advertisement: "An adventure and a chance to do a good deed await the one who builds the finest ship!" Responding to the ad, they meet the enigmatic Mr. Bass, who claims to have discovered the tiny planet Basidium-X, and to the boys' amazement, asks them to make a journey into outer space. They arrive on a moss-covered planet teeming with towering mushrooms of every imaginable shape and color. They soon meet their alien counterparts, the little green men Mebe and Oru, who explain the great plague affecting Basidium's crops and take them to meet planet's leader, the Great Ta. Through David and Chuck's ingenuity, along with the unusual items Bass instructed them to bring on their journey, they succeed in saving the planet and its mushroom people. Peppered throughout with charming illustrations of the fungi-filled world by American artist Robert Henneberger (1921–1999), who also created its cover, the book was a rousing success that spawned five sequels.

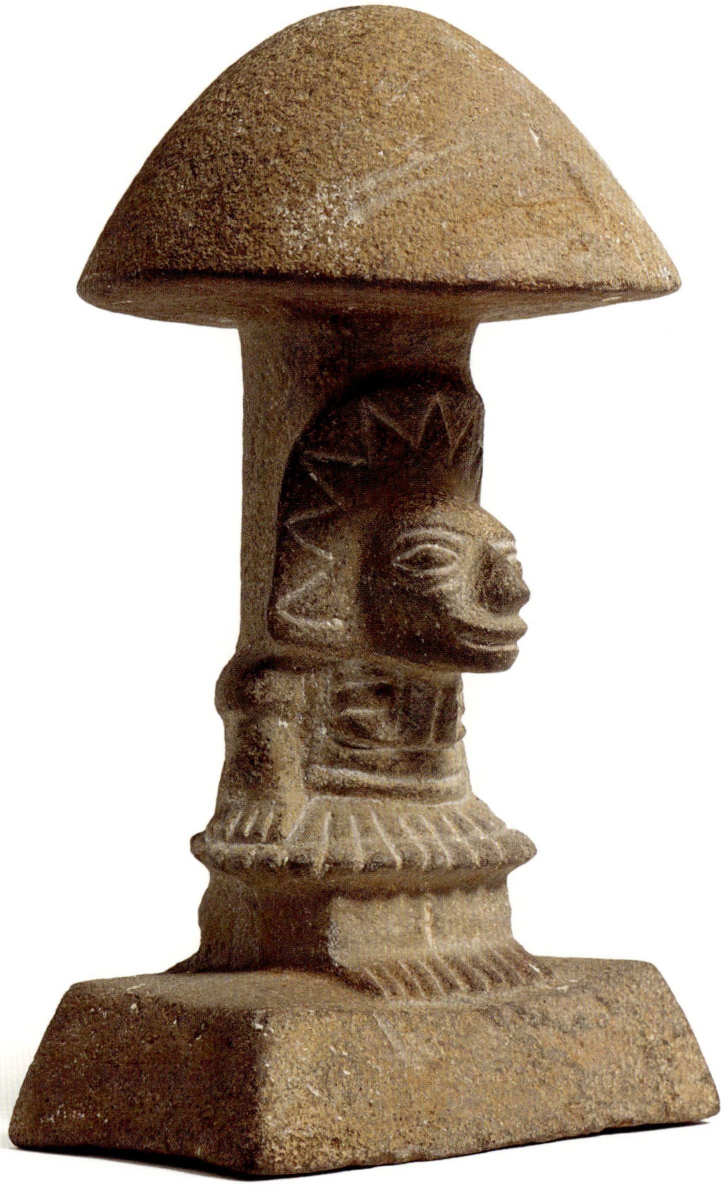

ANONYMOUS

Mushroom stone, 400–250 BCE
Stone, 13⅛ × 7½ × 6¾ in. / 33.5 × 19 × 17 cm
Museum Rietberg, Zürich

Carved by an unknown artisan more than two thousand years ago, this stone figure standing under a large, upright mushroom originated in Central America. Mushroom stones, as they are known, have been found in pre-Colombian cultures across present-day Mexico, El Salvador, Guatemala, and Honduras. The statues are evidence of the significant role psychoactive mushrooms played in ancient Mesoamerican civilizations, where they were an important part of sacred ceremonies. The Mayans and then the Aztecs who lived in present-day Central America and Mexico used a psychoactive mushroom called *k'aizalaj okox*, or teonanacatl, which contains the compounds psilocybin and psilocin, both of which cause visual hallucinations and can alter the perception of reality. The mushrooms would have been eaten only by high-ranking priests during religious ceremonies to commune with their gods, often accompanied by music and dancing. There is speculation that mushrooms would also have been consumed during other sacred events, such as the Mesoamerican ballgame and warfare. Another theory suggests the mushroom stones, which stand around 12 inches (30 cm) tall, were not purely decorative but were used to grind the psychoactive mushrooms into a fine powder ready for consumption during the ceremonies. Nearly all known mushroom stones have been uncovered in archaeological sites in former Mayan territory. After the Spanish conquest of 1492 and during the subsequent Spanish colonial rule, as the colonizers imposed Catholicism on the Indigenous people, the use of psychoactive mushrooms was outlawed.

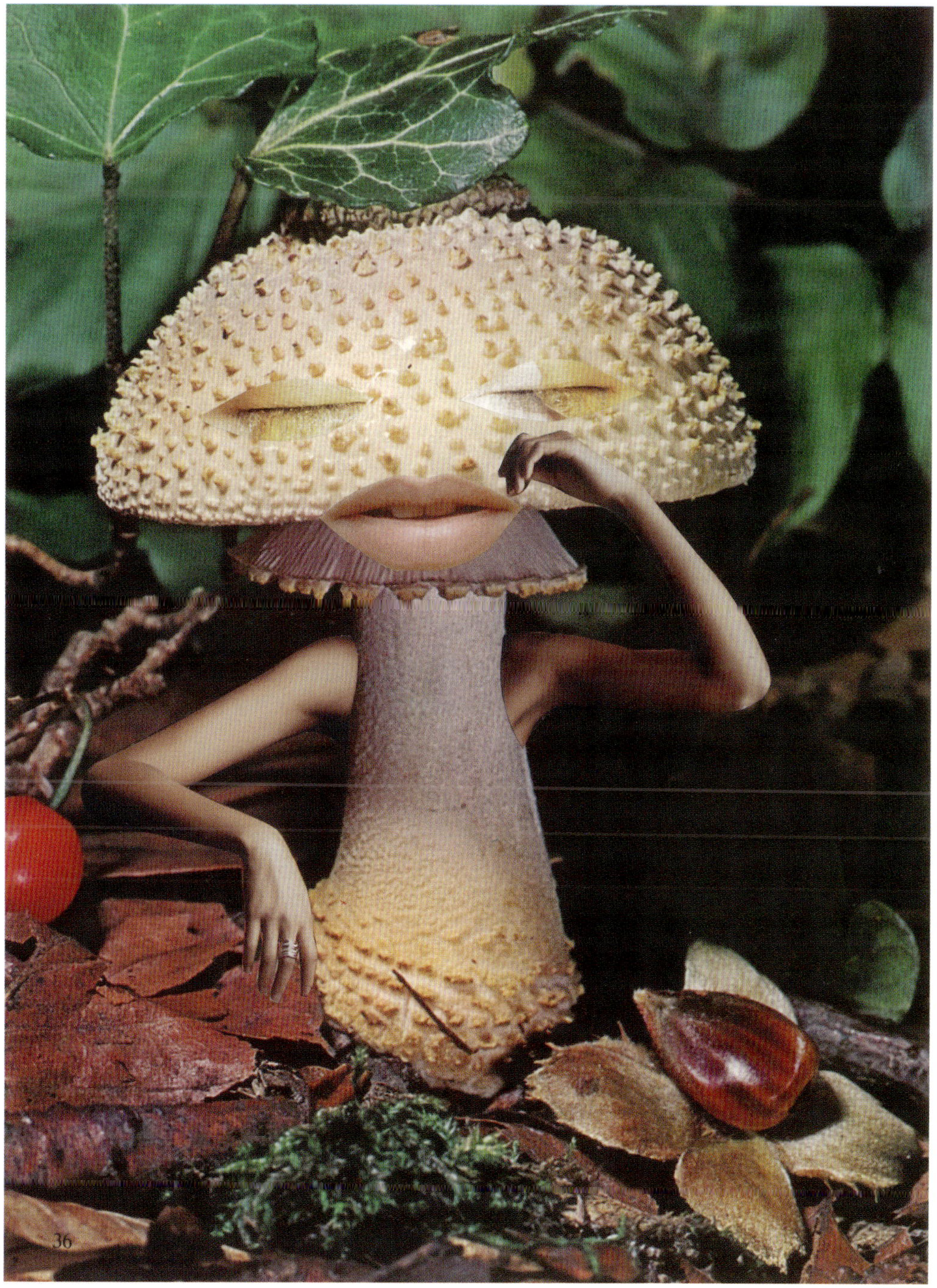

SEANA GAVIN

Mindful Mushroom, 2017
Hand-cut collage,
11¾ × 8¼ in. / 29.7 × 21 cm
Private collection

A photographic collage of a pale brown mushroom nestled in woodland undergrowth is transformed into something figurative with arms, closed eyes, and a mouth. This image of a meditative mushroom in nature by British artist Seana Gavin (b. 1977) has become one of the most recognized images of fungi in contemporary art, featured in exhibitions at London's Somerset House (2020), Fundação de Serralves in Porto, Portugal (2022–23), and Stockholm's Nobel Prize Museum (2023). Gavin's hand-cut collage works are composed of historic found images that she makes into intuitive, surreal landscapes, architectural spaces, or anthropomorphic figures. Her influences include works of science fiction and historical artists, and range from Dutch proto-Surrealist Hieronymus Bosch to English Romanticist painter, engraver, and illustrator John Martin. Gavin is also known for her photographic series devoted to a decade in the Free Party movement, traveling across Europe from 1994 to 2001, which manifested in the monograph *Spiralled*, published in 2020. Mushrooms, however, remain a central motif in her practice, originally a reflection of her wider interest in fantasy, fairy tales, and fiction. Gavin's work is also influenced by the psychedelic and other states of consciousness, a nod to her own experimentation with psychedelics, including LSD and mushrooms, and perhaps an indirect reference to her childhood growing up in hippie mecca Woodstock in Upstate New York. Playing with scale and color, in Gavin's fungi collages mushrooms can be the size of castles, represent architectural daydreams, or become metaphors for humanity's desire for connection.

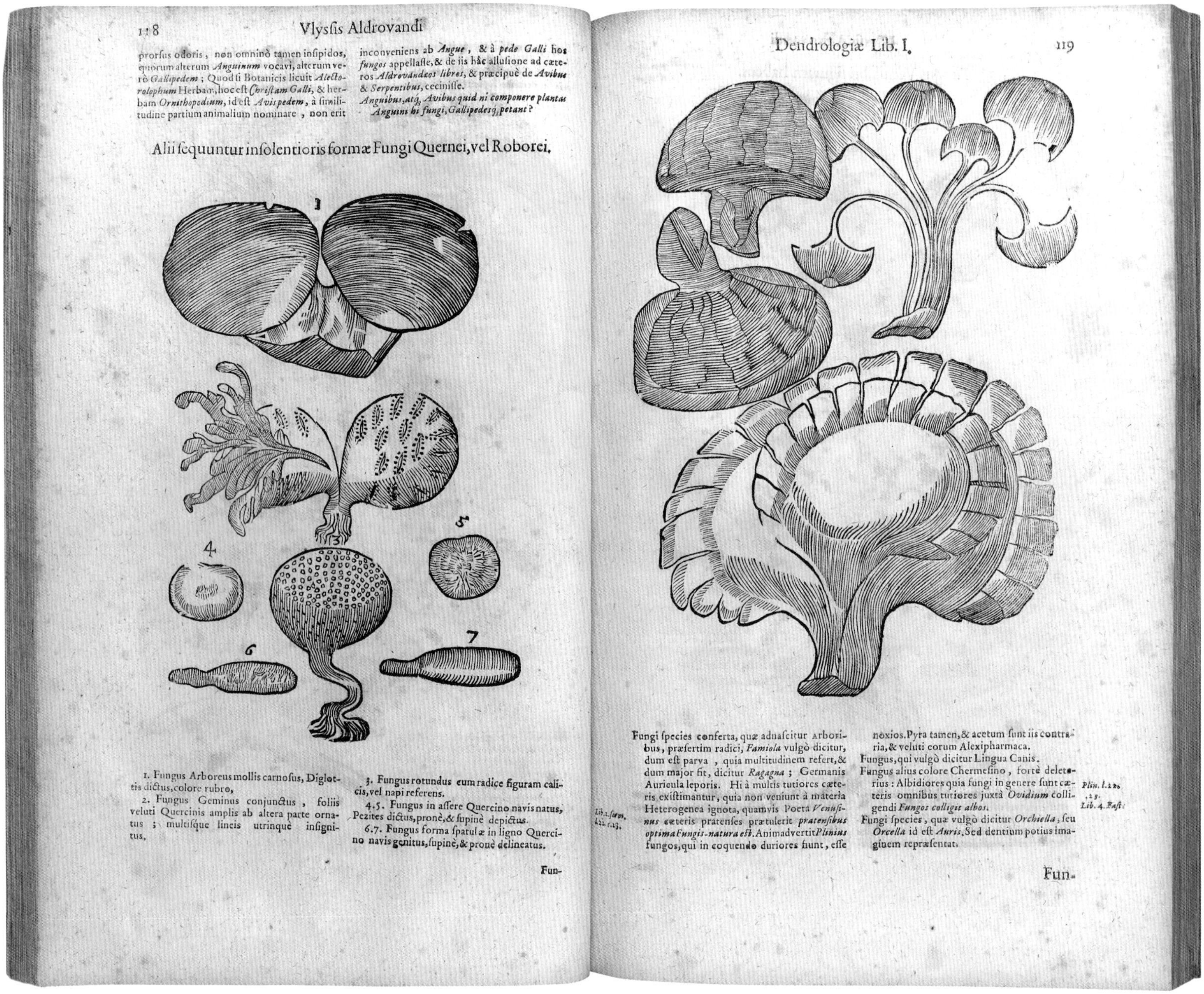

ULISSE ALDROVANDI

Dendrologiae Naturalis Scilicet Arborum Historiae (Natural Dendrology, Specifically the History of Trees), vol. 2, 1671
Woodcut, each page approx. 13¾ in. × 9 in. / 35.1 × 22.8 cm
Wellcome Collection, London

Rendered with bold outlines, these black-and-white woodcuts of mushrooms come from *Dendrologiae Naturalis Scilicet Arborum Historiae*, a history of trees by prolific Italian Renaissance polymath, naturalist, and physician Ulisse Aldrovandi (1522–1605). Aldrovandi approached this study of mushrooms—which was published posthumously—with the same exacting observation as his larger studies of the animal world, plants, and minerals. Appointed as the first professor of natural sciences at the University of Bologna—Europe's oldest university—in 1561, Aldrovandi became the leading light of sixteenth-century encyclopedic naturalism. An avid collector, he traveled widely, studying and collecting around eighteen thousand animal, plant, and mineral specimens, which he referred to as his "microcosm of nature." From this microcosm, Aldrovandi assembled more than seven thousand species of dried plants into a fifteen-volume herbarium and identified them for their medicinal properties. For the thirteen-volume encyclopedia of natural history, from which comes the volume featured here, he used drawings instead of actual specimens to illustrate his observations of, among others, bracket fungi, such as the fungus *Flammula* (top right), which causes trees to decay as its fruit grows in clusters at the base of the trunk, and the edible oyster mushroom (*Pleurotus ostreatus*, bottom right), which grows on dying deciduous trees. While the drawings are accurate, Aldrovandi's accompanying text, written in Latin, is less scientific, relying as it does on folklore and traditional beliefs, which were just beginning to be challenged as empirical science took hold.

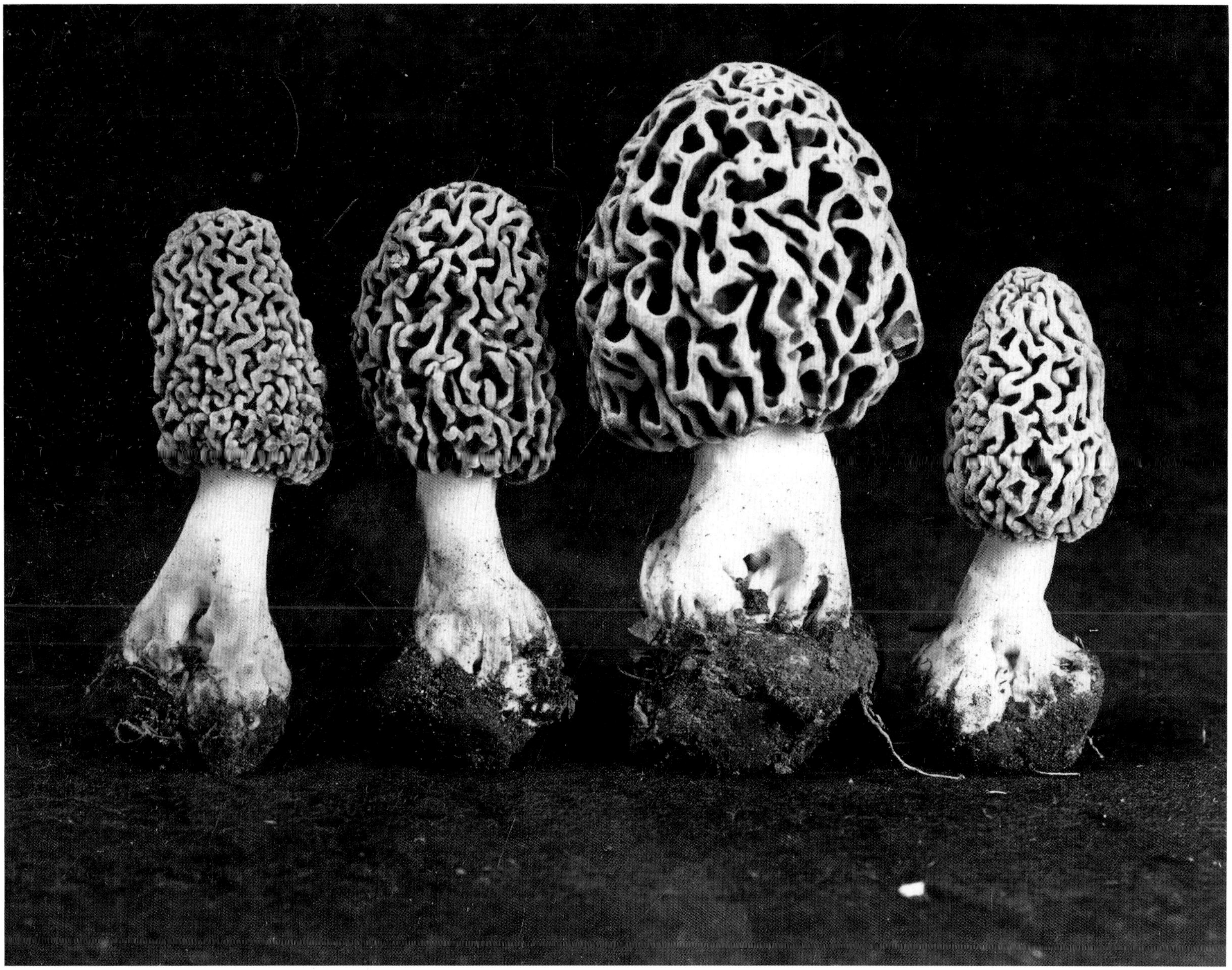

GEORGE FRANCIS ATKINSON

Morchella, c. 1892–1918
Photograph, dimensions variable
Cornell University Library, Ithaca, New York

Four specimens of *Morchella*, or true morels, sit neatly lined up, the texture of their brainlike caps starkly contrasted with the black backdrop. The image was taken by George Francis Atkinson (1854–1918), who was the first mycologist at Cornell University in Ithaca, New York. Between the late nineteenth century and his untimely death in 1918 from the Spanish Influenza, Atkinson collected almost forty-five thousand fungi specimens, principally from New York State. His rationale for such diligence in photographing his specimens was that a photograph and notes taken at the point of collection were key tools for mycologists since fungi change over time and lose some characteristics as they dry—a dried mushroom bears little similarity to its freshly picked equivalent. The first president of the American Botanical Society, Atkinson was an undergraduate at Cornell and later returned to the university to study plant pathology and mycology, becoming head of the Department of Botany in 1896. In addition to his mushroom collecting, he also collected mosses, and his two collections were the foundation of the Cornell Plant Pathology Herbarium. Here, Atkinson captured newly collected specimens of morels, with dirt still clinging to their stems. With more than seventy species of *Morchella* estimated worldwide, they are distinguished by their characteristic honeycombed caps. Morels thrive in a variety of habitats depending on the species but are mostly found in temperate regions of North America, Europe, and parts of Asia. With an earthy, nutty taste, they are prized by gourmands globally.

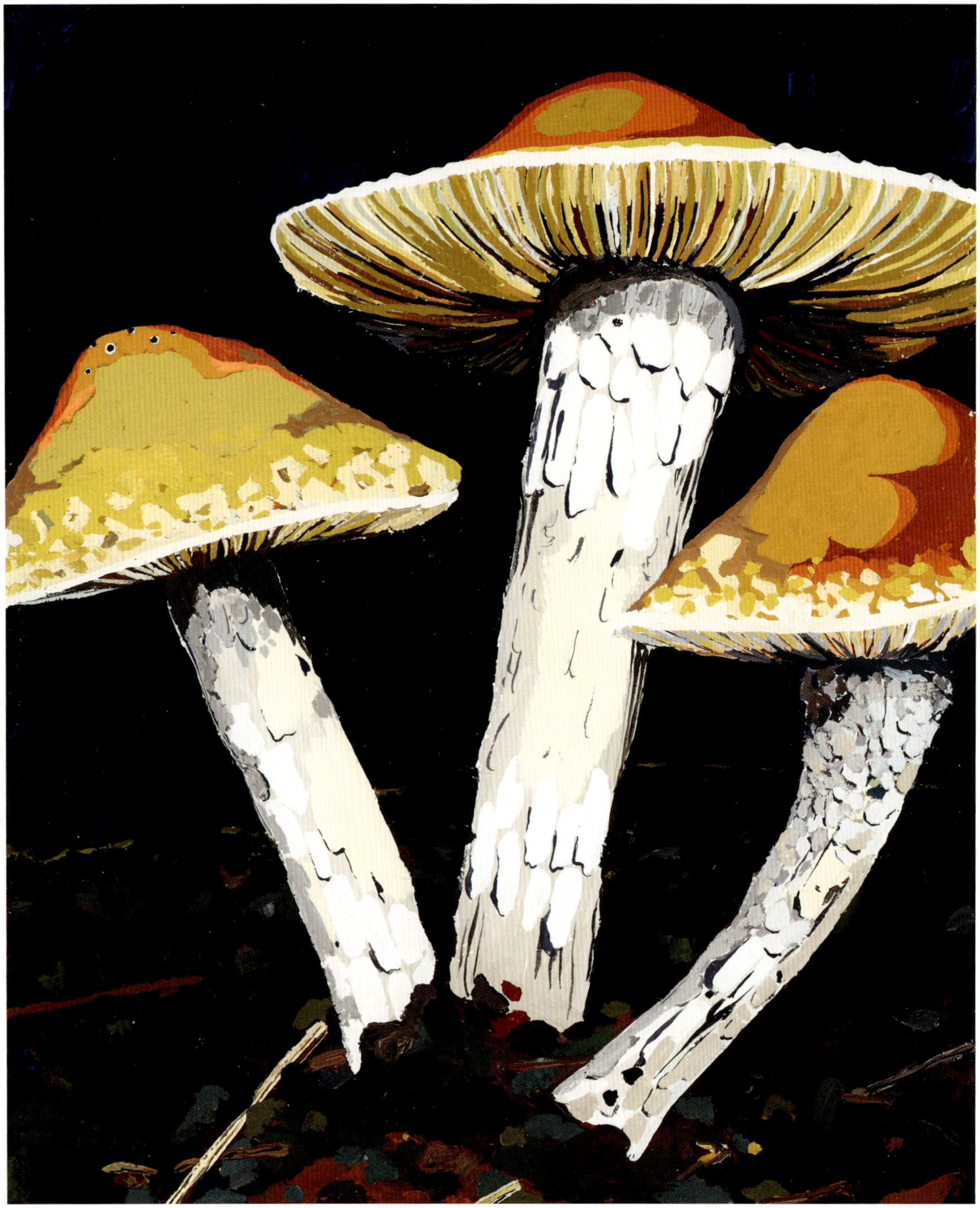

HILARY PECIS

Mushroom Patch, 2018
Acrylic on canvas,
20 × 16 in. / 50.8 × 40.6 cm
Private collection

American artist Hilary Pecis (b. 1979) is known for her color-splashed canvases depicting the familiar sun-drenched homes and landscapes of her native Southern California. She works mainly in acrylics, and her paintings reimagine the timeworn landscape, interior, and still-life traditions, imbuing them with a distinctly modern sensibility. Her canvases are typically marked by a riot of colors—deep reds, vibrant ochers, bright yellows, sparkling aquas—and focus on everyday scenes. From a book vendor's sidewalk stand and cluttered coffee tables to a seascape with wild California poppies, Pecis's bold works give glimpses into ordinary encounters from her daily life but rarely feature human figures. She has earned frequent comparisons to English painter David Hockney, who moved to Los Angeles in 1964 and often depicted the same California climes in his own vibrant style. Pecis's color work and representational imagery also recall the Fauvists, an undeniable influence. *Mushroom Patch* is a bit of a departure for Pecis, who this time follows her subject out of the sunshine and into the shadows, yet the artist's style is unmistakable. The titular mushrooms occupy the entirety of the canvas, bleeding off the edges and out of frame. Her bold brushstrokes give deep dimension to the fungi's scales and gills, setting the texture in sharp relief. Prized by contemporary art collectors, her work is also held in the permanent collections of the National Gallery of Art in Washington, DC; Los Angeles County Museum of Art; and Museum of Fine Arts, Boston, among others.

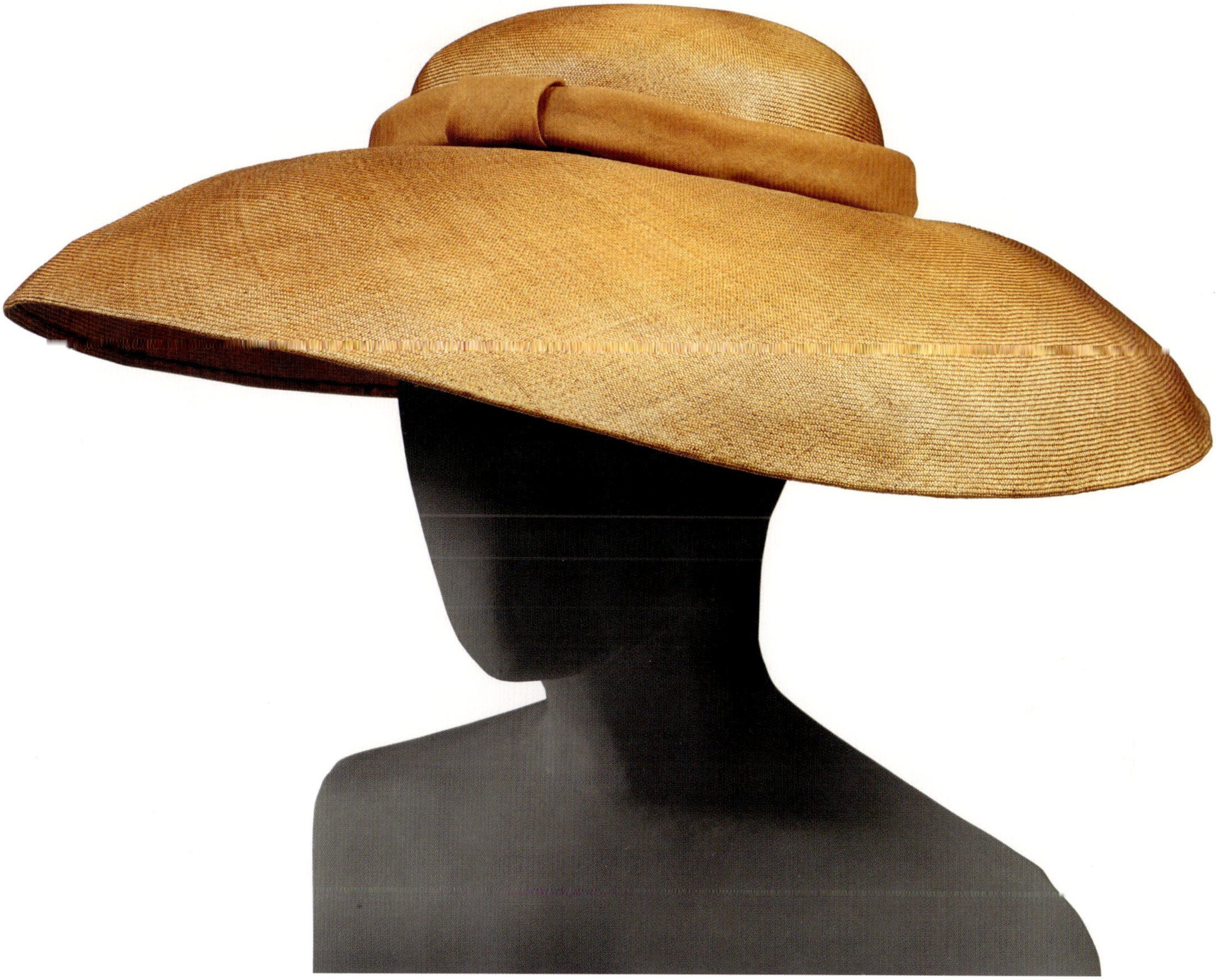

CRISTÓBAL BALENCIAGA

Mushroom hat, 1963
Hand-sewn fine straw hat with plastic haircomb and brown cotton headband, Victoria and Albert Museum, London

The oversize café-au-lait, fine straw "mushroom" hat lives up to its name, with its rounded top and wide downward-facing brim that resembles the cap of a mushroom. It is the work of the fashion house of Spanish couturier Cristóbal Balenciaga (1895–1972), who was famed for his innovative dress designs and exquisite tailoring. In 1963, when it was made, hats were still de rigueur for daily wear and at formal occasions, where they were intended to complement and transform a couture outfit. Balenciaga recognized their importance for his clients, and his Paris couture house had two workshops dedicated exclusively to millinery. While he did not design the hats himself, he worked closely with his in-house milliners, first Russo-French milliner Wladzio d'Attainville and then Spanish milliner Ramón Esparza.

Mushroom-shaped hats were first fashionable in the 1870s and 1880s, when they were typically made from straw. They reappeared in 1947, when Christian Dior launched his revolutionary New Look in Paris. The New Look featured nipped-in waists, fitted jackets, and a mushroom hat (also known as a dish hat), which was notable for its large, downturned brim. Balenciaga adopted the idea and reimagined it here as a chic accessory made from hand-sewn fine straw with a shiny finish and a matching silk organza band around the brim, which has proven a timeless fashion design.

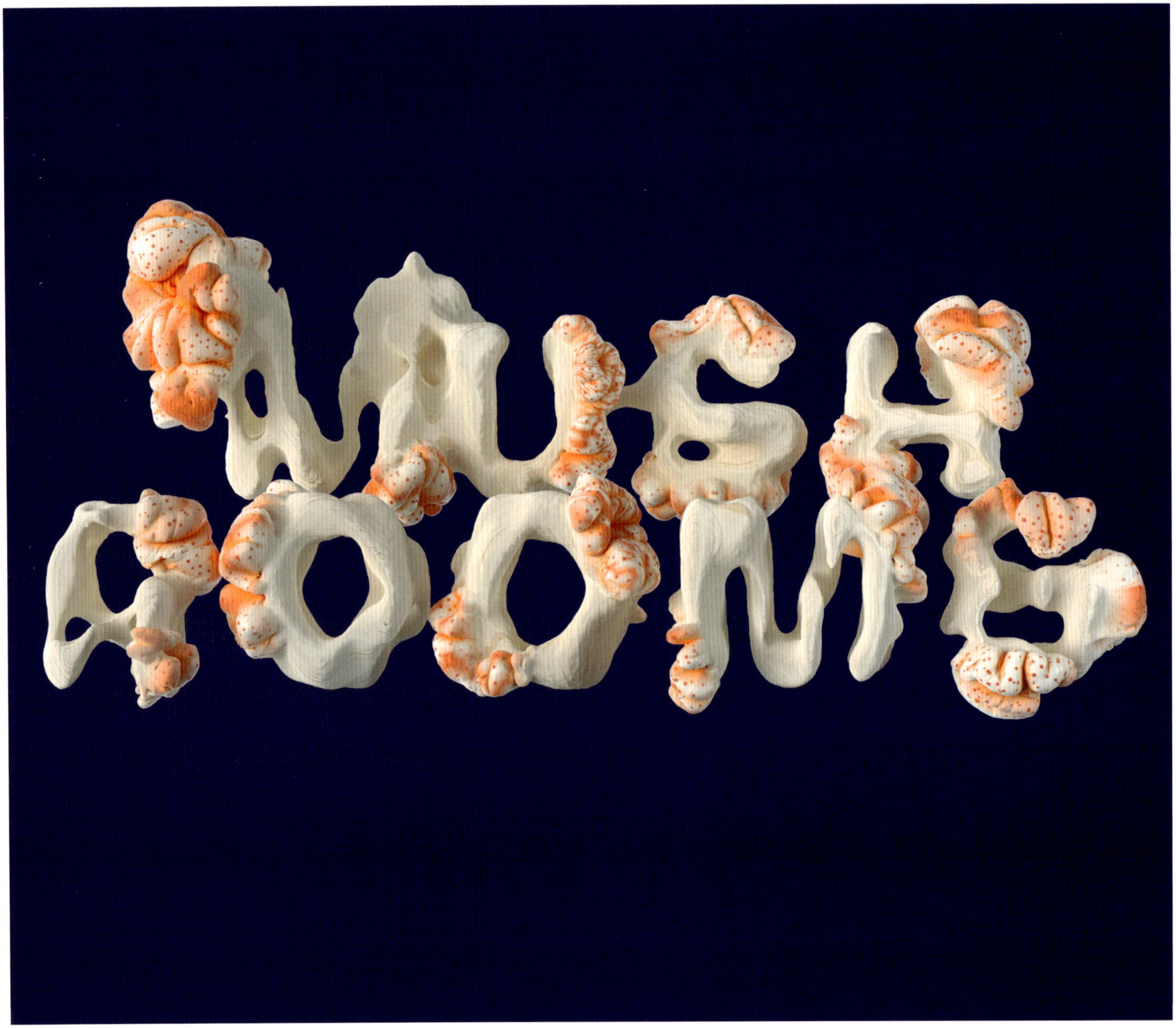

PENTAGRAM, COUNTERPOINT STUDIO, AND ROSIE EMERY

Hypha, 2023
Generative system and interactive web tool, dimensions variable

An alphabet grows and emerges from the ether, echoing the formation and organic expansion of mycelium and spore growth. Renowned design studio Pentagram developed *Hypha* when they were commissioned to create the design and artistic direction of the exhibition *Mushrooms: The Art, Design and Future of Fungi*, curated by Francesca Gavin at Somerset House in London. The project was a collaboration with 3D and technical artist Rosie Emery and generative typeface designers Counterpoint. Alongside the online tool, where any word could be transformed into a fungilike font, Pentagram's typeface also featured as 3D-printed exhibition tiles and growthlike sculptures that clung onto the sides of tables and the show's architecture. The tool was also presented in the exhibition as an interactive artwork in its own right, as well as in Fundação de Serralves's exhibition *The Art of Mushrooms*. *Hypha* balanced the psychedelic and magical with information and science. Viewers could watch a word grow and form in front of their eyes like the fruiting body of a mushroom in digital space. The piece was particularly informed by research on mycological field analysis and scientific cataloging archives. *Hypha* is just one of the many innovative designs by Pentagram since it was founded in London in 1972, sitting among a portfolio that includes the brand identity for iconic clients such as musicians Gorillaz, PJ Harvey, and Pink Floyd, and big brands such as Citibank, Coca-Cola, and Google. This project was led by the firm's partners, Luke Powell and Jody Hudson-Powell, and expands on their expertise regarding the intersection of the analog, virtual, and conceptual.

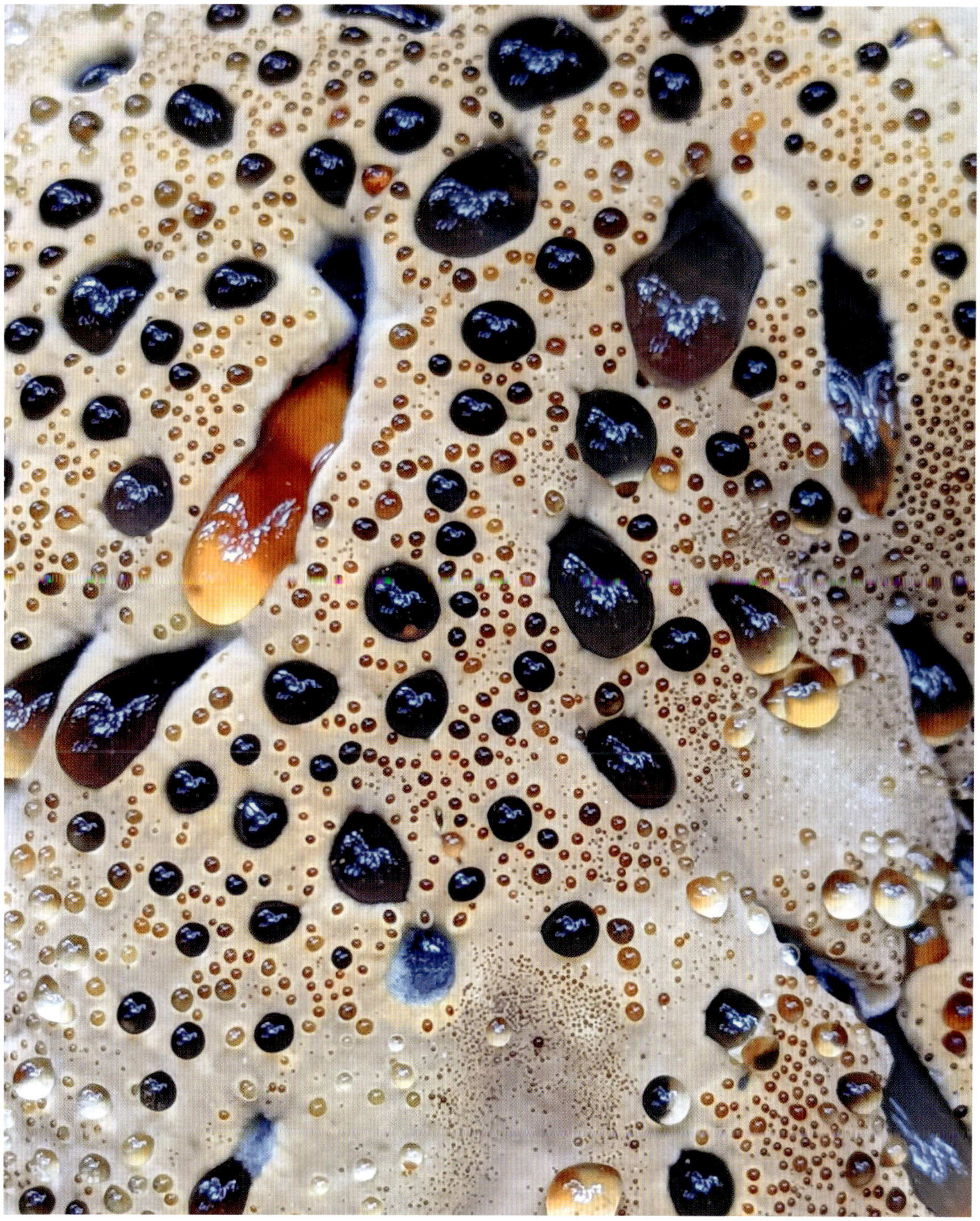

GREG T. BALLARD

The Weeping Polypore, 2022
Digital photograph, dimensions variable

Inonotus dryadeus is a species of many names—oak bracket, weeping polypore, warted oak polypore, weeping conk—but of singular purpose: to feast upon wood. It feeds indiscriminately and opportunistically. And although most often found on oak trees, it can grow on numerous broadleaf trees and even conifers. It is an insidious enemy to arborists precisely because most of its work is done stealthily. It is only after a tree falls, its core and root system weakened to the point that a wind could topple it completely, that the toll of oak bracket can truly be seen. The fungus starts at the base of the tree, attacking its root system from underground and penetrating its heartwood, causing white rot—named for the white-colored cellulose left behind by the fungi after breaking down the wood—and the tree trunk to decay. Perhaps the most distinctive feature of the oak bracket is its appearance: the telltale beads of amber liquid that drip from its surface. The emissions are the result of a process called guttation, in which the fungus releases water and waste during periods of rapid growth. Despite its destructive nature, there is no denying its beauty in this image by photographer Greg T. Ballard. Ballard came upon this specimen of oak bracket in the woods of Portland, Oregon, skillfully capturing the reflections seen in the "tears" of the fungus.

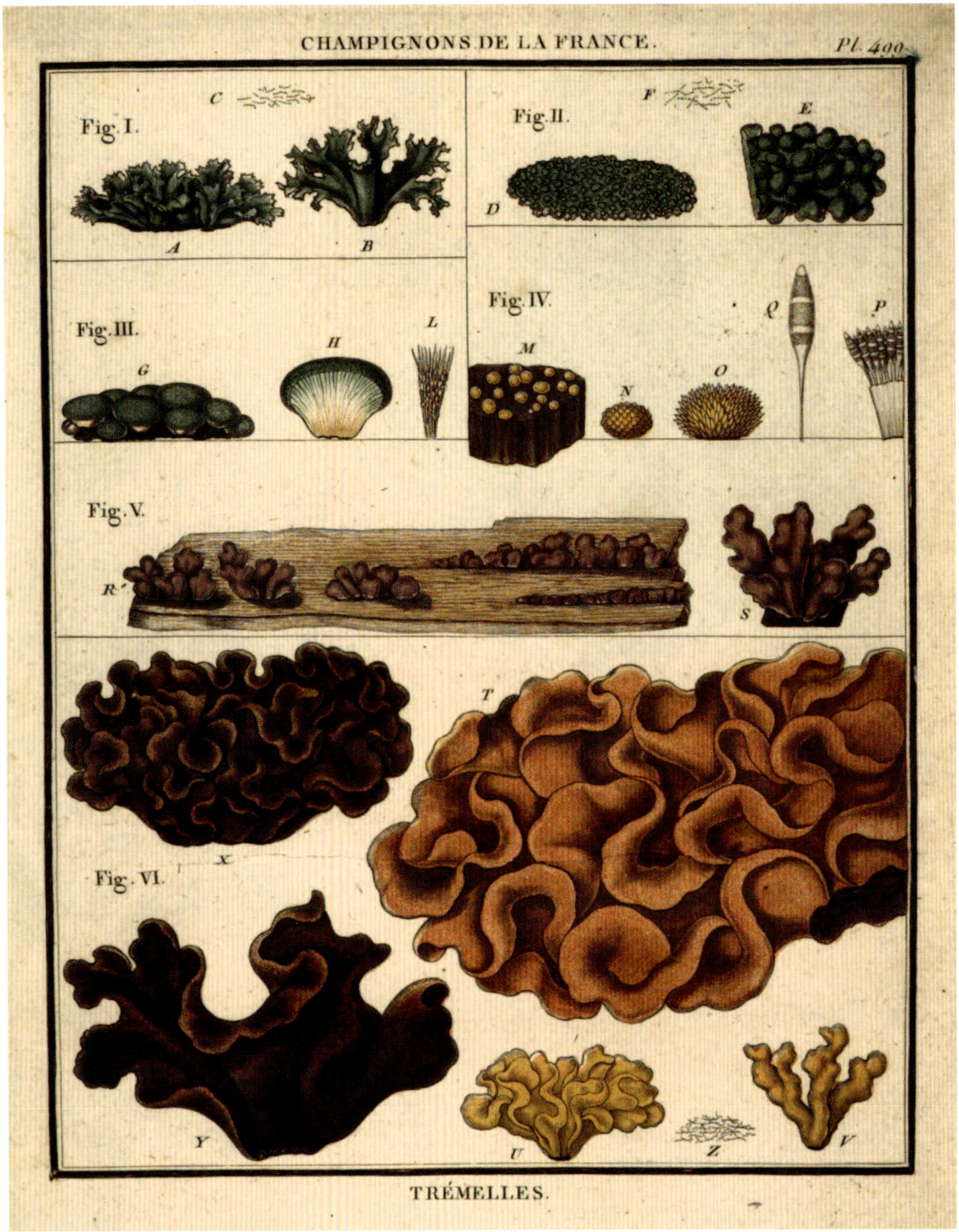

JEAN BAPTISTE FRANÇOIS PIERRE BULLIARD

Trémelles, from *Histoire des champignons de la France* (*History of the Mushrooms of France*), 1809–12
Colored engraving, H. 14½ in. / 37 cm
Harvard University, Cambridge, Massachusetts

Richly illustrated, *Histoire des champignons de la France* is one of the earliest French systematic mycological floras. Its pages reveal meticulous engravings of toadstools, bracket fungi, and other cryptogams, each rendered with delicate shading, naturalistic posture, and careful attention to structural details. In this example, *Tremella* fungi appear in rich browns, muted greens, and warm ochers, with identifying characteristics isolated and magnified. Despite their frilled and folded bodies, the dense weight of these parasitic jelly fungi is evident. The work was principally the creation of Jean Baptiste François Pierre Bulliard (1752–1793), a French physician-botanist whose hands-on approach was exceptional for the time. His ambitious project, begun in the late eighteenth century, was completed posthumously, with additional text by French botanist Étienne Pierre Ventenat. The book's purpose was representative of Bulliard's desire for wider dissemination of scientific knowledge, acting as a field guide and placing visual identification at the center of natural history. These images, although striking in their variety and skillfulness, were not decorative; viewed alongside their descriptions, they helped naturalists with the practical work of differentiating species. In the broader history of fungi in art and science, Bulliard's *Histoire* was an important step in the circulation of knowledge, both for its content and for the printing choices he made. Accuracy of color was crucial in assisting identification, and Bulliard drew, engraved, and even pioneered a color-printing technique for these plates, striving to capture the true hues of each fungus so that readers could recognize them in the field.

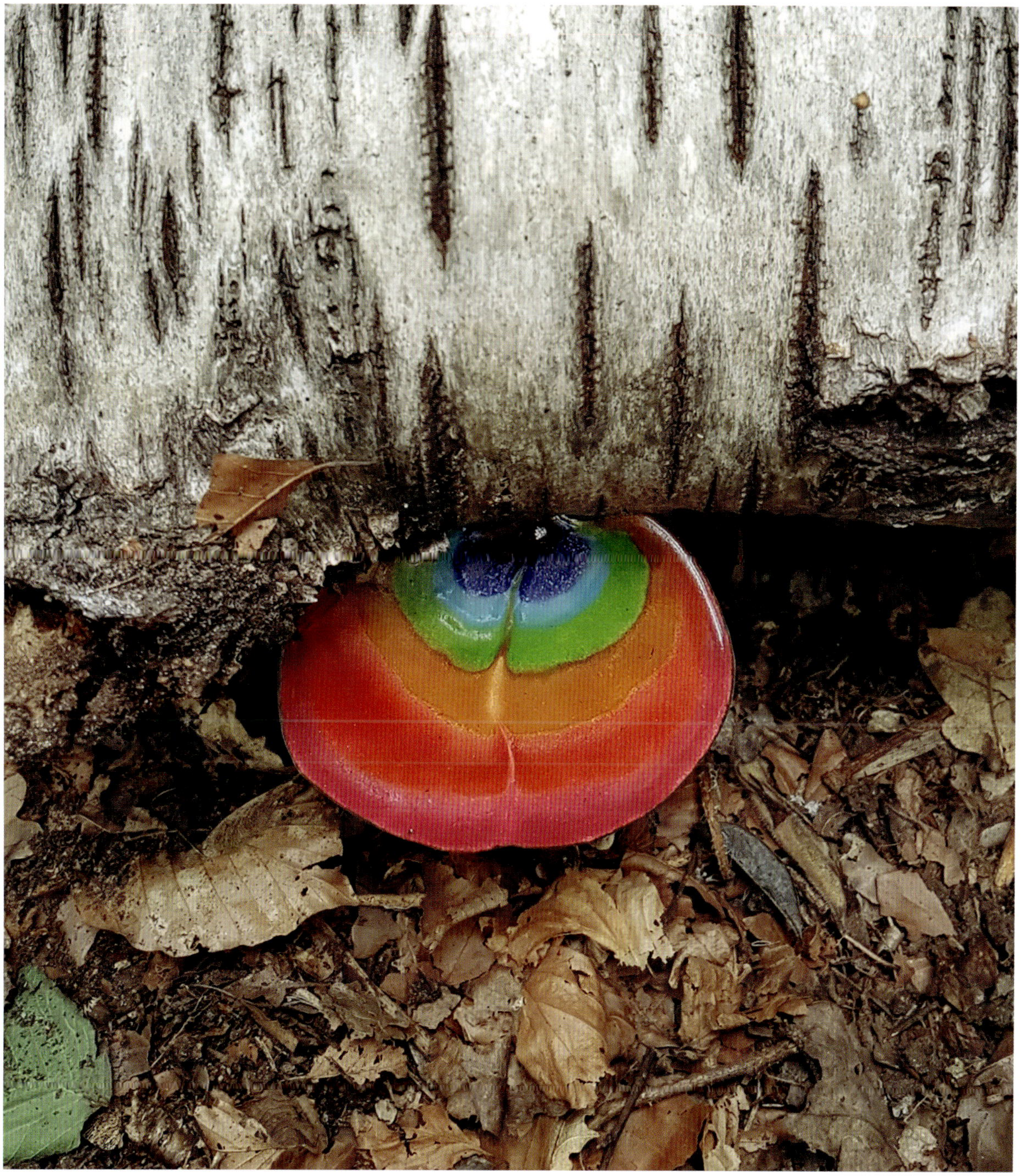

GEERTJE GEERTSMA

Where the Rainbow Ends, from the series *Cloud Flowers*, 2020
Installation with enamel on metal and wood, 4 × 4 in. / 10 × 10 cm
Private collection, the Netherlands

A rainbow-colored mushroom nestles beneath the rotting trunk of a white birch tree on a bed of decaying leaves, its vibrant color palette providing a stark contrast with its gloomy surroundings. Created by Dutch visual artist Geertje Geertsma (b. 1964), *Where the Rainbow Ends* was created as part of a series of mushrooms she made using enameled glass and copper plate. Titled *Cloud Flowers*, the series was inspired by a quote from novelist Margaret Atwood: "Mushrooms were the roses in the garden . . . because the real mushroom plant was underground. . . . The part you could see . . . was just a brief apparition. A cloud flower." For Geertsma, mushrooms have become something of an obsession. She is fascinated by what she sees as their "otherness": their different shapes, their ability to grow anywhere, and their seemingly infinite variety. She is equally inspired by the history of mycology and, more particularly, by the many—often unknown—nineteenth-century women who made mushroom hunting and drawing their life's work, benefiting from the scant regard their male contemporaries paid to fungi. Visiting her local park in the northern Dutch city of Groningen, where she lives and works, Geertsma drew in precise detail the different mushrooms she saw over the course of a year, amazed by the diversity found in the urban landscape. In an homage to the talented women mycologists of the past, Geertsma has gathered her illustrations into a book, *100 Paddenstoelen* (*100 Mushrooms*), published in 2026.

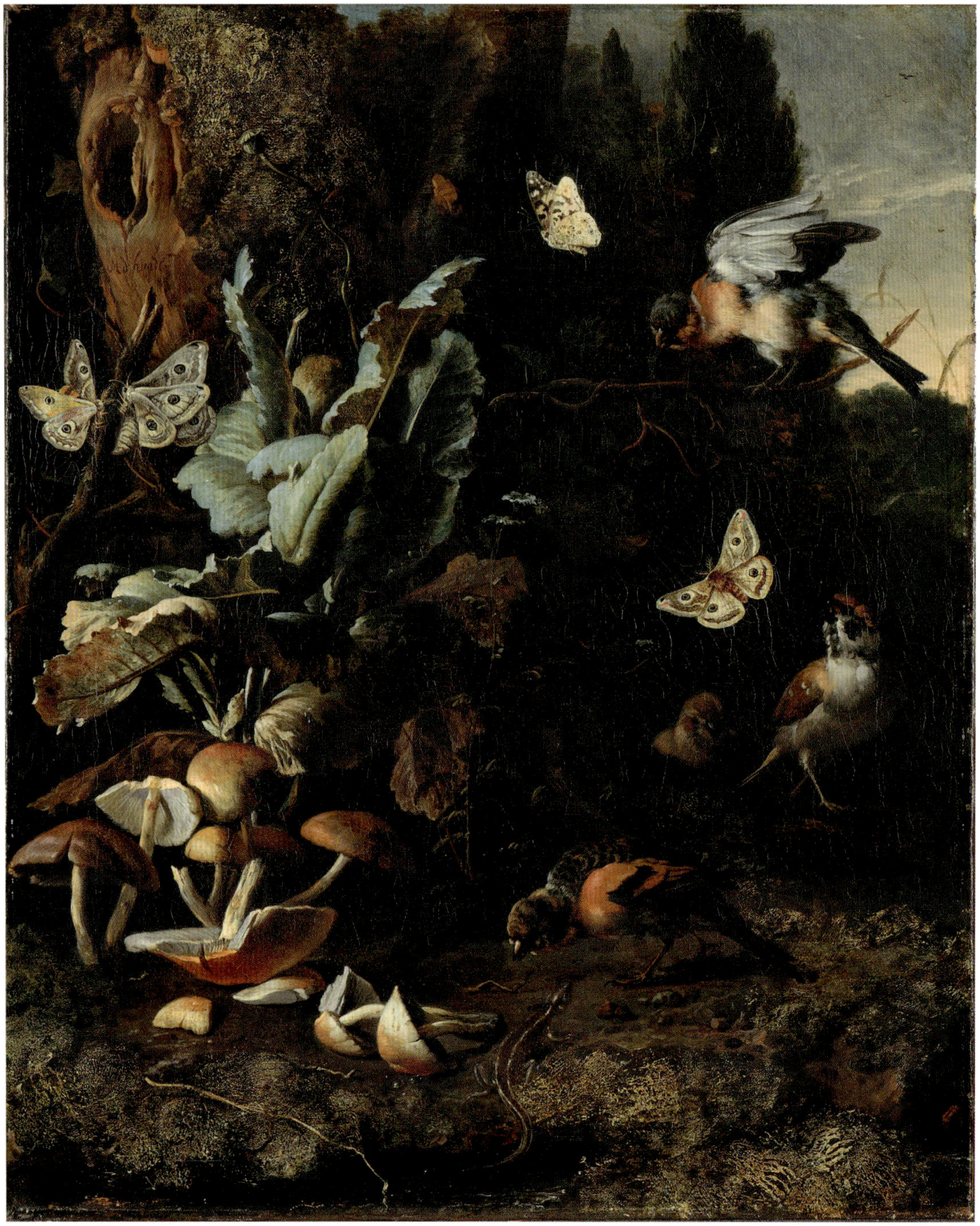

MELCHIOR D'HONDECOETER

Forest Floor Still Life with Birds, Butterflies, and a Lizard, c. 1668
Oil on canvas, 26 × 20⅝ in. / 66 × 52.5 cm
Rijksmuseum, Amsterdam

This oil painting of a dark forest floor illuminated by evening light teems with life: a common wall lizard slithers along the ground while a chaffinch looks on, perhaps intending to pounce. A closer inspection of the brooding landscape reveals more: trees and a moss-covered trunk, and luminous emperor moths that flutter while a painted lady flies above, created using real butterfly wings. In the bottom left corner, the dim forest light illuminates a cluster of mushrooms; some have been damaged and uprooted, the underside gills delicately rendered, while the others remain firmly in the ground. This densely filled canvas is the work of seventeenth-century Dutch painter Melchior d'Hondecoeter (c. 1636–1695), who is best known for his bird paintings. For most of his career, he specialized in painting often exotic species of birds, particularly either fighting or in flight, and attracted an international clientele for his work. For this early painting from around 1668, he included both insects and mushrooms, neither of which appeared in his later works. During the seventeenth century, there was a growing appetite for the natural sciences in Holland, and Dutch artists reflected this in their work. D'Hondecoeter probably painted the canvas in Amsterdam using specimens (some may well have been stuffed) that were close at hand, which explains the level of detail he imparts to both the birds and mushrooms.

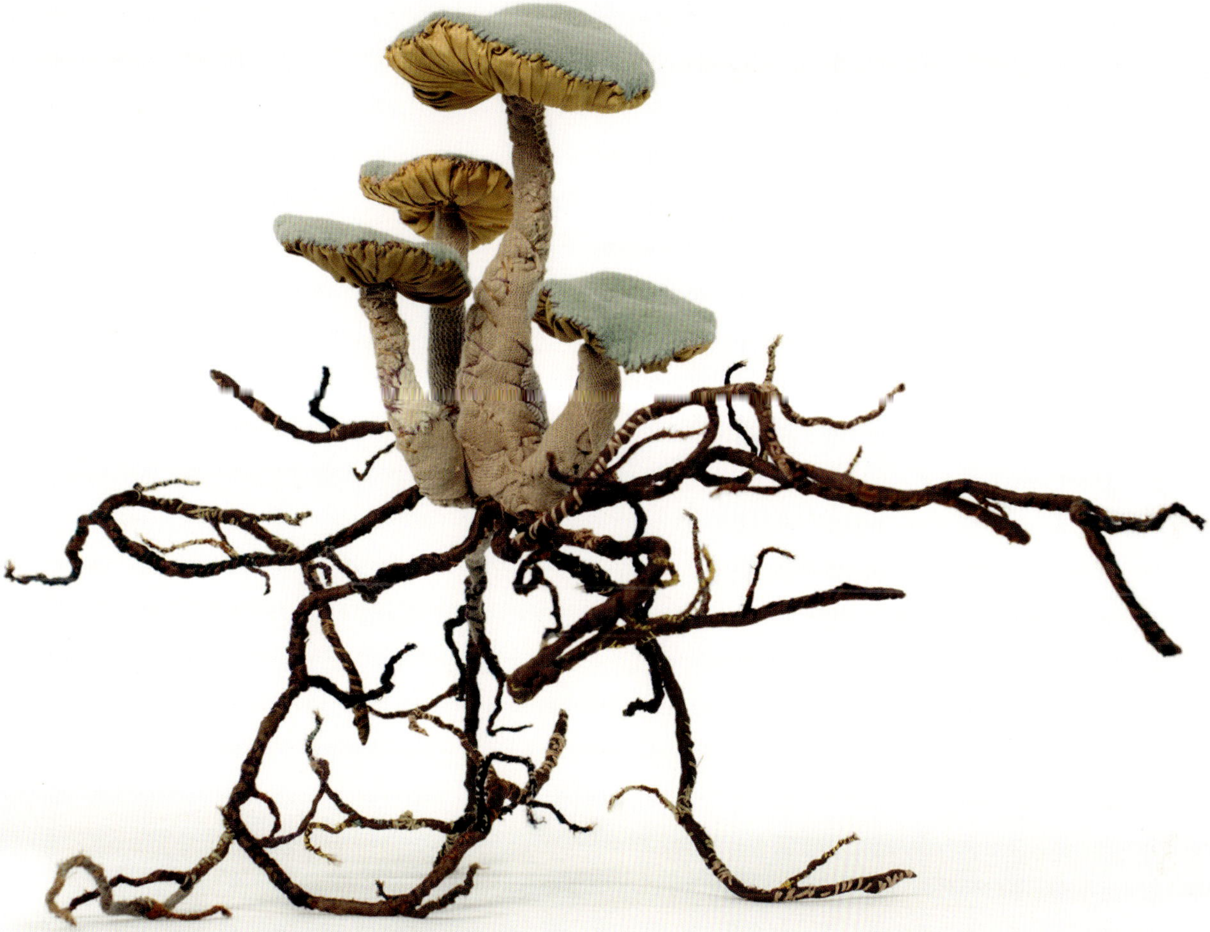

ANNE CARNEIN

Without Title, 2015
Textile, yarn, and wire,
7⅞ × 11¾ × 9⅞ in. / 20 × 30 × 25 cm

A cluster of four mushrooms, each with velvety, pale blue caps and golden undersides, perches atop a tangle of rootlike mycelial threads. While typically hidden underground, here artist Anne Carnein (b. 1982) makes the supporting mycelia visible in her intricate, seemingly floating textile sculpture. Carnein fashions her sculptures of fungi, flowers, and plants from pieces of worn clothing and other textiles, sewing together fabric, yarn, thread, and wire. Analyzing forms, colors, and biological characteristics, she translates botanical observations into her three-dimensional creations, selecting the materials that best match the essence of the mushroom or plant she's fabricating—an artistic interpretation of nature rather than imitation. Still, Carnein's works are typically shown at life size, and her inclusion of wiry roots dangling from mushroom stalks or plant stems give the appearance of having been recently plucked from the ground. While many of her sculptures are freestanding, some are attached to walls or perched on shelves, and when viewed in small narrative groupings, they give the impression of a vast herbarium. Born in Germany, Carnein went on to study sculpture at the State Academy of Fine Arts in Karlsruhe, working with German artist Stephan Balkenhol. Today, she works from her studio in the Allgäu region of southern Germany, in a self-admitted stop-and-start process, where she often puts down one sculpture and picks up again on a different partially finished piece. In each of her sculptures, she strives to explore our fundamental connection with the natural world.

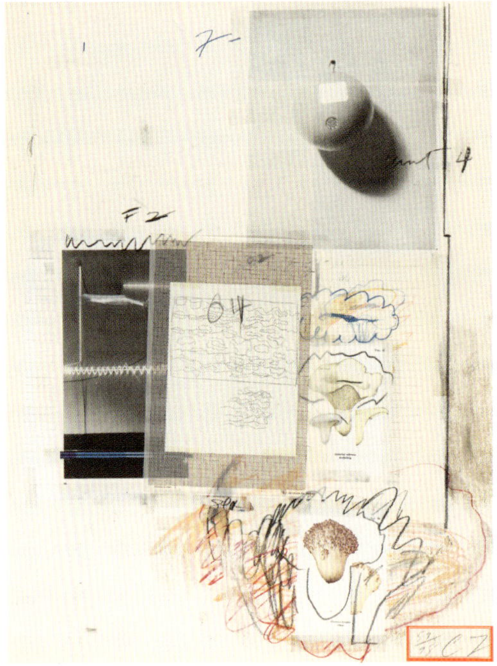

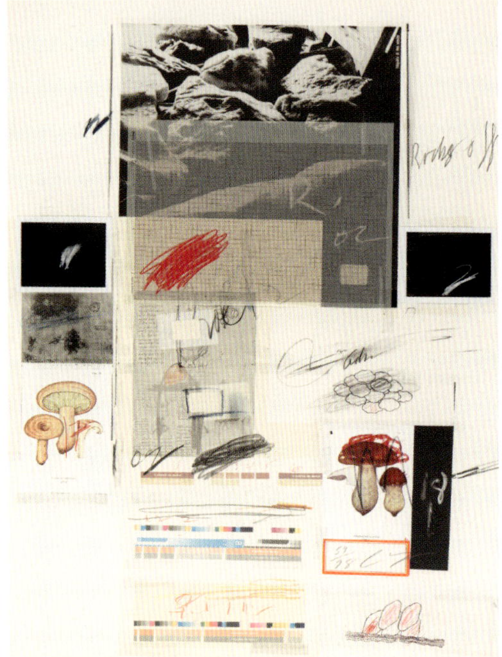

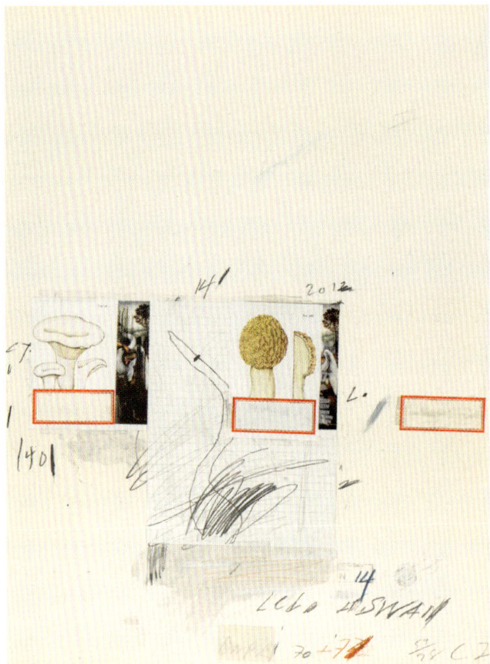

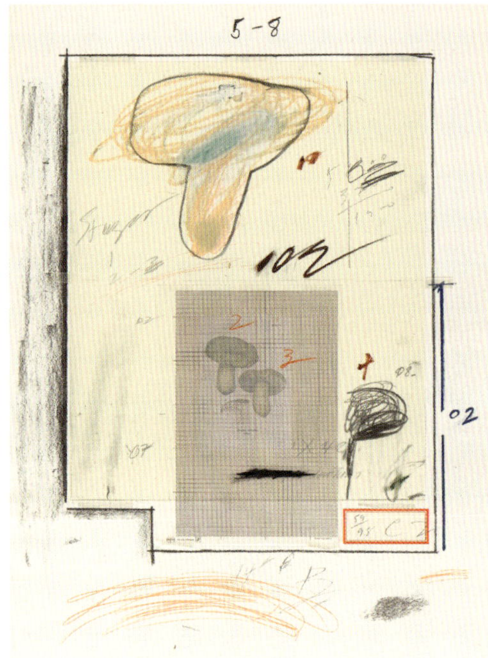

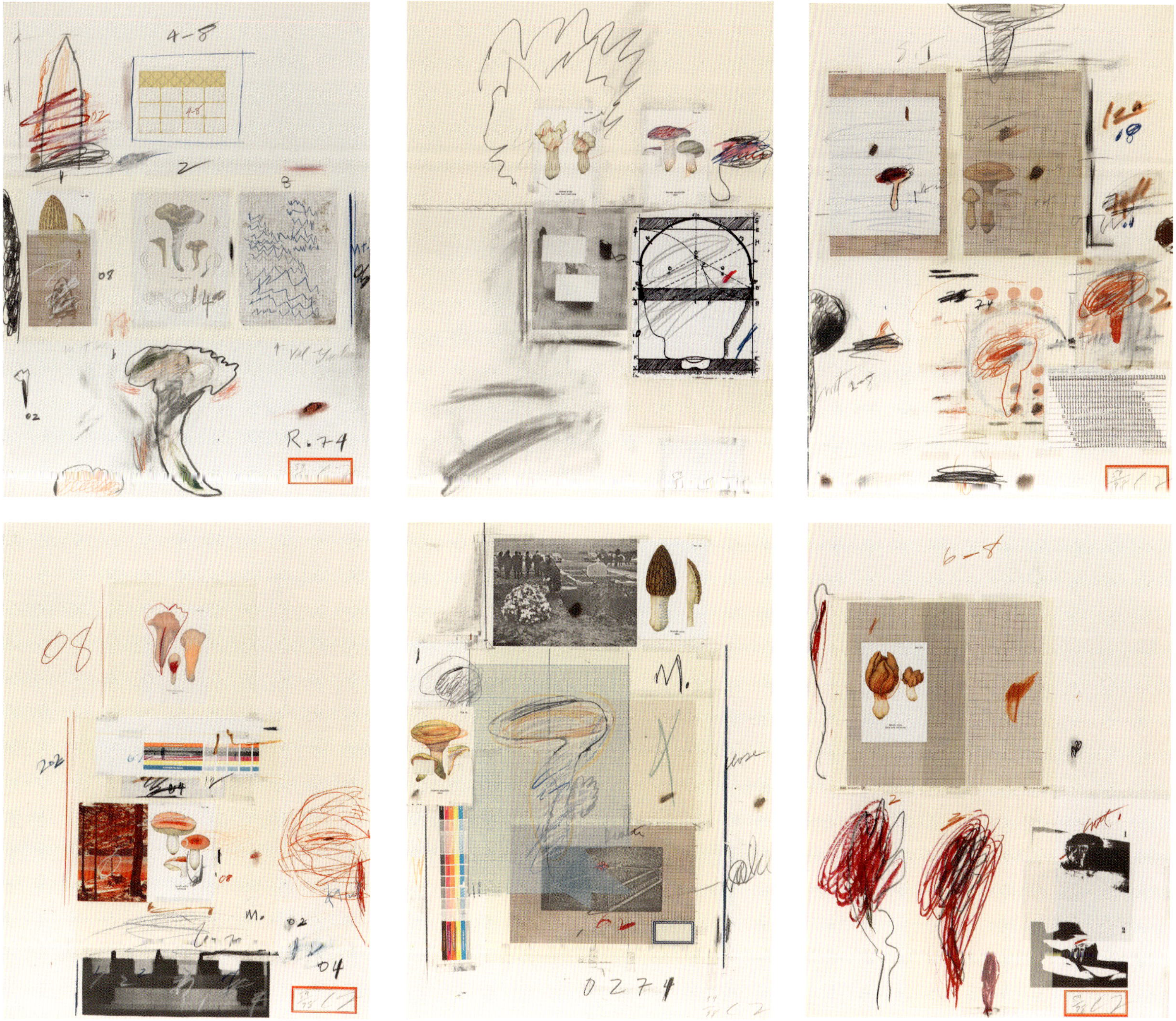

CY TWOMBLY

Natural History Part I: Mushrooms, 1974
A portfolio of ten mixed-media prints using lithograph, grano-lithograph, and collotype photochrome with collage and drawing, each sheet 29⅞ × 22 in. / 76 × 56 cm

American artist Cy Twombly (1928–2011) was known to hunt for mushrooms while he hiked in the Dolomite Mountains of South Tyrol in Italy. Such foraging trips provided the stimulus for this portfolio of ten lithograph and mixed-media prints that combine mycological illustrations and mushroom sketches with collaged elements and Twombly's idiosyncratic visual language of gestural marks, frenetic scribbles, and scrawled text. With the inclusion of recognizable fungi appropriated from old field guides, the prints represent a significant departure from the artist's more abstract works, dwelling on the rich, organic forms and textures of different species. Among the illustrated mushrooms are the fleecy milkcap (*Lactifluus vellereus*), goliath webcap (*Cortinarius praestans*), black morel (*Morchella elata*), and horn of plenty (*Craterellus cornucopioides*). Many are labeled with their binomial names and are partly obscured by translucent tracing and graph paper. Words are scrawled across a number of the prints, including in *No. VI* the name Val Gardena, a valley in the Dolomites where mushrooms are found during the late summer and autumn months, although hunting is highly regulated by provincial law. The portfolio's title references Pliny the Elder's *Historia Naturalis*, a wide-ranging, thirty-seven-volume encyclopedia completed around 79 CE. Twombly's inclusion of naturalist imagery similarly speaks to the human inclination for classifying and categorizing nature, though by partially obscuring his subject matter, he questions the extent to which the mycological world can be known through empirical means.

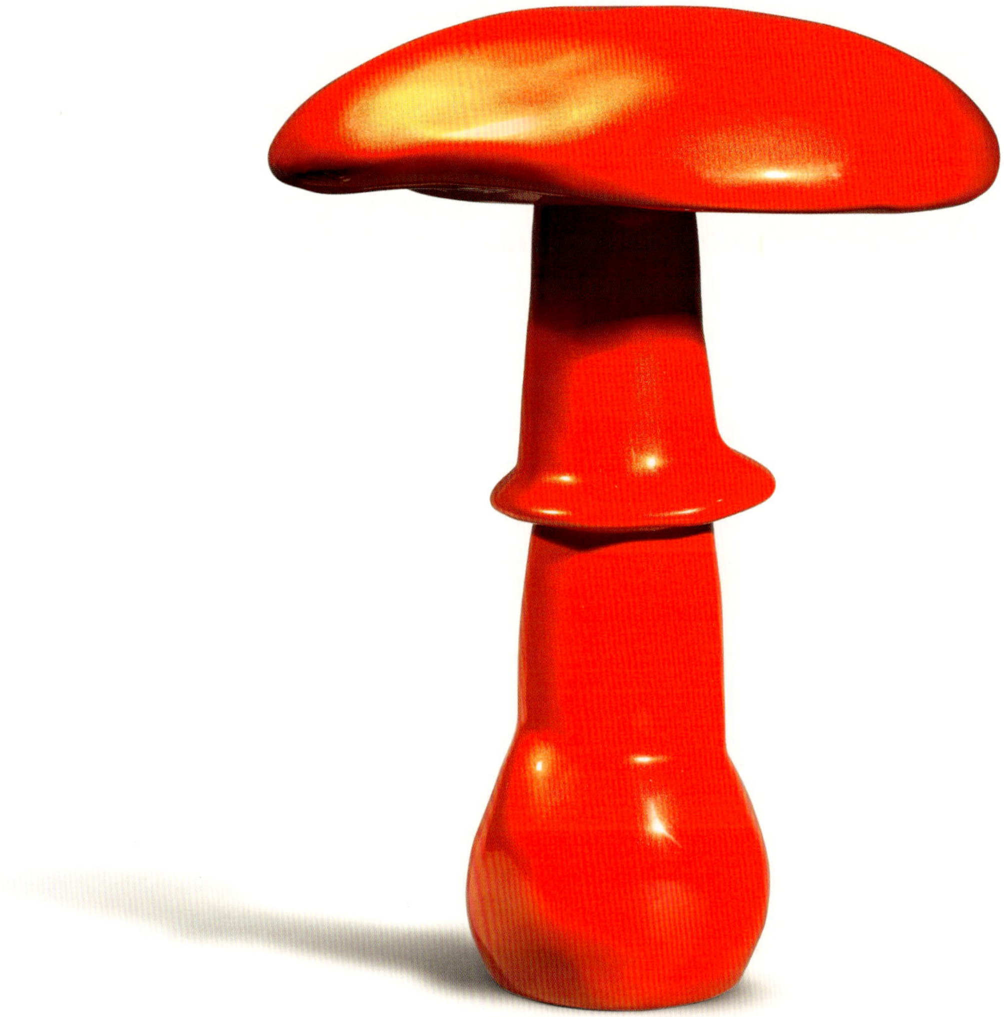

SYLVIE FLEURY

Mushroom (Colour Map E5 Rage xtc AG), 2006–11
Fiberglass and metallic car paint,
51⅛ × 43¼ × 43¼ in. / 130 × 110 × 110 cm
Private collection

A bright red fiberglass sculpture of a stylized mushroom stands boldly alone, emitting the glossy reflection of iridescent car paint. Titled with its detailed hue, *Mushroom (Colour Map E5 Rage xtc AG)* is part of a series of large mushroom sculptures, exhibited both inside galleries and as outdoor works, by contemporary Swiss artist Sylvie Fleury (b. 1961). Fleury emerged in the 1990s with works across sculpture, video, painting, and installation that satirize the male dominance of both art history and pop culture. Shopping, fashion, and car culture are all recurring themes in her often humorous work. Fleury balances conceptual references to iconic artworks with the complexity of consumerism and the experience of what it is to be a woman. She began exhibiting her mushroom series in the mid-2000s, often showing them in clusters of different sizes and shapes. Their colors are always deeply unnatural, echoing how fungi subconsciously register as magical and psychedelic in the contemporary consciousness. The creation of these pieces is intentionally industrial and echoes methods used in car production—a jarring contrast to their playful, natural forms. Fleury's mushrooms, which have been painted in blues, purples, and golds and often have an oil-slick sheen, have been exhibited at institutions that include MAMCO in Geneva, the Centre Pompidou in Paris, and Aranya Art Center in Beidaihe, China. Mixing high and low cultural references, the monolithic works intentionally nod to the relationship between psychotropic drugs and the experience of looking at art.

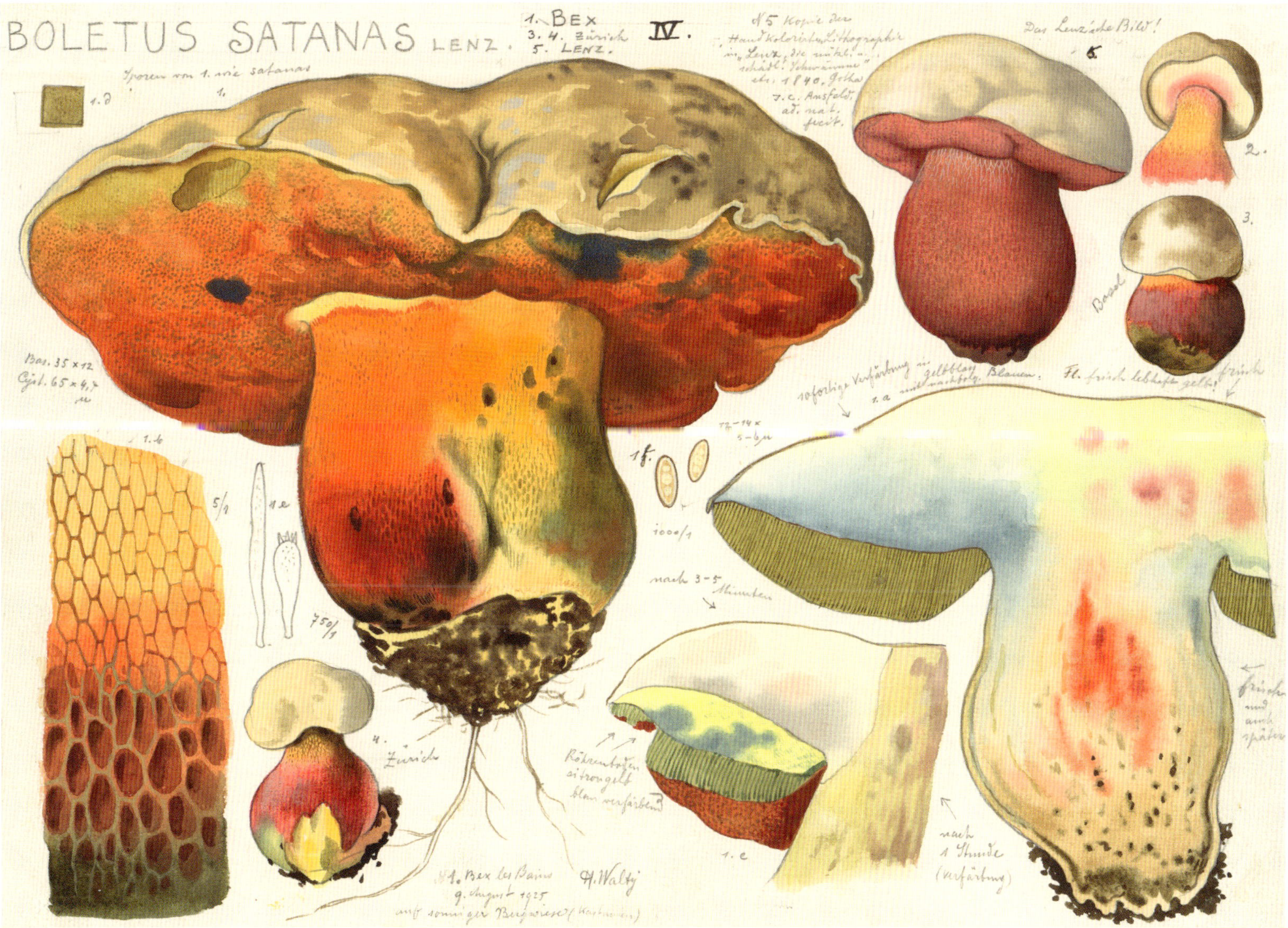

HANS WALTY

Boletus satanas, 1925
Pencil and watercolor on paper,
9¾ × 13½ in. / 24.8 × 34.2 cm
Swiss National Library, Bern, Switzerland

Emanuel Johann Anton Walty, known as Hans Walty (1868–1948), was a highly skilled artist who taught at the Zurich School of Applied Arts. Walty is known for his decorative paintings within buildings—including that in the vestibule of the Swiss National Museum in Zürich—and he also painted landscapes and portraits. But he is best known for his paintings of fungi, most famously depicted in a series of plates created over a period of more than thirty years. Swiss interest in edible fungi increased markedly toward the end of World War I, when food shortages encouraged people to search for edible items from nature. To meet this demand, guides on edible fungi appeared in local markets, and the tradition of foraging became firmly established. Though largely self-taught, Walty became very knowledgeable about fungi, and his vibrant watercolor illustrations graced the pages of a field guide to mushrooms and other fungi, *Schweizer Pilztafeln* (*Swiss Fungi Illustrated*), published in 1942. This handy, portable guide became the bible for mushroom enthusiasts. This plate shows specimens of the devil's bolete (*Boletus satanas*, now *Rubroboletus satanas*) at several stages of development, including details of structure, shown in cross section, and even the spores highly magnified. Walty added explanatory notes referring to locations of collection or key features of the anatomy. The secondary title here correctly states that the species is poisonous (*giftig* in German), although devil's bolete is thankfully rather rare. Several of its close relatives, however, are edible, and some, such as the cep (*Boletus edulis*) are highly prized for their culinary uses.

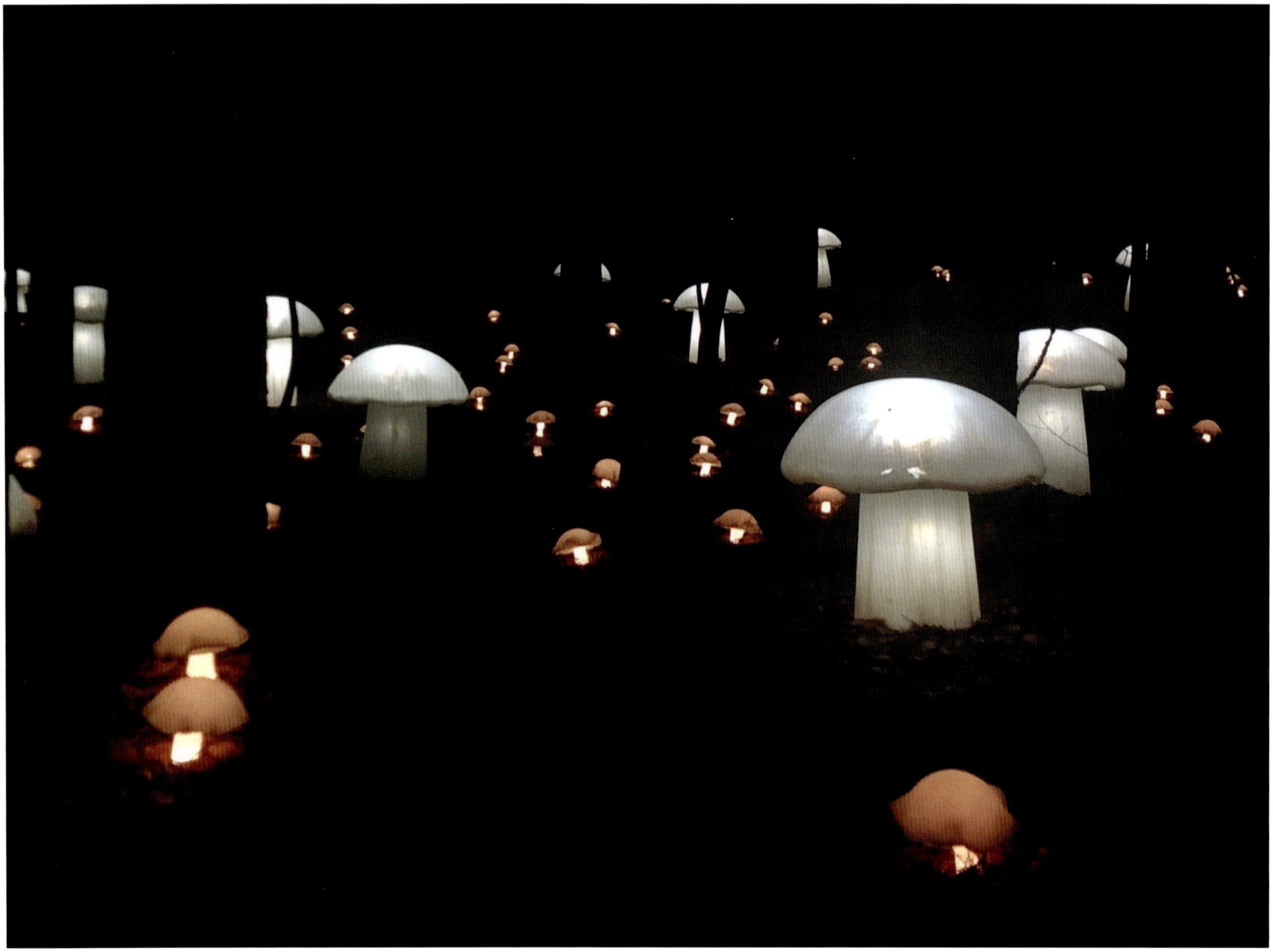

MEREDITH CONNELLY

Mushrooms II, 2022
Cast fiberglass, thermoplastic, and lighting, dimensions variable
US National Whitewater Center, Charlotte, North Carolina

Lighting up the woodland floor with an ethereal glow, these illuminated mushrooms charmed visitors with an after-dark walking trail at the US National Whitewater Center in Charlotte, North Carolina. With their domed caps, thick stems, and fine gills, the fiberglass fungi were among more than two hundred luminous objects—including crystals, tendrils of lights hanging from trees, and an incandescent stream—that local sculptor Meredith Connelly (b. 1984) created especially for the temporary installation, titled *Lights*. Although her oversize LED-lit mushrooms have an otherworldly appearance, they take inspiration from the familiar *Agaricus* genus, which includes common edible varieties as well as toxic ones. The project, which brought together the artist's interests in science, nature, and technology, took her and her team about two months to fabricate and install, and just like with real mushrooms, the resulting spectacle was temporary, open to the public for only four months. At the heart of Connelly's practice is an interest in connectivity and bringing people together. For her, mushrooms are the perfect metaphor for this, because research has suggested that floor-dwelling fungi communicate among themselves using electrical signals that travel through a complex network of subterranean filaments called hyphae, which comprise the mycelial network. Conceived as a way to encourage people to reconnect with nature and with each other, the artist's mesmerizing light installation attracted 250,000 visitors and was followed up the next year with a second iteration featuring three hundred additional elements, including two hundred smaller, hand-sculpted mushrooms.

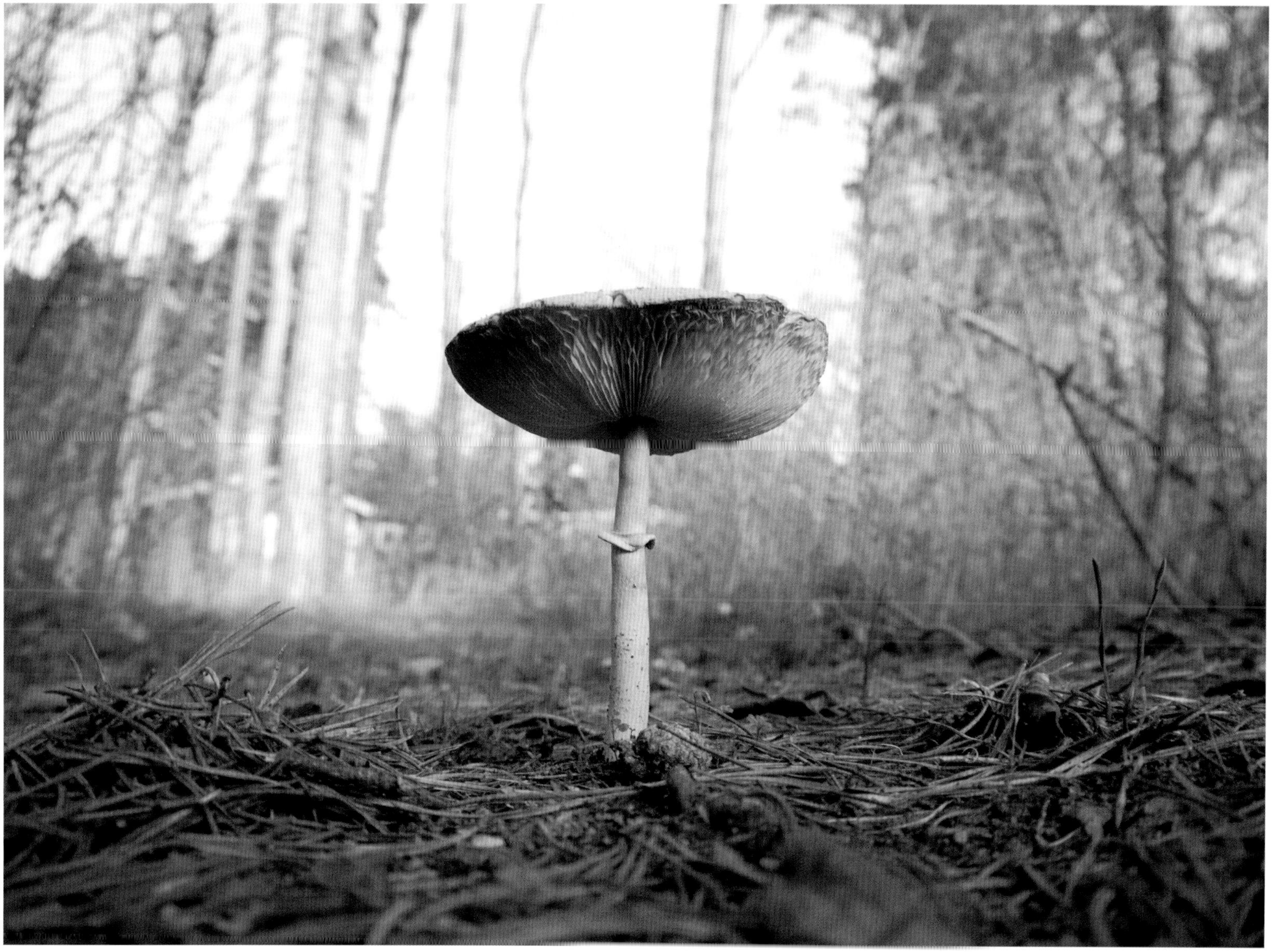

PAUL QUANT

The Mushroom, 2014
Digital photograph, fine art print on Hahnemühle Photo Rag Ultra Smooth, 8⅝ × 15¾ / 22 × 40 cm
Galleri Duerr, Stockholm

When studying the often stark, brooding imagery of Swedish photographer Paul Quant (b. 1959), it is important to consider his background. Quant's career began in 1977 in the lab of the Försvarets Materialverk (FMV)—the Swedish Defence Materiel Administration, a government agency responsible for, among other things, developing film for the Swedish Armed Forces. It was there that he acquired a mastery of the darkroom from veteran, old-school photographers as he developed glass plate negatives, 70mm World War II flight films, large-scale prints using specialized enlargers, and everything in between. This darkroom expertise has allowed him total control over exposure, contrast, mood, and other photographic elements. Next, in 1982 Quant took a job as a police photographer, shooting crime scenes as well as studio shots for case evidence. Many of his photos after this time seem to borrow from this world, and even his nature shots are imbued with a heavy solitude and a rawness that echoes crime scene photography. Since 1992 Quant has focused exclusively on commercial and fine-art photography, but the lessons he learned in previous decades have never left his work, often lending it a noir vérité quality. Here, for instance, a lone unidentified mushroom—most likely a parasol mushroom (*Macrolepiota procera*)—stands in perfect solitude, tense and alive against a monochromatic forest clearing in Öland, Sweden. Quant has worked extensively with Sinar large format Polaroid photography (Sinar stands for still, industrial, nature, architectural, reproduction), lending his images their signature blend of crystal clear detail and controlled depth of field.

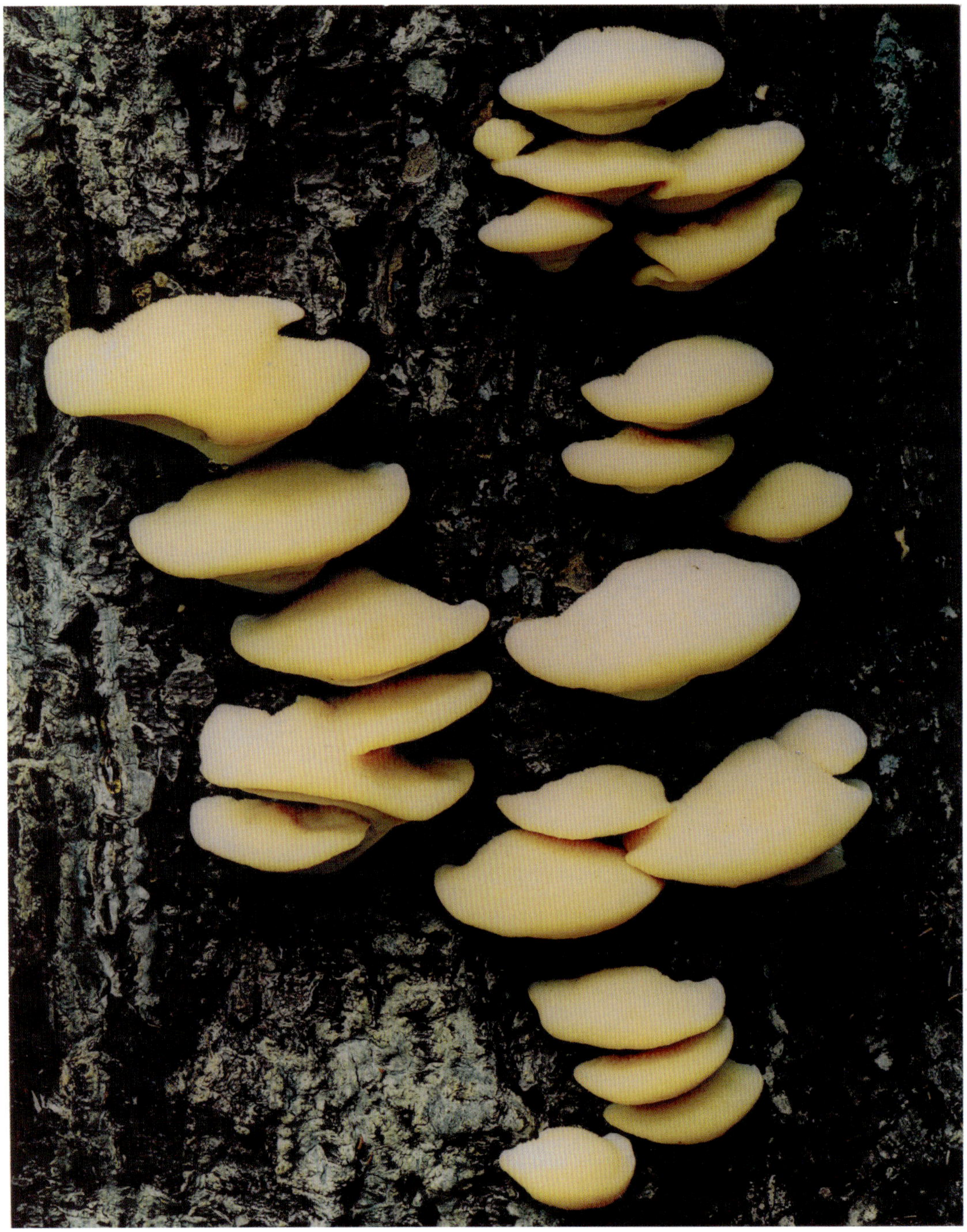

ELIOT PORTER

Yellow Fungi on Tree Trunk, Great Spruce Head Island, Maine, August 28, 1973, 1973
Dye imbibition print, 10½ × 8⅛ in. / 26.5 × 20.6 cm
Amon Carter Museum of American Art,
Fort Worth, Texas

In an almost otherworldly photograph, the bright, honey yellow fungi *Climacocystis borealis* contrasts strikingly with the dark, mossy bark of a spruce tree. The visually arresting woodland scene proved irresistible to American photographer Eliot Porter (1901–1990), who was among the earliest artists to work extensively with color photography. This picture was taken on his family's private island in the Penobscot Bay of Maine—a significant source of inspiration for Porter, whose photographs frequently captured the flora and fauna of its expansive spruce forests, showcasing details such as lichen, moss, and, as here, fungi. Porter is celebrated for his vibrant images of the natural world that reflect his passionate concern for nature and environmental conservation. With his three-color dye imbibition prints, which are noted for their image stability and permanence, he set the standard for color landscape and nature photography, elevating it from its association with commercial advertising to an expressive artistic medium—even in a photograph of a humble forest mushroom. *Climacocystis borealis* is a species of polypore fungus, also called bracket or shelf fungi, that plays a significant role in wood decay, nutrient cycling, and aiding carbon dioxide absorption by forest ecosystems. Commonly found in coniferous forests across the northern hemisphere, the velvety, wedge-shaped fungi usually grow in large colonies of up to fifty pieces in overlapping shelves, turning lighter in color as they age. Depending on regional conditions, *Climacocystis borealis* appears from late spring to autumn, thriving on old, decaying trees and stumps, though its prevalence varies greatly from year to year.

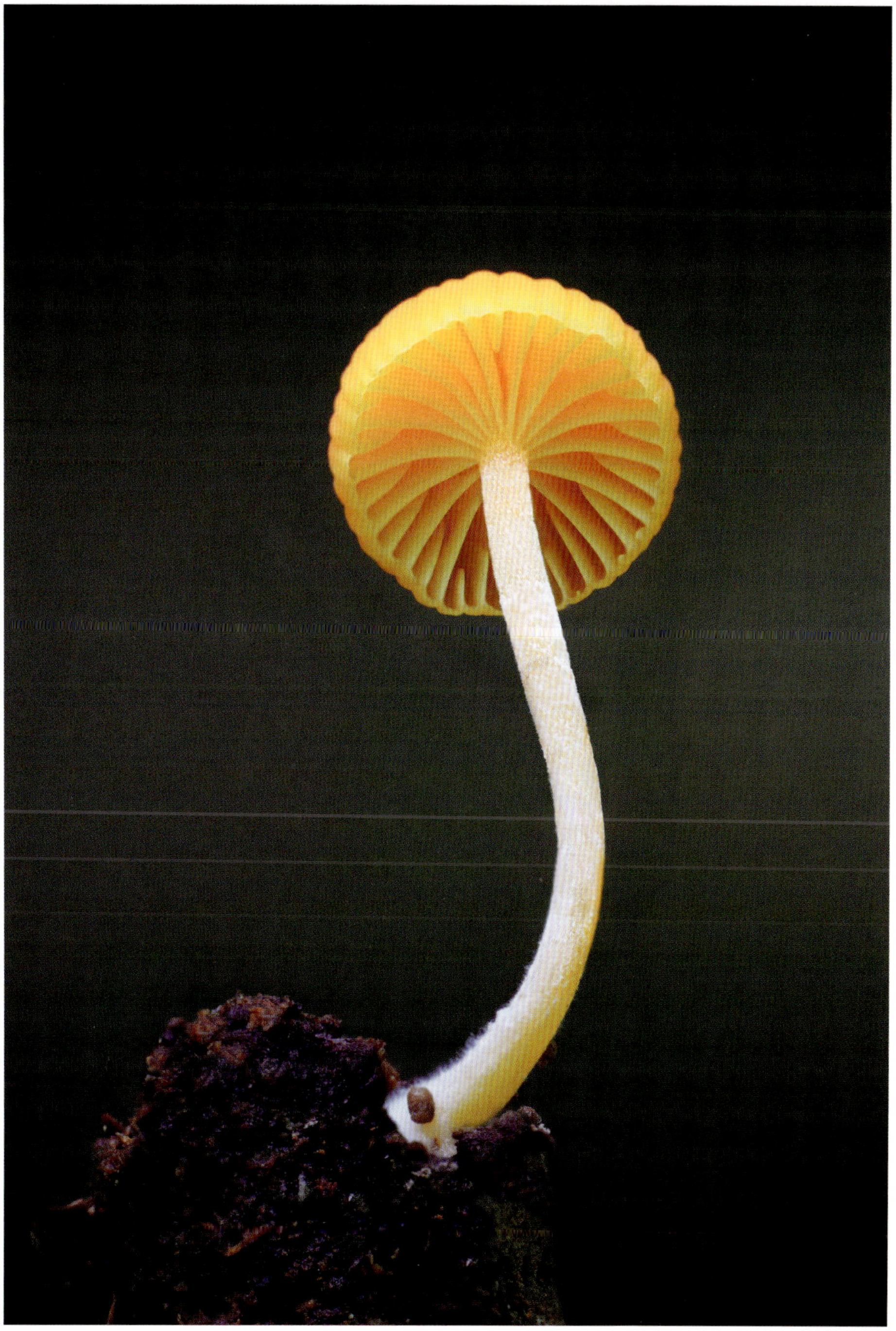

STEVE AXFORD

Mycenas p., 2016
Digital photograph,
dimensions variable

Until his late fifties, Australian photographer Steve Axford (b. 1951) was, by his own admission, "an ordinary bloke" whose career had been spent in IT. Retirement saw the New South Wales–based Axford take up photography—more specifically, photographing the unusual forms of fungi, slime molds, and lichens he finds in the forested areas close to his home. Working with a macro 90mm lens, Axford photographs different species in situ, many of which he suspects have yet to be identified by the scientific world. This fungus, which is a species of bonnet from the genus *Mycena*, was captured in the nearby Nightcap National Park. Despite its bright yellow gills and long stem, this mushroom would be easy to miss by the casual observer because of its small size, typical of the estimated five hundred species of *Mycena*, which, in some cases, only reach about a quarter of an inch (5 mm) wide. *Mycenas* are saprotrophic, meaning that they feed on decaying organic material, an essential process for breaking down the litter layer of a forest floor. Axford's photographs catalog the seemingly endless permutations of color, shape, and size found in the fungi kingdom. His passion is shared by his partner, filmmaker Catherine Marciniak, and together they have made time-lapse films of the mushrooms they have cataloged. Their work has garnered international recognition, appearing on television programs, including British naturalist David Attenborough's award-winning BBC documentary series *Planet Earth 2* (2016).

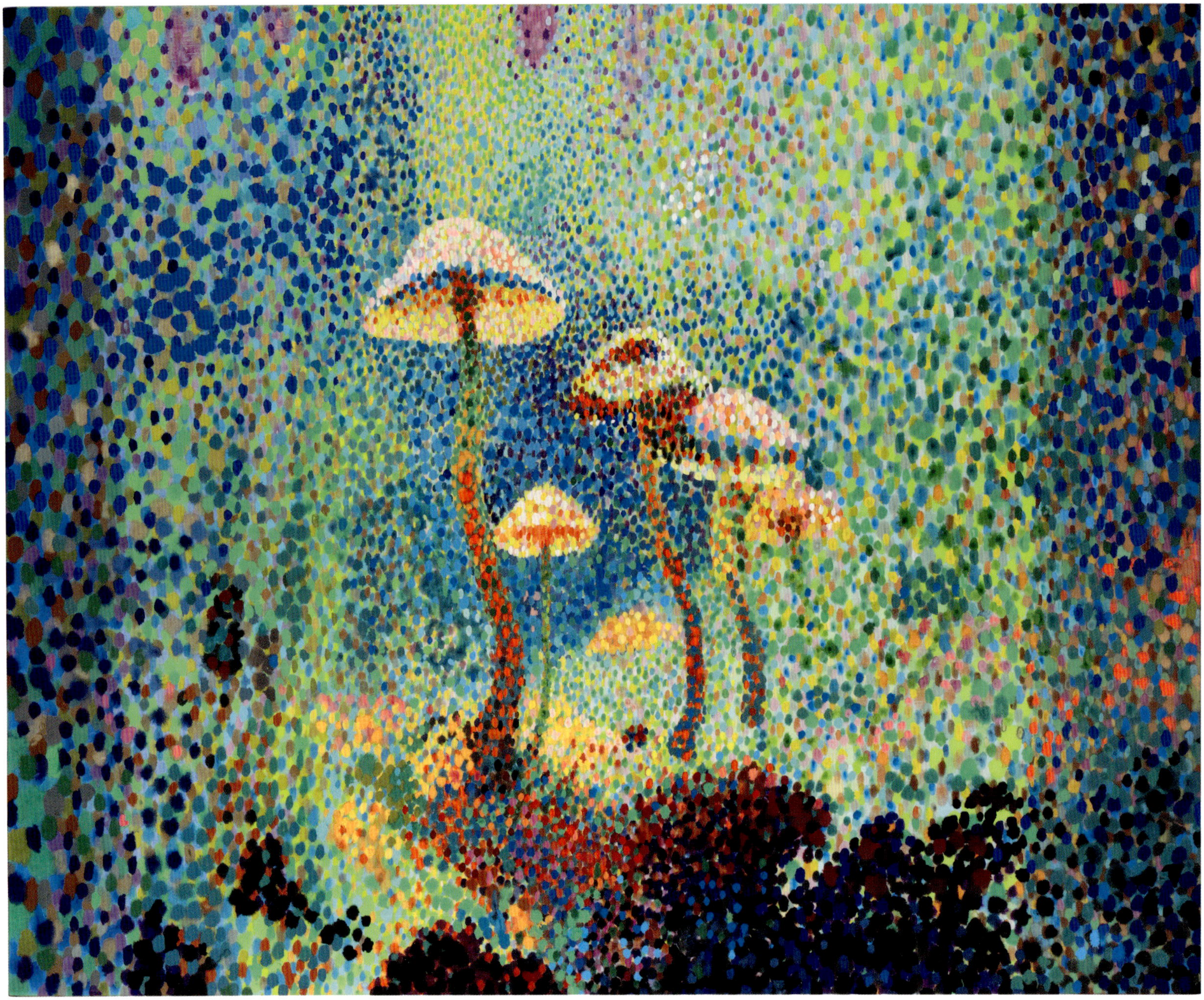

HULDA GUZMÁN

Joy in the Rain, 2025
Acrylic gouache on linen,
50 × 60 in. / 127 × 152.4 cm
Bisonó León Family Collection,
Santo Domingo, Dominican Republic

Contemporary artist Hulda Guzmán (b. 1984) hails from Santo Domingo, Dominican Republic, and now splits her time between her hometown and the mountainous rainforests of Samaná, a coastal town in the northeast of the island nation. Influenced by Mexican muralism and Caribbean folk art, Guzmán's surrealist-tinged Technicolor artworks explore the dichotomy of waking and dreaming, reality and fiction. But they also tackle themes of postcolonialism, imperialism, and ecology, often celebrating nature and questioning humanity's relationship to the natural world and our place within it. Guzmán, who says she "finds a lot of comfort in the meditative process of contemplating nature," welcomes us into the world she scrutinizes. Through the lens of her own brand of magical realism, she juxtaposes imaginary or anthropomorphized creatures with realistic ones, plays with size and spatial relationships, and introduces an otherworldly color palette. Her surrounding tropical landscape is a constant source of inspiration, with many of her vibrant paintings depicting the flora and fauna of the rainforests and the people who live among them. In *Joy in the Rain*, Guzmán uses an idiosyncratic style of pointillism to reveal a group of fantastically colored mushrooms that could just as easily be underwater as in a forest glen. The mushrooms in question bear a strong resemblance to *Psilocybe semilanceata*, or liberty caps—known for containing the psychoactive compound psilocybin—perhaps adding yet another dimension to how we internalize and understand the natural world.

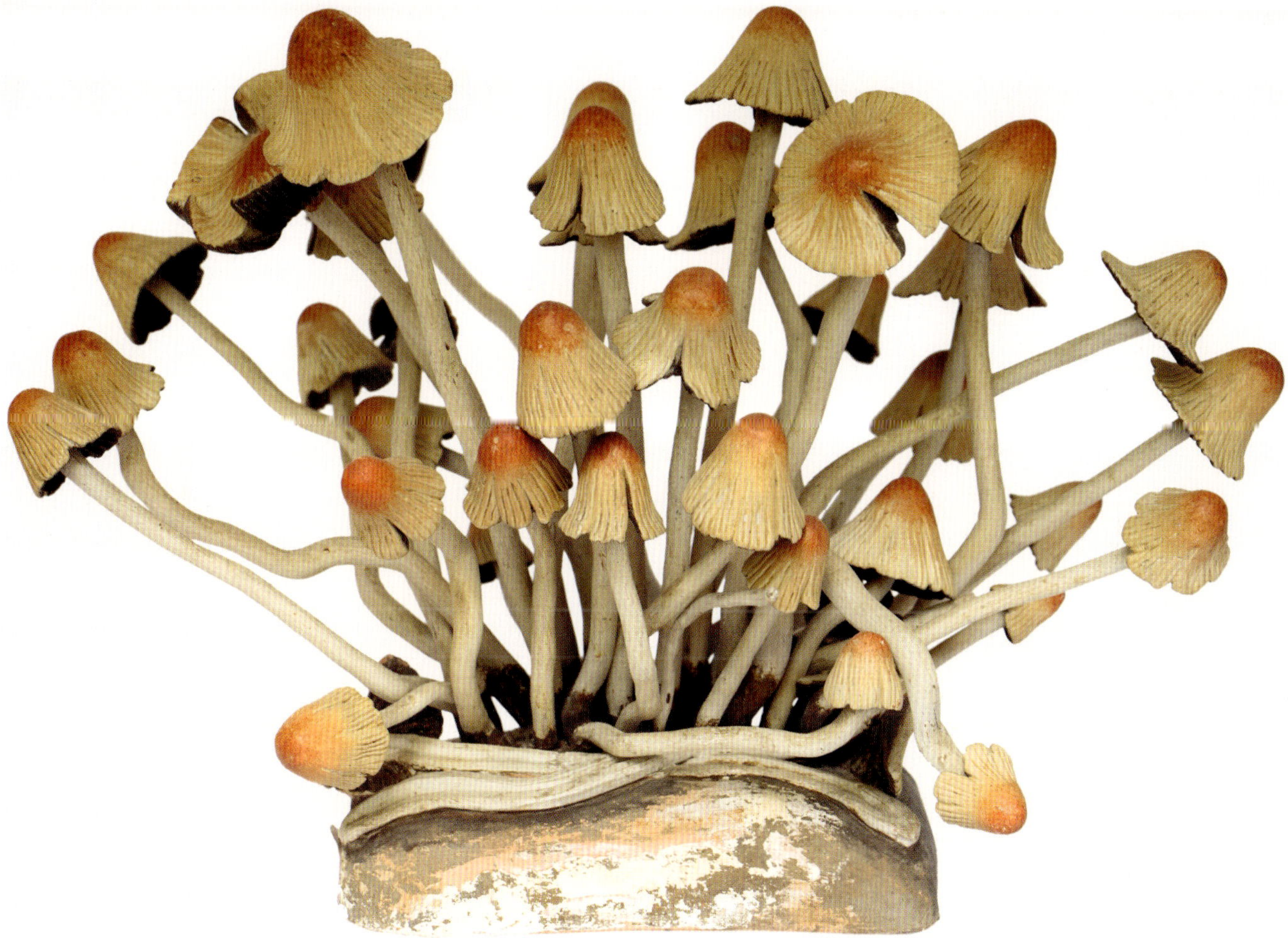

FRANCESCO VALENTI SERINI

Coprinellus micaceus, 1838–72
Terra-cotta, 14½ × 10¼ × 10⅝ in. / 37 × 26 × 27 cm
Museo di Storia Naturale dell'Accademia dei Fisiocritici, Siena, Italy

A cluster of spindly mushrooms with delicately rendered caps emerges in a tangle in this remarkable terra-cotta model by Francesco Valenti Serini (1795–1872). Born and educated in Siena, Italy, Serini graduated with a degree in medicine and a particular interest in botany and mycology. When practicing medicine in hospitals in Florence, he became much concerned about illnesses caused by the consumption of poisonous fungi. On returning to Siena, he began studying fungi, with the aim of educating the public about which species were edible and which were poisonous. To this end, he proposed five categories: edible, edible and good, edible and delicious, suspect, and poisonous. As part of his endeavor, Serini created accurate terra-cotta models of many of the fungi he collected—often, as here, with the fruiting bodies seeming to grow out from clay tablets. Fungi often disintegrate soon after collection, and capturing their characteristics in clay enabled him to display their features permanently. Such models were very suitable for educating the public about their appearance and identification. This model is of *Coprinellus micaceus*, a rather fragile fungus known as the mica cap or glistening inkcap, which typically grows in dense clusters on rotting wood, such as logs or tree stumps. It is common and widespread in Europe and is also found in the Americas and in Australasia. Edible, with a delicate flavor, it is sometimes added to omelets or used as a flavoring for sauces.

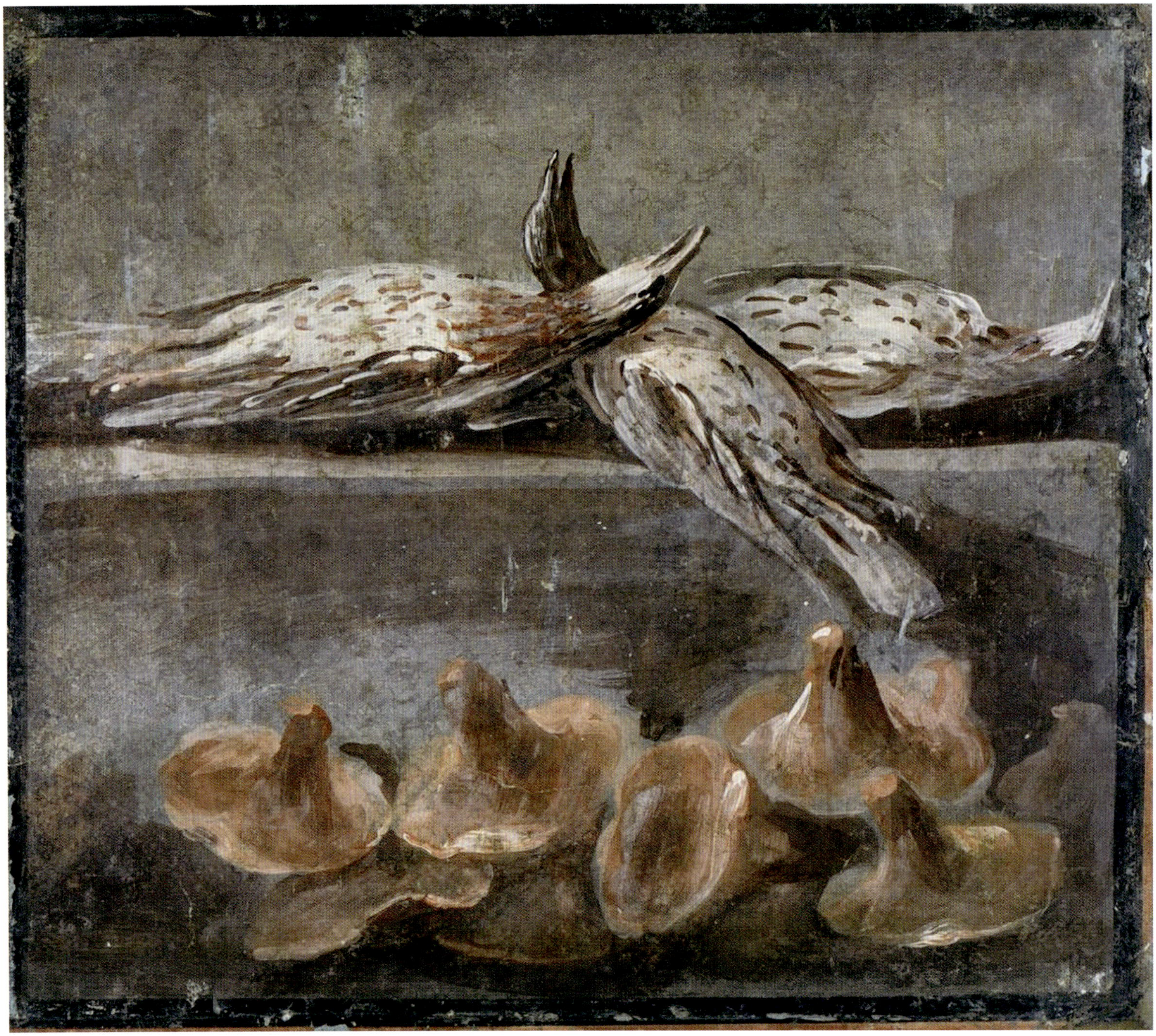

ANONYMOUS

Still Life with Thrushes and Mushrooms, c. 62–69 CE
Fresco, 14 × 13½ in. / 35.6 × 34.3 cm
National Archaeological Museum, Naples, Italy

Remarkably preserved, this fresco comes from the House of Stags, an opulent ancient Roman domus decorated with sweeping mosaics, fountains, and massive marble statues of the deer for which it is named. Located in Herculaneum, a seaside retreat for many of the richest Romans just north of Pompeii, it was decimated by the eruption of Mount Vesuvius in 79 CE and rediscovered in the eighteenth century. Frescoes were common practice at the time, and paintings such as this can tell us much about the culinary habits of the ancient civilization, as artists often depicted everyday scenes, including food, chores, and marketplaces. Created around 62 to 69 CE, this fresco is one panel of a larger series of still lifes in the Fourth Style of Roman wall painting that were commissioned by Q. Granius Vero, a wealthy merchant and the owner of the House of Stags, to decorate the twenty-five rooms of the house. At top are three song thrushes (*Turdus philomelos*), which were prized in Ancient Rome for their beautiful song as well as their delectable taste. Below lies a pile of mushrooms, which were commonly foraged for culinary use in the Campania region. A staple of the Roman diet, mushrooms were referred to as "the food of the gods" and, depending on the variety, could indicate status—as with the edible *Amanita caesarea*, or Caesar's mushroom—or be used for more nefarious purposes, such as the poisoning of Emperor Claudius in 54 CE.

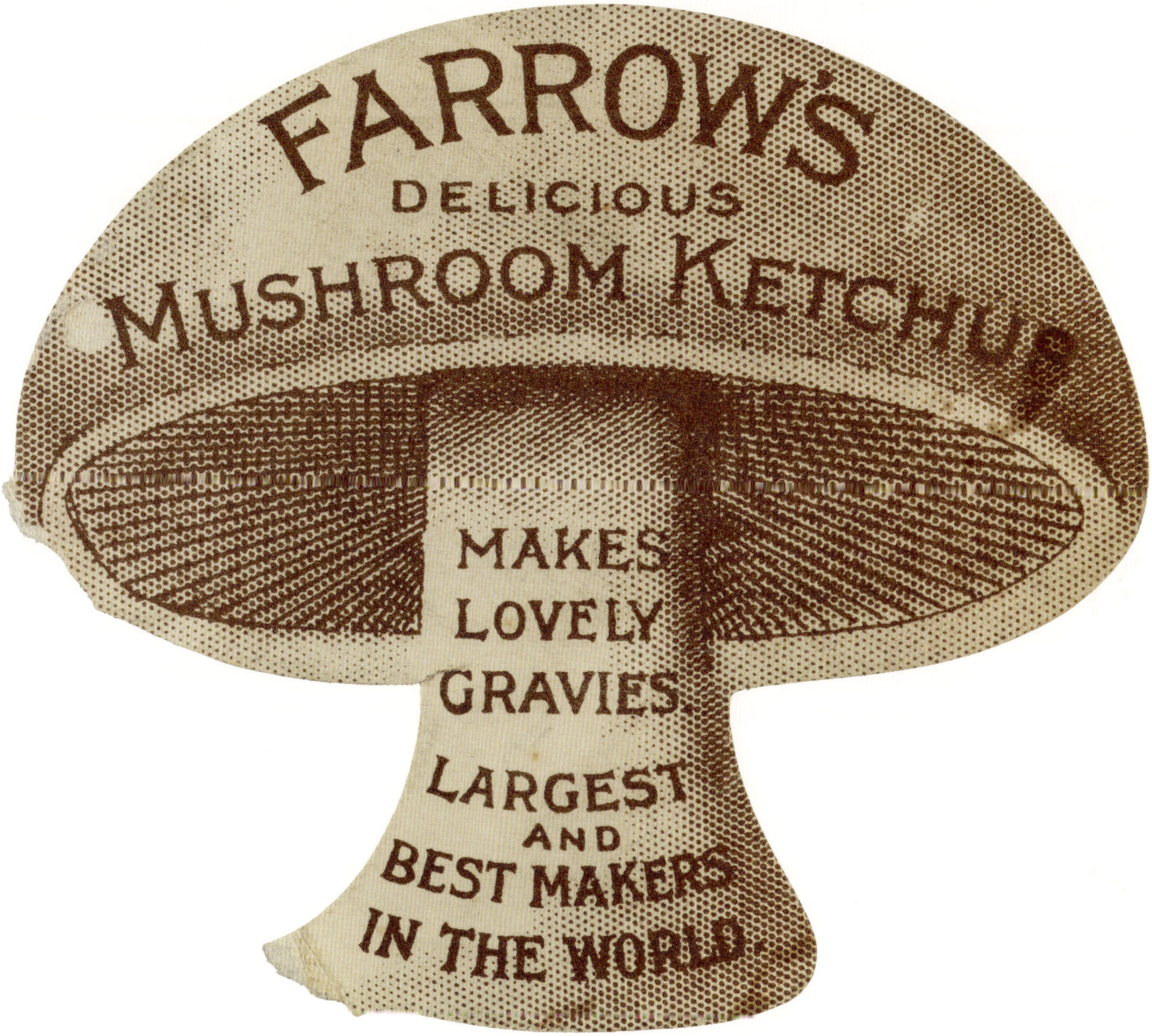

FARROW'S

Advertising card, c. 1890–99
Printed die-cut card, H. 2¾ in. / 7 cm
Wellcome Collection, London

Before there was tomato ketchup, there was mushroom ketchup. The word *ketchup* is actually an anglicization of the Chinese word *ke-tsiap*, a fermented fish sauce that arrived in Britain in the seventeenth century. Mushroom ketchup was the result of a homegrown attempt to re-create the pricey fish sauce, although the result was more of a liquid than a sauce. Made principally by stewing fungi—although some recipes included walnuts and oysters—mushroom ketchup later traveled with the early British settlers to the American colonies. Commercial production of this condiment, which had been strictly a homemade item until then, began in 1840 with a small Cambridgeshire, England–based condiment maker, Joseph Farrow & Co., Ltd., better known as Farrow's. Appreciated for its distinctive umami flavor, which was created by slow-cooking and salting the mushrooms before adding a secret blend of spices, the strained, dark savory liquid sold by Farrow's was typically used to flavor soups, pies, and gravies. Farrow's launched an advertising campaign at the end of the nineteenth century with this beige-colored mushroom shape, hinting at the major constituent of the sauce. Highly successful at first, the thin ketchup began to fall out of favor in the late nineteenth century as people's tastes changed. In 1876 the American H. J. Heinz Company launched its tomato ketchup, made from real tomatoes, which only hastened the demise of mushroom ketchup. With its added sugar and thicker consistency, tomato ketchup appealed to more "modern" palates. Recently, however, foodies have rediscovered mushroom ketchup, resulting in a revived cult following.

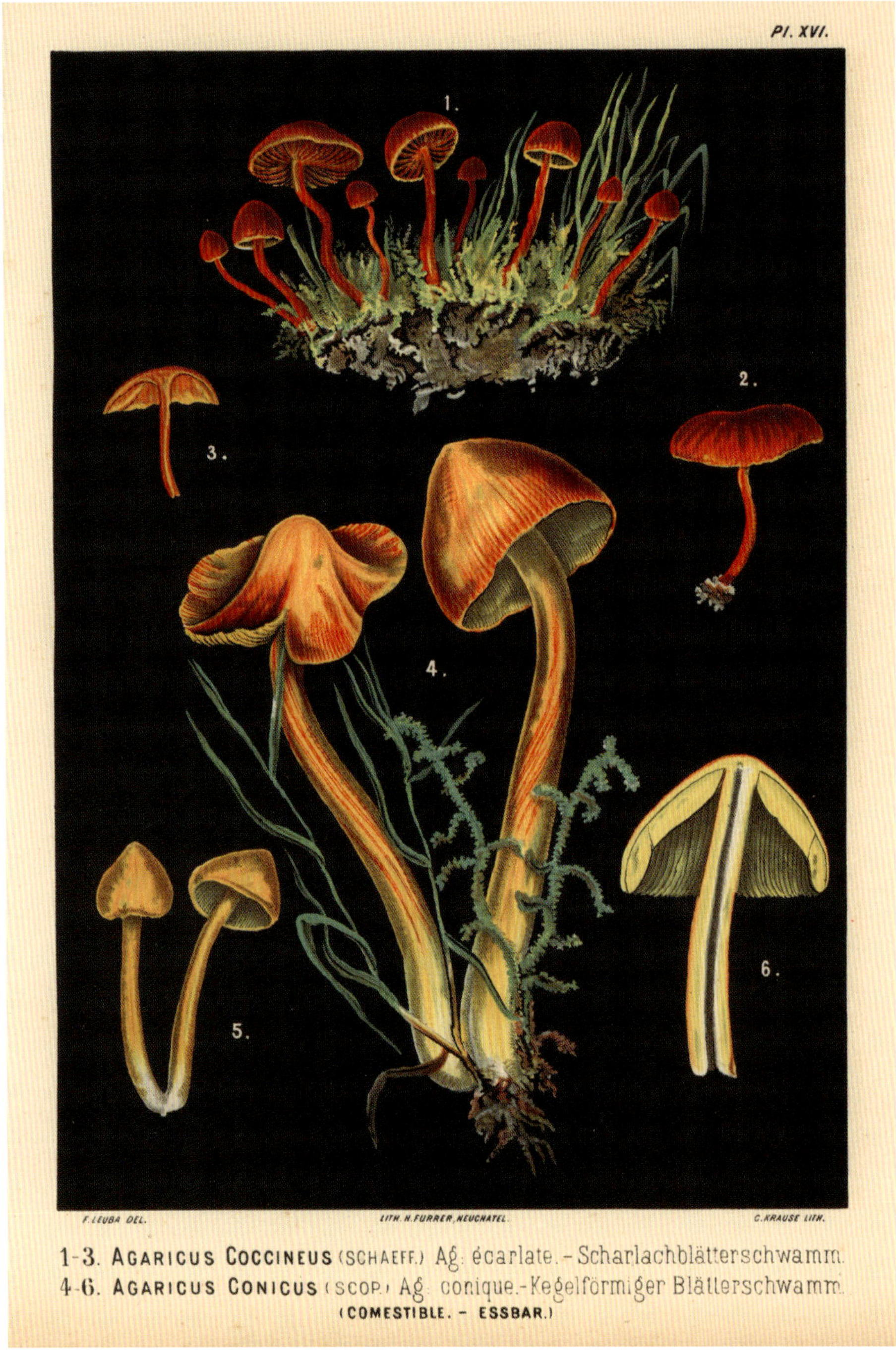

FRITZ LEUBA

Plate XVI, from Les champignons comestibles et les espèces vénéneuses avec lesquelles ils pourraient être confondus (Edible Mushrooms and the Poisonous Species They Should Not Be Confused With), 1890
Chromolithograph, H. 14⅛ in. / 36 cm
Private collection

Against a dramatic black background are illustrations of two species of fungi, one of which is harmless and edible—the scarlet hood, or scarlet waxcap (*Agaricus coccineus*, now *Hygrocybe coccinea*) at top—and the other toxic—the witch's hat (*Agaricus conicus*, now *Hygrocybe conica*) at bottom. Both species are drawn as they can be found growing in nature, then in relief, and, finally, isolated on their own. These illustrations are the creation of Swiss artist and mycologist Fritz Leuba (1848–1910) from his 1890 work *Les champignons comestibles et les espèces vénéneuses avec lesquelles ils pourraient être confondus* (*Edible Mushrooms and the Poisonous Species They Should Not Be Confused With*). Mycology was then a relatively new discipline because, despite being part of the human diet since prehistoric times, fungi were thought not to be worthy of study, as they were considered to lack organs. It was only in 1836, after the English clergyman Miles Berkeley, one of the founders of plant pathology, wrote a description of British fungi that was included in Sir William Jackson Hooker's seminal *The British Flora*, that mushrooms began to be taken seriously as a subject to be investigated and understood. By the time Leuba produced his book, the medicinal qualities of fungi and—just as importantly—their toxicity, were becoming better understood. The intention of Leuba's book was to show the reader how easy it was to mistake poisonous and edible mushrooms.

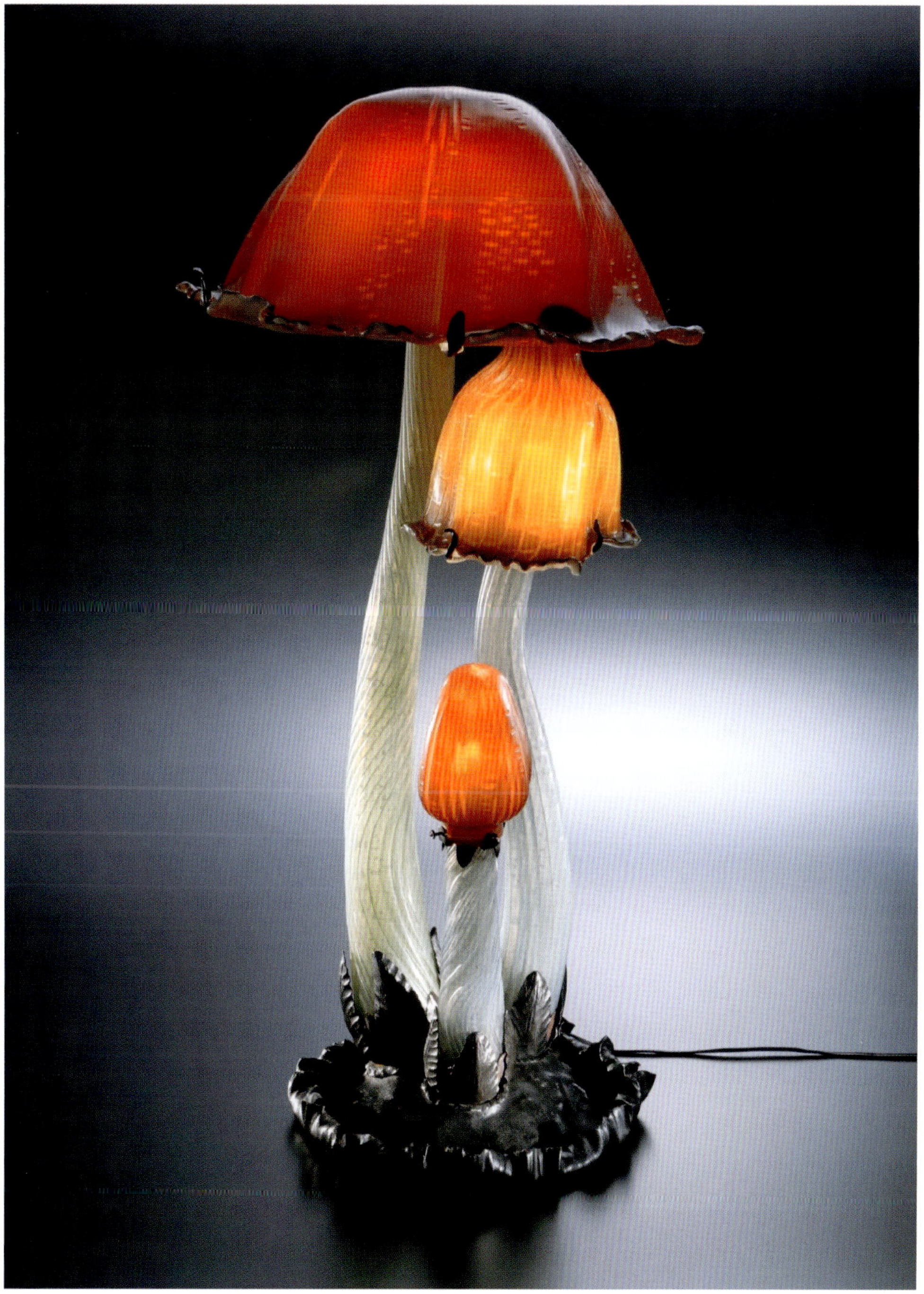

ÉMILE GALLÉ

Ink Cap Lamp, c. 1902
Glass, 33 × 12⅜ in. / 83.8 × 31.4 cm
Suntory Museum of Art, Tokyo

Three ink cap mushrooms (*Coprinopsis atramentaria*) in varying stages of maturity emit an ethereal glow, something most definitely not seen in nature. This table lamp was created by innovative French designer Émile Gallé (1846–1904), a leading light of the Art Nouveau movement, toward the end of his life. Intended for use in the corner of a sitting room, each mushroom has a glass cap sitting atop a long, milky white glass stem: the smallest cap is tightly shut, the middle-size cap is partially opened, and the largest cap is fully expanded. The ink cap is a fungus that grows in a single night, wilting as dawn approaches, although the cap is typically a grayish-brown color, not the vibrant reddish-brown Gallé chose for his lamp. A master glassmaker, Gallé used three layers of glass to make the caps—an opaque, cream-colored glass that was covered with transparent glass and then coated with reddish-brown glass—so that when the lamp is illuminated, the three caps give out a red color, suggesting the fungus's brief burst of energy. Although the lamp's color deviates from reality, Gallé—who was a trained botanist and required artists at his workshop to work from actual plants as models—was at pains to make the caps as realistic as possible in every other way, marking each cap with lines and turning up their edges. The dramatic color that would flood a room with brilliant red light was a spectacular twist on the lamp's "realism."

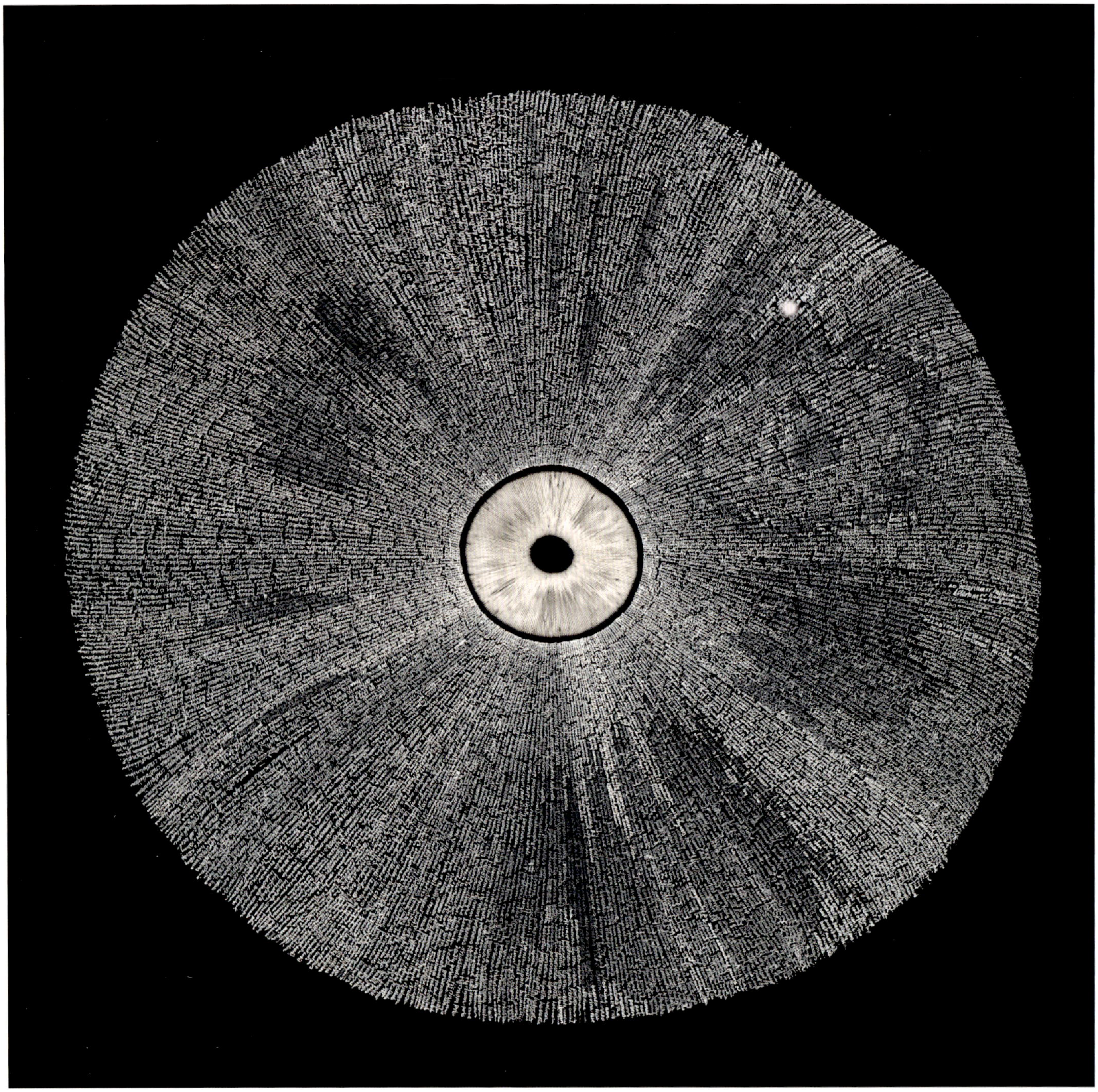

CHRIS DRURY

Poison Pie I, 2000
Spore print on black card with text in white ink, 32¼ × 32¼ in. / 82 × 82 cm
Private collection, London

To describe Sri Lankan–born British artist Chris Drury (b. 1948) as a Land artist would be an oversimplification. While he has created numerous three-dimensional Land installations, he is also a performance artist, a printmaker, and a video artist. Drury himself says he seeks to "make connections between nature and culture, inner and outer, microcosm and macrocosm." He collaborates with scientists and scholars across a broad field of studies to bring his artistic vision into being. *Poison Pie I* exemplifies Drury's twining of nature and mankind. It is an etching that depicts at its center a spore print, made by leaving a mushroom cap, underside down, on wet paper overnight, so that as the spores dropped, they left an impression on the paper. The mushroom Drury chose is the *Amanita muscaria*, commonly called the fly agaric, which is known to cause (usually nonfatal) poisoning and is prized by some for its psychoactive properties. The fly agaric has been used in art and literature since the Renaissance, and typifies mushroom iconography; it has appeared in various fairy tales, for example, as well as the *Super Mario* video games and Disney's *Fantasia* (see pp.209, 213). Although this print seems abstract at first blush, as one looks closer it becomes apparent that the white lines radiating out from the spore print nucleus are lines of handwritten text that detail the Latin names of poisonous fungi and their effects on the human body. And then when viewed in aggregate, the image comes full circle and resembles the underside of a mushroom cap.

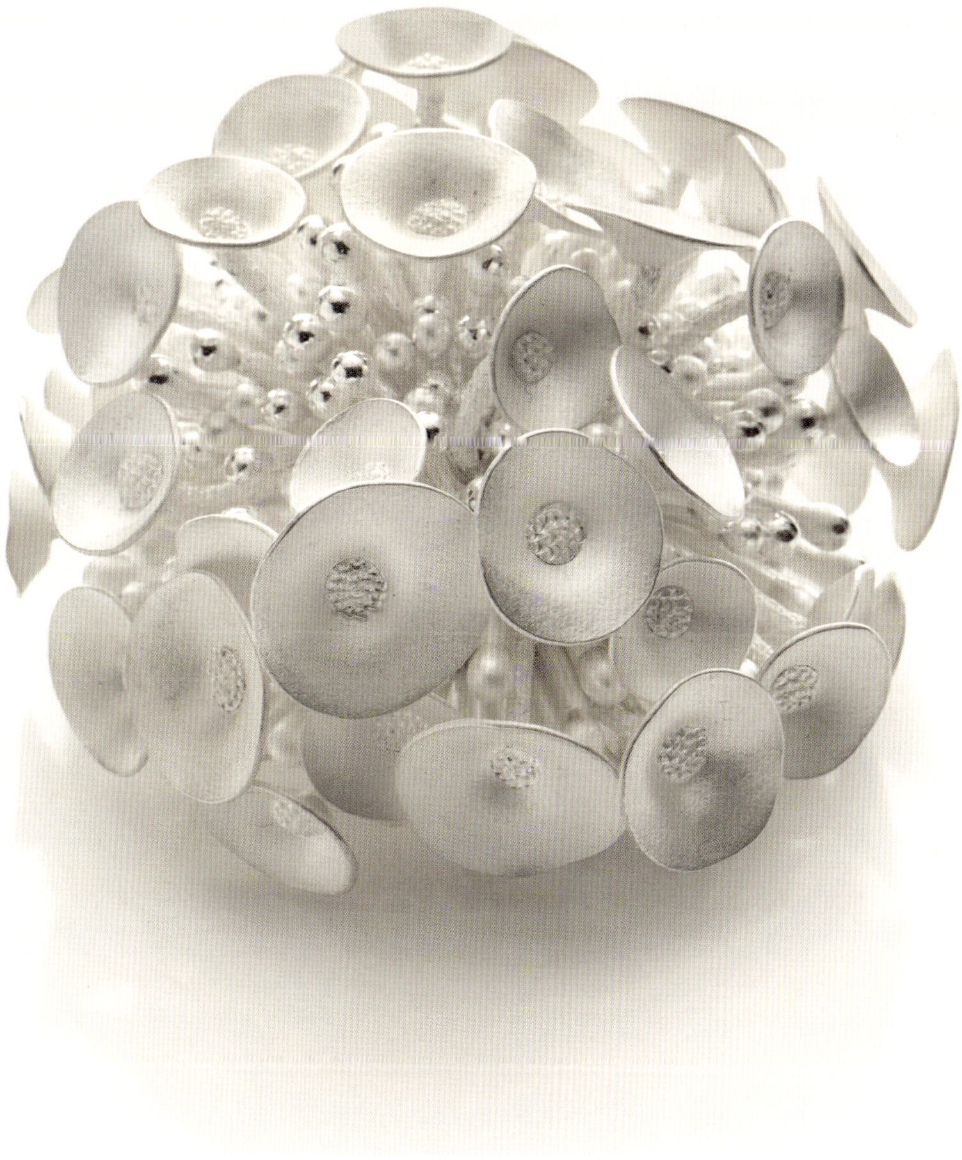

JUNKO MORI

Silver Organism, Fungi Dew, 2024
.999 fine silver, 2⅜ × 3½ × 3⅜ in. / 6 × 9 × 8.5 cm
Private collection

The science of evolutionary biology has long fascinated Japanese artist Junko Mori (b. 1974), whose sculptural creations are often inspired by nature. Junko is particularly intrigued by forms composed of multiple aggregate parts, and many of her sculptures echo the complex structures of natural organisms. *Silver Organism, Fungi Dew* is a petite work—small enough to be held in the palm of one's hand—but with maximum detail. Constructed from .999 fine silver, each element is individually forged and cast before being hand-hammered and assembled into a whole. The artist lets the final form develop organically, just as nature takes its course. Her technical prowess is on full display, the tiny dew droplets glittering among clusters of mushrooms, arranged as precisely as gems on a piece of high jewelry. Born in Yokohama, Japan, Junko was trained at Musashino Art University in Tokyo before earning a bachelor's degree in silversmithing and metalwork at Camberwell College of Arts in London. She lives and practices in Wales, and her work is in the permanent collection of dozens of institutions around the world, including many in the UK, London's Victoria and Albert Museum and the British Museum among them. Whether working in small scale, as here, or creating large-scale sculptures constructed from forged steel, Junko's metalsmithing skills have established her as one of the most esteemed and exhibited artists working in the medium today.

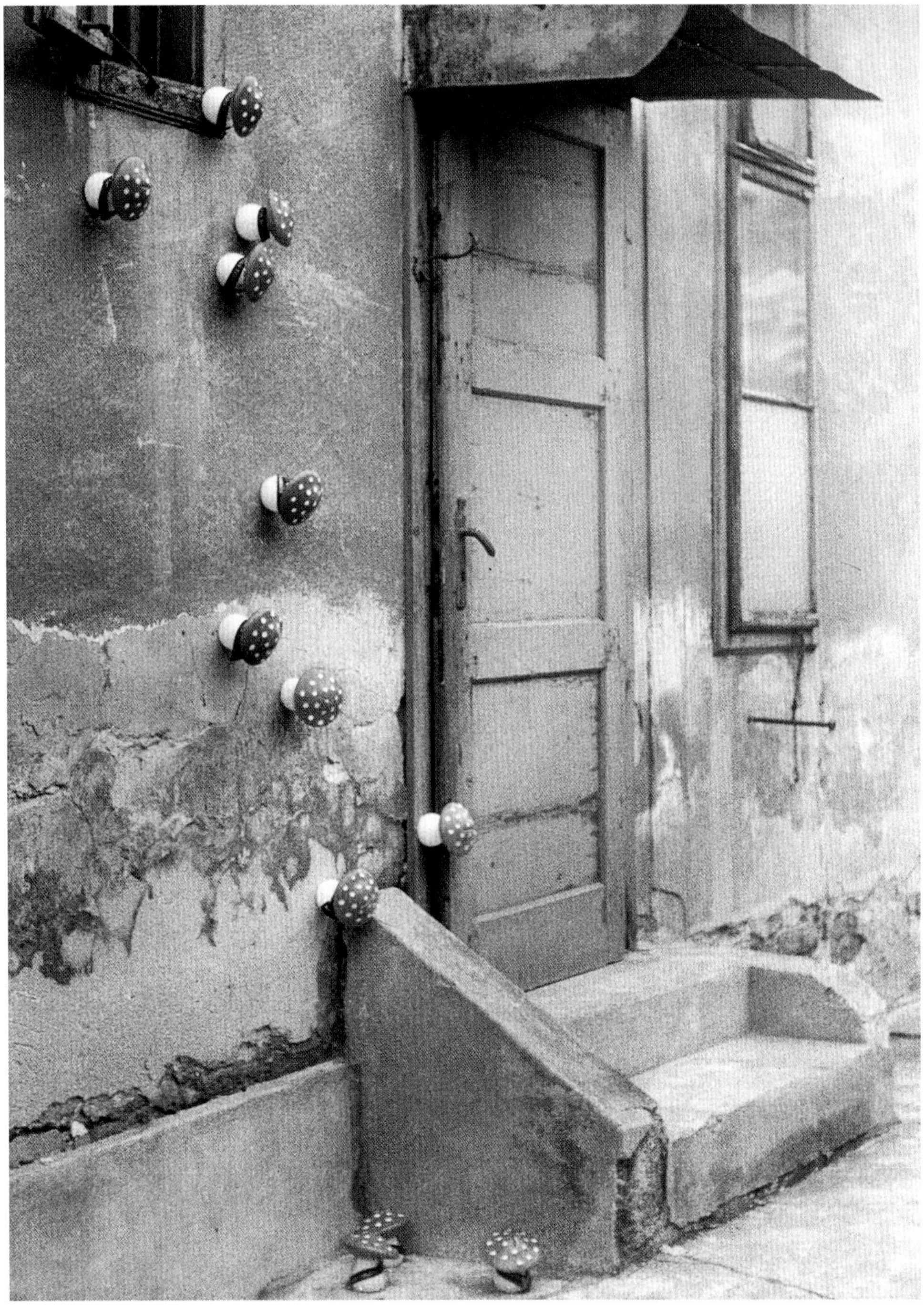

JANA ŽELIBSKÁ

Amanita muscaria—The Possibility of Saving Throughout the Year, 1970
Installation with plastic coin banks and black ribbons, dimensions variable
1st Open Studio, House of Rudolf Sikora, Tehelná Street, Bratislava, Czechoslovakia (Slovakia)

A series of toylike facsimiles of fly agaric mushrooms climb playfully up the facade of a house in Bratislava, the capital city of what is now Slovakia. One of a series of mushroom installations by Slovak artist Jana Želibská (b. 1941), these whimsical fungi are in fact plastic coin banks, which informed the title, *The Possibility of Saving Throughout the Year*. Želibská's mushrooms, each with a black ribbon identifying it as *Amanita muscaria*, appeared in galleries, on house walls—as the one here, which belonged to fellow Slovak artist Rudolf Sikora—and later as stand-alone models in vitrines. Želibská studied at the Academy of Fine Arts and Design in Bratislava, where she is still based, before attending a residency in Paris and witnessing the May 1968 protests there. A notable figure in the Czechoslovakian conceptual art scene of the 1960s, she drew from Pop art and French New Realism. She encouraged viewer interaction with her assemblages and installations, and much of her work focused on the female body, often depicted symbolically or in fragments. In the 1970s and '80s, unable to exhibit in Communist Czechoslovakia for political reasons, Želibská focused on experimental Land art, performance, and video art. In recent years, her work has been exhibited in group exhibitions at the Walker Art Center in Minneapolis, Mumok in Vienna, Ludwig Museum in Budapest, and the Tate Modern in London, and she represented Slovakia at the 57th Venice Biennale in 2017. Her early mushroom interest and experimental Pop sensibility continues in her current focus on climate change and environmental challenges.

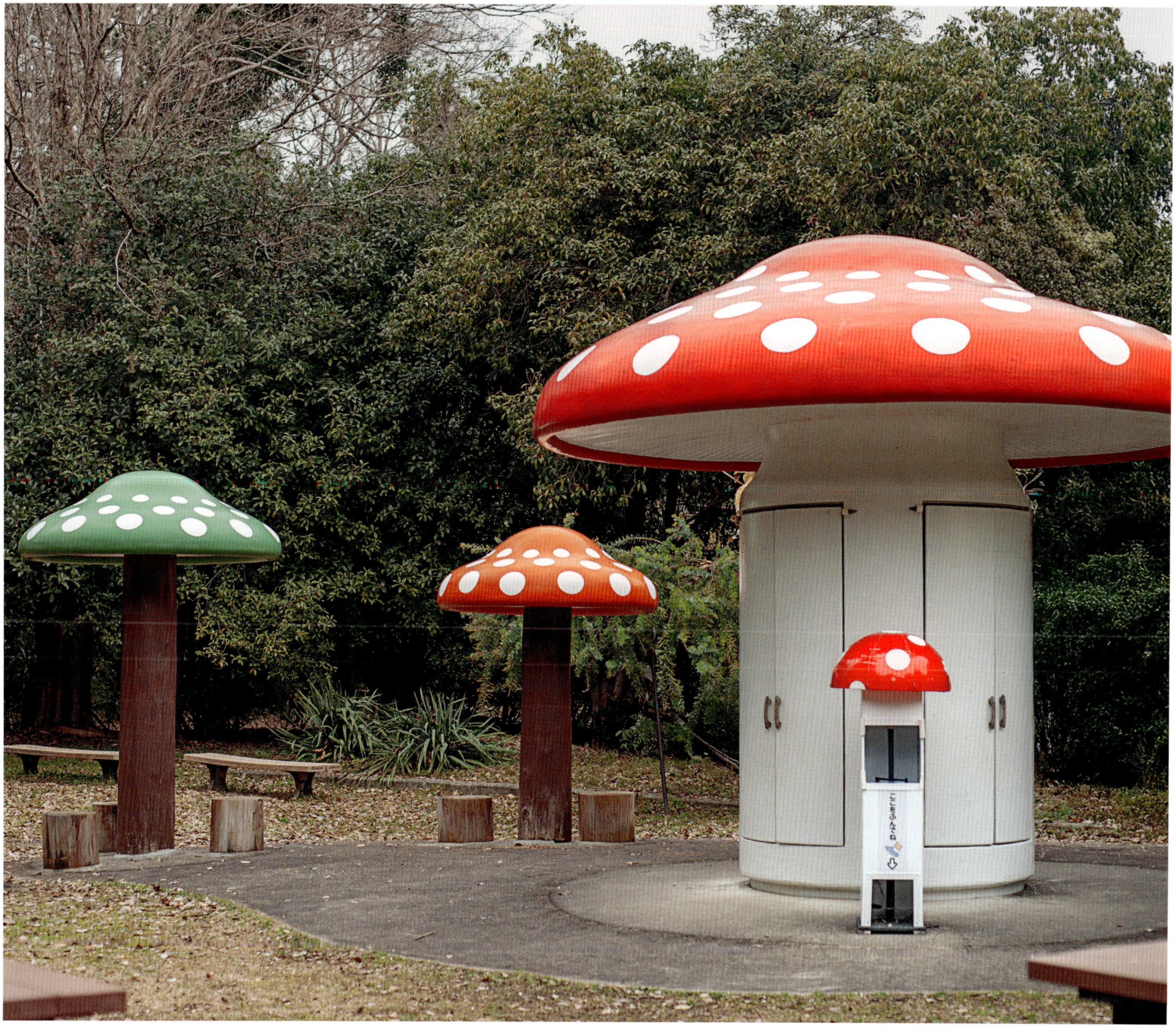

ANONYMOUS

Kinoko Library, 2022
Installation, dimensions variable
Kyoto Botanical Gardens, Japan

Nestled within Kyoto Botanical Gardens is a small, outdoor library spread among several structures, each in the shape of a brightly colored mushroom. The collection, which includes children's books and other reading materials, is available free of charge to all garden visitors. Opened in 1924, and later restored and reopened in 1961, Kyoto's was the first public botanical garden to be established in Japan. The garden's more than 59 acres (24 hectares) boasts 12,000 plant species that flower through the seasons, from spring cherry blossom to the deep reds and golds of autumn. Fungal architecture is a fashionable, if fanciful, idea, and mushroom-inspired structures can be found worldwide on a variety of scales, from full-size houses to ornamental follies. Dwellings often have curvaceous, sometimes thatched canopies that spread over several small structures like creeping fungus, with windows peeping out under low-spreading roofs. Taking its name from the Japanese word for mushroom, the Kinoko Library in Kyoto is designed for children: several large round-roofed wooden structures are set apart from each other in a woodland play area. Each giant-size mushroom "roof," or cap, is painted in a vibrant hue—red, green, and orange, among others. Here, the sturdy wooden "stem" on which the red cap rests has double doors that open to reveal shelves tightly packed with books, at a height that a child can reach. This main "library" is familiar as the oft-depicted fly agaric toadstool (*Amanita muscaria*), with its white-spotted red cap. Thanks to its clever design, the giant fungus keeps both books and readers dry throughout the year.

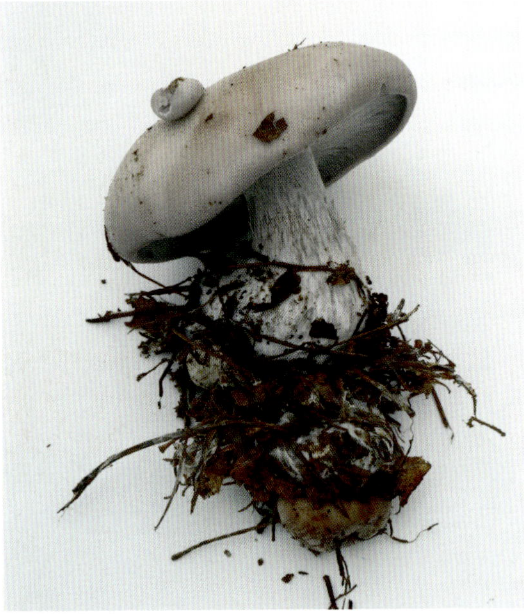
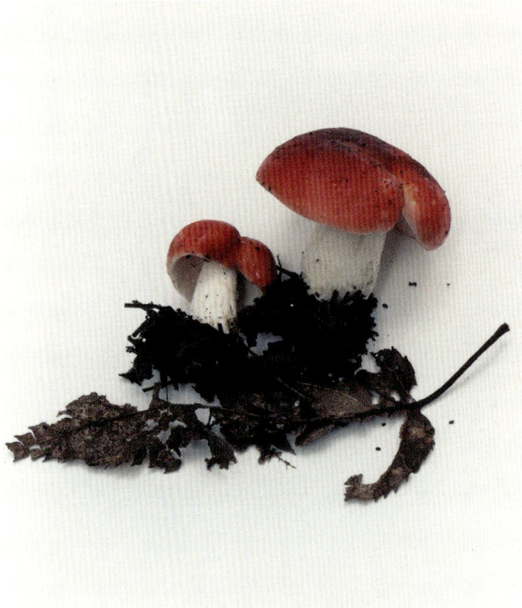
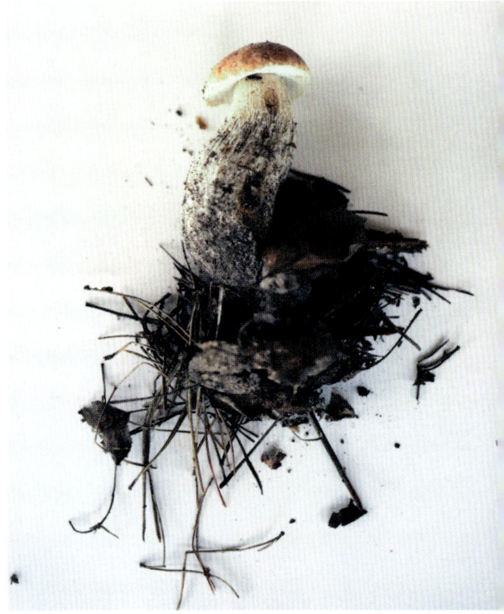
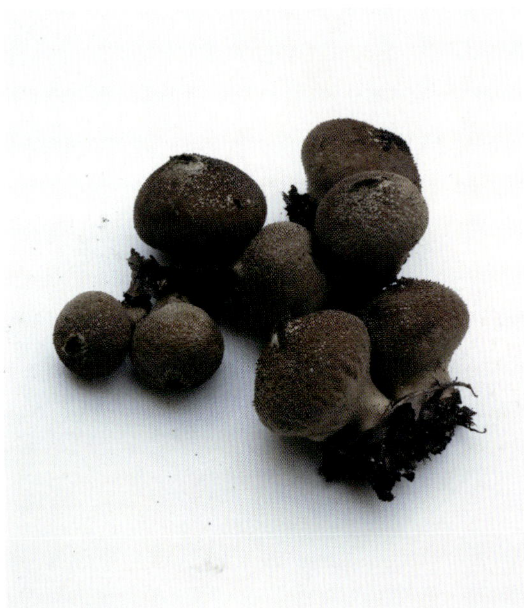
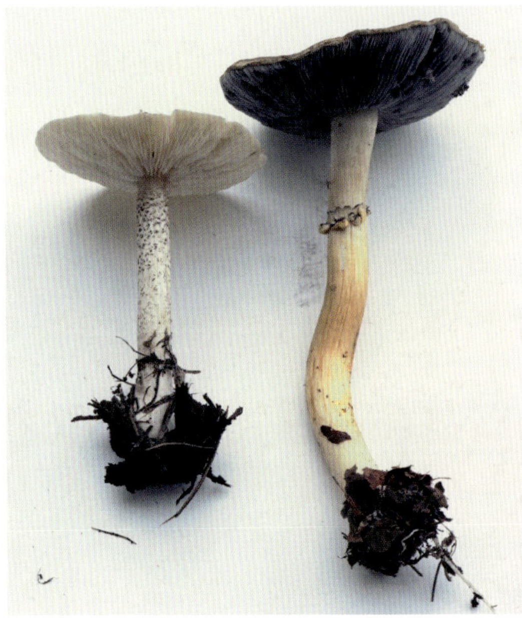
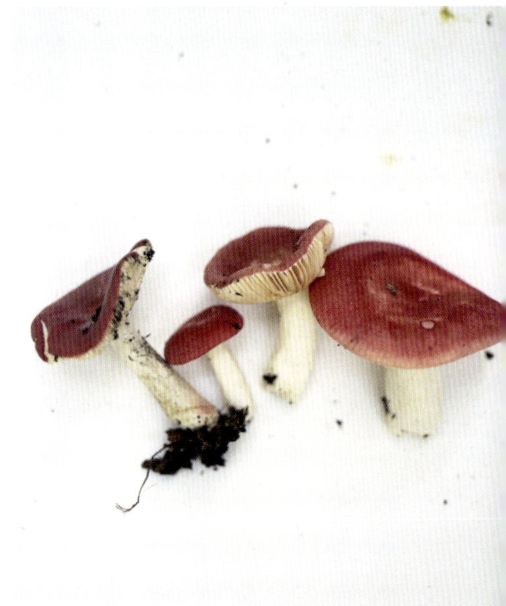

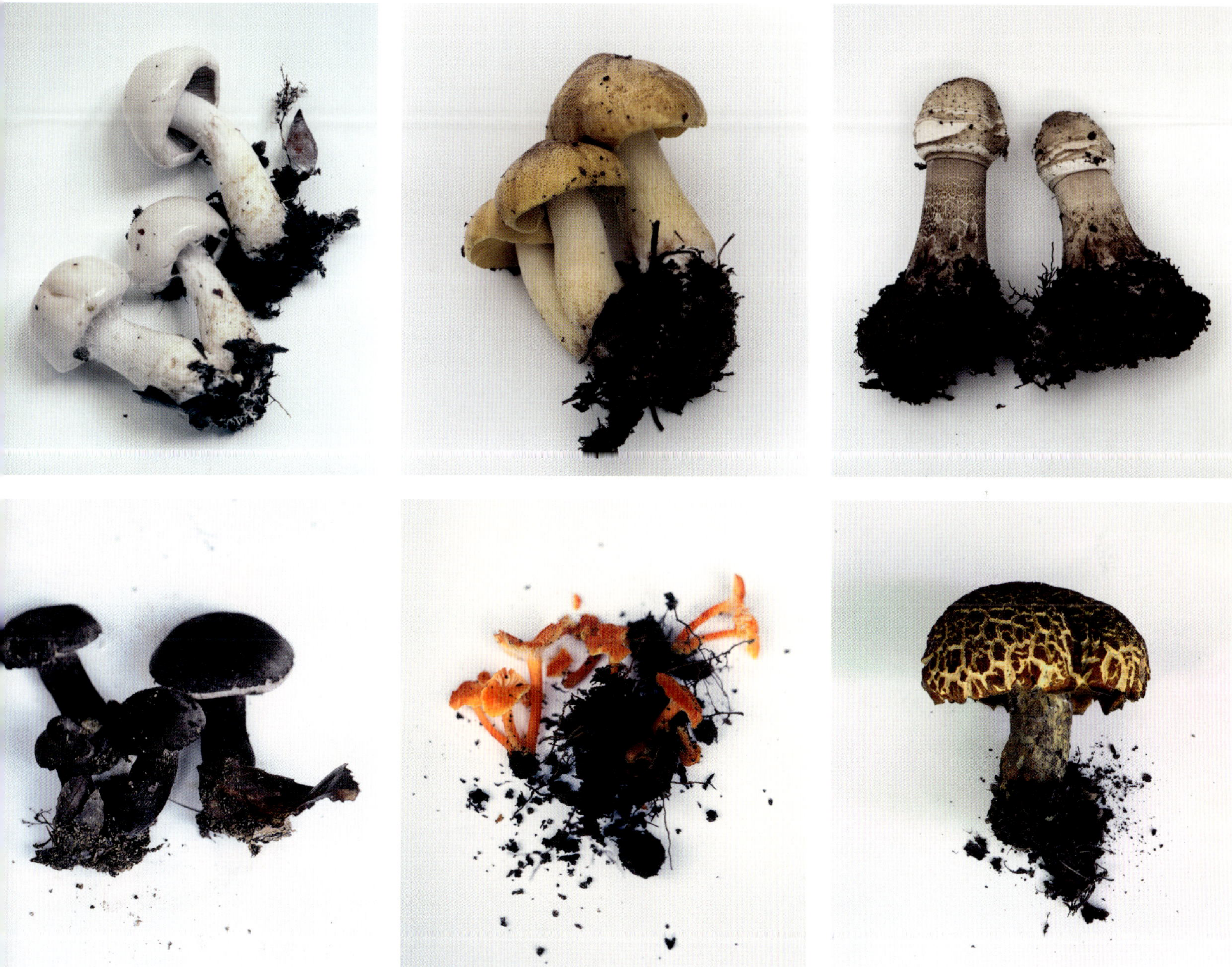

TAKASHI HOMMA

mushrooms from the forest, 2011–25
C-type print, each approx. 13¼ × 10⅜ in. / 33.5 × 26.5 cm

The series of photographs by Japanese artist Takashi Homma (b. 1962), *mushrooms from the forest*, comprises more than just benign portraits of nature. Each specimen is laid against a white background, soil still clinging to the bottom of its stem. Yet there is a more complex story behind this series—the mushrooms Homma photographed were sourced from sites of radioactive aftermath. First were the fungi of the forests contaminated by the Fukushima nuclear accident following the tsunami and earthquake in March 2011. Then Homma expanded to document mushroom specimens, mycelium, and host soil from other contaminated zones, such as the Scandinavian forests affected by Chernobyl in 2017 and, the following year, Stony Point, New York, where composer and mycologist John Cage would forage fungi (see p.83). Homma's mushrooms highlight our changing relationship with forests and fungi, from items of decay and death toward rebirth and regeneration. Interestingly, in the wake of Homma's images, scientists have discovered that fungi and mycelium are the first organism to help clean oil spills and that they notably absorb radiation. Homma began his career in 1984, working as a professional photographer in the advertising industry. He later shot street style images for *i-D* magazine in the 1990s in London, before returning to Japan. He won the respected Kimura Ihei Commemorative Photography Award in 1999 for his artist book *Tokyo Suburbia* and has become one of the medium's most respected artists for his unique place between documentary objectivity and the expression of poetic emotions.

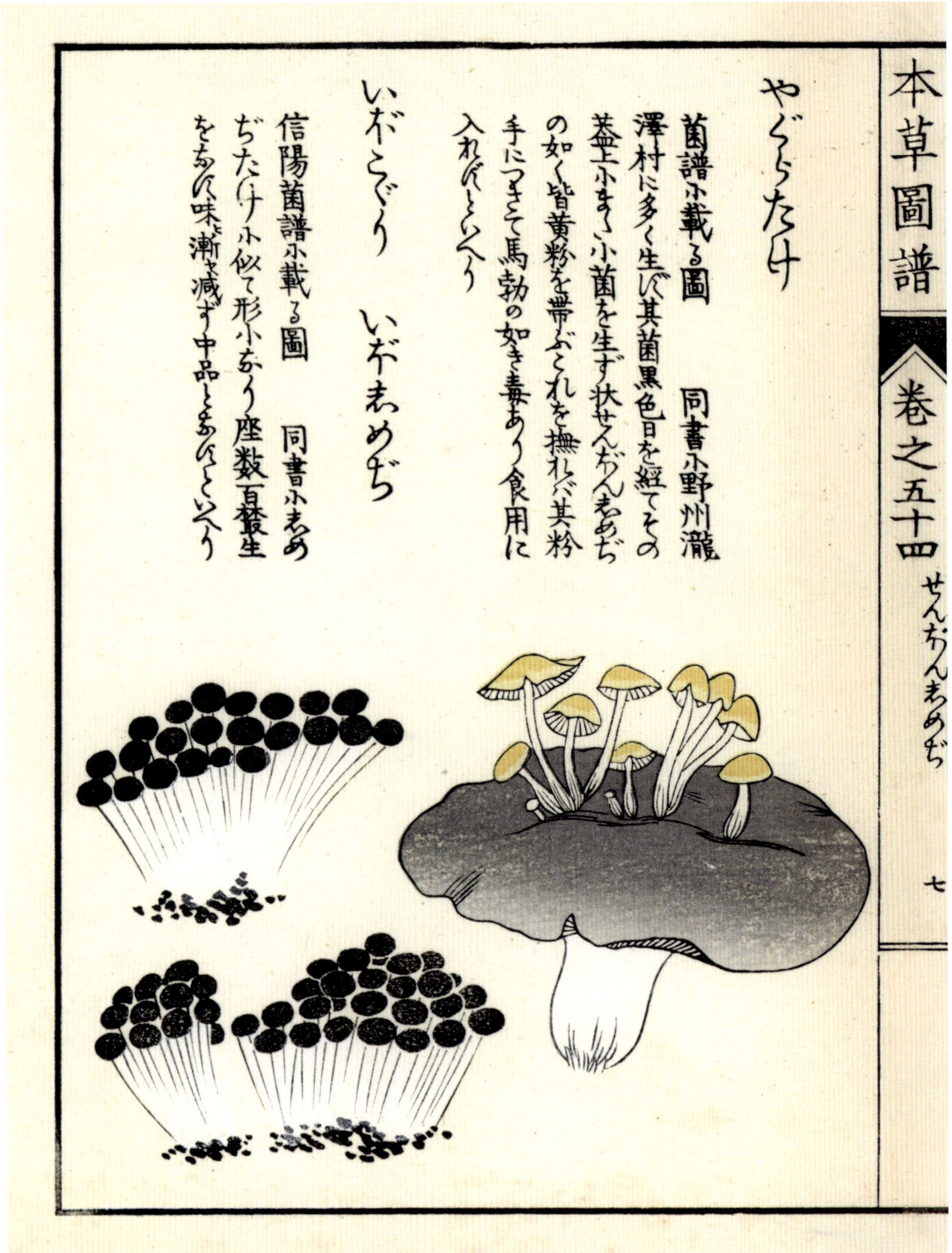

KAN'EN (IWASAKI TSUNEMASA)

Page from *Honzō Zufu* (*Illustrated Book of Medicinal Herbs*), c. 1916–20
Color printed woodblock engraving, approx. 10 × 6⅞ in. / 25.6 × 17.4 cm

Clusters of delicately rendered mushrooms sit side by side on this page from *Honzō Zufu*, a monumental illustrated botanical work produced in nineteenth-century Japan. It is the creation of Japanese botanist, zoologist, entomologist, and samurai Iwasaki Tsunemasa (1786–1842), who worked under the pseudonym Kan'en and served the Tokugawa shogunate during the Edo period (1603–1867). A massive undertaking, the *Honzō Zufu* was intended to be an illustrated guide to medicinal plants, and it detailed through illustrations and written descriptions—as seen here—some two thousand species, including some fungi. Here, Kan'en depicts *Lyophyllum decastes* on the left—also known as shimeji, chicken fried mushroom, or clustered domecap—which is a popular culinary mushroom widely eaten across Japan. On the right is *Asterophora lycoperdoides*, or the powdery piggyback mushroom, a parasitic fungus that grows only on other mushrooms. The precision of Kan'en's illustrations was enabled by developments in block printing during the Edo period. Woodcut printing had originated in China as early as the eighth century, but the Japanese developed a way to produce multiple blocks to allow prints to be created using different colors, paving the way not only for botanically accurate illustrations but also printed text. Kan'en's prints were originally distributed as hand-colored manuscripts, and it was only around 1916 to 1920 that his work was gathered and printed. His research extended to more than ninety volumes, of which two volumes were indices.

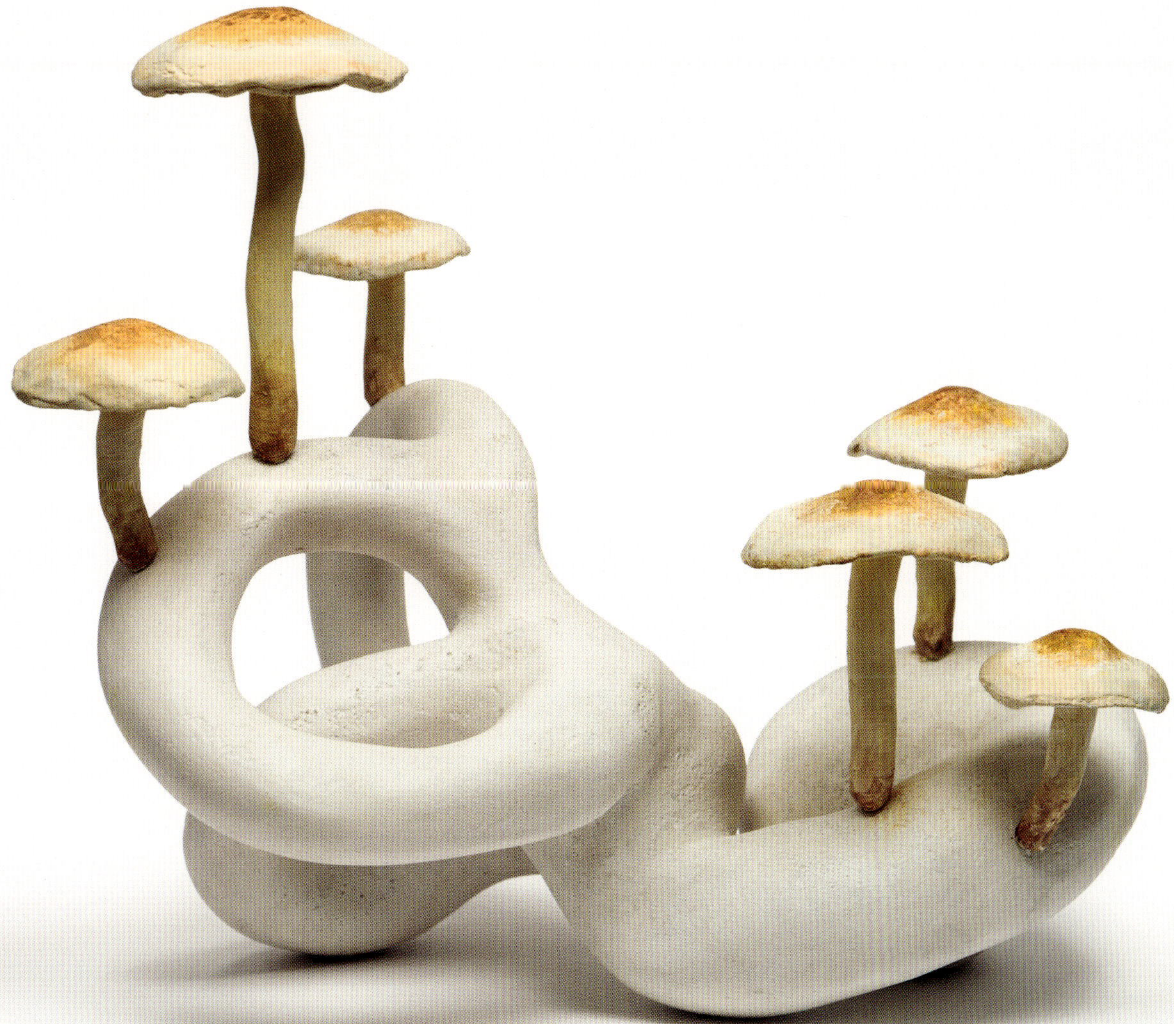

HAMISH PEARCH

Cochlea Brick Tuft, 2019
Plaster polymer, epoxy putty, and oil paint, 9⅞ × 9⅞ × 6 in. / 25 × 25 × 15 cm

Six facsimile mushrooms with painted warm-toned heads sprout from a twisting, cream-colored plaster base, at once a stylized echo of mycelium and a representation of the spiral-shaped cochlea of the human inner ear. This artwork is one of the first of an ongoing series of fungi sculptures by British artist Hamish Pearch (b. 1993), which he began while completing his master's degree at the Royal Academy Schools in London. These lifelike objects are made from a fusion of epoxy, plaster, and oil paints, and they intentionally draw uncanny juxtapositions that depend on fungi's ability to transform and grow out of existing materials. Pearch has created mushrooms appearing to grow out of everything from burnt toast and piles of discarded newspapers to aluminum cooking pots. The British artist is drawn particularly to the bodily and psychological aspect of mushrooms. Here, fungi form a metaphor for creative inspiration itself, a force emerging from a void. In his practice, which includes installation, drawing, and sound, Pearch is drawn toward the strange liminal spaces created by man on the edge of nature. There is a tension between nature and the manmade in his creations. Increasingly working with bronze and metal, among other materials, other recurring themes in Pearch's work include frogs, turtles, and what appear as twisted branches that function as pseudo-chairs or bags.

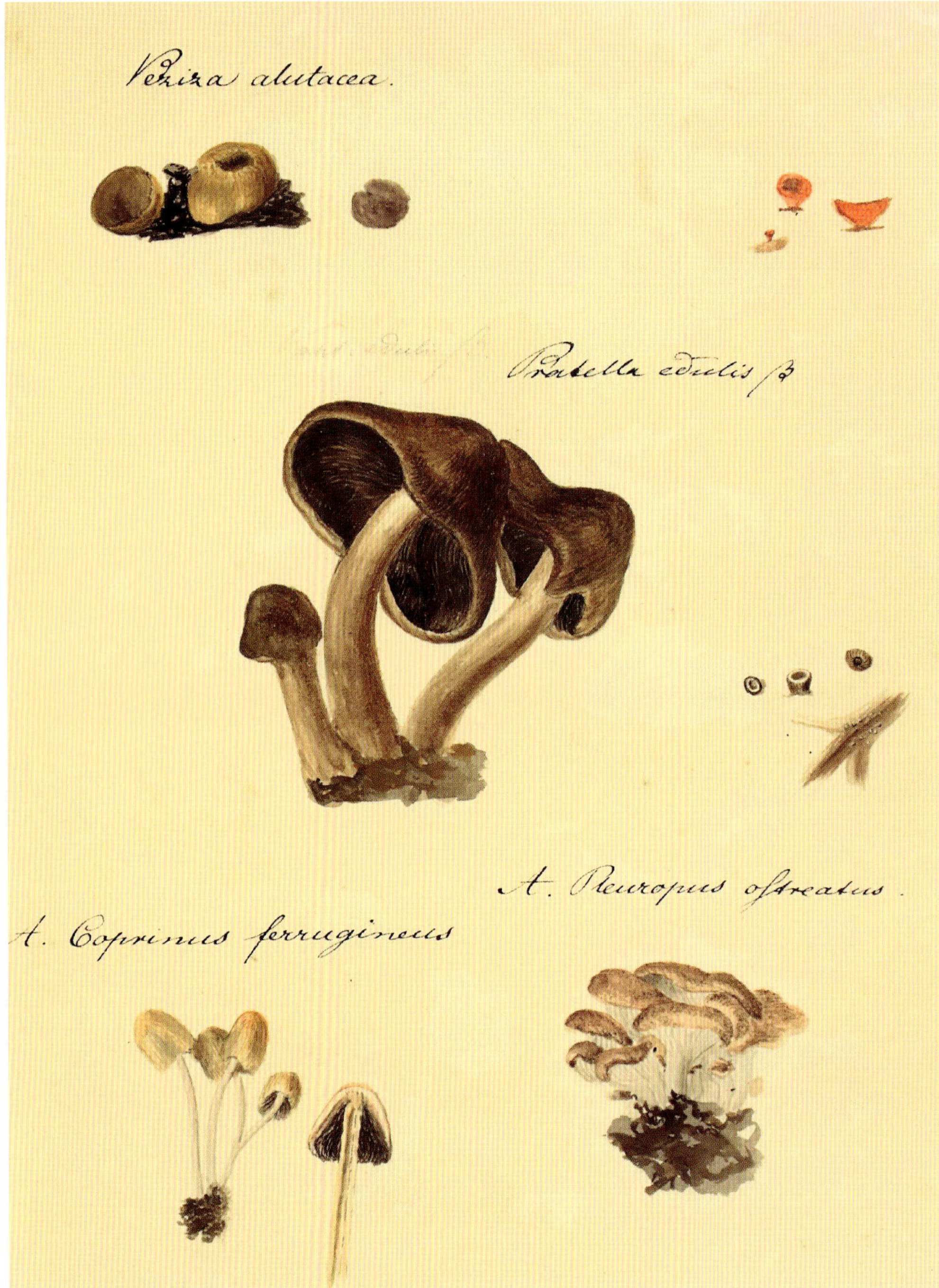

LEWIS DAVID VON SCHWEINITZ

Icones fungorum Niskiensium (Illustrations of the Fungi of Niesky), c. 1798–1805
Watercolor on paper, H. 10¼ in. / 26 cm
Harvard University Botany Libraries, Cambridge, Massachusetts

Included in a volume of unpublished watercolor paintings of fungi, these precisely rendered illustrations show the drawing skills of German American mycologist Lewis David von Schweinitz (1780–1834). Now regarded as the father of North American mycology, Schweinitz collected his illustrations in the undated tome *Icones fungorum Niskiensium*, one of seven that served as a reference for his groundbreaking *Conspectus Fungorum in Lusatiae*. The *Conspectus*, published in 1805 with the collaboration of his friend and mentor, professor Johannes Baptista von Albertini, is considered a classic text of mycology. Schweinitz's masterwork, it illustrates more than one thousand species of fungi, around a hundred of which were previously unrecorded species, many of which now bear his name. To ensure the accuracy of his watercolors, Schweinitz drew from recently collected fresh specimens—a crucial undertaking, as fungi dramatically change shape as they start to dry out—or referenced his existing watercolors. Schweinitz produced the illustrations for *Icones fungorum Niskiensium* while living in a theological seminary in Niesky in Upper Lusatia in Saxony (present-day eastern Germany). The book contains some 249 watercolors of various mushroom species, from agarics of the *Amanita* genus and *Peziza* cup fungi to *Clavaria* coral fungi and truffles of the *Tuber* genus. Although Schweinitz's name does not appear on any of the illustrations, mycologists have recognized his distinctive handwriting in the labels that provide each species's scientific name.

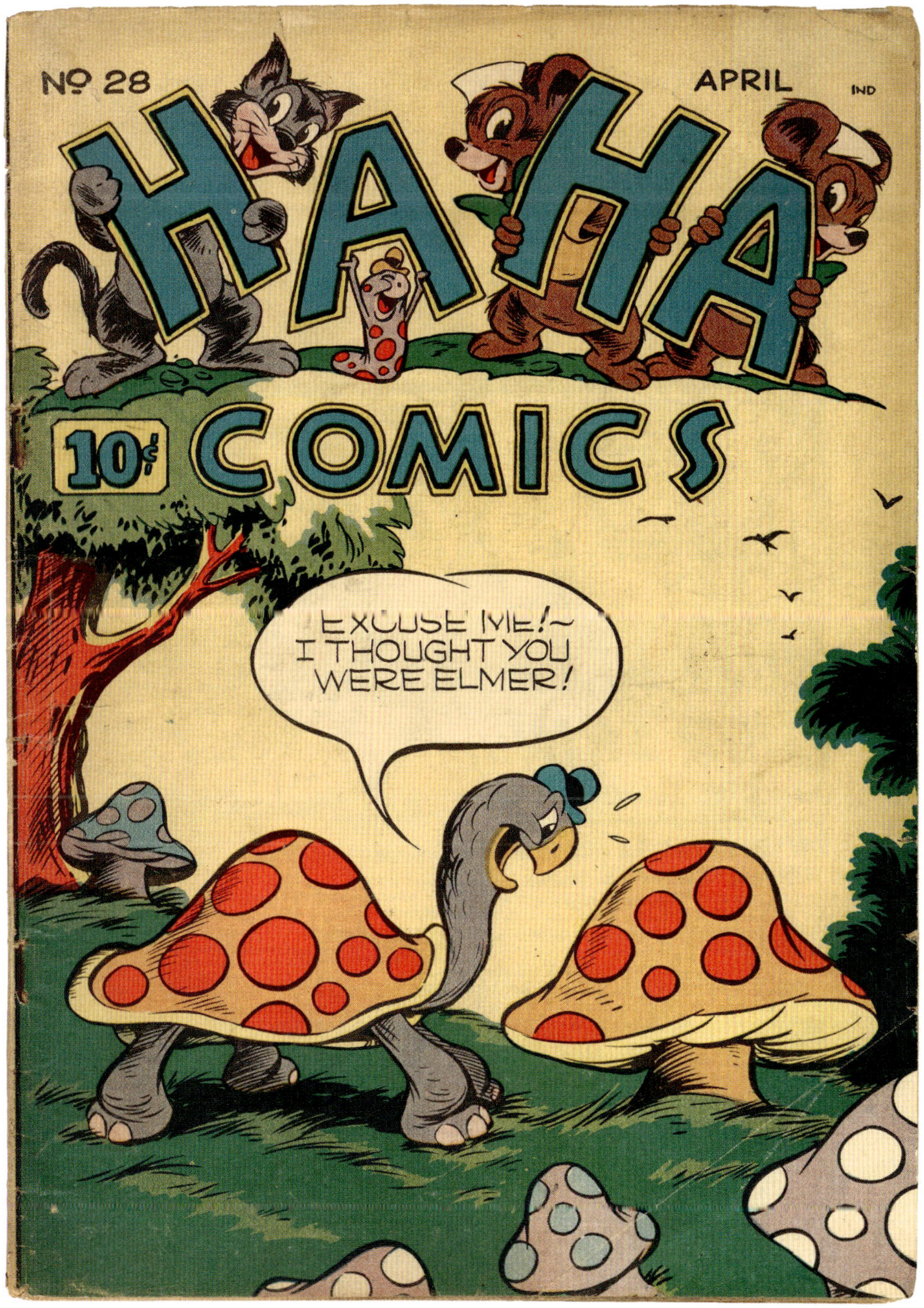

KEN HULTGREN

Cover of *Ha Ha Comics* no. 28, 1946
Printed book, 10 × 7 in. / 25.4 × 17.8 cm
Private collection

In this whimsical 1946 cover for *Ha Ha Comics*, illustrator Ken Hultgren (1915–1968) amusingly depicts a turtle with a polka-dot shell mistaking a similarly patterned mushroom for his friend Elmer. The brainchild of the Creston Publishing Corp. in New York, *Ha Ha Comics* published its first issue in October 1943, offering much-needed laughs to a readership mired in the stresses of World War II. A native of Minneapolis, Hultgren studied at his hometown school of art before departing for California in 1935, where he soon found work as an animator for Disney Studios. He quickly rose to become one of the studio's top animators and worked on films such as *Snow White and the Seven Dwarfs* (1937) and *Bambi* (1942). He began drawing comics when he was tapped to ink an adaptation of *Bambi*, as well as several spin-off comic stories featuring some of Bambi's friends. After leaving Disney in 1943, Hultgren worked for Sangor Studios, which produced comic adaptations of films by Disney, Warner Brothers, Screen Gems, MGM, and others. He specialized in drawing anthropomorphized animals, and illustrated recurring comic titles, including "Robespierre" (a cat) and "Izzy & Dizzy" (two bear cubs). Sangor closed in 1949, and the following year Hultgren published *The Art of Animal Drawing*, which is still considered a seminal text for any would-be illustrator. He returned to Disney from 1955 to 1957 and then worked a variety of jobs in the last decade of his life, including seven years at Archie Comics and as an animator for *Mr. Magoo*, *Popeye the Sailor*, and other television shows.

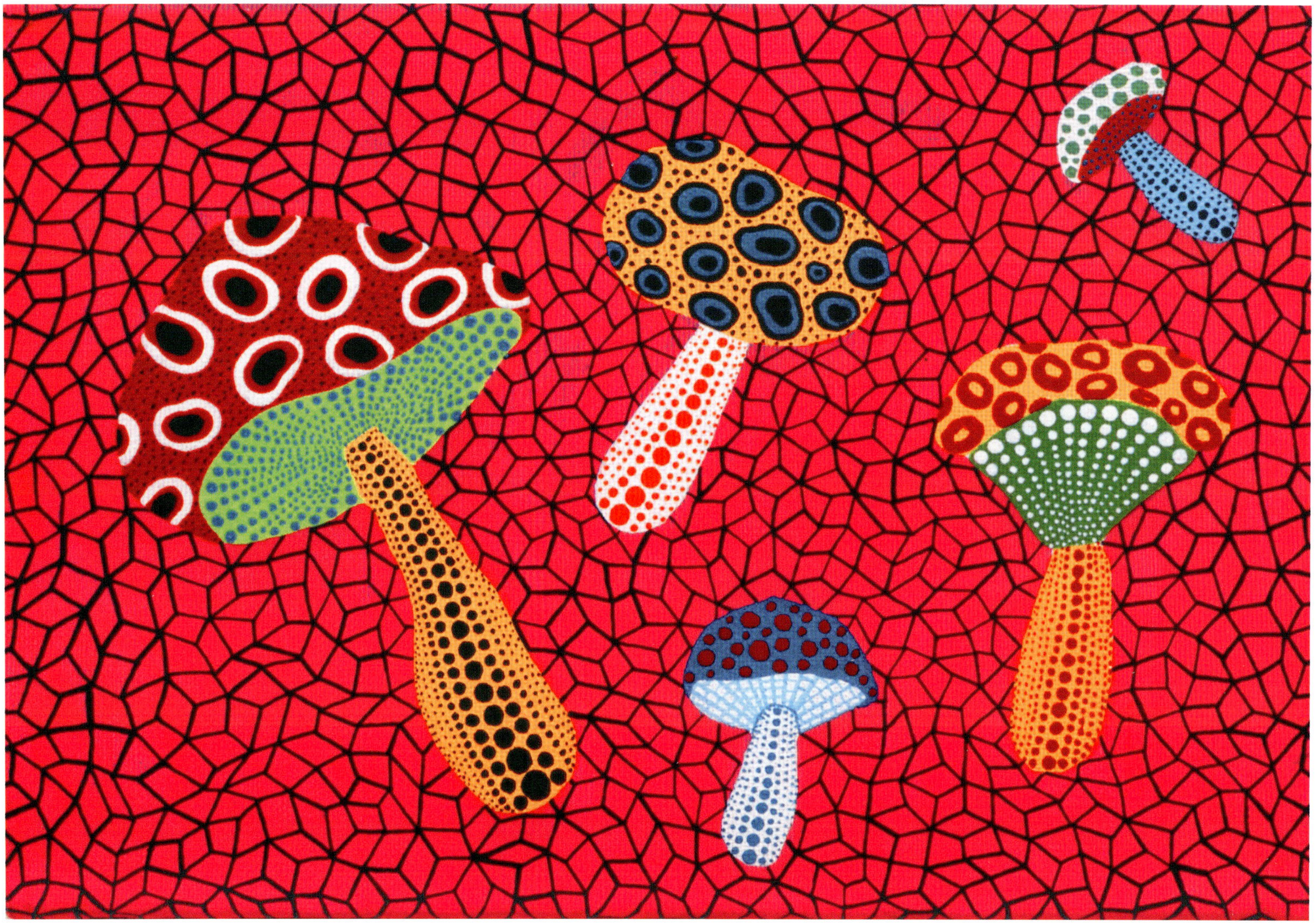

YAYOI KUSAMA

Mushrooms, 1995
Acrylic on canvas,
6¼ × 9 in. / 15.8 × 22.7 cm
Private collection

While Yayoi Kusama (b. 1929) may be better known for depicting pumpkins and flowers in her paintings, sculptures, and immersive installations, the internationally celebrated Japanese artist has also featured mushrooms in her art for several decades. In 1979 she produced a naturalistic felt pen-and-ink drawing of *kinoko* (Japanese for *mushrooms*) floating in a sea of small red dots. Kusama has since returned repeatedly to the subject, creating compositions of varying scales, complexities, and styles, from a 5 by 4 foot (1.6 by 1.3 m) red-and-black painting of a single mushroom enlarged to gigantic proportions created in 1980 to this small, kaleidoscopic canvas featuring five different types of mushrooms. *Mushrooms* is typical of her approach to painting: surfaces are flattened and covered with patterns of grids and dots, and the vibrant palette is far removed from reality. These characteristics add to an overall sense of the psychedelic and hallucinatory, concepts that are also associated with mushrooms on account of the perception-altering properties of certain varieties—such as those that contain the compound psilocybin—also called "magic mushrooms." For Kusama, however, such hallucinations and visual disturbances have been part of her life since childhood. Her compulsive drive to create art is well-documented as being her strategy to cope with the intrusive visions of repetitive forms and patterns: translations of her inner experiences into physical media.

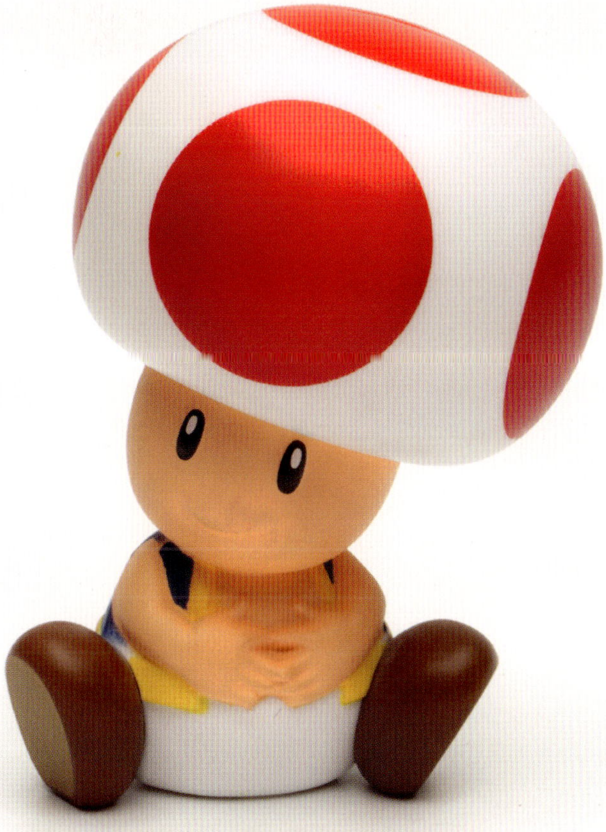

SHIGERU MIYAMOTO FOR NINTENDO

Toad figurine, 2013
Plastic, H. approx. 3¼ in. / 8.3 cm
Private collection

This small, cheery character with short arms, stubby feet, and a large, red-spotted, mushroom-like head has appeared in virtually every iteration of Nintendo's Mario video game franchise, either as a playable or nonplayable character. Known as Toad, he is a citizen of the Mushroom Kingdom and one of the most loyal attendants to Princess Peach, who he tries to protect with varying degrees of success. Known in Japan as Kinopio, Toad was created by Shigeru Miyamoto (b. 1952), who developed the original *Super Mario Bros.* platform game for the Nintendo Entertainment System in 1985. While he typically provides assistance to Mario and friends, Toad takes center stage in spin-off games such as *Wario's Woods* (1994) and *Captain Toad: Treasure Tracker* (2014)—the latter being the first official Nintendo title to feature Toad as the titular character. Toad's mushroom head has historically been the subject of some ambiguity as to whether it is, in fact, a hat. However, in 2018, Nintendo stated categorically that it is, indeed, a part of his body. Toad belongs to a species of mushroom people known collectively as Toads, all of whom have a very similar appearance, with spotted mushroom heads and vests of varying colors. They feature prominently in *The Super Mario Bros. Movie* (2023), in which Toad, voiced by Keegan-Michael Key, also plays a major role, heroically working alongside Mario and Princess Peach to keep Bowser from invading their peaceful kingdom.

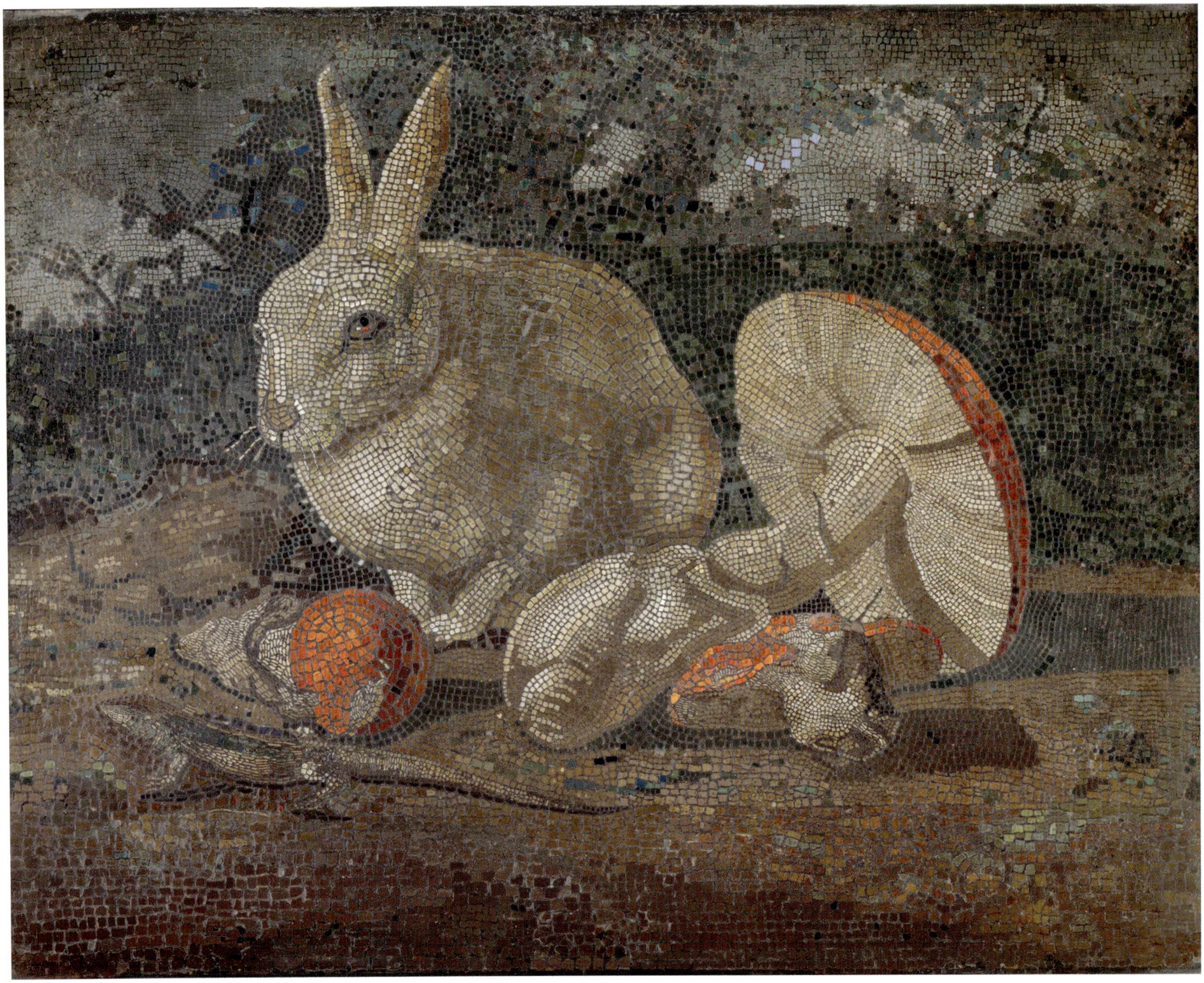

ANONYMOUS

Mosaic with rabbit, lizard, and mushroom, 19th or early 20th century
Tile, 15⅜ × 19⅜ × ½ in. / 39.1 × 49.1 × 1.3 cm
Metropolitan Museum of Art, New York

Mosaic is an artistic tradition that dates back millennia. The earliest examples can be traced to ancient Mesopotamia in the second half of the third millennium BCE, and the practice proliferated in ancient Rome and Greece, becoming the intricate art form we know it to be today. This particular mosaic was created by an unknown artist around the late nineteenth to early twentieth century. Although little is known about the work, it was a gift to New York's Metropolitan Museum of Art from famed industrialist John Pierpont Morgan, who was one of the Gilded Age's most prominent collectors of art and antiquities. Executed in a style to resemble Greco-Roman mosaic, the intricate assemblage is rendered in fine detail by the deft hand of its maker. One can easily observe the contours of the rabbit's musculature, the craggy scales of the lizard's body, and even the gills of the overturned red-capped mushroom. Mosaics are formed from hundreds—sometimes thousands, or even millions—of tesserae, tiny pieces of colored stone, ceramic, or glass, usually in a roughly square shape. With a meticulous artistic vision, the tesserae are painstakingly adhered to ceilings, floors, wall panels, and other surfaces to eventually form a larger pattern or tableau when fully assembled. While its maker's intention is unknown, the scene in this work could be a reference to Lewis Carroll's 1865 novel, *Alice's Adventures in Wonderland*, in which the White Rabbit, gargantuan mushrooms, and the workman Bill the Lizard all feature prominently (see p.79).

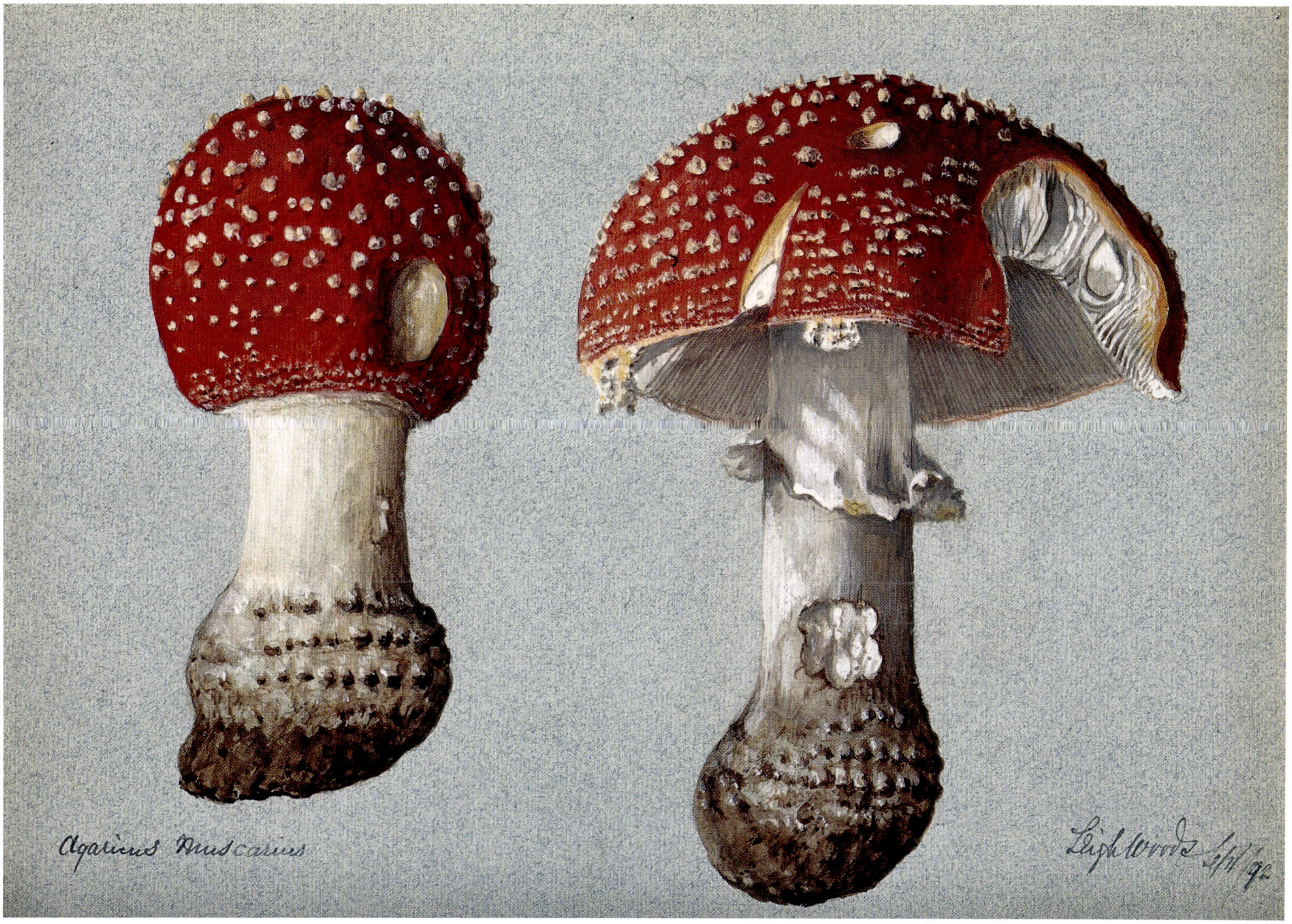

ANONYMOUS

Agaricus muscarius, 1892
Watercolor, approx. 26 × 34 in. / 66 × 86.4 cm
Wellcome Collection, London

This watercolor by an unknown nineteenth-century British artist captures two *Amanita muscaria* specimens. What is remarkable is not just the detail the artist paints—the iconic red cap dotted with protruding white spores, the gills, and the bulbous stem—but also the imperfections found in each mushroom, indicating that these illustrations were almost certainly drawn from fresh, real-life specimens. On the left, the mushroom is missing a small chunk from its cap, while the sections missing from the mushroom on the right reveal more of its white gills, or lamellae. As noted in the bottom right, the mushrooms were found in Leigh Woods, a woodland close to the city of Bristol in the west of England known for its ancient trees, which make it a perfect habitat for mushrooms. The *Amanita muscaria*, or fly agaric, is perhaps the most easily recognizable of all the fungi species and has featured in everything from emojis to garden statues to fairy tales. Toxic in large doses, in smaller quantities the mushroom is known to produce hallucinogenic effects that can lead to the blurring of reality and dreams. Used in Siberia centuries ago in shamanic rites, its hallucinogenic qualities are said to have inspired Lewis Carroll's 1865 novel *Alice's Adventures in Wonderland* (see p.79). The mushroom has a prominent role in popular culture, from video games such as *Super Mario Brothers* to movies such as Walt Disney's *Fantasia* (see pp.209, 213).

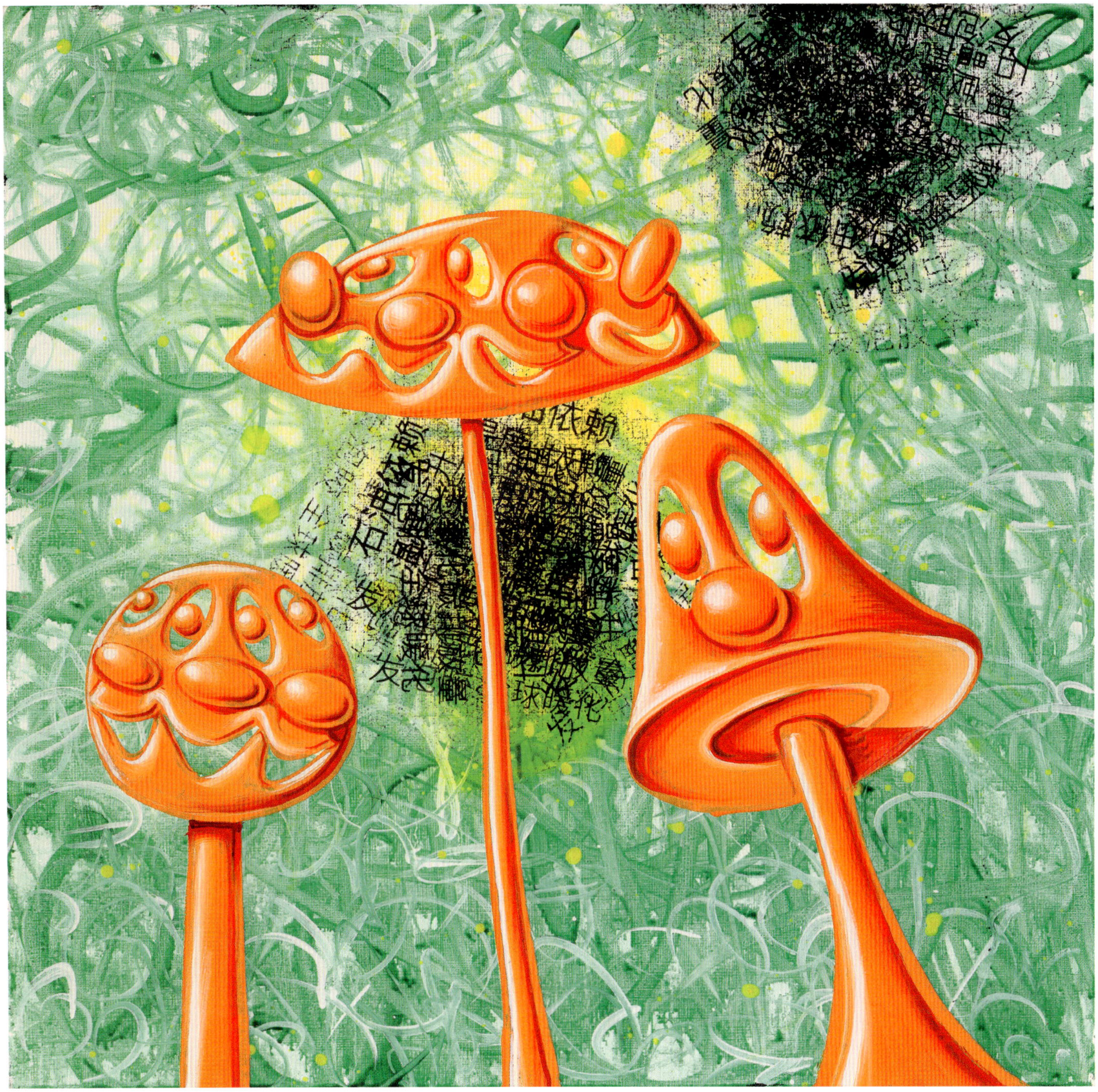

KENNY SCHARF

Jingu, 2021
Oil and silk-screen ink on linen,
24 × 24 in. / 61 × 61 cm
Private collection

Growing up in Los Angeles, Kenny Scharf (b. 1958) aspired to be an artist from the time he could grip a pencil. After attending the prestigious Beverly Hills High School, he moved up the coast to study at the University of California at Santa Barbara before entering New York's School of Visual Arts two years later in 1978. There he soon met, and became roommates with, Keith Haring. Tight friends, the two spent their nights in New York's art-punk underground, especially in the East Village. Scharf made art across a variety of media during this time, primarily painting but also sculpture, performance art, video, and street art. He was known to creep around the tenements of the Lower East Side at night, spray-painting buildings with large murals of the Jetsons and other Hanna-Barbera characters. Scharf's bright, psychedelic, often cartoon-influenced artworks (now championed as prime examples of Pop Surrealism) didn't resonate widely at the time, and after he and Haring graduated from SVA in 1980, the latter's star began a brisk ascent while Scharf's stalled. By 1985 that had changed, and Scharf's impact was being felt across the contemporary art scene at large. The mushrooms pictured in *Jingu* are classically Scharf-ian, with their exaggerated, cartoony shapes; bright colors; and faces smiling with a combination of too many eyes, mouths, and noses. And while Scharf has stated that he never painted under the influence of psychedelic mushrooms, he has said that he took profound inspiration from them in creating his perfectly unique artistic universe.

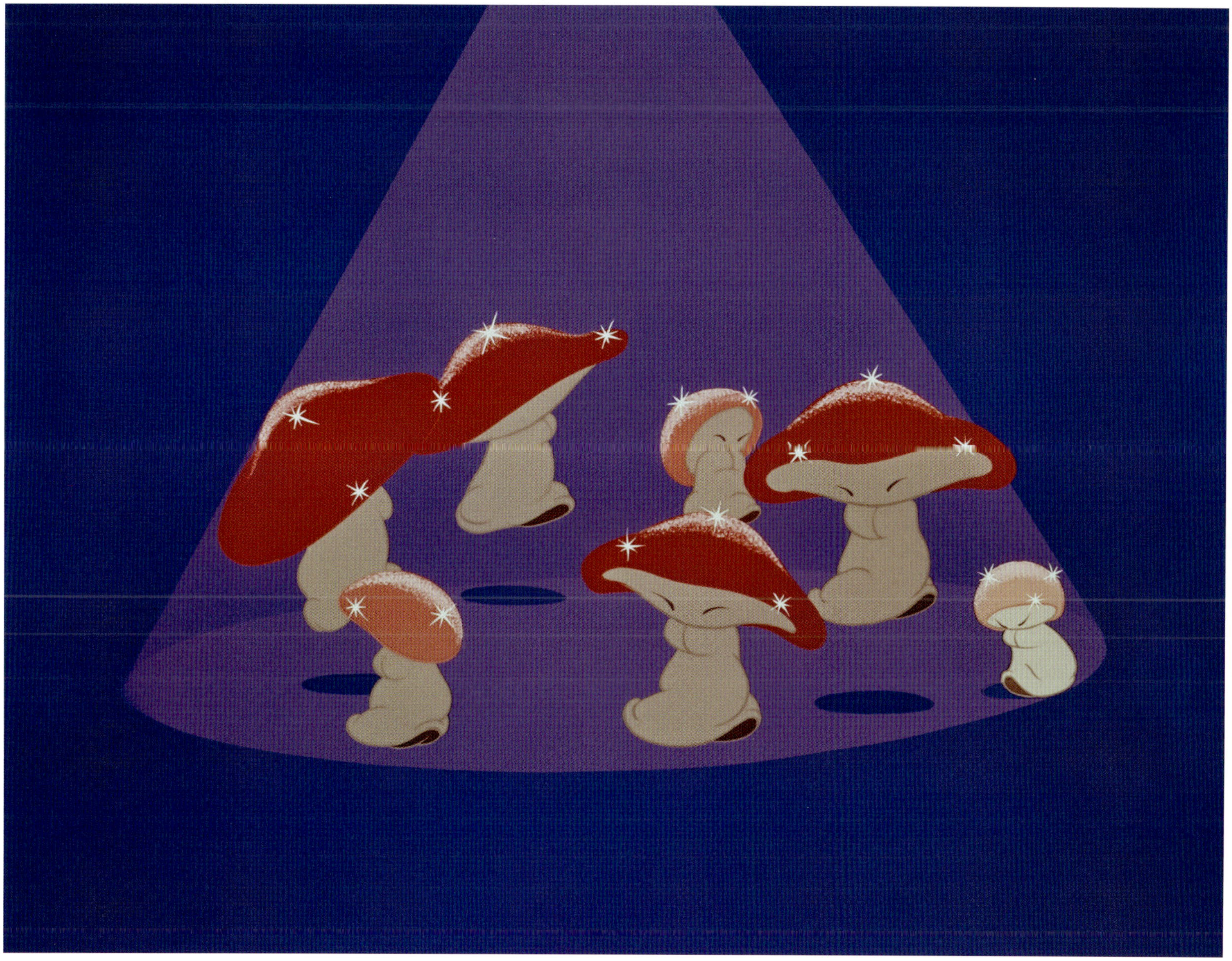

WALT DISNEY PRODUCTIONS

Fantasia, 1940
Film still, dimensions variable

An enduring classic and one of the most visionary animated films of all time, *Fantasia* began with a chance encounter. One evening, Walt Disney and famed conductor Leopold Stokowski happened to be dining at the same Hollywood restaurant. According to Stokowski, after striking up a conversation, the two men arrived at the kernel of an idea—to create a film where audiences could see what they heard. The collaboration between Disney—casting his studio's marquee character, Mickey Mouse, as the sorcerer's apprentice—and Stokowski, who selected the music and recorded seven of the eight pieces with his Philadelphia Orchestra, resulted in the now-iconic 1940 film. In the film's opening scene, pre–Abstract Expressionist colors and shapes accompany Bach's "Toccata and Fugue in D Minor," while one of *Fantasia*'s more lighthearted entries is set to Tchaikovsky's *Nutcracker Suite*, music from the wildly popular ballet that would have been familiar to audiences at the time. Each piece is its own short vignette, in which fairies and woodland flora, fauna, and fungi variably serve as stand-ins for the ballet's dancers. As Tchaikovsky's "Chinese Dance" begins, a group of mushrooms, dotted by morning dew, springs to life. Inspired by the traditional Chinese fan dance, the mushrooms twirl and whirl in unison, while the smallest of the troupe tries desperately to keep up with its bigger counterparts. While more than a thousand artists, animators, sound engineers, and other technicians worked on the film, animator Art Babbitt was the lead on the mushroom dance sequence and a number of the other most memorable scenes in all of *Fantasia*.

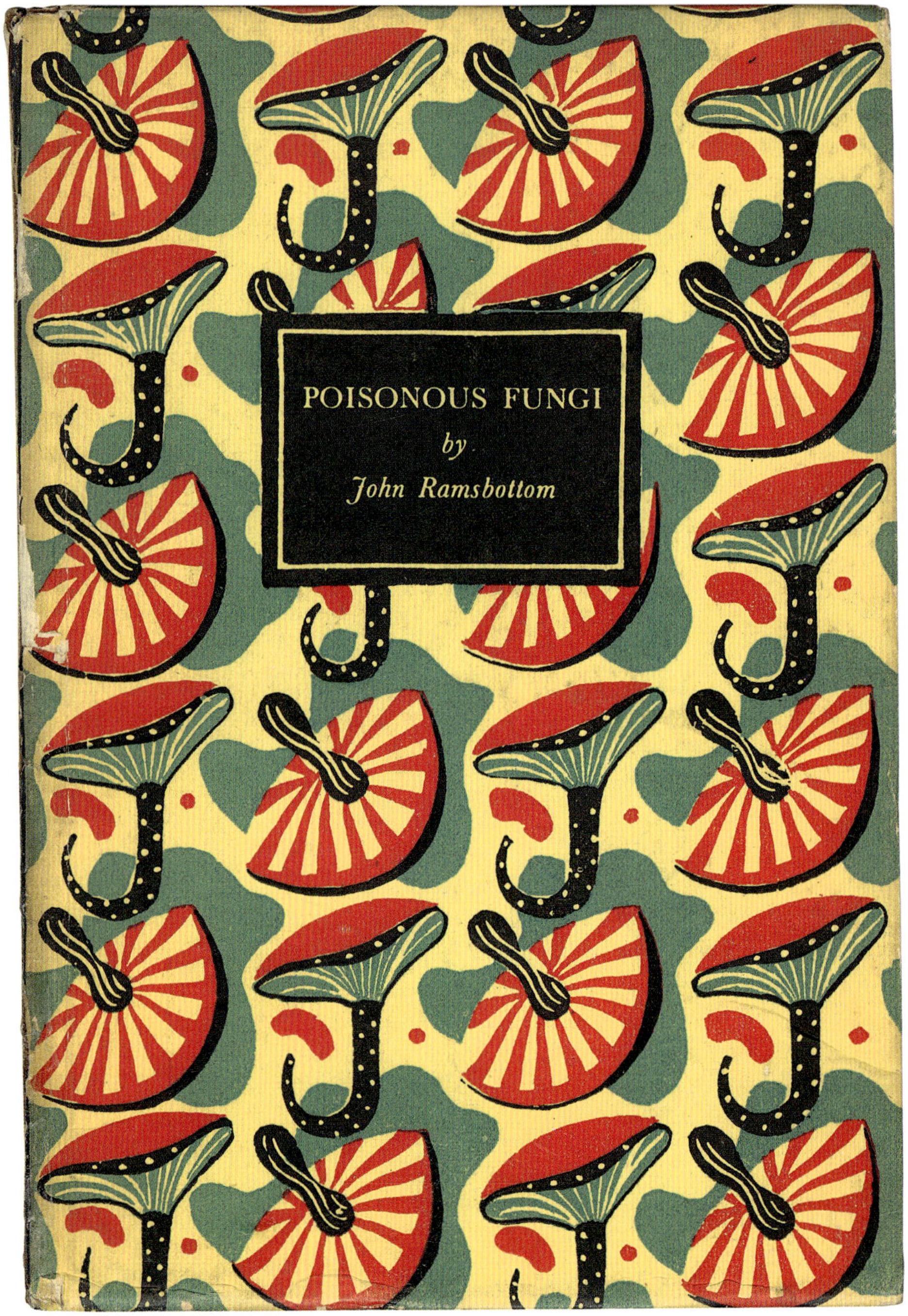

JOY JARVIS

Cover of *Poisonous Fungi*
by John Ramsbottom, 1945
Printed book, 7 × 4¾ in. / 18 × 12 cm
Private collection

In nature, the colors red and yellow often signal danger, so it is apt that these hues predominate on the cover of John Ramsbottom's *Poisonous Fungi*, published in 1945. Ramsbottom, a leading British mycologist, was the author of several specialist publications on fungi, as well as popular books, such as this one, for a more general readership. The book includes sixteen botanical illustrations by artist Rose Ellenby, but the cover is the work of British designer Joy Jarvis (active 1940s). Jarvis's design repeats two stylized mushroom motifs. They are not identifiable species, but their bright colors serve to remind the reader that some of the most visually appealing mushrooms are also the most toxic. This cover is in marked contrast to that of the companion volume *Edible Fungi*, published two years earlier, which is printed in more muted tones and shows mushrooms in a skillet. While much about Jarvis's career remains unknown, she developed her skill in illustrating repeating images through her work as a textile designer in the 1940s, producing patterns—mostly for furnishing fabrics—for clients such as London Transport (see p.72) as well as independent retailers. Thanks to the bold cover design, *Poisonous Fungi* is one of the most eye-catching of the King Penguin series. Beginning in 1939, these collectable picture books (described as "illustrated keepsakes" by Penguin's founder Allen Lane) covered an eclectic range of subjects, ranging from art, architecture, fashion, and toys to myriad subjects of natural history.

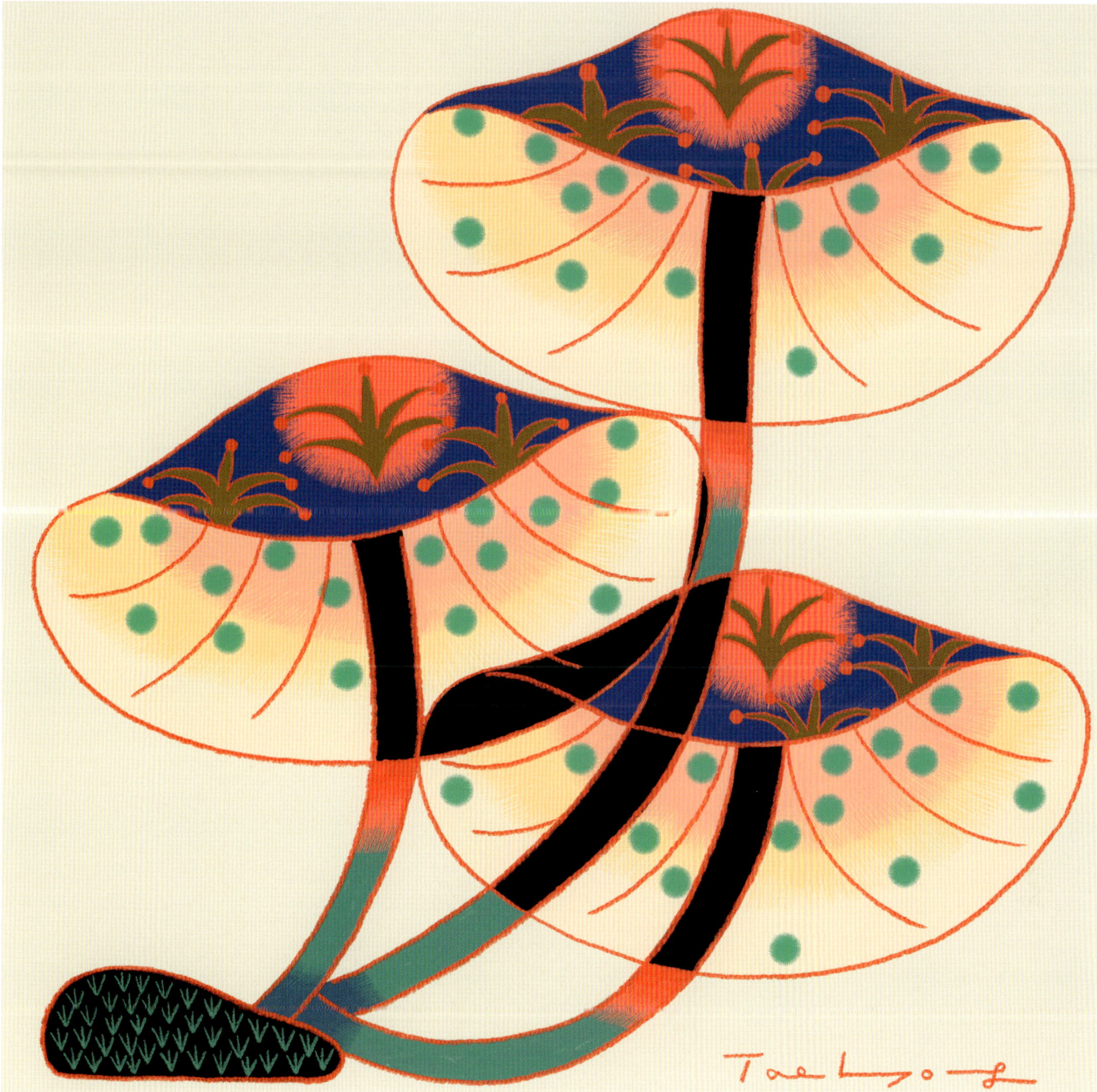

TAEHYOUNG JEON

Plant 137, 2023
Digital painting,
15¾ × 15¾ in. / 40 × 40 cm

Three brightly colored mushrooms writhe and flow, their umbrella-like caps fanning out from long, sinuous stems in this digital drawing by Korean artist Taehyoung Jeon (b. 1992). In this piece—one of a series inspired by botanical life—Taehyoung uses a combination of traditional printmaking and digital animation. With a printmaker's eye, her construction of layered images leads her to create artwork in separate slices. While taking inspiration from nature, Taehyoung does not directly represent a particular species but instead interprets the mushroom's form in her own artistic perspective. Working digitally allows her to try out color combinations until she finds those that best represent her vision: in this case brilliant blue, green, and pink, with gills decorated in polka dots. Taehyoung was born in South Korea, and her formative years were spent in New Zealand and the UK, bringing a multicultural quality to her work that blends traditional Eastern and Western aesthetics. With her background in printmaking and computer composition, Taehyoung's practice combines abstract forms with plant life in simple and balanced images. Working from her studio in Seoul, and inspired by the potted plants that fill her apartment, in 2022 she began her *Plant* series, painted digitally on her iPad, which has since grown and developed into her green leaf period, followed by her flower period, mushroom period, and insect period. No matter what she depicts, at its core Taehyoung's art is founded on the belief that vegetation and humans have a close and enduring relationship.

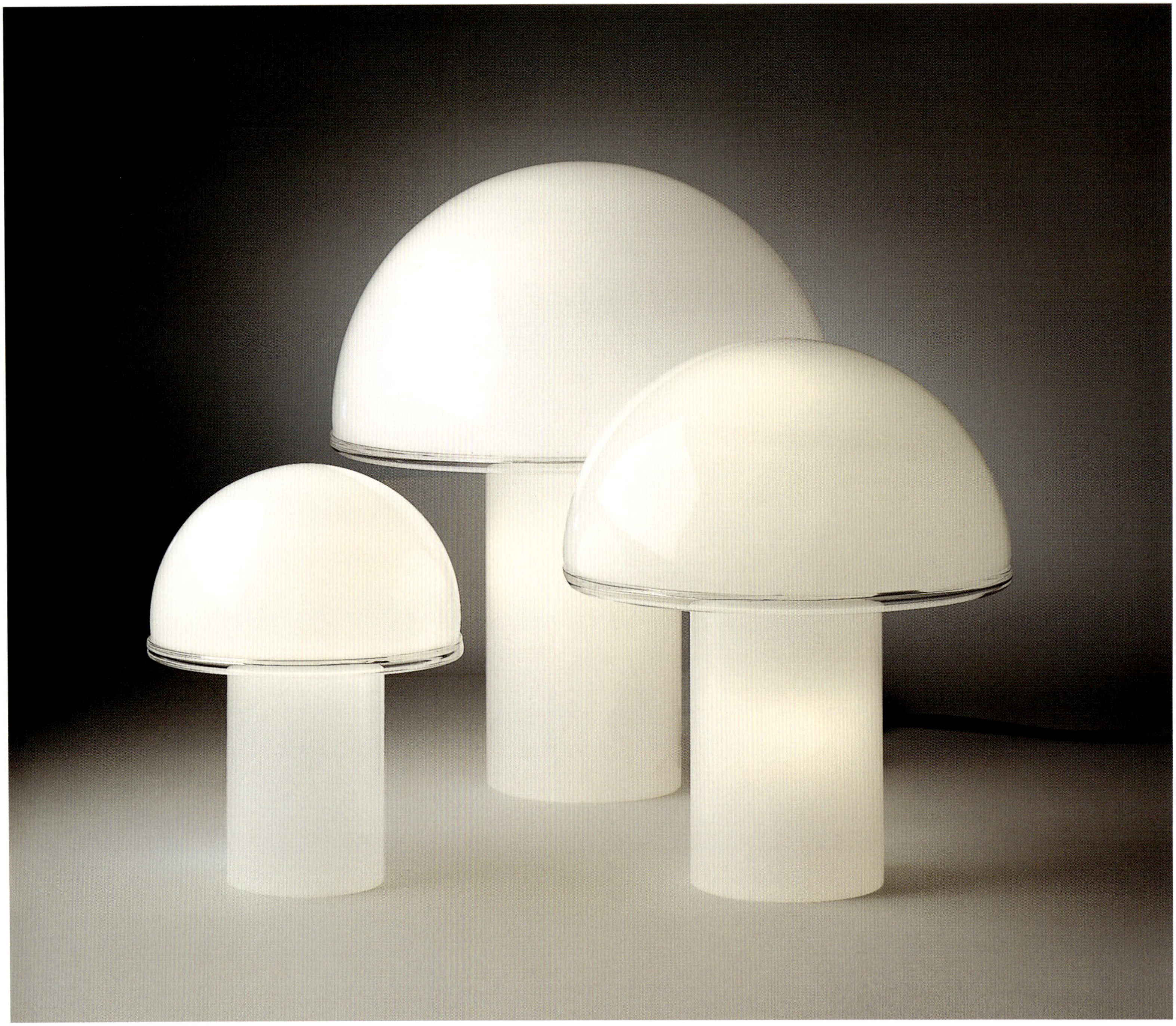

LUCIANO VISTOSI

Onfale, 1978
Glass and lamp components available in three sizes,
Diam. 7⅞ in. / 20 cm; 11 in. / 28 cm; 14⅛ in. / 36 cm
Private collection

Growing up around his family's glassworks, which produced industrial glass products, Venice-born artist and designer Luciano Vistosi (1931–2010) was perhaps destined to work in the medium of glass. When his father died in 1952, Vistosi dropped out of university and, along with his brother and their uncle, repurposed the family glassworks into a new company, Vetreria Vistosi (meaning Vistosi Glassworks). With a future-minded outlook and unrestrained creativity, the company sought to make design-forward glassware and move Venetian glass from the realm of tourist curiosity to that of refined art and design. The *Onfale* lamp, designed for Italian lighting company Artemide—unmistakably owing its form to the mushroom, and available in three sizes—represents one such object Vistosi designed during this period. But by the mid-1970s, his interest in design objects had started to wane, and he was more interested in glass sculpture of an increasingly large stature. After the death of his brother and uncle, Vistosi sold the glassworks in 1980 to concentrate on his career as an artist. The subject of many solo and group exhibitions, Vistosi flourished in the art world, winning awards and creating many large-scale glass sculptures, such as his model for the Accademia Bridge, measuring about 26 feet (8 m), and his large, lagoon-green glass cross for St. Mark's Basilica. But despite Vistosi's shift from the world of design to the larger stage of the art world, many of his creations, including the *Onfale* lamp, remain in high demand by collectors to this day.

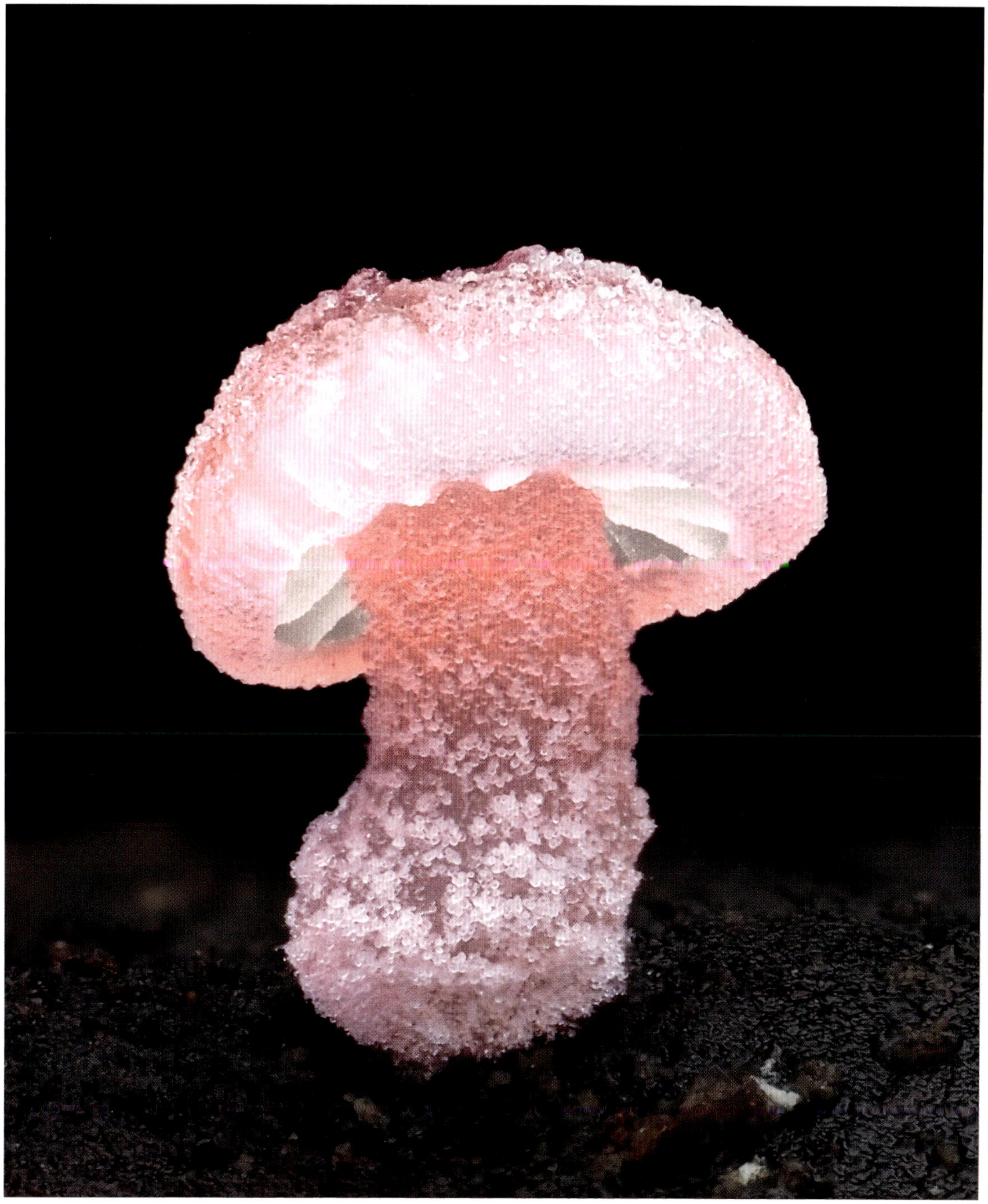

JAY LICHTER

Mycena roseoflava, 2024
Photograph, dimensions variable

Resembling an enticing, sugar-coated candy, this minuscule pink mushroom is easily overlooked by day, but at night its gentle glow illuminates the gloomy forest floor with tiny pinpricks of light. Discovered in 1964, *Mycena roseoflava* is a species of agaric mushroom native to New Zealand that reaches only about a quarter to half an inch (5 to 10 mm) in height. It grows on dying plants and rotting wood and is one of the world's few naturally bioluminescent fungi. Its glow, which is produced by a chemical reaction between luciferin molecules and the enzyme luciferase, was first recorded in 2021, having previously been overlooked by mycologists. It is believed that the mushroom's luminescence attracts small insects, which aid in the spreading of its spores. This specimen was photographed by Jay Lichter (b. 1996), who specializes in capturing macroimages of tiny mushrooms and other fungi in the forests of his native New Zealand. His journey into photography began casually, using just an iPhone with a cracked screen, although he has since upgraded to a sophisticated macro setup that allows him to capture the beauty of his tiny subjects in stunning close-up detail. Lichter regularly hunts for fungi in the damp woodlands of Auckland's Hūnua Falls and Blockhouse Bay, often spending many hours looking for the ideal specimens, which can be hard to spot because of their extremely small size. He has also photographed mushrooms in his own back garden and even discovered some of his subjects growing on houseplants.

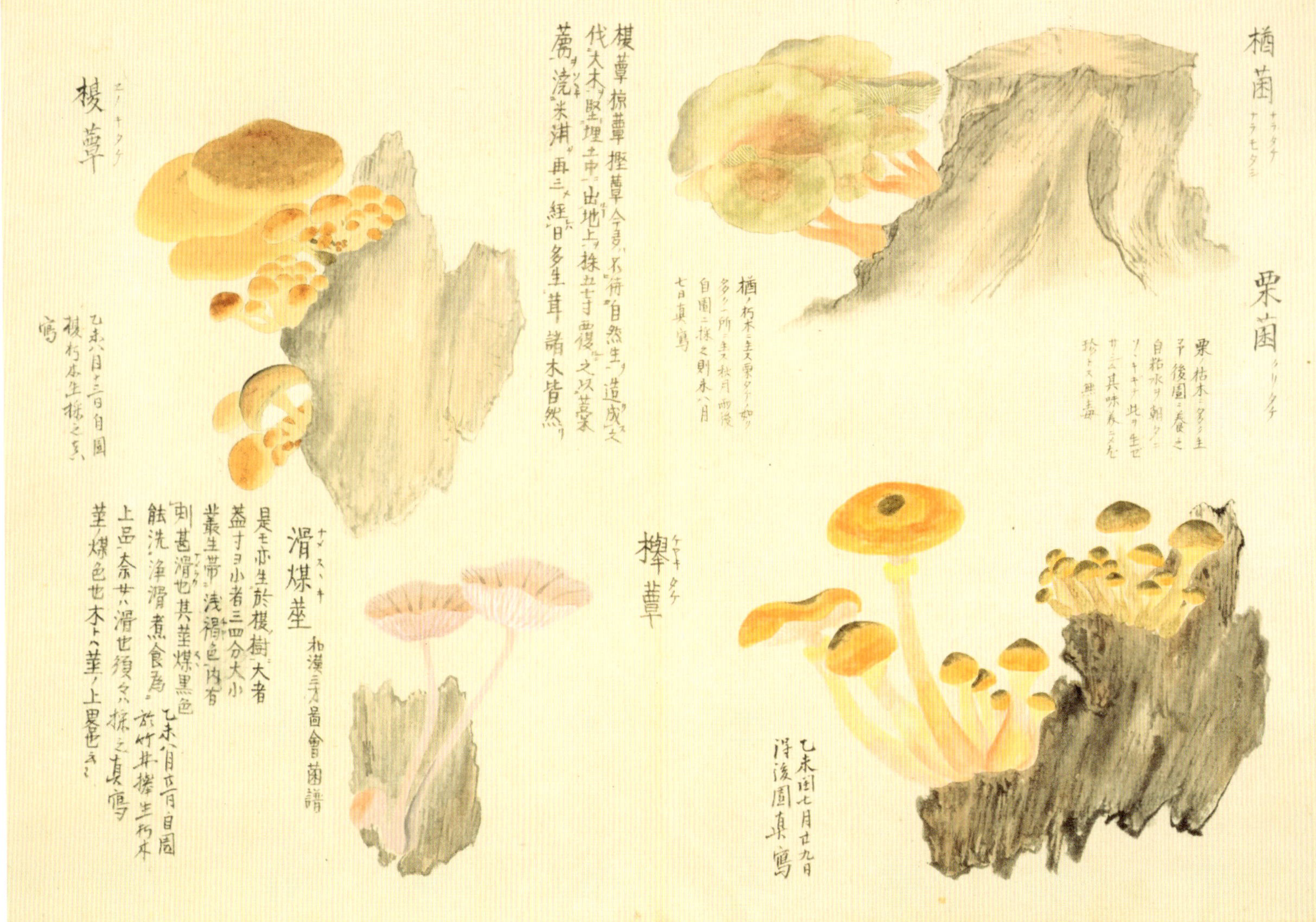

MŌRI BAIEN

Pages from *Baien kinpu*
(*Baien's Treatise on Mushrooms*), 1836
Illustrated manuscript,
each page 7½ × 11 in. / 19.2 × 27.9 cm
National Diet Library, Tokyo

Four different species of native Japanese mushrooms rise delicately from wooden branches and tree stumps in this painting by Japanese samurai and naturalist Mōri Baien (1798–1851). One of the most famous nature painters of the Edo period (1603–1867), Baien included these illustrations in *Baien kinpu* (*Baien's Treatise on Mushrooms*), part of a series of albums the artist produced to illustrate Japanese flowers, fungi, fishes, marine life, and birds. The high level of accurate botanical detail in the mushroom paintings reflects Baien's dedication to producing a pictorial record of Japanese flora and fauna, which was helped by his use of photographs from which he painted. *Baien kinpu* covers edible, medicinal, and poisonous mushrooms, and Baien added a written text to each image to tell his readers as much as possible about the species depicted. Japan has more than five thousand varieties of fungi, of which only around a hundred are edible. Species that grow on trees include the popular shiitake, which grows on the shii tree and is used widely in Japanese cuisine. The nameko grows in small clusters; its orange caps and gelatinous coating make it easy to identify. Another popular culinary ingredient, it has a mildly nutty flavor. Perhaps the most famous and highly sought-after edible forest fungus is the matsutake, which signals the arrival of autumn. Found under the Japanese red pine, on the forest floor, its strong aroma and pine flavor—and rarity—make it especially prized.

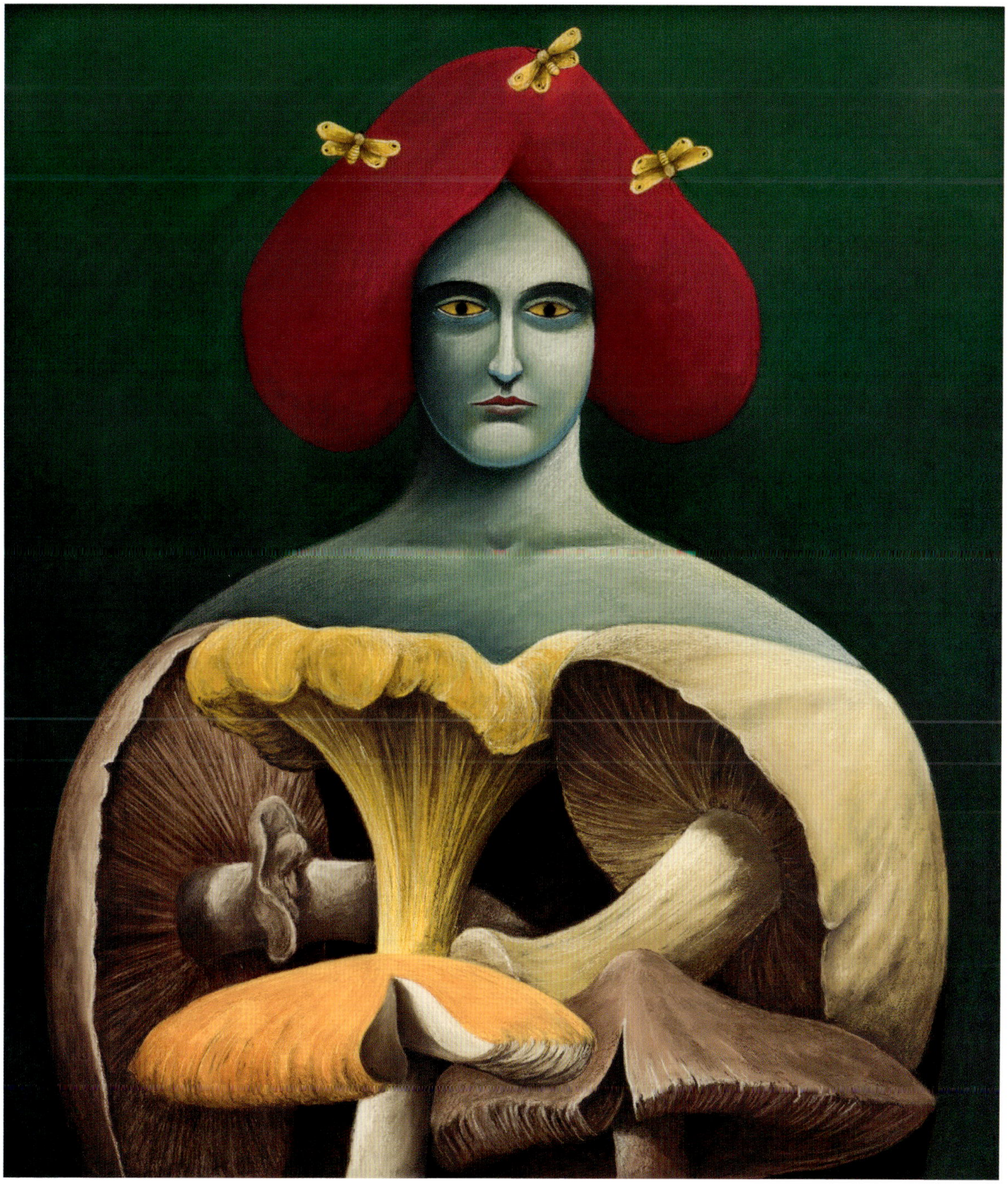

NICOLAS PARTY

Portrait with Mushrooms, 2019
Soft pastel on linen,
59 × 50 in. / 149.9 × 127 cm
Private collection

The dreamy paintings of Swiss figurative artist Nicolas Party (b. 1980) draw on both art traditions and surreal imagery to create uncanny other worlds. Party's early career as a graffiti artist and muralist established in him a confidence in working on a large scale and with bold color, lending a contemporary feel to his work, which is also inspired by traditional techniques and materials. He pays tribute to the historical art canon by using pastels—more commonly associated with the whimsical rococo style from the eighteenth century—and referencing the work of the Old Masters to create a body of work distinctive for saturated tones and overemphasized forms. At first glance, his portraits, landscapes, and still lifes have a sense of the familiar that, upon closer inspection, is belied by otherworldly detailing. His figures are often hard to place—frequently androgynous, they can appear in masks or with features that are almost, but not quite, human, juxtaposing style cues from the Renaissance and the clean lines of graphic design. Party also draws upon motifs from the natural world, including the forest, mushrooms, and insects, most notably in his *Sottobosco* series, where he explores the mysteries of the forest floor. In *Portrait with Mushrooms*—part of the *Sottobosco* series—the oversize fungi encompass hidden ecosystems and a vivid undercurrent of life. Party also uses the mushrooms as symbolic interpretations of metamorphosis, alongside the butterflies, frogs, and flowers referenced throughout the series.

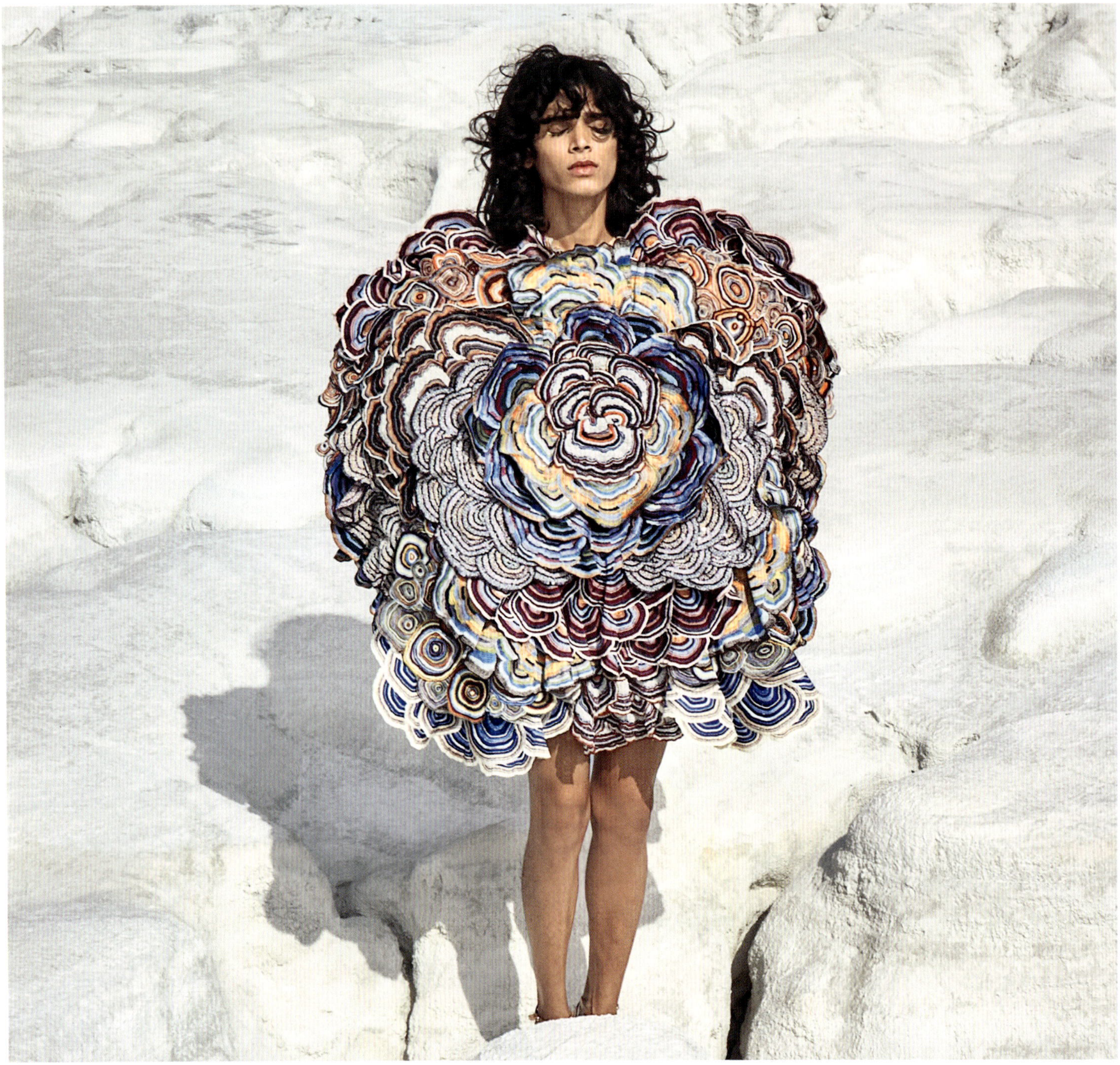

RAHUL MISHRA

Radial Mushroom Dress, from *The Dawn* collection, 2021
Tuille and silk organza with 3D hand embroidery, dimensions variable

Watching the British natural historian David Attenborough's television documentary *A Life on Our Planet* five times during the COVID-19 pandemic of 2020 provided the inspiration for this dress by Indian fashion designer Rahul Mishra (b. 1979). Debuting at his Spring 2021 couture collection at Paris Haute Couture Week, this minidress is made up entirely of 3D mushrooms, each one painstakingly cut out by hand, hand-embroidered, and then tacked onto a tuille and silk organza garment. The optimistic message Mishra took from Attenborough's documentary was the planet's ability to regenerate itself if left undisturbed, despite the human-made destruction it has suffered. That regeneration is something he witnessed firsthand. Growing up in the north Indian state of Uttar Pradesh, he watched a much-loved tree that had been felled come alive again after fungi colonized it, creating a whole new ecosystem. The ability of the sprouting fungi to regenerate stayed with him and crystallized in his 2021 show in mushroom minidresses and layered, flowing gowns that mimicked shelves of fungi sprouting from trees. These mushrooms most resembled turkey tail mushrooms, so-called because they look like the bird's tail as they grow in tiers on tree trunks. While inedible, the turkey tail mushroom is known for its medicinal properties: in Japan it is used in cancer treatment. Weaving together traditional Indian aesthetics with contemporary fashion, Mishra has become globally renowned for his intricately crafted garments.

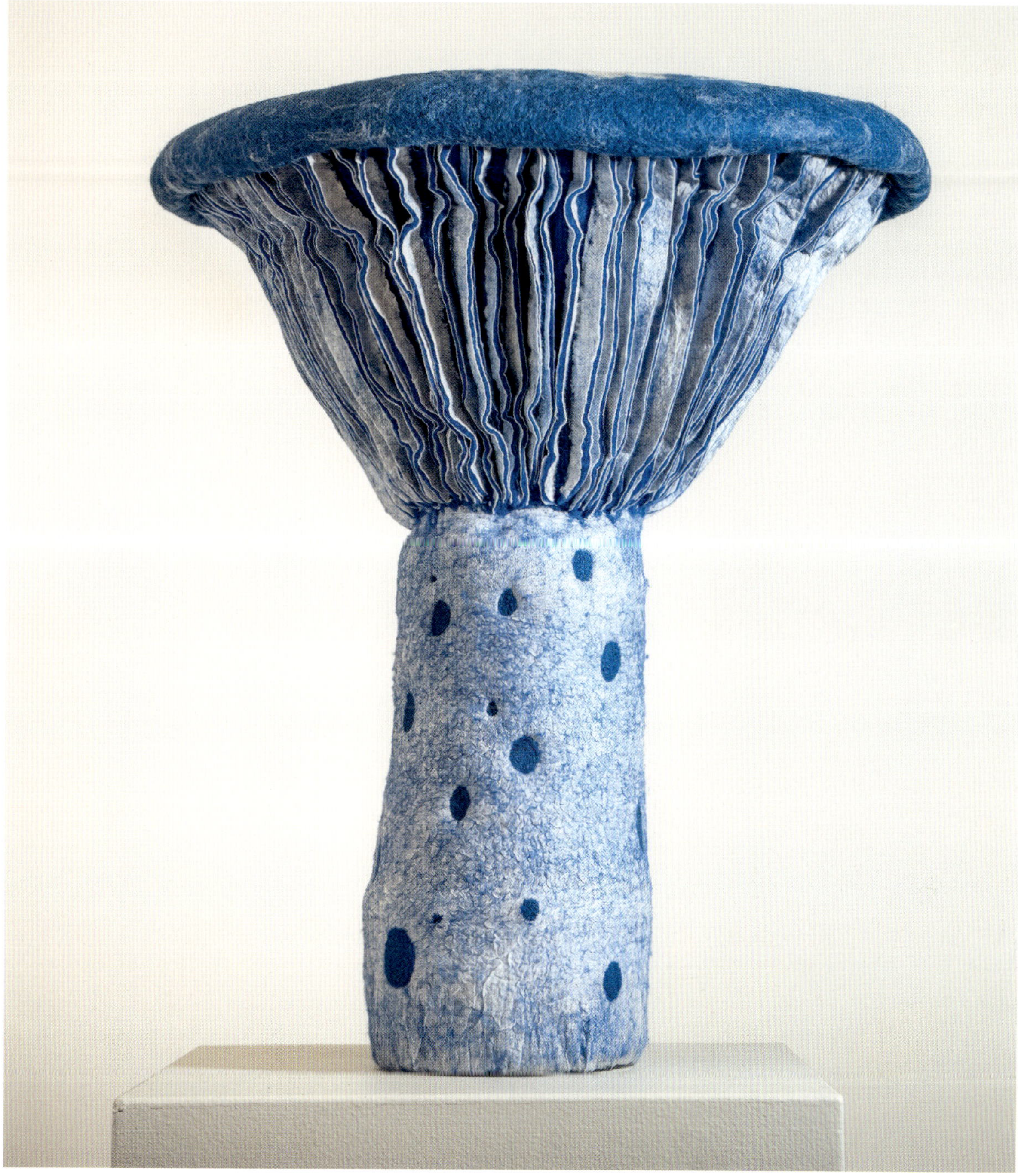

ELISE CAMPBELL

Lactarius Indigo (Indigo Milkcap), 2024
Wool, silk fabric, Kozo paper, and dowel,
17¾ × 14½ × 14½ in. / 45 × 37 × 37 cm
Private collection

For Canadian fiber artist Elise Campbell (b. 1985), who lives and works in Nova Scotia, mushrooms are an essential symbol of connectivity. Through vast mycelial networks that are connected to nearly everything in the natural world, fungi have also earned the title of the biggest living organism on Earth. And while the real-life indigo milk cap (*Lactarius indigo*) only grows up to around 3 inches (8 cm) tall, Campbell has exaggerated its proportions in this giant mushroom sculpture. The indigo milk cap is highly distinctive, thanks to its vibrant blue, and is, perhaps surprisingly, edible. In fact, its color varies depending on its freshness: a dark blue fungus is fresh, while a blue-gray mushroom is an older, drier specimen. To create her oversize sculpture, Campbell worked with a combination of natural wool and silk fibers and dowels to fabricate the flat cap and gills, while the stem is made from layers of Kozo paper sculpted into a three-dimensional shape. In making her artwork, Campbell uses different contemporary and traditional techniques of needle and wet felting, as well as nuno felting, all of which involve teasing the fibers into the required shape (some with and others without water), building up layer upon layer into the final solid form. Campbell's choice of materials and techniques reflects her belief in the need to respect the environment by using eco-friendly methods. This is also echoed in the subjects of her sculptures, which in addition to myriad mushrooms, including morels, chanterelles, and oysters, often include flora and fauna from her surrounding Nova Scotian landscape.

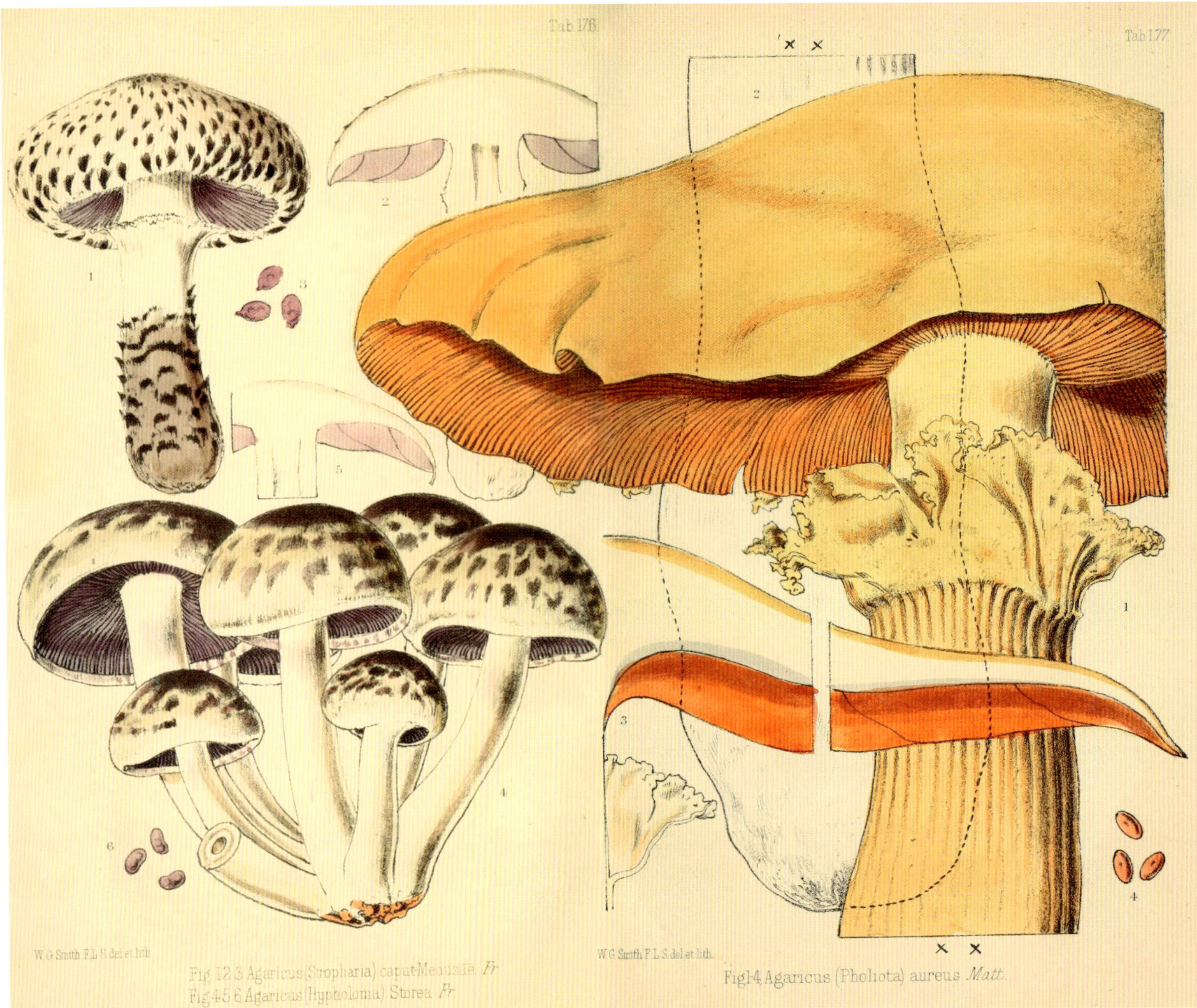

WORTHINGTON GEORGE SMITH

New and Rare Hymenomycetous Fungi, from *Journal of Botany, British and Foreign*, vol. 14, 1876
Hand-colored lithograph, H. 9 in. / 23 cm
Peter H. Raven Library, Missouri Botanical Garden, St. Louis

This striking plate presents fungi as both monumental and intimate, combining close observation with dramatic scale. On the left, smaller mushrooms (*Psathyrella caput-medusae* and *P. storea*) are presented singly and clustered, their creamy white stems and caps dotted with shades of brown and mauve, and revealing violet gills. On the right, a large golden bootleg mushroom (*Phaeolepiota aurea*) dominates the composition, its cap stretched across the page and its deep ocher gills radiating outward in precise, rhythmic lines, contrasting with the fringed edge of the stem's ring. Each species is presented with cross sections and areas of magnification, including spores shown in shades matching the respective gills from which they will fall. This innovative and graphic approach is the work of Worthington George Smith (1835–1917), an English illustrator, architect, and mycologist. Smith was an expert in fungi, writing more than two hundred articles and papers as well as books, and he was a founding member of the British Mycological Society. This example of "New and Rare Hymenomycetous Fungi" was written and illustrated for the *Journal of Botany, British and Foreign*. The tone of the journal was practical and scholarly, intended for other naturalists and scientists. Importantly, it gathered contributions from many field botanists and naturalists across Britain and beyond at a time when the understanding of fungi was rapidly evolving. Throughout history, detailed illustrations were essential to scientific discourse, and periodicals such as *Journal of Botany, British and Foreign* were key forums for that exchange.

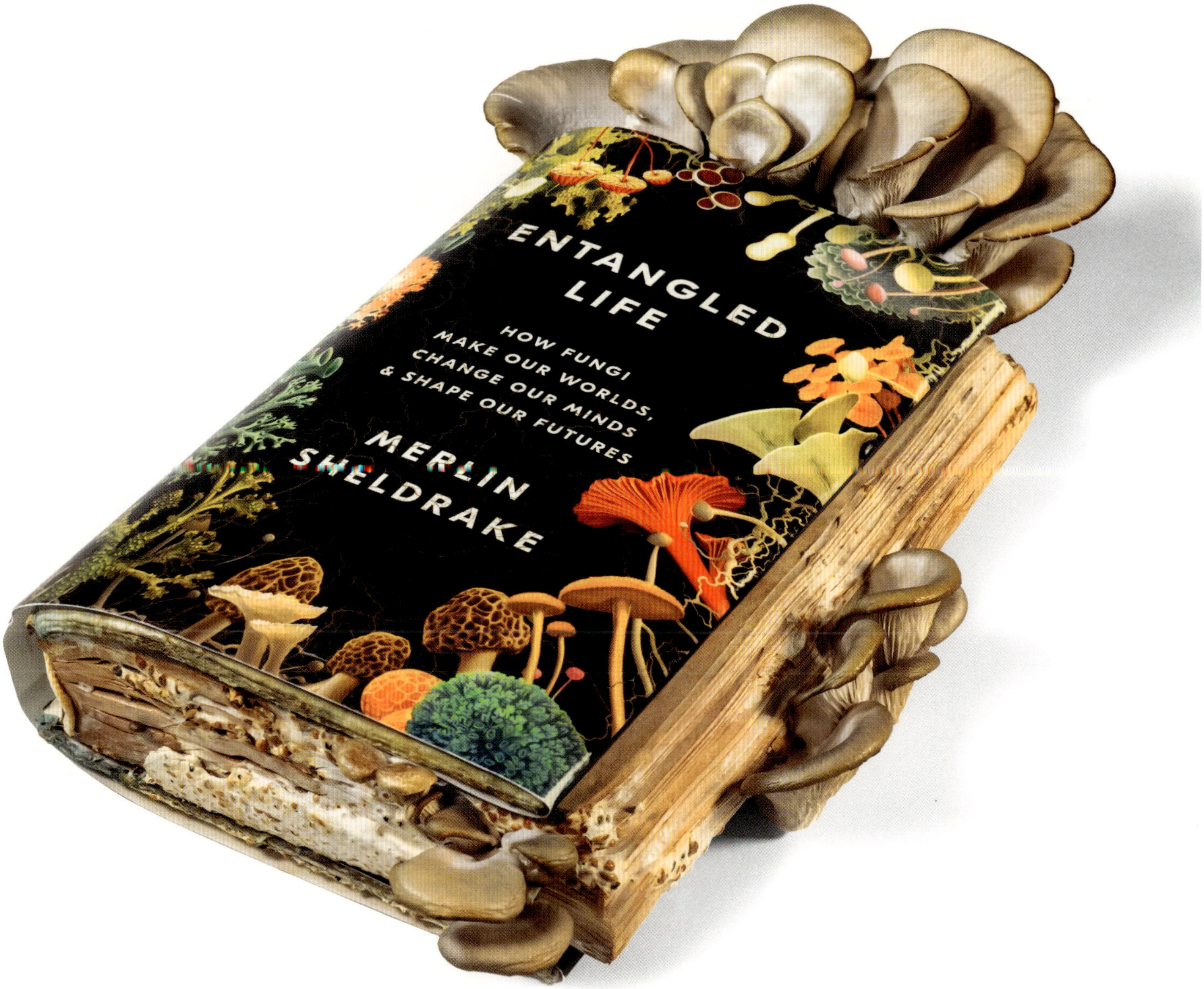

DAN ROEBUCK

Entangled Life, Partially Consumed by Pleurotus, 2020
Photograph, dimensions variable

Overtaken by nature, shell-like oyster mushrooms grow out of the bulging pages of the book *Entangled Life: How Fungi Make Our Worlds, Change Our Minds & Shape our Future* by scientist and author Merlin Sheldrake (b. 1987). Created as a marketing image for Sheldrake's deeply influential 2020 debut book, the photograph by Dan Roebuck was shared on social media to bring the microbiologist's vibrant take on fungi to life. (Another video depicted the author eating the mushrooms.) Sheldrake is the son of author and parapsychology researcher Rupert Sheldrake, who introduced him to notable psychedelic and botany specialists Terence McKenna and Paul Stamets in his youth. Merlin studied biology at Cambridge, and *Entangled Life* was the outcome of his PhD in tropical ecology. His interest in fungi was initially driven by a fascination with symbiosis, how humanity and organisms live in tandem and collaborate with each other. The award-winning popular science book looks at the impact of fungi on philosophy, design, agriculture, geology, medicine, food, and the mind, and in 2023 was turned into the documentary film *Fungi: Web of Life*. Sheldrake's interest in mycology has expanded into a music project with his brother, ecological sound artist and composer Cosmo, and he has also collaborated and consulted with Björk and Stella McCartney (see p.42). In addition to his creative projects, Sheldrake also works with lawyers and other mycologists for the legal protection of fungi, which are a vital component of ecosystems.

INDEX

Page numbers in *italics* refer to illustrations

N

O

P

Q

R

S

ACKNOWLEDGMENTS

A project of this size requires the commitment and expertise of many people. Special thanks are due to the following for their knowledge, passion, and advice in the selection of the works for inclusion:

Matthew Biggs
Gardener, author, and broadcaster, Hertfordshire, UK

Eugenia Bone
Food and nature author and journalist, New York

Francesca Gavin
Author, curator, and Editor in Chief of *EPOCH Review*, London

Tony Kirkham
Former Head of Arboretum, Gardens, and Horticultural Services at the Royal Botanic Gardens, Kew, London

Lawrence Millman
Author and mycologist, Cambridge, Massachusetts

Carrie Roy
Curator of Art and Research Scholar, Hunt Institute for Botanical Documentation, Carnegie Mellon University, Pittsburgh

Gill Saunders
Honorary Senior Research Fellow, Victoria and Albert Museum, London

Giuliana Furci
Founding Director, Fungi Foundation, Santiago, Chile

We are particularly grateful to Rosie Pickles and Sara Bader for researching and compiling the long list of entries for inclusion. Additional thanks are due to Sarah Bell and Rosie Pickles for their picture research, and to Laine Morreau, Theresa Bebbington, Vanessa Bird, Olivia Clark, João Mota, and Andrew Mott, for their invaluable assistance.

Finally, we would like to thank all the artists, illustrators, photographers, collectors, libraries, institutions, and museums who have given us permission to include their images.

TEXT CREDITS

The publisher is grateful to Francesca Gavin for writing the foreword to the book and the following writers for the individual entry texts:

Matthew Biggs: 26, 137, 159

Anita Croy: 14, 16, 23, 27, 30, 31, 32, 34, 35, 36, 48, 49, 50, 55, 58, 59, 60, 64, 65, 77, 80, 81, 86, 87, 89, 95, 96, 97, 100, 120, 124, 126, 128, 131, 138, 146, 148, 149, 153, 154, 158, 168, 169, 172, 174, 175, 177, 181, 182, 191, 195, 196, 197, 204, 206, 211, 218, 220, 221

Francesca Gavin: 6, 21, 38, 42, 79, 83, 90, 103, 118, 132, 136, 143, 147, 151, 163, 173, 178, 186, 200, 202, 205, 223

Katy Sprinkel Morreau: 7, 8, 12, 18, 20, 22, 29, 40, 52, 54, 57, 67, 74, 75, 101, 110, 113, 114, 116, 119, 122, 123, 129, 134, 139, 140, 142, 145, 155, 157, 160, 165, 171, 176, 179, 194, 199, 210, 213

Laine Morreau: 19, 24, 37, 43, 46, 53, 56, 66, 73, 88, 93, 98, 115, 127, 135, 141, 162, 164, 167, 189, 192, 198, 207, 212, 216

Rebecca Morrill: 69, 99, 106, 208

Clare Preston: 63

Carrie Roy: 28, 102, 112, 166, 180, 222

Gill Saunders: 45, 72, 214

Hannah Silver: 39, 68, 108, 111, 219

David Trigg: 9, 10, 13, 17, 44, 70, 76, 82, 91, 117, 133, 144, 152, 161, 184, 188, 190, 209, 217

Martin Walters: 11, 33, 41, 51, 61, 62, 71, 78, 84, 92, 104, 105, 107, 109, 125, 130, 150, 170, 183, 187, 193, 201, 215

PICTURE CREDITS

Every reasonable effort has been made to identify owners of copyright. Errors and omissions will be corrected in subsequent editions.

Accademia dei Fisiocritici (Siena, Italy)/Photo Gianmaria Bonari: 193; © 2026 ADAGP, Paris and DACS, London: 159; Christopher Adams: 98; akg-images/ Florilegius: 196; Greg Allen: 14–15; The Armitt Trust: 6; © 2026 ARS, NY and DACS, London/Image Courtesy Rago Wright: 135; Artemide: 216; Stephen Axford, Planet Fungi: 191; Greg T. Ballard: 179; Julie Beeler: 30; Courtesy Beeple: 133; David Benjamin/The Living: 91; Ms. Mediceo Palatino 218–20, Biblioteca Medicea Laurenziana, Florence, MiBACT, 1577: 96; Biophoto Associates/Science Photo Library: 66; Artist: Björk, Photographer: Vidar Logi, Styling: Edda Gudmundsdottir, Look/Dress: Daniel Del Core, Mask: Daniel Del Core in collaboration with David Åberg: 123; Bonhams: 146, 186; © Ross Bowden, PhD/Image courtesy National Gallery of Victoria, Melbourne: 48; From the British Library archive/Bridgeman Images: 79; © The Trustees of the British Museum: 36, 77; Photo by John Bryson/ Getty Images: 110; © Buccellati/Courtesy Rago/Wright: 39; John Cage, Lois Long, and Alexander H. Smith, *Mushroom Book* (New York: Hollanders Workshop, Inc., 1972), © John Cage Trust: 83; © 2026 Calder Foundation, New York/DACS, London: 161; *Wonderful Flight to the Mushroom Planet* by Eleanor Cameron, copyright © 1954/Reprinted by permission of Little, Brown, an imprint of Hachette Book Group, Inc.: 171; Arian Camilleri for Maison Gerard: 115; Courtesy Elise Campbell/Photo: Prateek Tamta: 221; © Canada Post Corporation, 1989. Reproduced with permission/Canada Modern, canadamodern.org: 13; Anne Carnein: 183; Amon Carter Museum of American Art, Fort Worth, Texas, Bequest of the artist, P1990.51.3476 © 1990 Amon Carter Museum of American Art: 190; © Gusmano Cesaretti: 37; Courtesy Chisholm Larsson Gallery: 74; Christie's Images/Bridgeman Images: 51; Christie's Images, London/Scala, Florence: 23; Christie's Images/Scala, Florence: 35, 53, 126, 139; Amanda Cobbett: 143; Meredith Connelly, *Mushrooms II*, 2022, cast fiberglass, thermoplastic, and lighting/ Courtesy the artist: 188; Corey Corcoran: 169; Cornell University Library/Biodiversity Heritage Library: 86, 92; Cornell University, Plant Pathology Herbarium (CUP), G.F. Atkinson Herbarium: 175; Courtesy Pearl Cowan and Good Naked Gallery info@goodnakedgallery.com: 38; © 2026 DACS: 45; Courtesy Nike Davies-Okundaye and kó, Lagos: 40; Gift of G.A. de Graag: 154; © 1940 Disney: 213; Photo by Stephan Doitschinoff Studio: 151; Chris Drury: 198; Edward C. Moore Collection, Bequest of Edward C. Moore, 1891/ Metropolitan Museum of Art: 107; Guy Edwardes/ naturepl.com: 62; E.R. Butler & Co: 49; Courtesy the Ernest Bloch Society: 93; © Florilegius/Bridgeman Images: 26; Reproduced by permission of Floris Books, Edinburgh from *Children of the Forest* by Elsa Beskow. All rights reserved: 33; Artwork: FoldHaus Collective/ Photography: Rene Smith: 18; Courtesy Adam Fuss: 136; © 2025 Estate of Ingeborg Gabriel/Galerie Nagel Draxler, Berlin, Cologne, Meseber/Digital image, the Museum of Modern Art, New York/Scala, Florence/The Judith Rothschild Foundation Contemporary Drawings Collection Gift: 101; *Champignons*, drawing by Anne Gavarni d'Autray and Françoise de la Perrière © Hermès Paris, 2026: 64; *Mindful Mushroom*, 2017, unique hand-cut collage by Seana Gavin: 173; Geertje Geertsma: 181; Digital image courtesy Getty's Open Content Program: 65; Research Library, the Getty Research Institute/Biodiversity Heritage Library: 124; *Blue Moon Mushroom* by Alex Grey, secret writing by Allyson Grey: 80; Amy Gross: 95; Selma Gurbuz: 90; © Hulda Guzmán/Courtesy the artist, Alexander Berggruen, and Stephen Friedman Gallery: 192; Harvard University Botany Libraries/Biodiversity Heritage Library: 206; Harvard University, courtesy HathiTrust: 180; © Sarah Hearn: 9; Jaymi Heimbuch/ naturepl.com: 29; Erich Heinemann: *Wichtelgeschichten*. With pictures by Fritz Baumgarten © 2024 Esslinger Verlag in the Thienemann-Esslinger Verlag GmbH, Stuttgart: 109; Courtesy Hemmerle and Private Collection: 43; © Hergé/Tintinimaginatio 2026: 58; Hironoyama/Wikimedia Commons: 201; © 2026 Damien Hirst and Science Ltd. All rights reserved, DACS/Christie's Images, London/Scala, Florence: 68; Courtesy Carsten Höller and Fondazione Prada, Milan © Attilio Maranzano: 111; Courtesy Takashi Homma, Nonaka-Hill, and Taro Nasu: 202–3; Courtesy Max Hooper Schneider and Maureen Paley, London/Photo: Paul Salveson: 103; © Stacy Hsu, All rights reserved 2026/Bridgeman Images: 160; Huff, Ned L., 1876–1964, *Hydnum*, University of Minnesota Libraries, University Archives: 142; © Courtesy the Huntington Art Museum, San Marino, California: 89; Courtesy Hunt Institute for Botanical Documentation, Carnegie Mellon University, Pittsburgh: 105; Nigel Hutchings/Science Photo Library: 162; © Alex Hyde/naturepl.com: 94; *Chanterelle*, 2009 © Riitta Ikonen and Anni Koponen: 7; Photo © The Israel Museum, Jerusalem: 131; Taehyoung (Maria) Jeon: 215; Photo Brian Jones: 67; *La Vie en Rose* by Stéphanie Kilgast: 81; © Estate of Martin Kippenberger, Galerie Gisela Capitain, Cologne: 76; © Yayoi Kusama: 208; Lamyai/Getty Images: 164; "Magic" design is copyright © Liberty Fabric Limited: 52; Jay Lichter, @cyanesense: 217; Liszt Collection/ Bridgeman Images: 27; ©TfL from the London Transport Museum collection: 72; Carol Long: 73; © Look and Learn/Bridgeman Images: 56b; LuEsther T Mertz Library, New York Botanical Garden/Biodiversity Heritage Library: 60, 78, 166, 222; Phyllis Ma: 145; Marshmallow Laser Feast, *Poetics of Soil: Fly Agaric I*, 2025, London, courtesy the artist: 84–85; © Mary Evans Picture Library: 56t; Stella McCartney and Aoi Yamada Mushroom Campaign Summer 2022 collection: 42; Meow Wolf (multiple artists): 102; Metropolitan Museum of Art: 34; © Rahul Mishra: 220; Monsieur Plant - @monsieur_plant: 134; Gift of J. Pierpont Morgan, 1917/Metropolitan Museum of Art: 210; The Morgan Library & Museum, New York: 16; Morisawa Studio: 140; © Estate of Cedric Morris. All rights reserved 2025/Bridgeman Images: 138; Max Mudie: 148; © 1999 Takashi Murakami/Kaikai Kiki Co., Ltd. All Rights Reserved: 10; Photo: Museum Rietberg Zürich, Rainer Wolfsberger: 172; Christiaan Nagel Studio: 144; Photo © Peter Nahum at the Leicester Galleries, London/Bridgeman Images: 57; © 2025 Tina Napangardi Martin/Copyright Agency. Licensed by DACS: 106; © Yoshitomo Nara/Image courtesy Yoshitomo Nara Foundation: 63; © National Diet Library, Japan: 218; National Museum of Nature and Science, Tokyo: 128; Naturhistoriska Riksmuseet, Stockholm. M.A. Lindblad Archive: 22; New York Public Library: 50, 132; New York Public Library Digital Collections: 114; Courtesy New York State Museum, Albany: 168; Tony North: 130; Österreichische Nationalbibliothek: 55; Otto's Antiques & Interiors: 129; Paris Musées/Maisons de Victor Hugo: 75; © Nicolas Party/Courtesy the artist and Hauser & Wirth: 219; Courtesy Hamish Pearch: 205; Sculpture by Jo Pearl/Photo by Elsa Pearl: 8; © Hilary Pecis, Courtesy Rago/Wright: 176; Cover of *Poisonous Fungi* by John Ramsbottom used by permission of Penguin Random House UK: 214; Kristel Peters, Studio Cojak: 147; © Peyo, 2025, Licensed through I.M.P.S. (Brussels), www.smurf.com: 108; piemags/Alamy: 194; © 2026 The Estate of Sigmar Polke, Cologne/DACS/ACT Art Collection, Deutschland: 152; Courtesy Simon Popper: 118; Luke Powell, Jody Hudson-Powell, Nav Bhatia, Luis Gutiérrez, Jack Llewellyn, David Loy, Albert Sanjuán: 178; Courtesy Robert Pruitt and Salon 94: 17; Public domain: 70; © Paul Quant: 189; Courtesy Rago Arts: 46–47; Courtesy Annie Ratti/Photo: Agostino Osio: 163; Chris Ritson: 20; © Dale M Reid: 97; © 2026 Albert Renger-Patzsch/Archiv Ann und Jürgen Wilde/DACS/Image copyright the Metropolitan Museum of Art/Art Resource/Scala, Florence: 44; The Rijksmuseum: 182; RMN-Grand Palais/Georges Meguerditchian/Dist. Foto Scala, Florence12; RMN-Grand Palais/Patrick Gries, Valérie Torre/Dist. Foto Scala, Florence: 100; RMN-Grand Palais/Thierry Ollivier/Dist. Foto Scala, Florence: 127; Dan Roebuck: 223; © Mika Rottenberg/Courtesy the artist and Hauser & Wirth: 167; © Kathleen Ryan: 69; Image courtesy Adrian Sassoon, London/Photography by Sylvain Deleu: 199; © Margaret Saylor: 82; Photo Scala, Florence: 125; © Kenny Scharf/Courtesy the artist and Almine Rech/Photo: Alessandro Wang: 212; Schweizerische Nationalbibliothek/Wikimedia Commons: 187; Science Stock Photography/Science Photo Library: 104; Marianne Seiman: 116; Courtesy Alake Shilling and Jeffrey Deitch, New York and Los Angeles/Collection of Shio Kusaka and Jonas Wood/ Photo by Evan Bedford: 59; © David Shrigley. All Rights Reserved, DACS Images: 99; Smithsonian American Art Museum, Gift of the artist, 1970.355.215: 112; Smithsonian Libraries and Archives/Biodiversity Heritage Library: 141; Courtesy Sotheby's: 120–21; Collaboration with Dr. Danielle Stevenson: 31; Suntory Museum of Art: 197; Sustainable Design (Material) Museum: 155; © Cy Twombly Foundation: 184–85; University of Connecticut Libraries/Biodiversity Heritage Library: 54; University Library, University of Illinois Urbana Champaign/Biodiversity Heritage Library: 28; University of Michigan, courtesy HathiTrust: 149; University of Michigan Museum of Art, Gift of Mr. and Mrs. James Agah, Class of 1989 (BBA), 2012/2.14.1: 61; University of Wisconsin-Milwaukee Libraries: 158; US Department of Agriculture, National Agricultural Library/Biodiversity Heritage Library: 41; Courtesy Myung Urso: 87; Courtesy Iris van Herpen: 165; Arie Van 'T Riet/Science Photo Library: 24–25; Courtesy Sage Vaughn: 157–58; Vincent Verroust: 88; © Alexander Viazmensky: 71; Image © Victoria and Albert Museum, London: 177; Courtesy Megan Walsh: 113; © 2026 The Andy Warhol Foundation for the Visual Arts, Inc./ Licensed by DACS, London/Digital image Whitney Museum of American Art/Licensed by Scala: 21; Barry Webb FRPS: 153; Wellcome Collection: 32, 137, 150, 174, 195, 211; Wellcome Library/Biodiversity Heritage Library: 117; Edward Weston © Center for Creative Photography, Arizona Board of Regents/Photo: Scala, Florence: 19; White Images/Scala, Florence: 170, 204; Wikimedia Commons: 11; Melvyn Yeo/Science Photo Library: 122; Courtesy Jana Želibská and Gandy Gallery/ Photo by Miloš Vančo: 200; Boris Zhitkov/Alamy: 209.

Phaidon Press Limited
2 Cooperage Yard
London E15 2QR

Phaidon Press Inc.
111 Broadway
New York, NY 10006

Phaidon SARL
55, rue Traversière
75012 Paris

phaidon.com

First published 2026

ISBN 978 1 83729 206 6

A CIP catalogue record for this book is available from the British Library and the Library of Congress.

Commissioning Editor: Victoria Clarke
Project Editor: Lynne Ciccaglione
Production Controller: Zuzana Cimalova
Cover Design: Julia Hasting
Layout Design: João Mota
Typesetting: Daniel Chehade

Printed in China